MW00800571

A Resource and Strategies Guide for Chemistry

www.apstrategies.org

Acknowledgments

Funding for the *Laying the Foundation* series was provided through a grant from the O'Donnell Foundation.

Advanced Placement* Strategies gratefully acknowledges the tireless efforts of the following educators to write and edit the *Laying the Foundation* series.

Project Directors

René McCormick
AP* Strategies, Inc.
Dallas, Texas

Lisa McGaw
AP* Strategies, Inc.
Dallas, Texas

Authors

Carol Brown
Saint Mary's Hall
San Antonio, Texas

Lynn Kirby
Kealing Junior High
Austin, Texas

Lisa McGaw
AP* Strategies, Inc.
Dallas, Texas

Hugh Henderson
Plano Senior High School
Plano, Texas

Carol Leibl
James Madison High School
San Antonio, Texas

Mary Payton
AP Strategies, Inc.
Dallas, Texas

Jason Hook
Kealing Junior High
Austin, Texas

René McCormick
AP Strategies, Inc.
Dallas, Texas

Debbie Richards
Bryan High School
Bryan, Texas

Editor

Mary Payton
Editor In Chief

Contributing Authors

Randy Baskin
Rider High School
Wichita Falls, Texas

Judy Cordell
Nolan High School
Fort Worth, Texas

Brian Kaestner
Saint Mary's Hall
San Antonio, Texas

Adrian Carrales
Kealing Junior High
Austin, Texas

Ron Esman
Plano Senior High School
Plano, Texas

René Moses
Carroll High School
Southlake, Texas

Andrew Cordell
Fort Worth Country Day
School
Fort Worth, Texas

Jeff Funkhouser
Northwest High School
Justin, Texas

Mary Anne Potter
Micron Systems
Dallas, Texas

*AP and Advanced Placement Program are registered trademarks of the College Entrance Examination Board, which is not involved in the production of this product.

Other Contributors and Reviewers

Pat Chriswell
Fort Bend Baptist Academy
High School
Sugar Land, Texas

Kristen Jones
A&M Consolidated High
School
College Station, Texas

Dennis Ruez, Jr.
UT Austin, Department of
Geology
Austin, Texas

Charlotte Taggart
Abilene High School
Abilene, Texas

Syllabi Contributors

Rhonda Alexander
Robert E. Lee High School
Tyler, Texas

Carol Brown
St. Mary's Hall
San Antonio, Texas

Tom Campbell
Keeling Junior High
Austin, Texas

Lynn Cook
Putnam City Schools
Oklahoma City, Oklahoma

Luiz DeCarvalho
Carroll High School
Southlake, Texas

Denise DeMartino
Westlake High School
Austin, Texas

Chuy Garcia
Hyde Park Baptist
Austin, Texas

Lawanna Jenkins
Hodges Bend Middle School
Houston, Texas

Sharon Hamilton
Fort Worth Country Day
Fort Worth, Texas

Nancy Nixon
Covington Middle School
Austin, Texas

Patti O'Conner
James Madison High School
San Antonio, Texas

Camie Fillpot
O'Henry Middle School
Austin, Texas

Nancy Ramos
Northside Health Careers
High School
San Antonio, Texas

Jackie Snow
Troy High School
Troy, Texas

Charlotte Taggart
Abilene High School
Abilene, Texas

Production

Sonya Pullen
AP* Strategies, Inc.
Dallas, Texas

Table of Contents

Assessment

Appendixes

Introduction to the Laying the Foundation Series

The Laying the Foundation Series in Science is designed to support classroom teachers in better preparing students for Advanced Placement* science courses. We believe this goal is readily accomplished through well-designed science programs that begin in the middle grades. These guide books are also designed to assist a school or school district in building a strong science vertical team. Each guide is designed to provide the teacher with insight into the process skills, content skills, and assessment strategies that will better prepare students as they pursue Advanced Placement science courses and other advanced coursework.

Each guide begins with a set of Foundation Lessons. The Foundation Lessons target the process skills needed to provide a solid foundation for further scientific study. These lessons should serve as a spring board for establishing the basic expectations for a science vertical team and should be reinforced at each grade level.

Pre-AP* teachers often say, "If I just knew what to teach, I would teach it!" Content skills specific to each course are outlined in the appropriate LTF guide. The Pre-AP course should have a greater depth and breadth of content. Therefore, a course scope along with sample syllabi from successful Pre-AP teachers is provided to aid teachers in designing their course. It is also our belief that connections between math and science should be demonstrated and emphasized to students from an early age. The middle grades content has been divided into two LTF guides, one for life and earth sciences and one for chemistry and physics. Some middle schools have found an integrated chemistry and physics course offered to students in Algebra I to be very successful. Middle grade teachers with an integrated science course covering all four disciplines should utilize both of the middle school guides. The remaining three guides in the series focus on first-year courses in biology, chemistry and physics.

Currently, there are no Pre-AP science textbooks. The lessons found in the LTF guides model how to add the fullness to each course that is currently lacking in most widely adopted textbooks. Teacher pages included with each lesson provide correlations to the TEKS and the National Science Standards as well as a connection to the relevant AP course outline. In addition, these pages provide helpful hints for setting up laboratory activities and insightful teaching strategies. Finally, the teacher pages offer content assistance for the lessons dealing with topics that are not typically found in the on-level science textbook.

We hope that you will use these guides to enhance instruction in your classroom, formulate strong horizontal and vertical teams, and better prepare students to succeed on AP science exams. It is also our hope that you use the lessons from the guides with all of your students. Ultimately, we believe student achievement depends upon the strong expectation that all of your students are preparing for Advanced Placement science classes or advanced course work.

— René McCormick and Lisa McGaw

Overview of Laying the Foundation in Chemistry

All of life is dependent upon chemistry. Chemical reactions are everywhere. A thorough grounding in basic chemical principles is necessary in order to understand the other sciences such as modern biology. We readily recognize that society is increasingly dependent upon technology. In order to be competitive in the 21st century, communities and nations will need citizens with an understanding of the natural world. Today's students must understand basic scientific principles in order to be intelligent consumers and informed citizens.

The first year of high school chemistry is designed to lay the foundation for further study in the sciences. Content and skills acquired in this course provide the basis for students to become lifelong learners with a love and appreciation for the natural world. While much emphasis in high school is placed on developing patterns of learning that will continue beyond formal education, there is also a body of scientific knowledge to be acquired that enriches life and enables the student to progress in the study of science.

Unfortunately, students often find science courses difficult, not because the material is unmanageable, but rather because they do not have the skills needed to be successful. If teachers, schools, and districts work together to provide a seamless curriculum that moves students from one level to the next, all students will be able to continue in their study of science. Furthermore, by presenting Pre-AP strategies at an early age, many more students will to enroll in and be successful in advanced science courses.

The teacher in a first-year chemistry class serves as a mentor, guide, and coach to lead students to competency in scientific thinking and understanding of basic chemical principles. To reach these goals, the teacher should make available an enriched atmosphere of academic challenge. Engagement in laboratory activities, problem solving, and guided inquiry provide a background for a students to achieve success in advanced science courses, whether these courses are AP courses taken on the high school campus or university science courses.

All teachers want their students to achieve, but it is often difficult to assess exactly what material should be covered and to what depth. It would be wonderful to assume that this guide will address all of those issues; however, no one guide or book will ever be able to accomplish such an ideal. Instead, much of the responsibility rests upon the shoulders of the individual teacher to develop creative methods to engage the interest of students.

Laying the Foundations in Chemistry provides resources for the teacher to implement Pre-AP strategies in a first-year chemistry course. The lessons and learning activities are appropriate for any beginning chemistry course and should provide students with skills needed to develop into critical thinkers. It should be noted that many of the lessons require mathematical skills to analyze data. Attention must be given to mathematics because an inability to apply appropriate mathematics to scientific problems is often a major stumbling block to success in science. This guide provides activities to help students employ technology by using data collection devices and computers, thereby integrating science and mathematics. Problem solving strategies such as factor-label (dimensional analysis) will be emphasized.

The ultimate purpose of this guide is to provide a launching pad for teachers to develop new and creative methods for engaging students. It is our hope that the entire series of guides will encourage teachers and administrators to be engaged constantly in refining, aligning, and defining the curriculum in all schools.

— Carol Brown

Process Skills
Progression Chart

PROCESS SKILLS PROGRESSION CHART

	Factual Knowledge	Conceptual Understanding	Reasoning and Analysis
Acquire Data By Experimentation and Observation	Identify scientific equipment, instruments, and technology Know and observe safety precautions Follow a procedure	Choose appropriate equipment and technology	Work collaboratively to obtain scientific data
Record and Manipulate Data	Measure and record data in SI units Make and record observations	Determine variables to be measured Estimate and approximate quantities Solve mathematical equations using data	Design a data table or chart as appropriate Create appropriate graphical representations of data Analyze error Apply statistical analysis such as standard deviation, percent error, and chi square
Graph and Analyze Data	Plot data points Label the axes Title the graph	Translate graph into words Scale axes Calculate slope, area, and intercepts Construct line of best fit, curve fits, regression equations	Evaluate line of best fit, curve fit, and regression equation Detect patterns in data Interpret physical meaning of slope, area, and intercepts Interpolate, extrapolate and predict from a graph Transform data into linear form Recognize cause and effect relationships Draw appropriate conclusions Apply conclusions to new situations and further investigations

	Factual Knowledge	Conceptual Understanding	Reasoning and Analysis
Communicate and Share Results		Translate data into words Read and understand scientific articles	Defend results and conclusions in both written and oral format Relate concepts to unifying themes
Design Experiments	State the purpose Practice identifying variables	Design and use models to explain scientific concepts Understand the importance of controls Apply steps of scientific method to solve a problem Formulate a feasible and practical procedure	Formulate testable questions and hypotheses Critique experimental designs Predict outcomes Make environmentally friendly choices when designing experiments
Demonstrate Mathematical Problem-solving Skills	Identify relevant given information	Substitute values into an equation and solve Use dimensional analysis Estimate reasonable answers	
Use Technology	Recognize useful data tools such as graphing calculators, probes, data collection device and computers	Use data collection tools such as graphing calculators, probes, data collection device and computers	

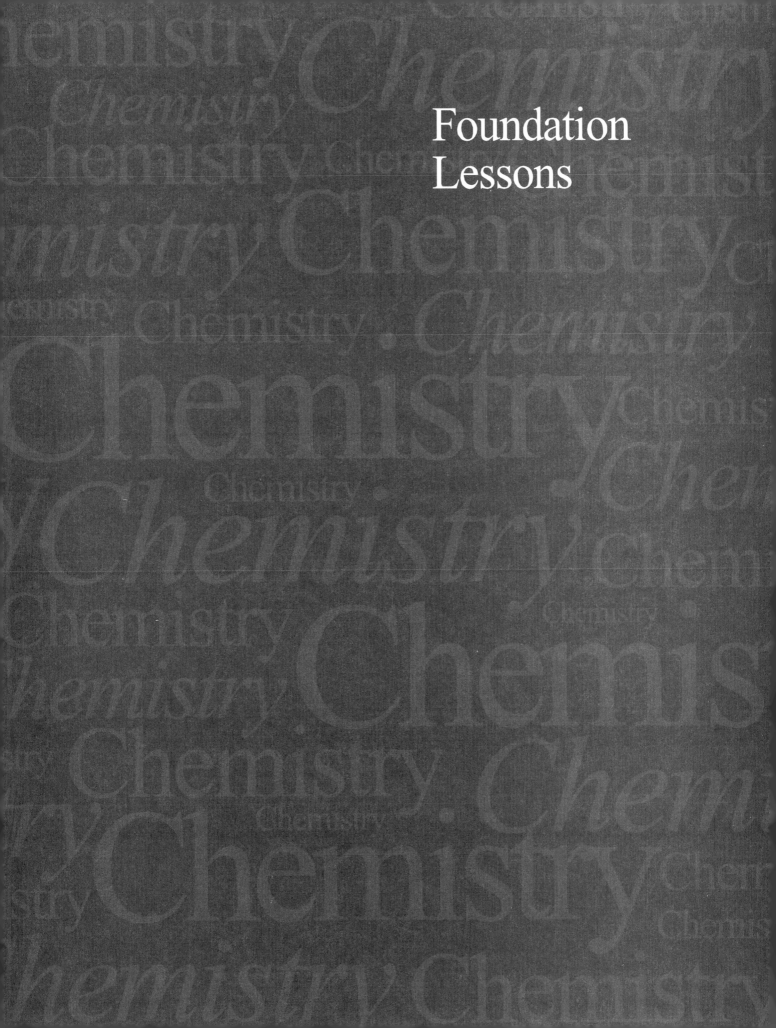

Foundation
Lessons

TEACHER PAGES

The Scientific Method
Exploring Experimental Design

Unit Overview

OBJECTIVE
Students will identify and apply the steps of the scientific method.

LEVEL
All

NATIONAL STANDARDS
UCP.1, UCP.2, UCP.3, A.1, A.2, G.2

TEKS
6.1(A), 6.2(A), 6.2(B), 6.2(C), 6.2(D), 6.2(E), 6.3(A)
7.1(A), 7.2(A), 7.2(B), 7.2(C), 7.2(D), 7.2(E), 7.3(A)
8.1(A), 8.2 (A), 8.2(B), 8.2(C), 8.2(D), 8.2(E), 8.3(A)
IPC: 1(A), 2(A), 2(B), 2(C), 2(D), 3(A)
Biology: 1(A), 2(A), 2(B), 2(C), 2(D), 3(A)
Chemistry: 1(A), 2(A), 2(B), 2(C), 2(D), 2(E), 3(A)
Physics: 1(A), 2(A), 2(B), 2(C), 2(D), 2(E), 2(F), 3(A)

CONNECTIONS TO AP
AP Science courses all contain a laboratory component where the scientific method will be used.

TIME FRAME
Two 45 minute class periods

MATERIALS

Come Fly With Us student pages   *Scientific Method Practice 1* student pages
Penny Test Lab student pages   *Scientific Method Practice 2* student pages

TEACHER NOTES

Modern scientific inquiry or science (from *scientia*, Latin for knowledge) is generally attributed to the historical contributions of Galileo Galilei and Roger Bacon, though some historians believe that their practices were inspired by earlier Islamic tradition. In spite of the rich human tradition of scientific inquiry, there is, today, no single or universal method of performing science. According to the National Science Teachers Association, science is "characterized by the systematic gathering of information through various forms of direct and indirect observations and the testing of this information by methods including, but not limited to, experimentation." Although this definition is helpful to explain the process of science, it does not specify a list of experimental steps that one should logically progress through to perform an experiment. (An experiment can be defined as an organized series of steps used to test a probable solution to a problem, commonly called a hypothesis, or educated guess.) Despite the absence of a *standard* scientific method, there is a generally agreed upon model that describes how science operates.

Steps of the Scientific Method

1. State the problem: What is the problem? This is typically stated in a question format.
   - *EXAMPLE: Will taking one aspirin per day for 60 days decrease blood pressure in females ages 12-14?*

2. Research the problem: The researcher typically will gather information on the problem. They may read accounts and journals in the subject or be involved in communications with other scientists.
   - *EXAMPLE: Some people relate stories to doctors that they feel relief from high blood pressure after taking one aspirin per day. It is not scientific if the idea is untested or if one person reports this (called anecdotal evidence).*

3. Form a probable solution, or hypothesis, to your problem. Make an educated guess as to what will solve the problem. Ideally this should be written in an *if-then* format.
   - *EXAMPLE: If a female aged 12-14 takes one aspirin per day for 60 days, then it will decrease her blood pressure.*

4. Test your hypothesis: **Do an experiment**.
   - *EXAMPLE: Test 100 females, ages 12-14, to see if taking one aspirin a day for 60 days lowers blood pressure in those females.*

   > **Independent Variable (I.V.)**: The variable you change, on purpose, in the experiment. To help students remember it suggest the phrase "**I** change it" emphasizing the **I**ndependent variable.
   > - *EXAMPLE: In this described experiment, taking an aspirin or not would be the independent variable. This is what the experimenter changes between his groups in the experiment.*

Dependent Variable (D.V.): The response to the I.V.
- *EXAMPLE: The blood pressure of the individuals in the experiment, which may change from the administration of aspirin.*

Control: The group, or experimental subject, which does not receive the I.V.
- *EXAMPLE: The group of females that does not get a dose of aspirin.*

Constants: Conditions that remain the same in the experiment.
- *EXAMPLE: In this scenario some probable constants would include: only females were used, only females at around the same age, the same dosage of aspirin was given to all the individuals in the experimental group for the same defined time interval—60 days, the same brand of aspirin was given, the same type of diet was ideally given to the members of the experimental group as well as the same activity level prescribed.*

5. Recording and analyzing the data: What sort of results did you get? Typically data is organized into data tables. The data is then graphed for ease of understanding and visual appeal.
   - *EXAMPLE: Out of 100 females, ages 12-14 yrs., 76 had lower blood pressure readings after taking one aspirin per day for 60 days.*

6. Stating a conclusion: What does all the data mean? Is your hypothesis correct?
   - *EXAMPLE: It appears that taking one aspirin per day for 60 days decreases blood pressure in 76% of the tested females ages 12-14, therefore the original hypothesis has been verified, that taking aspirin can decrease blood pressure.*

7. Repeating the work: Arguably, the most important part of scientific inquiry! When an experiment can be repeated and the same results obtained by different experimenters, that experiment is validated.

Included in this unit is a hands-on lab, the *Penny Test Lab*, that can be modified for use at any level (although it was initially designed for middle school use), to teach the steps of the scientific method. Students are given the simple task of determining the number of drops that can fit on the "Lincoln" side of a penny. As the lab is designed, the students quickly learn that even the most simple of experiments can contain many hidden variables that decrease the validity of the experiment.

Another student-centered activity has been included called *Come Fly with Us*. This activity makes a great first day activity to get kids warmed up to the scientific method. They will examine what happens to the spin direction of a paper helicopter when you fold the blades in different directions. Students construct a paper helicopter to test their hypothesis about how the helicopter will fly upon folding the blades in different directions.

Suggested Teaching Procedure

Day 1

1. Present notes on the steps of the scientific method as you see fit. Although this part is teacher-directed, ideally the steps should be presented as more of a discussion. Some questions to ask during your discussion are:
   - "What is the variable that the scientist changes?"
   - "What makes a valid experiment?"
   - "Why is it important to have detailed procedures for other scientists to repeat your experiment?"
   - "Why is the control such an important part of the experiment?"

   Also, students can be asked to imagine a scientific problem while the teacher asks students, "What is the independent variable in your problem?" and so on. An example of a scientific question was just presented in these teacher notes. You can use this example or make up another example to illustrate the prescribed steps. Students should record the steps in their notebooks.

2. After students take notes, pass out the student activity pages for *Come Fly With Us*.

3. Students should read the directions and perform the prescribed tasks in the procedure, applying new-found scientific method knowledge to the activity. Students should complete and turn in *Come Fly with Us* before leaving.

4. Assign *Scientific Method Practice 1* reading and questions for homework. Students are to return the completed questions the following class period.

Day 2

5. Use the answers that follow to review *Scientific Method Practice 1* after collecting the students' papers.

6. Pass out the student activity pages for the *Penny Test Lab*. Students should read the instructions and perform the lab during class.

7. After completing the lab, students should turn in a lab write-up at the end of the period.

8. Assign *Scientific Method Practice 2* reading and questions for homework. Students should answer questions and return the completed assignment the following class period.

Day 3

9. After collecting the students' papers, use the answers that follow to review *Scientific Method Practice 2* with students.

10. At this point you can begin an introduction to *Can Mosquitoes Transmit HIV Roleplay*, a complex student inquiry-based activity also found in this *Laying the Foundation* guide.

T E A C H E R   P A G E S

The Scientific Method
Exploring Experimental Design

Come Fly With Us

OBJECTIVE
Students will practice applying the steps of the scientific method to a problem.

LEVEL
All levels

NATIONAL STANDARDS
UCP.1, UCP.2, UCP.3, A.1, A.2, G.2

TEKS
6.1(A), 6.2(A), 6.2(B), 6.2(C), 6.2(D), 6.2(E), 6.3(A)
7.1(A), 7.2(A), 7.2(B), 7.2(C), 7.2(D), 7.2(E), 7.3(A)
8.1(A), 8.2 (A), 8.2(B), 8.2(C), 8.2(D), 8.2(E), 8.3(A)
IPC: 1(A), 2(A), 2(B), 2(C), 2(D), 3(A)
Biology: 1(A), 2(A), 2(B), 2(C), 2(D), 3(A)
Chemistry: 1(A), 2(A), 2(B), 2(C), 2(D), 2(E), 3(A)
Physics: 1(A), 2(A), 2(B), 2(C), 2(D), 2(E), 2(F), 3(A)

CONNECTIONS TO AP
Using the scientific method by acquiring data through experimentation and design of experiments are all fundamental skills needed for the AP Science courses.

TIME FRAME
45 minutes

MATERIALS

28 models of helicopter (provided)   28 pairs of scissors
28 pens or pencils

TEACHER NOTES
Come Fly With Us is an effective way for students to experimentally test a variable in a simple activity. This activity is designed to be the first activity that students do after learning the steps of the scientific method. The students can apply their newfound knowledge in a meaningful way.

Students cut out and fold a paper helicopter according to the instructions on the lab. After constructing the simple helicopter, students are instructed to fold the blades of the helicopter in opposing directions. Students generate a hypothesis as to how they think it will affect the direction of spin. The students then test their hypothesis and fly the helicopter after folding the blades in each direction. The students will discover that folding the blade one way will produce a clockwise spin of the helicopter. Folding the blades in the opposite direction will produce a counterclockwise spin of the helicopter. The students will ideally discover that applying the independent variable (folding the blades in opposing directions) causes the dependent variable to change (the direction of spin clockwise or counterclockwise). The students must also take into consideration the constants in their experiment: holding the helicopter at the same initial height, maintaining a stable wind environment, no other external forces acting on the helicopter, and holding the helicopter at the T each time. Although a control setup is not at first apparent, it is illustrated later.

The control is best illustrated in the second half of *Come Fly With Us*. A fictitious student, Bonita, believes that adding mass (paper clips) will stabilize her paper helicopter and increase the flight time. The independent variable is the presence or absence of the added mass. The control, by definition, does not receive the independent variable. Therefore, the control in Bonita's experiment is a paper helicopter with no paper clips added. The students can typically clearly envision the idea of a control. Further discussion of a control can describe how the experimental subject can only be truly tested when the results of the control setup are compared to the results of the experimental setup. Then the effect of the independent variable upon the dependent variable can clearly be seen. An extension of this lab could be to have the students actually try testing the extra weight and seeing how it affects the flight time.

POSSIBLE ANSWERS TO THE CONCLUSION QUESTIONS

1. In the helicopter experiment, what was the independent variable?
   - Folding the blades in different directions, with the black circle up and the white square down, or with the black circle down and the white square up.

2. What was the dependent variable?
   - The dependent variable is the direction of spin, clockwise or counterclockwise.

3. List three things you should try to keep constant each time you try this experiment.
   - There are many correct answers for this question. Possible answers include:
     - holding the helicopter in the same place (on the body versus the wing)
     - holding it at the same height
     - making sure there is no cross breeze each time
     - using the same helicopter
     - adding no extra force when letting it go each time

4. What is the problem question in Bonita's experiment?
   - Will adding extra mass in the form of paper clips to the helicopter stabilize it, making it stay in the air longer?

5. What is Bonita's hypothesis?
   - If additional paperclips are added to the helicopter, then the helicopter will be stabilized resulting in a longer flight time.

6. What is her independent variable?
   - Bonita's independent variable is the addition of paper clips (weight) to the helicopter.

7. What is her dependent variable?
   - Bonita's dependent variable is the amount of time the helicopter stays in the air.

8. What should her constants be?
   - Her constants should be the same as those listed in #3, plus: use the same size paper clips, attach the paper clips to the same place on the helicopter each time, etc...

9. What can she use for a control?
   - Her control is the same helicopter with no added mass.

10. Why should Bonita retest her experiment between 5-10 times?
    - Bonita should retest to make sure her results are reasonable and valid.

TEACHER PAGES

The Scientific Method
Exploring Experimental Design

Scientific Method Practice 1

POSSIBLE ANSWERS TO THE CONCLUSION QUESTIONS

NOTE: Problem, hypothesis and conclusion should all match in wording!

1. What was Erika's problem? [The problem should be stated as a question.]
   - Is the oven heating to the correct temperature? OR
   - Why didn't the cake rise?

2. What was Erika's hypothesis? [This is an answer to your problem question.]
   - No, the oven is not heating to the correct temperature. OR
   - The cake did not rise because the oven was not heating to the correct temperature.

3. What was Erika's conclusion? [This states whether your hypothesis was correct.]
   - The oven is heating to the correct temperature. OR
   - The oven was heating to the correct temperature and therefore could not have been the cause of the cake's failure to rise.

4. Which step in the scientific method do you think Erika should do next? Explain your reasoning.
   - Form a new hypothesis OR gather more information OR repeat the experiment.

5. List two other hypotheses which might explain why the cake did not rise.
   - Answers will vary

The Scientific Method
Exploring Experimental Design

Penny Test Lab

OBJECTIVE
Students will learn about controls and variables in an experiment. Additionally, they will learn what constitutes valid experimental procedure.

LEVEL
All levels

NATIONAL STANDARDS
UCP.1, UCP.2, UCP.3, A.1, A.2, G.2

TEKS
6.1(A), 6.2(A), 6.2(B), 6.2(C), 6.2(D), 6.2(E), 6.3(A)
7.1(A), 7.2(A), 7.2(B), 7.2(C), 7.2(D), 7.2(E), 7.3(A)
8.1(A), 8.2 (A), 8.2(B), 8.2(C), 8.2(D), 8.2(E), 8.3(A)
IPC: 1(A), 2(A), 2(B), 2(C), 2(D), 3(A)
Biology: 1(A), 2(A), 2(B), 2(C), 2(D), 3(A)
Chemistry: 1(A), 2(A), 2(B), 2(C), 2(D), 2(E), 3(A)
Physics: 1(A), 2(A), 2(B), 2(C), 2(D), 2(E), 2(F), 3(A)

CONNECTIONS TO AP
Using the scientific method by acquiring data through experimentation and design of experiments are all fundamental skills needed for the AP Science courses.

TIME FRAME
50 minutes

MATERIALS

28 pennies	28 calculators
28 eyedroppers	28 pieces of graph paper (in appendix)
28 small beakers of water	28 metric rulers
28 paper towel	28 pieces of notebook paper
28 pencils or pens	

TEACHER NOTES

The *Penny Test Lab* takes a simple problem, how many drops of water will fit onto the "Lincoln" side of a penny, and expands it into an excellent lab that can be used to study the steps of the scientific method. To summarize, students use an eyedropper to determine how many drops of water will fit onto the penny before it spills over. This seemingly simple task generates many diverse results. By definition, a valid experiment is one that can be repeated by anyone else with the same results obtained. The lack of similar student results verifies that there are hidden variables that are unaccounted for in the procedure. This lab procedure is designed to show the students what an invalid experiment looks like.

Some of the hidden variables include pennies of different ages and conditions and different droppers (some plastic, some glass). Also, no exact procedure is given to the students about how to hold the dropper, how much pressure to put on the dropper, how to make a drop, how to drop it onto the penny, or from what height the drop should be released.

Before class, draw on the chalkboard or the overhead a data table with three rows and as many columns as there are students in the classroom. The first row should be labeled "student initials". The second row will be labeled "predicted" and the bottom row be labeled "observed".

One of the first steps of the procedure is for students to make a prediction about the number of drops they believe will fit on the "Lincoln" side of a penny. If students have never done a lab like this before or have no knowledge of the cohesive properties of water, they tend to underestimate the number of drops that will actually fit. Have students write their initials and their predicted number of drops on the chalkboard. Students then perform three trials, take an average from these three trials, and round it to the nearest whole number. Once they have completed this, they should write their average whole number of drops in the space provided on the class data table.

student initials	jh	hs	if	hg	sc	br
predicted #	12	8	13	20	26	11
actual #	23	16	35	12	22	13

A partial data table is shown here.

After everyone has completed their trials, students will analyze the class data by graphing the frequency of ranges of drops. Together as a class, you and the students will count the number of people that averaged 0-10 drops. Repeat this procedure for the ranges of 11-20, 21-30, 31-40, 41-50, etc. Instruct students to make a bar graph (technically a histogram since the intervals are equivalent) showing the number of people on the *y*-axis versus the range of drops on the *x*-axis.

Here is an example of a possible student histogram.

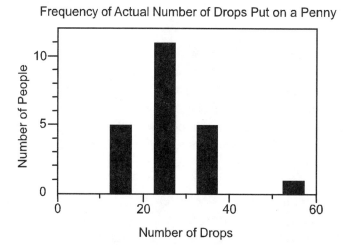

Frequency of Actual Number of Drops Put on a Penny

Before the students answer the questions, lead a discussion with students regarding what is considered a valid experiment versus an invalid experiment. Most students come to the conclusion that this is an invalid experiment due to the diverse results. Also, the discussion can include student ideas of hidden variables. These will include: different droppers, different pennies, no exact procedure for dropping, no definition of "drop," unstable table, etc. You can reveal to the students that this lab was purposefully designed to produce invalid results so that students could begin to understand that even in a simple task there can be many hidden variables.

POSSIBLE ANSWERS TO THE CONCLUSION QUESTIONS AND SAMPLE DATA

A typical data table for the student's 3 trials:

	Trial #1	Trial #2	Trial #3	Average
Number of Water Drops	_____	_____	_____	_____

1. Using your bar graph, determine if the average number of drops for each experimenter is about the same?
   - No, the results are not the same.

2. List four reasons why the actual number of drops for each experimenter was similar or dissimilar.
   - Student answers will vary
   - Four possible reasons:
     a. different types pennies
     b. no standard way of administering drops
     c. the table could be uneven
     d. each dropper is physically different and delivers drops of varying volume

3. Are the results of this experiment "valid"? Why or why not? Be sure to think about what makes an experiment valid.
   - Results are not valid in this experiment. To make an experiment valid, the results should be repeatable regardless of experimenter.

4. In this experiment, there were a limited number of constants. Name two of them.
   - Water (as opposed to alcohol or some other fluid)
   - Using pennies (as opposed to using nickels and pennies, etc…)

5. What was the independent variable in this experiment?
   - Student answers will vary. Some variation or factor that could affect the outcome (the dependent variable) could be accepted as the independent variable: height of dropper, size of dropper hole, pressure used when squeezing the dropper, size of drops, and so on.

6. What was the dependent variable in this experiment?
   - The dependent variable is the number of drops that fit on the head side of a penny.

7. Is it possible to state definitively how many drops of water will fit on the "Lincoln" side of a penny with this lab procedure? Why or why not?
   - Using this procedure it is not possible to state exactly how many drops fit onto a penny. This is not a valid experiment. There are too many hidden variables.

The Scientific Method
Exploring Experimental Design

Scientific Method Practice 2

POSSIBLE ANSWERS TO THE CONCLUSION QUESTIONS

1. State the problem in the form of a question.
   - What causes fresh water to freeze at a higher temperature than sea water?

2. Form a hypothesis to answer the problem question above based on the fact that fresh water does not contain salt.
   - The salt in sea water lowers the temperature at which water freezes.

3. According to the data table above, at what temperature did the experiment begin?
   - The experiment began at 25°C.

4. At what time intervals were the temperature measurements taken?
   - The time intervals were 5 minutes.

5. What conclusions can you draw from these graphs about the effect of salt on the freezing point of water?
   - Salt lowers the freezing point of water.

6. What can you say about the rate at which the temperature in the fresh water container dropped compared to the rate at which the temperature in the salt water container dropped?
   - The rate at which the temperature in the fresh water container dropped was the same as the rate at which the temperature in the salt water container dropped.

7. What was the independent variable in Stephanie and Amy's experiment?
   - The independent [manipulated] variable was the addition of salt.

8. What was the dependent variable?
   - The dependent [responding] variable was the temperature at which the water froze.

9. Explain why detailed, step-by-step written procedures are an essential part of any scientific experiment.
   - When a scientist writes a report on his or her experiment, it must be detailed enough so that scientists throughout the world can repeat the experiment for themselves. In many cases, it is only when an experiment has been repeated by scientists worldwide that it is considered to be accurate.

10. The following hypothesis is suggested to you: Water will heat up faster when placed under the direct rays of the sun than when placed under indirect, or angled, rays of the sun. Design an experiment to test this hypothesis. Be sure to number each step of your procedure. Identify your independent variable, dependent variable and control. Identify those things which will remain constant during your experiment.

    • Answers will vary.

The Scientific Method
Exploring Experimental Design

Overview

PURPOSE

Through this series of activities you will identify and apply the steps of the scientific method.

MATERIALS

Come Fly With Us   *Scientific Method Practice 1*
Penny Test Lab   *Scientific Method Practice 2*

PROCEDURE

Day 1

1. Take notes in your notebook from your teacher's discussion about the steps of the scientific method.

2. Do the *Come Fly With Us* activity in class. Turn in at the end of the period.

3. Do the *Scientific Method Practice 1* activity for homework. Be ready to turn in the write-up at the beginning of the next period.

Day 2

4. Turn in *Scientific Method Practice 1* to your teacher at the beginning of class. Your teacher will review the correct answers.

5. Begin *Penny Test Lab* after collecting your materials.

6. Turn in the *Penny Test Lab* write-up to your teacher after completing it.

7. Do the *Scientific Method Practice 2* activity for homework. Be ready to turn in the write-up at the beginning of the next period.

Day 3

8. Turn in *Scientific Method Practice 2* to your teacher at the beginning of class. Your teacher will review the correct answers.

The Scientific Method
Exploring Experimental Design

Come Fly With Us

This assignment is intended to be a quick and easy guide to the methods scientists use to solve problems. It should also give you information about how to "wing your way" through your own experiments. You are going to start by making a model helicopter with the attached instructions. You will be given a problem question, and it is your job to write a suitable hypothesis. Remember, your hypothesis should be a possible answer to the problem question and it should be based upon what you already know about a topic.

GLOSSARY OF WORDS USED IN CONDUCTING EXPERIMENTS

- **problem**: scientific question that can be answered by experimentation.
- **hypothesis**: an educated prediction about how the independent variable will affect the dependent variable stated in a way that is testable. This should be an "If…then…" statement.
- **variable**: a factor in an experiment that changes or could be changed
- **independent variable**: the variable that is changed on purpose.
- **dependent variable**: the variable that responds to the independent variable.
- **control**: the standard for comparison in an experiment; the independent variable is not applied to the control group.
- **constant**: a factor in an experiment that is kept the same in all trials.
- **repeated trials**: the number of times an experiment is repeated for each value of the independent variable.

PURPOSE
In this assignment you will practice applying the steps of the scientific method to a problem by experimentation.

MATERIALS

model of helicopter (provided)   scissors
pen

PROCEDURE

1. Find the section labeled Hypothesis on your student answer page. Read the problem question and respond with an appropriate hypothesis. Remember to use an "If...then..." format.

2. Once you have made your hypothesis, you should test it for accuracy. Stand on a chair and hold your helicopter by the "T" at shoulder level.

3. Drop the helicopter and note whether it spins clockwise or counterclockwise. Repeat this test several times.

4. Refold the blades so that the square on section Y shows when you look down on top of the helicopter.

5. Stand on a chair and hold your helicopter by the top of the "T" at shoulder level. Drop the helicopter and note whether it spins clockwise or counterclockwise.

6. Repeat this test several times.

Name _____

Period _____

The Scientific Method
Exploring Experimental Design

Come Fly With Us

PROBLEM

How will changing the direction that the paper helicopter blades are folded affect the "flight" of the helicopter?

HYPOTHESIS

ANALYSIS

You have just performed an experiment. Experiments involve changing something to see what happens. In this case, you refolded the helicopter blades. You made this change on purpose to learn about its effect on the flight of the helicopter. The parts of an experiment that change are called *variables*.

When designing an experiment, you should choose one variable that you will purposely change. You will measure the effect of this *independent variable* on another variable that you think will respond to the change. The responding variable is called the *dependent variable*.

If you kept every variable except the folds the same in each test, you were making it a fair test. Why? Only the variable you changed could be causing the dependent variable to change because everything else was kept constant.

To have a fair test, you also need a *control*, or a standard for comparison. A control for the helicopter experiment would be an "unchanged" helicopter against which you could compare the results. You could make another helicopter as your standard for comparison and not refold its blades.

It is important to note that in some experiments, it is impossible to have a control that is completely unchanged. For example, let's say you are trying to determine the effect of light from different light sources on plant growth. The control plant needs some kind of light in order to live through the experiment. So, you have to choose one light source — any one say, normal sunlight — to be the standard of comparison.

After you refolded the blades of the helicopter, you dropped the helicopter several times and observed the results. These repeated trials enable you to be more confident of your results. If you conducted your experiment only once, the results could be due to an error or a chance event, such as a draft. But, when you repeat your experiment many times and each time achieve similar results, you can be more confident that your findings are not due to an error or chance.

Read the following paragraph and then answer the conclusion questions that follow using complete sentences:

> Bonita wanted to know if adding mass to her paper helicopter would affect how long it would stay in the air. She predicted that adding some mass would help to stabilize the helicopter and keep it in the air longer than a helicopter without extra mass. She experimented with different numbers of paper clips attached to her helicopter.

CONCLUSION QUESTIONS

1. In the helicopter experiment, what was the independent variable?

2. What was the dependent variable?

3. List three things you should try to keep constant each time you try this experiment.
   a.
   b.
   c.

4. What is the problem question in Bonita's experiment?

5. What is Bonita's hypothesis?

6. What is her independent variable?

7. What is her dependent variable?

8. What should her constants be?

9. What can she use for a control?

10. Why should Bonita retest her experiment between 5-10 times?

PAPER HELICOPTER MODEL

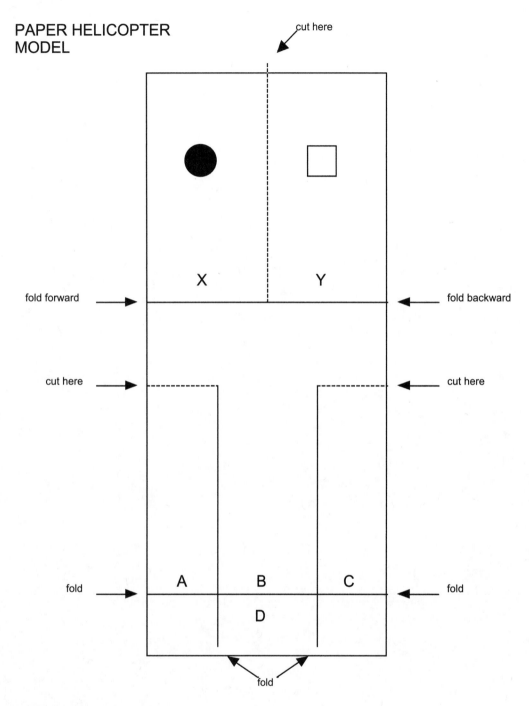

INSTRUCTIONS:

1. Cut out the rectangular helicopter (above).
2. Now cut along dotted lines.
3. Fold along the solid lines: section C behind section B, section A behind section B, and section D behind section B.
4. Complete the helicopter by folding blade X with the dot up and blade Y in the opposite direction with the square down.

Name _____

Period _____

The Scientific Method
Exploring Experimental Design

Scientific Method Practice 1

DIRECTIONS: *Read the following paragraphs and then answer the questions that follow on a separate sheet of paper. Use complete sentences to answer all questions. Be certain to restate the question in your answer.*

Science differs from other subject areas in the way it seeks to answer questions. This approach to problem solving is called the scientific method. The scientific method is a systematic approach to problem solving. The basic steps, in no particular order, of the scientific method are:

- Stating the problem
- Gathering information on the problem
- Forming a hypothesis
- Performing experiments to test the hypothesis
- Recording and analyzing data
- Stating a conclusion
- Repeating the work

Erika baked a cake for her mother's birthday. When the cake was taken from the oven, Erika noticed that the cake had not risen. She guessed that the oven had not heated to the correct temperature. She set up the following experiment to test her hypothesis.

First, Erika put a thermometer in the oven. She then turned the oven dial to 375°F. She noticed that the preheating light came on when she turned the oven on. She waited until the preheating light went out, indicating that the oven was up to temperature. Erika then read the thermometer within the oven. It read 400°F. Erika concluded that the oven was heating properly.

CONCLUSION QUESTIONS

1. What was Erika's problem? [The problem should be stated as a question.]

2. What was Erika's hypothesis? [This is an answer to your problem question.]

3. What was Erika's conclusion? [This states whether your hypothesis was correct.]

4. Which step in the scientific method do you think Erika should do next? Explain your reasoning.

5. List two other hypotheses which might explain why the cake did not rise.

Name _____

Period _____

The Scientific Method
Exploring Experimental Design

Penny Test Lab

PURPOSE
In this activity you will learn about controls and variables in an experiment. You will also learn what constitutes valid experimental procedure.

MATERIALS

penny	calculator (optional)
eyedropper	graph paper
water	ruler
paper towel	pen

DEFINITIONS
- **variable**: things in an experiment that change or could be changed.
- **independent variable**: variable that is changed on purpose.
- **dependent variable**: variable that responds to the independent variable.
- **constant**: things in an experiment that are kept the same in all trials.

PROCEDURE
1. Answer each of the following questions using complete sentences. For fill-in-the blank statements, copy the entire sentence.

2. Copy the lab purpose onto your paper.

3. Your task is to guess how many drops of water will fit on the "Lincoln" side of a penny.

   [Copy the following statement.] PROBLEM: How many drops of water will fit onto the "Lincoln" side of a penny?

4. [Copy the following statement and make a prediction by filling in the blank.] HYPOTHESIS: I predict that _____ drops of water will fit on the head side of a penny.

5. After you have made your hypothesis and have written it down on your lab paper, you will write it on the chalkboard under the heading "Predicted Number of Drops."

6. Copy the following chart onto your paper. Be neat and use a ruler!

TEST RESULTS:

	Trial #1	Trial #2	Trial #3	Average
Number of Water Drops	_____	_____	_____	_____

7. Test to see if your hypothesis is correct. Place your penny on a paper towel and, using the eyedropper, add water to the "Lincoln" side of the penny, one drop at a time, counting each drop until the water spills over. Do not count the drop that causes the water to spill over. Write the number of drops you counted under Trial #1 on your chart. Repeat this procedure two more times, filling in the number of drops you counted for each trial under the appropriate heading on your Test Results chart.

8. Find the average of your three trials, round your answer to a whole number (no decimals), and write the average number of drops on your Test Results chart. Then write your average on the chalkboard under the heading "Actual Average".

9. Write a sentence that will serve as your conclusion for this experiment. Remember that your conclusion should state whether your hypothesis was correct.

10. Make a bar graph of the class test results using the data from "Actual Average" on the chalkboard. The *x*-axis (horizontal line) should be titled "Average Number of Drops" and the *y*-axis (vertical line) should be titled "Number of Tests." Before graphing, you will need to organize the class data into ranges — make a chart that shows how many people got averages between 0-10, 11-20, 21-30, etc. When you have finished your bar graph, give it an appropriate title.

Answer the conclusion questions on your paper. Be sure to use complete sentences.

CONCLUSION QUESTIONS

1. Using your bar graph, determine if the average number of drops for each experimenter is about the same.

2. List four reasons why the actual number of drops for each experimenter was similar or dissimilar.

3. Are the results of this experiment "valid"? Why or why not? Be sure to think about what makes an experiment valid.

4. In this experiment, there were a limited number of constants. Name two of them.

5. What was the independent variable in this experiment?

6. What was the dependent variable in this experiment?

7. Is it possible to state definitively how many drops of water will fit on the "Lincoln" side of a penny with this lab procedure? Why or why not?

Name _____

Period _____

The Scientific Method
Exploring Experimental Design

Scientific Method Practice 2

DIRECTIONS: *Answer these questions on a separate sheet of paper using a black or dark blue ink pen. Use complete sentences to answer all questions. Be certain to restate the question in your answer!*

Stephanie and Amy were vacationing in Canada. Bundled up in warm clothing, they walked along the beach. Glistening strips of ice hung from the roofs of the beach houses. Only yesterday, Stephanie commented, these beautiful icicles had been a mass of melting snow. Throughout the night, the melted snow had continued to drip, freezing into lovely shapes. Near the ocean's edge, Amy spied a small pool of sea water. Surprisingly, she observed, it was not frozen as were the icicles on the roofs. What could be the reason, they wondered?

A scientist might begin to solve the problem by gathering information. The scientist would first find out how the sea water in the pool differs from the fresh water on the roof. This information might include the following facts: The pool of sea water rests on sand, while the fresh water drips along a tar roof. The sea water is exposed to the cold air for less time than the fresh water. The sea water is saltier than the fresh water.

Using all of the information that has been gathered, the scientist might be prepared to suggest a possible solution to the problem. A proposed solution to a scientific problem is called a hypothesis. A hypothesis almost always follows the gathering of information about a problem. Sometimes, however, a hypothesis is a sudden idea that springs from a new and original way of looking at a problem.

A scientist (or a science student) does not stop once a hypothesis has been suggested. In science, evidence that either supports a hypothesis or does not support it must be found. This means that a hypothesis must be tested to show whether it is correct. Such testing is usually done by performing experiments.

Experiments are performed according to specific rules. By following these rules, scientists can be confident that the evidence they uncover will clearly support or not support a hypothesis. For the problem of the sea water and freshwater, a scientist would have to design an experiment that ruled out every factor but salt as the cause of the different freezing temperatures. Stephanie and Amy, being excellent science students, set up their experiment in the following as follows.

First, they put equal amounts of fresh water into two identical containers. Then Stephanie added salt to only one of the containers. [The salt is the variable. In any experiment, only one variable should be tested at a time. In this way, scientists can be fairly certain that the results of the experiment are caused by one and only one factor — in this case the variable of salt.] To eliminate the possibility of hidden or unknown variables, Stephanie and Amy conducted a control experiment. A control experiment is set up exactly like the one that contains the variable. The only difference is that the control experiment does not contain the variable. Scientists compare the results of an experiment to a control experiment.

In the control experiment, Stephanie and Amy used two containers of the same size with equal amounts of water. The water in both containers was at the same starting temperature. The containers were placed side by side in the freezing compartment of a refrigerator and checked every five minutes. But only one container had salt in it. In this way, they could be fairly sure that any differences that occurred in the two containers were due to the single variable of salt. In such experiments, the part of the experiment with the variable is called the experimental setup. The part of the experiment with the control is called the control setup.

Stephanie and Amy collected the following data: the time intervals at which the containers were observed, the temperatures of the water at each interval, and whether the water in either container was frozen or not. They recorded the data in the tables below and then graphed their results.

WATER (control setup)

Time (min)	0	5	10	15	20	25	30	
Temperature (°C)	25	20	15	10	5	0*	-10	

   *Asterisk means liquid has frozen

WATER WITH SALT (experimental setup)

Time (min)	0	5	10	15	20	25	30	
Temperature (°C)	25	20	15	10	5	0	-10*	

Stephanie and Amy might be satisfied with their conclusion after just one test. For a scientist, however, the results from a single experiment are not enough to reach a conclusion. A scientist would want to repeat the experiment many times to be sure the data were reproducible. So, a scientific experiment must be able to be repeated. And before the conclusion of a scientist can be accepted by the scientific community, other scientists must repeat the experiment and check the results. Consequently, when a scientist writes a report on his or her experiment, that report must be detailed enough so that scientists throughout the world can repeat the experiment for themselves. In most cases, it is only when an experiment has been repeated by scientists worldwide that it is considered to be accurate and worthy of being included in new scientific research.

By now it might seem as if science is a fairly predictable way of studying the world. After all, you state a problem, gather information, form a hypothesis, run an experiment, and determine a conclusion. Well, sometimes it isn't so neat and tidy.

In practice, scientists do not always follow all the steps in the scientific method. Nor do the steps always follow the same order. For example, while doing an experiment a scientist might observe something unusual or unexpected. That unexpected event might cause the scientist to discard the original hypothesis and suggest a new one. In this case, the hypothesis actually followed the experiment.

As you already learned, a good rule to follow is that all experiments should have only one variable. Sometimes, however, scientists run experiments with several variables. Naturally, the data in such experiments are much more difficult to analyze. For example, suppose scientists want to study lions in their natural environment in Africa. It is not likely they will be able to eliminate all the variables in the environment and concentrate on just a single lion. So, although a single variable is a good rule and you will follow this rule in almost all of the experiments you design or perform, it is not always practical in the real world.

There is yet another step in the scientific method that cannot always be followed. Believe it or not, many scientists search for the truths of nature without ever performing experiments. Sometimes the best they can rely on are observations and natural curiosity. Here is an example. Charles Darwin is considered the father of the theory of evolution, how living things change over time. Much of what we know about evolution is based on Darwin's work. Yet Darwin did not perform a single controlled evolutionary experiment! He based his hypotheses and theories on his observations of the natural world. Certainly it would have been better had Darwin performed experiments to prove his theory of evolution. But as the process of evolution generally takes thousands, even millions of years, performing an experiment would be a bit too time consuming.

CONCLUSION QUESTIONS

1. State Stephanie and Amy's problem in the form of a question.

2. Form a hypothesis (to answer the problem question above) based on the fact that fresh water does not contain salt.

3. According to the data table above, at what temperature did the experiment begin?

4. At what time intervals were the temperature measurements taken?

5. What conclusions can you draw from these graphs about the effect of salt on the freezing point of water?

6. What can you say about the rate at which the temperature in the fresh water container dropped compared to the rate at which the temperature in the salt water container dropped?

7. What was the independent variable in Stephanie and Amy's experiment?

8. What was the dependent variable?

9. Explain why detailed, step-by-step written procedures are an essential part of any scientific experiment.

10. The following hypothesis is suggested to you: Water will heat up faster when placed under the direct rays of the sun than when placed under indirect, or angled, rays of the sun. Design an experiment to test this hypothesis. Be sure to number each step of your procedure. Identify your independent variable, dependent variable and control. Identify those things which will remain constant during your experiment.

Numbers in Science
Exploring Measurements, Significant Digits, and Dimensional Analysis

OBJECTIVE
Students will be introduced to correct measurement techniques, correct use of significant digits, and dimensional analysis.

LEVEL
All

NATIONAL STANDARDS
UCP.1, UCP.3, A.1, G.2

TEKS
6.1(A), 6.2(A), 6.2(B), 6.2(C), 6.2(D), 6.4(A)
7.1(A), 7.2(A), 7.2(B), 7.2(C), 7.2(D), 7.4(A)
8.1(A), 8.2 (A), 8.2(B), 8.2(C), 8.2(D), 8.4(A)
IPC: 1(A), 2(A), 2(B), 2(C), 2(D)
Biology: 1(A), 2(A), 2(B), 2(C), 2(D)
Chemistry: 1(A), 2(A), 2(B), 2(C), 2(E)
Physics: 1(A), 2(A), 2(B), 2(C), 2(D), 2(F)

CONNECTIONS TO AP
Students are expected to report measurements and perform calculations with the correct number of significant digits.

TIME FRAME
180 minutes, depending on level

MATERIALS

small cube	spherical object
metric ruler	tweezers
200 mL beaker	flexible tape measure
large graduated cylinder	balance

TEACHER NOTES
Small wooden alphabet blocks or dice should be inexpensive and easy to obtain. Cubic shaped ice could also be used. Be sure to find a cube/graduated cylinder combination that ensures total submersion of the cube since its volume will be determined by water displacement.

Spherical objects might be a large marble or small rubber ball. Again, be sure to check the sphere/cylinder size to ensure that total submersion of the sphere is possible.

For the flexible tape measure, photocopy the metric side of a tape measure and have students cut it out on paper. Or, provide students with a length of string and metric ruler. The string can be wrapped around the sphere, marked, and then removed and measured.

This lesson is designed to introduce or reinforce accurate measurement techniques, correct usage of significant digits, and dimensional analysis. Dimensional analysis is also called the Factor-Label method or Unit-Label method and is a technique for setting up problems based on unit cancellations. Lecture as well as guided and independent practice of these topics should precede this activity. Students should be provided with reference tables containing metric and English conversion factors.

The purpose of significant digits is to communicate the accuracy of a measurement as well as the measuring capacity of the instrument used. Remind students repeatedly to take measurements including an estimated digit and to perform their calculations with the correct number of significant digits. Emphasize that points will be deducted for answers containing too many or too few significant digits. The correct number of significant digits to be reported by your students will depend entirely upon your equipment.

POSSIBLE ANSWERS TO CONCLUSION QUESTIONS AND SAMPLE DATA

Introduction
Left: _____5.75 mL_____

Middle: _____3.0 mL_____

Right: _____0.33 mL_____

DATA AND OBSERVATIONS

Data Table		
Cube		
Mass: 15.05 g (4sd)		
Dimensions	length: 3.68 cm (3sd)	width: 3.65 cm (3 sd)	height: 3.67 cm (3 sd)
Volume	Beaker initial volume: 100 mL (1sd)	Beaker Final volume: 150 mL (2 sd)
	Graduated cylinder initial volume: 175.0 mL (4 sd)	Graduated cylinder final volume: 225.1 mL (4 sd)
Sphere		
Mass: 19.38 g (4sd)		
Dimensions	Circumference: 7.62 cm (3sd)	
Volume	Beaker initial volume: 100 mL (1sd)	Beaker final volume: 110 mL (2sd)
	Graduated cylinder initial volume: 175.0 mL (4sd)	Graduated cylinder final volume: 182.3 mL (4sd)
Formula for calculating the volume of a cube:	V = length x width x height	
Formula for calculating the circumference of a circle:	C = πd	
Formula for calculating the diameter of a circle:	d = 2r	
Formula for calculating the volume of a sphere:	$V = \frac{4}{3}\pi r^3$	

ANALYSIS

- Remember to follow the rules for reporting all data and calculated answers with the correct number of significant digits.
- You may need tables of metric and English conversion factors to work some of these problems.

1. For each of the measurements you recorded above, go back and indicate the number of significant digits in parentheses after the measurement. Ex: 15.7 cm (3sd)
   - The number of significant digits will be determined by the equipment you are using.

2. Use dimensional analysis to convert the mass of the cube to (a) mg and (b) ounces.

   (a) $\quad 15.05 \, \cancel{g} \times \dfrac{1000 \, mg}{1 \, \cancel{g}} = 15,050 \, mg$

   (b) $\quad 15.05 \, \cancel{g} \times \dfrac{1 \, \cancel{lb}}{454 \, \cancel{g}} \times \dfrac{16 \, oz}{1 \, \cancel{lb}} = 0.5304 \, oz$

3. Calculate the volume of the cube in cm^3.
   - V=l×w×h

     $V = 3.68 \, cm \times 3.65 \, cm \times 3.67 \, cm = 49.3 \, cm^3$

4. Use dimensional analysis to convert the volume of the cube from cm^3 to m^3.

- $49.3\,cm^3 \times \dfrac{1\,m}{100\,cm} \times \dfrac{1\,m}{100\,cm} \times \dfrac{1\,m}{100\,cm} = 4.93 \times 10^{-5}\,m^3$

5. Calculate the volume of the cube in mL as measured in the beaker. Convert to cm^3 knowing that $1\ cm^3 = 1$ mL.

- $V = V_{final} - V_{initial}$

   $V = 150\,mL - 100\,mL$

   $V = 50\,mL = 50\,cm^3$

6. Calculate the volume of the cube in mL as measured in the graduated cylinder. Convert to cm^3 knowing that $1\ cm^3 = 1$ mL.

- $V = V_{final} - V_{initial}$

   $V = 225.1\,mL - 175.0\,mL$

   $V = 50.1\,mL = 50.1\,cm^3$

7. Using the density formula $D = \dfrac{mass}{volume}$, calculate the density of the cube as determined by the (a) ruler (b) beaker (c) graduated cylinder.

(a) $D = \dfrac{15.05\,g}{49.3\,cm^3}$

   $D = 0.305\ \dfrac{g}{cm^3}$

(c) $D = \dfrac{15.05\,g}{50.1\ cm^3}$

   $D = 0.300\ \dfrac{g}{cm^3}$

(b) $D = \dfrac{15.05\,g}{50\ cm^3}$

   $D = 0.3\ \dfrac{g}{cm^3}$

8. Use dimensional analysis to convert these three densities into kg/m^3.

(a) $0.305\,\dfrac{g}{cm^3} \times \dfrac{1\,kg}{1000\,g} \times \dfrac{100\,cm}{1\,m} \times \dfrac{100\,cm}{1\,m} \times \dfrac{100\,cm}{1\,m} = 305\,\dfrac{kg}{m^3}$

(b) $0.3\,\dfrac{g}{cm^3} \times \dfrac{1\,kg}{1000\,g} \times \dfrac{100\,cm}{1\,m} \times \dfrac{100\,cm}{1\,m} \times \dfrac{100\,cm}{1\,m} = 300\,\dfrac{kg}{m^3}$

(c) $0.300\,\dfrac{g}{cm^3} \times \dfrac{1\,kg}{1000\,g} \times \dfrac{100\,cm}{1\,m} \times \dfrac{100\,cm}{1\,m} \times \dfrac{100\,cm}{1\,m} = 300\,\overline{}\dfrac{kg}{m^3}$

- The bar above the last zero of the number 300 communicates it is a significant zero transforming the recorded answer from one significant digit to three. It is equally appropriate to teach your students to use scientific notation to effectively communicate three significant digits. The number would be correctly written as 3.00 x 10^2. Another way to communicate a number accurate to the ones position is to use a decimal at the end of the number. The number could be written as 300. representing that this measurement is accurate to the last digit.

9. Convert the mass of the sphere to (a) kg and (b) lbs.

   (a) $19.38 \, g \times \dfrac{1 \, kg}{1000 \, g} = 0.01938 \, kg$

   (b) $19.38 \, g \times \dfrac{1 \, lb}{454 \, g} = 0.04269 \, lbs$

10. Using the measured circumference, calculate the diameter of the sphere.
    - $C = \pi d$

    $d = \dfrac{C}{\pi}$

    $d = \dfrac{7.62 \, cm}{3.14} = 2.43 \, cm$

11. Calculate the radius of the sphere.
    - $d = 2r$

    $r = \dfrac{d}{2}$

    $r = \dfrac{2.43 \, cm}{2} = 1.22 \, cm$

12. Calculate the volume of the sphere from its radius.
    - $V = \dfrac{4}{3} \pi r^3$

    $V = \dfrac{4}{3} (\pi)(1.22)^3$

    $V = 7.61 \, cm^3$

13. Calculate the volume of the sphere in mL as measured in the beaker. Convert to cm^3 knowing that 1 cm^3 = 1 mL.
    - $V = V_{final} - V_{initial}$

    $V = 110 \, mL - 100 \, mL$

    $V = 10 \, mL = 10 \, cm^3$

14. Calculate the volume of the sphere in mL as measured in the graduated cylinder. Convert to cm^3 knowing that $1\ cm^3 = 1\ mL$.

- $V = V_{final} - V_{initial}$

  $V = 182.3\,mL - 175.0\,mL$

  $V = 7.3\,mL = 7.3\,cm^3$

15. Using the density formula $D = \dfrac{mass}{volume}$, calculate the density of the sphere as determined by the (a) tape measure (b) beaker (c) graduated cylinder.

(a) $D = \dfrac{19.38\,g}{7.60\,cm^3}$

$D = 2.55\ \dfrac{g}{cm^3}$

(c) $D = \dfrac{19.38\,g}{7.3\,cm^3}$

$D = 2.7\ \dfrac{g}{cm^3}$

(b) $D = \dfrac{19.38\,g}{10\,cm^3}$

$D = 2\ \dfrac{g}{cm^3}$

16. Use dimensional analysis to convert the densities into lbs/ft^3.

(a) $\dfrac{2.55\,g}{cm^3} \times \dfrac{1\,lb}{454\,g} \times \left(\dfrac{2.54\,cm}{1\,in}\right)^3 \times \left(\dfrac{12\,in}{1\,ft}\right)^3 = 159\,\dfrac{lbs}{ft^3}$

(b) $\dfrac{2\,g}{cm^3} \times \dfrac{1\,lb}{454\,g} \times \left(\dfrac{2.54\,cm}{1\,in}\right)^3 \times \left(\dfrac{12\,in}{1\,ft}\right)^3 = 100\,\dfrac{lbs}{ft^3}$

(c) $\dfrac{2.7\,g}{cm^3} \times \dfrac{1\,lb}{454\,g} \times \left(\dfrac{2.54\,cm}{1\,in}\right)^3 \times \left(\dfrac{12\,in}{1\,ft}\right)^3 = 170\,\dfrac{lbs}{ft^3}$

TEACHER PAGES

CONCLUSION QUESTIONS

1. Compare the densities of the cube when the volume is measured by a ruler, beaker and graduated cylinder. Which of the instruments gave the most accurate density value? Use the concept of significant digits to explain your answer.
   - The density of the cube had 3 significant digits when measured with the ruler. After subtracting to find the difference between the initial and final water levels in the graduated cylinder and beaker there are 2 significant digits when measured with the graduated cylinder but only 1 significant digit when measured with the beaker.
   - The ruler is the more accurate measure of the volume when compared to the volume obtained by water displacement using the graduated cylinder. The tweezers used to submerge the cube will contribute a small amount to the volume recorded since they contribute to the TOTAL amount of water displaced. See if you can get your students to come up with this concept!
   - Student answers may vary in significant digits depending on equipment used.

2. A student first measures the volume of the cube by water displacement using the graduated cylinder. Next, the student measures the mass of the cube before drying it. How will this error affect the calculated density of the cube? Your answer should state clearly whether the calculated density will increase, decrease or remain the same and must be justified.
   - The calculated density of the cube would increase.
   - Measuring a wet block will make the mass appear greater. Since mass is in the numerator of the equation $D = \dfrac{mass}{volume}$, the density value reported will be too large.

3. A student measures the circumference of a sphere at a point slightly higher than the middle of the sphere. How will this error affect the calculated density of the sphere? Your answer should state clearly whether the calculated density will increase, decrease, or remain the same and must be justified.
   - The density of the sphere would increase.
   - If the student measured the circumference at any point other than the center, the circumference would be reported as too low.
   - If the circumference is too small then the diameter will be too small.
   - If $\dfrac{C_\downarrow}{\pi} = d \therefore d_\downarrow$ the diameter is reported as too small, the radius will also be reported as too small.
   - If $\dfrac{d_\downarrow}{2} = r \therefore r_\downarrow$ the radius is reported as too small then the volume will be reported as too small.
   - If $V = \dfrac{4}{3}\pi r_\downarrow^3 \therefore V_\downarrow$ the volume is reported as too small the density will be reported as too large.

   $Density = \dfrac{m}{V_\downarrow} \therefore Density_\uparrow$

Laying the Foundation in Chemistry

Numbers in Science
Exploring Measurements, Significant Digits, and Dimensional Analysis

TAKING MEASUREMENTS

The accuracy of a measurement depends on two factors: the skill of the individual taking the measurement and the capacity of the measuring instrument. When making measurements, you should always read to the smallest mark on the instrument and then estimate another digit beyond that.

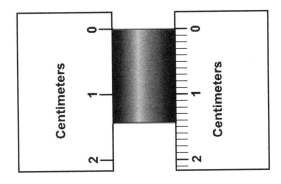

For example, if you are reading the length of the steel pellet pictured above using only the ruler shown to the left of the pellet, you can confidently say that the measurement is between 1 and 2 centimeters. However, you MUST also include one additional digit estimating the distance between the 1 and 2 centimeter marks. The correct measurement for this ruler should be reported as 1.5 centimeters. It would be incorrect to report this measurement as 1 centimeter or even 1.50 centimeters given the scale of this ruler.

What if you are using the ruler shown on the right of the pellet? What is the correct measurement of the steel pellet using this ruler? 1.4 centimeters? 1.5 centimeters? 1.40 centimeters? 1.45 centimeters? The correct answer would be 1.45 centimeters. Since the smallest markings on this ruler are in the tenths place we must carry our measurement out to the hundredths place.

If the measured value falls exactly on a scale marking, the estimated digit should be zero.

The temperature on this thermometer should read 30.0°C. A value of 30°C would imply this measurement had been taken on a thermometer with markings that were 10° apart, not 1° apart.

When using instruments with digital readouts you should record all the digits shown. The instrument has done the estimating for you.

When measuring liquids in narrow glass graduated cylinders, most liquids form a slight dip in the middle. This dip is called a *meniscus*. Your measurement should be read from the bottom of the meniscus. Plastic graduated cylinders do not usually have a meniscus. In this case you should read the cylinder from the top of the liquid surface. Practice reading the volume contained in the 3 cylinders below. Record your values in the space provided.

Left:_____

Middle: _____

Right:_____

SIGNIFICANT DIGITS

There are two kinds of numbers you will encounter in science, exact numbers and measured numbers. *Exact numbers* are known to be absolutely correct and are obtained by counting or by definition. Counting a stack of 12 pennies is an exact number. Defining 1 day as 24 hours are exact numbers. Exact numbers have an infinite number of significant digits.

Measured numbers, as we've seen above, involve some estimation. Significant digits are digits believed to be correct by the person making and recording a measurement. We assume that the person is competent in his or her use of the measuring device. To count the number of significant digits represented in a measurement we follow 2 basic rules:

1. If the digit is NOT a zero, it is significant.

2. If the digit IS a zero, it is significant if
   a. It is a sandwiched zero
      OR

   b. It terminates a number containing a decimal place

Examples:
>   3.57 mL has 3 significant digits (Rule 1)
>   288 mL has 3 significant digits (Rule 1)
>   20.8 mL has 3 significant digits (Rule 1 and 2a)
>   20.80 mL has 4 significant digits (Rules 1, 2a and 2b)
>   0.01 mL has only 1 significant digit (Rule 1)
>   0.010 mL has 2 significant digits (Rule 1 and 2b)
>   0.0100 mL has 3 significant digits (Rule 1 and 2b)
> 3.20×10^4 kg has 3 significant digits (Rule 1 and 2b)

SIGNIFICANT DIGITS IN CALCULATIONS

A calculated number can never contain more significant digits than the measurements used to calculate it.

Calculation rules fall into two categories:

1. Addition and Subtraction: answers must be rounded to match the measurement with the least number of decimal places.
   >   37.24 mL + 10.3 mL = 47.54 (calculator value), report as 47.5 mL

2. Multiplication and Division: answers must be rounded to match the measurement with the least number of significant digits.
   >   1.23 cm x 12.34 cm = 15.1782 (calculator value), report as 15.2 cm^2

DIMENSIONAL ANALYSIS

Throughout your study of science it is important that a unit accompanies all measurements. Keeping track of the units in problem can help you convert one measured quantity into its equivalent quantity of a different unit or set up a calculation without the need for a formula.

In conversion problems, equality statements such as 1 ft. = 12 inches, are made into fractions and then strung together in such a way that all units except the desired one are canceled out of the problem. Remember that defined numbers, such as the 1 and 12 above, are exact numbers and thus will not affect the number of significant digits in your answer. This method is also known as the Factor-Label method or the Unit-Label method.

To set up a conversion problem follow these steps.

1. Think about and write down all the "=" statements you know that will help you get from your current unit to the new unit.

2. Make fractions out of your "=" statements (there could be 2 fractions for each "="). They will be reciprocals of each other.

3. Begin solving the problem by writing the given amount with units on the left side of your paper and then choose the fractions that will let a numerator unit be canceled with a denominator unit and vice versa.

4. Using your calculator, read from left to right and enter the numerator and denominator numbers in order. Precede each numerator number with a multiplication sign and each denominator number with a division sign. Alternatively, you could enter all of the numerators, separated by multiplication signs, and then all of the denominators, each separated by a division sign.

5. Round your calculator's answer to the same number of significant digits that your original number had.

Example:
How many inches are in 1.25 miles?

Solution:

$$1\,ft=12\,in \qquad \frac{1\,ft}{12\,in} \ \ OR \ \ \frac{12\,in}{1\,ft}$$

$$5280\,ft.=1\,mile \qquad \frac{5280\,ft.}{1\,mile} \ \ OR \ \ \frac{1\,mile}{5280\,ft.}$$

$$1.25\,\cancel{miles} \times \frac{5280\,\cancel{ft.}}{1\,\cancel{mile}} \times \frac{12\,in.}{1\,\cancel{ft.}} = 79{,}200\,in.$$

As problems get more complex the measurements may contain fractional units or exponential units. To handle these problems treat each unit independently. Structure your conversion factors to ensure that all the given units cancel out with a numerator or denominator as appropriate and that your answer ends with the appropriate unit. Sometimes information given in the problem is an equality that will be used as a conversion factor.

Example: Suppose your automobile tank holds 23 gal and the price of gasoline is 33.5¢ per L. How many dollars will it cost you to fill your tank?

Solution: From a reference table we will find,
   1 L = 1.06 qt
   4 qt = 1 gal

We should recognize from the problem that the price is also an equality, 33.5¢ = 1 L and we should know that 100¢ = 1 dollar

Setting up the factors we find,

$$23\,\cancel{gal} \times \frac{4\,\cancel{qt}}{1\,\cancel{gal}} \times \frac{1\,\cancel{L}}{1.06\,\cancel{qt}} \times \frac{33.5\,\cancel{¢}}{1\,\cancel{L}} \times \frac{\$1}{100\,\cancel{¢}} = \$29$$

In your calculator you should enter $23 \times 4 \div 1.06 \times 33.5 \div 100$ and get 29.0754717. However, since the given value of 23 gal has only 2 significant digits, your answer must be rounded to $29.

Squared and cubed units are potentially tricky. Remember that a cm^2 is really cm x cm. So, if we need to convert cm^2 to mm^2 we need to use the conversion factor 1 cm = 10 mm twice so that both centimeter units cancel out.

Example: One liter is exactly 1000 cm^3. How many cubic inches are there in 1.0 L?

Solution:
We should know that

$$1000 \text{ cm}^3 = 1 \text{ L}$$

From a reference table we find,

$$1 \text{ in.} = 2.54 \text{ cm}$$

Setting up the factors we find,

$$1.0 \, \cancel{L} \times \frac{1000 \, \cancel{cm} \times \cancel{cm} \times \cancel{cm}}{1 \, \cancel{L}} \times \frac{1 \text{ in}}{2.54 \, \cancel{cm}} \times \frac{1 \text{ in}}{2.54 \, \cancel{cm}} \times \frac{1 \text{ in}}{2.54 \, \cancel{cm}} = 61 \text{ in}^3$$

(The answer has 2 significant digits since our given 1.0 L contained two significant digits.)

As you become more comfortable with the concept of unit cancellation you will find that it is a very handy tool for solving problems. By knowing the units of your given measurements, and by focusing on the units of the desired answer you can derive a formula and correctly calculate an answer. This is especially useful when you've forgotten, or never knew, the formula!

Example: Even though you may not know the exact formula for solving this problem, you should be able to match the units up in such a way that only your desired unit does not cancel out.

What is the volume in liters of 1.5 moles of gas at 293 K and 1.10 atm of pressure?

The ideal gas constant is $\dfrac{0.0821 \text{ L} \bullet \text{atm}}{\text{mol} \bullet \text{K}}$

Solution: It is not necessary to know the formula for the ideal gas law to solve this problem correctly. Working from the constant, since it sets the units, we need to cancel out every unit except L. Doing this shows us that moles and kelvins need to be in the numerator and atmospheres in the denominator.

$$\frac{0.0821 \text{ L} \bullet \cancel{atm}}{\cancel{mol} \bullet \cancel{K}} \times \frac{1.5 \, \cancel{mol}}{} \times \frac{293 \, \cancel{K}}{} \times \frac{}{1.10 \, \cancel{atm}} = 33 \text{ L}$$

(2 significant digits since our least accurate measuement has only 2 sig.digs.)

**NOTE: NEVER consider the number of significant digits in a constant to determine the number of significant digits for reporting your calculated answer. Consider ONLY the number of significant digits in given or measured quantities.

PURPOSE

In this activity you will review some important aspects of numbers in science and then apply those number handling skills to your own measurements and calculations.

MATERIALS

small cube	spherical object
metric ruler	tweezers
200 mL beaker	flexible tape measure
large graduated cylinder	balance

PROCEDURE

*Remember when taking measurements it is your responsibility to estimate a digit between the two smallest marks on the instrument.

1. Mass the small cube on a balance and record your measurement in the data table on your student page. *5.60 grams*

2. Measure dimensions (the length, width and height) of the small cube in centimeters, being careful to use the full measuring capacity of your ruler. Record the lengths in your data table. *1.42 cm .*

3. Fill the 200 mL beaker with water to the 100 mL line. Carefully place the cube in the beaker and use the tweezers to gently submerge the cube. The cube should be just barely covered with water. Record the new, final volume of water. *100ml 102.0 ml*

4. Fill the large graduated cylinder ¾ of the way full with water. Record this initial water volume. Again, use the tweezers to gently submerge the cube and record the final water volume. *75 ml → 78.0 .*

5. Mass the spherical object on a balance and record your measurement in the data table. *25.25 grams*

6. Use the flexible tape measure to measure the widest circumference of the sphere in centimeters. Be careful to use the full measuring capacity of the tape measure. *6 cm. 6.0*

7. Fill the 200 mL beaker with water to the 100 mL line. Carefully place the spherical object in the beaker and, if needed, use the tweezers to gently submerge the sphere. Record the final volume of water from the beaker.

$100 \longrightarrow$ 101.0 ml

8. Fill the large graduated cylinder ¾ of the way full with water. Record this initial water volume. If needed, use the tweezers to gently submerge the sphere and record the new water volume.

75 ml $\longrightarrow$ 78.8 ml .

9. Dry the cube and sphere and clean up your lab area as instructed by your teacher.

$$\frac{1.00 L \mid 1000\ ml \mid 1cm^3 \mid 1in^3}{\mid 1 L \mid 1ml \mid 2.54 cm^3}$$

16.4

Name _____

Period _____

Numbers in Science
Exploring Measurements, Significant Digits, and Dimensional Analysis

DATA AND OBSERVATIONS

Data Table		
CUBE DATA		
Mass: 5.6		
Dimensions	length: 1.42 cm width: 1.42 cm	height: 1.42 cm
Volume	Beaker initial volume: 100 mL	Beaker final volume:
	Graduated cylinder initial volume:	Graduated cylinder final volume:
SPHERE DATA		
Mass:		
Dimensions	Circumference:	
Volume	Beaker initial volume: 100 mL	Beaker final volume:
	Graduated cylinder initial volume:	Graduated cylinder final volume:
Formula for calculating the volume of a cube:		
Formula for calculating the circumference of a circle:		
Formula for calculating the diameter of a circle:		
Formula for calculating the volume of a sphere:		

ANALYSIS

- Remember to follow the rules for reporting all data and calculated answers with the correct number of significant digits.
- You may need tables of metric and English conversion factors to work some of these problems.

1. For each of the measurements you recorded above, go back and indicate the number of significant digits in parentheses after the measurement. Ex: 15.7 cm (3sd)

2. Use dimensional analysis to convert the mass of the cube to

    a. mg .00560mg ,

    b. ounces .198

3. Calculate the volume of the cube in cm^3.   2.86 cm3

    cm3
    2.86

4. Use dimensional analysis to convert the volume of the cube from cm^3 to m^3.

$$2.86 \times 10^{-6}$$

5. Calculate the volume of the cube in mL as measured in the beaker. Convert the volume to cm^3 knowing that 1 cm^3 = 1 mL.

    2.0 mL   2c   2.0 cm^3 .

6. Calculate the volume of the cube in mL as measured in the graduated cylinder. Convert to cm^3 knowing that 1 cm^3 = 1 mL.

    2.86   1 kg = 2.21 lbs
                    1 lb

7. Using the density formula $D = \dfrac{mass}{volume}$, calculate the density of the cube as determined by the

   a. ruler

   b. beaker

   c. graduated cylinder

8. Use dimensional analysis to convert these three densities into kg/m^3.

9. Convert the mass of the sphere to
   a. kg

   b. lbs.

10. Using the measured circumference, calculate the diameter of the sphere.

11. Calculate the radius of the sphere.

12. Calculate the volume of the sphere from its radius.

13. Calculate the volume of the sphere in mL as measured in the beaker. Convert to cm^3 knowing that 1 cm^3 = 1 mL.

14. Calculate the volume of the sphere in mL as measured in the graduated cylinder. Convert to cm^3 knowing that 1 cm^3 = 1 mL.

15. Using the density formula $D = \dfrac{mass}{volume}$, calculate the density of the sphere as determined by the

    a. tape measure

    b. beaker

    c. graduated cylinder

16. Use dimensional analysis to convert these three densities into lbs/ft^3.

CONCLUSION QUESTIONS

1. Compare the densities of the cube when the volume is measured by a ruler, beaker and graduated cylinder. Which of the instruments gave the most accurate density value? Use the concept of significant digits to explain your answer.

2. A student first measures the volume of the cube by water displacement using the graduated cylinder. Next, the student measures the mass of the cube before drying it. How will this error affect the calculated density of the cube? Your answer should state clearly whether the calculated density will increase, decrease or remain the same and must be justified.

3. A student measures the circumference of a sphere at a point slightly higher than the middle of the sphere. How will this error affect the calculated density of the cube? Your answer should state clearly whether the calculated density will increase, decrease, or remain the same and must be justified.

Literal Equations
Manipulating Variables and Constants

OBJECTIVE
Students will review how to solve literal equations for a particular variable.

LEVEL
Middle Grades: Chemistry/Physics, Chemistry I, Physics I

NATIONAL STANDARDS
UCP.1, UCP.2, G.2

TEKS
IPC: 4(A)
Chemistry: 2(C)
Physics: 3(B)

CONNECTIONS TO AP
In AP science courses, particularly physics and chemistry, the student is often given an equation and asked to solve it for a particular variable.

TIME FRAME
45 minutes

TEACHER NOTES
Throughout chemistry and physics courses, and at times in IPC, the students will need to be able to solve an equation for a particular variable to see how that variable depends on other variables and constants. Manipulation of variables without the substitution of numbers is an important skill in helping students understand that the variables depend on each other in a certain way regardless of any particular numbers which may be substituted into the equation. For example, in the equation $F_{net} = ma$ (Newton's second law), the acceleration a is always proportional to the net force F_{net} regardless of the value of the mass. In the equation $P_1V_1 = P_2V_2$ (Boyle's law), pressure P and volume V are always inversely proportional to each other.

The following examples and exercises illustrate the manipulation of many of the literal equations that commonly appear in physics and chemistry courses. Although the students may not understand the meaning of many of the equations during this practice exercise, practicing solving the equations will sharpen their algebra skills. When they ultimately learn the meaning of the equations, they will be more likely to feel comfortable with the conceptual understanding behind the equations rather than losing the meaning of the relationships among the variables in the algebraic manipulation.

A *literal equation* is one which is expressed in terms of variable symbols (such as d, v, and a) and constants (such as R, g, and π). Often in science and mathematics the students are given an equation and asked to solve it for a particular variable symbol or letter called the *unknown*.

The symbols, which are not the particular variable we are interested in solving for, are called *literals*, and may represent variables or constants. Literal equations are solved by isolating the unknown variable on one side of the equation, and all of the remaining literal variables on the other side of the equation. Sometimes the unknown variable is part of another term. A *term* is a combination of symbols such as the products *ma* or πr^2. In this case the unknown (such as r in πr^2) must be factored out of the term before we can isolate it.

The following rules, examples, and exercises will help you review and practice solving literal equations from physics and chemistry.

Suggested Teaching Procedure:

1. Review the procedure section with students. Emphasize keeping their equations neat and orderly.

2. Choose several of the listed examples to work with students on the overhead or chalkboard.

3. Instruct students to complete the remaining exercises in the space provided on their student answer pages.

ANSWERS TO EXERCISES

Directions: For each of the following equations, solve for the variable in **bold** print.

1. $v = \mathbf{a}t$

   - $\dfrac{v}{t} = \mathbf{a}$

2. $P = \dfrac{F}{\mathbf{A}}$

   - $P\mathbf{A} = F$

     $\mathbf{A} = \dfrac{F}{P}$

3. $\lambda = \dfrac{\mathbf{h}}{p}$

   - $\lambda p = \mathbf{h}$

4. $F(\mathbf{\Delta t}) = m\Delta v$

   - $\mathbf{\Delta t} = \dfrac{m\Delta v}{F}$

5. $U = \dfrac{G\mathbf{m_1}m_2}{r}$

   $Ur = G\mathbf{m_1}m_2$

   - $\dfrac{Ur}{Gm_2} = \mathbf{m_1}$

6. $C = \dfrac{5}{9}(\mathbf{F} - 32)$

   • $\dfrac{9}{5}C = \mathbf{F} - 32$

   $\mathbf{F} = \dfrac{9}{5}C + 32$

7. $v^2 = v_0{}^2 + 2\mathbf{a}\Delta x$

   $v^2 - v_0{}^2 = 2\mathbf{a}\Delta x$

   • $\dfrac{v^2 - v_0{}^2}{2\Delta x} = \mathbf{a}$

8. $K_{avg} = \dfrac{3}{2}k_B\mathbf{T}$

   $\dfrac{2}{3}K_{avg} = k_B\mathbf{T}$

   • $\dfrac{2K_{avg}}{3k_B} = \mathbf{T}$

9. $K = \dfrac{1}{2}m\mathbf{v}^2$

   $2K = m\mathbf{v}^2$

   • $\dfrac{2K}{m} = \mathbf{v}^2$

   $\mathbf{v} = \sqrt{\dfrac{2K}{m}}$

10. $v_{rms} = \sqrt{\dfrac{3RT}{\mathbf{M}}}$

   $v^2{}_{rms} = \dfrac{3RT}{\mathbf{M}}$

   • $\mathbf{M}v^2{}_{rms} = 3RT$

   $\mathbf{M} = \dfrac{3RT}{v^2{}_{rms}}$

11. $v_{rms} = \sqrt{\dfrac{3\mathbf{k}_B T}{\mu}}$

$v_{rms}{}^2 = \dfrac{3\mathbf{k}_B T}{\mu}$

- $\mu v_{rms}{}^2 = 3\mathbf{k}_B T$

$\dfrac{\mu v_{rms}{}^2}{3T} = \mathbf{k}_B$

12. $F = \dfrac{1}{4\pi\varepsilon_0}\dfrac{Kq_1 q_2}{\mathbf{r}^2}$

$4\pi\varepsilon_o \mathbf{r}^2 F = Kq_1 q_2$

- $\mathbf{r}^2 = \dfrac{Kq_1 q_2}{4\pi\varepsilon_o F}$

$\mathbf{r} = \sqrt{\dfrac{Kq_1 q_2}{4\pi\varepsilon_o F}}$

13. $\dfrac{1}{s_i} + \dfrac{1}{s_o} = \dfrac{1}{\mathbf{f}}$

- $\mathbf{f} = \dfrac{1}{\dfrac{1}{s_o} + \dfrac{1}{s_i}}$

14. $\dfrac{1}{C_{EQ}} = \dfrac{1}{\mathbf{C}_1} + \dfrac{1}{C_2}$

$\dfrac{1}{C_{EQ}} - \dfrac{1}{C_2} = \dfrac{1}{\mathbf{C}_1}$

- $\mathbf{C}_1 = \dfrac{1}{\dfrac{1}{C_{EQ}} - \dfrac{1}{C_2}}$

15. $V = \dfrac{4}{3}\pi \mathbf{r}^3$

$\dfrac{3}{4}V = \pi \mathbf{r}^3$

• $\dfrac{3V}{4\pi} = \mathbf{r}^3$

$\mathbf{r} = \sqrt[3]{\dfrac{3V}{4\pi}}$

16. $P + \mathbf{D}gy + \dfrac{1}{2}\mathbf{D}v^2 = C$

$P + \mathbf{D}\left(gy + \dfrac{1}{2}v^2\right) = C$

• $C - P = \mathbf{D}\left(gy + \dfrac{1}{2}v^2\right)$

$\mathbf{D} = \dfrac{C - P}{\left(gy + \dfrac{1}{2}v^2\right)}$

17. $P + Dgy + \dfrac{1}{2}D\mathbf{v}^2 = C$

$\dfrac{1}{2}D\mathbf{v}^2 = C - P - Dgy$

• $D\mathbf{v}^2 = 2(C - P - Dgy)$

$\mathbf{v}^2 = 2D(C - P - Dgy)$

$\mathbf{v} = \sqrt{2D(C - \mathbf{P} - Dgy)}$

18. $x = x_0 + v_0 t + \dfrac{1}{2}\mathbf{a}t^2$

$x - v_0 t = \dfrac{1}{2}\mathbf{a}t^2$

• $2\left(x - v_0 t\right) = \mathbf{a}t^2$

$\mathbf{a} = \dfrac{2\left(x - v_0 t\right)}{t^2}$

19. $n_1 \sin \theta_1 = n_2 \sin \theta_2$

$$\frac{n_1 \sin\theta_1}{n_2} = \sin\theta_2$$

- $$\theta_2 = \sin^{-1}\left[\frac{n_1 \sin\theta_1}{n_2}\right]$$

20. $mg \sin \theta = \mu mg \cos \theta \left(\dfrac{M+m}{m}\right)$

$$\frac{mg \sin\theta}{mg \cos\theta} = \mu\left(\frac{M+m}{m}\right)$$

$$\frac{\sin\theta}{\cos\theta} = \mu\left(\frac{M+m}{m}\right)$$

- $$\tan\theta = \mu\left(\frac{M+m}{m}\right)$$

$$\theta = \tan^{-1}\left[\mu\left(\frac{M+m}{m}\right)\right]$$

Literal Equations
Manipulating Variables and Constants

A *literal equation* is one which is expressed in terms of variable symbols (such as d, v, and a) and constants (such as R, g, and π). Often in science and mathematics you are given an equation and asked to solve it for a particular variable symbol or letter called the *unknown*.

The symbols which are not the particular variable we are interested in solving for are called *literals*, and may represent variables or constants. Literal equations are solved by isolating the unknown variable on one side of the equation, and all of the remaining literal variables on the other side of the equation. Sometimes the unknown variable is part of another term. A *term* is a combination of symbols such as the products ma or πr^2. In this case the unknown (such as r in πr^2) must factored out of the term before we can isolate it.

The following rules, examples, and exercises will help you review and practice solving literal equations from physics and chemistry.

PROCEDURE
In general, we solve a literal equation for a particular variable by following the basic procedure below.

1. Recall the conventional order of operations, that is, the order in which we perform the operations of multiplication, division, addition, subtraction, etc.:
   a. Parenthesis
   b. Exponents
   c. Multiplication and Division
   d. Addition and Subtraction

   This means that you should do what is possible within parentheses first, then exponents, then multiplication and division from left to right, then addition and subtraction from left to right. If some parentheses are enclosed within other parentheses, work from the inside out.

2. If the unknown is a part of a grouped expression (such as a sum inside parentheses), use the distributive property to expand the expression.

3. By adding, subtracting, multiplying, or dividing appropriately,

   (a) move all terms containing the unknown variable to one side of the equation, and

   (b) move all other variables and constants to the other side of the equation. Combine like terms when possible.

4. Factor the unknown variable out of its term by appropriately multiplying or dividing both sides of the equation by the other literals in the term.

5. If the unknown variable is raised to an exponent (such as 2, 3, or ½), perform the appropriate operation to raise the unknown variable to the first power, that is, so that it has an exponent of one.

EXAMPLES

1. $F = m\mathbf{a}$. Solve for **a**.
   $$F = ma$$

   Divide both sides by m:

   $$\frac{F}{m} = \mathbf{a}$$

   Since the unknown variable (in this case a) is usually placed on the left side of the equation, we can switch the two sides:

   $$\mathbf{a} = \frac{F}{m}$$

2. $P_1V_1 = P_2\mathbf{V}_2$. Solve for $\mathbf{V}_2$.
   $$P_1V_1 = P_2\mathbf{V}_2$$

   Divide both sides by P_2:

   $$\frac{P_1V_1}{P_2} = \mathbf{V}_2$$

   $$\mathbf{V}_2 = \frac{P_1V_1}{P_2}$$

3. $v = \dfrac{d}{\mathbf{t}}$. Solve for **t**.
   Multiply each side by **t**:

   $$\mathbf{t}v = d$$

   Divide both sides by v:

   $$\mathbf{t} = \frac{d}{v}$$

4. $PV = n\mathbf{R}T$. Solve for $\mathbf{R}$.

$PV = n\mathbf{R}T$

Divide both sides by n:

$$\frac{PV}{n} = \mathbf{R}T$$

Divide both sides by $\mathbf{T}$:

$$\frac{PV}{nT} = \mathbf{R}$$

$$\mathbf{R} = \frac{PV}{nT}$$

5. $R = \dfrac{\rho \mathbf{L}}{A}$. Solve for $\mathbf{L}$.

$$R = \frac{\rho \mathbf{L}}{A}$$

Multiply both sides by A:

$$RA = \rho \mathbf{L}$$

Divide both sides by ρ:

$$\frac{RA}{\rho} = \mathbf{L}$$

$$\mathbf{L} = \frac{RA}{\rho}$$

6. $A = h(a + \mathbf{b})$. Solve for $\mathbf{b}$.

Distribute the h:

$$A = ha + h\mathbf{b}$$

Subtract ha from both sides:

$$A - ha = h\mathbf{b}$$

Divide both sides by h:

$$\frac{A - ha}{h} = \mathbf{b}$$

$$\mathbf{b} = \frac{A - ha}{h}$$

7. $P = P_0 + \rho \mathbf{g} h$. Solve for $\mathbf{g}$.

Subtract P_0 from both sides:

$$P - P_0 = \rho \mathbf{g} h$$

Divide both sides by ρh:

$$\frac{P - P_0}{\rho h} = \mathbf{g}$$

$$\mathbf{g} = \frac{P - P_0}{\rho h}$$

8. $U = \frac{1}{2} \mathbf{Q} V$. Solve for $\mathbf{Q}$.

Multiply both sides by 2:

$$2U = \mathbf{Q} V$$

Divide both sides by V:

$$\frac{2U}{V} = \mathbf{Q}$$

$$\mathbf{Q} = \frac{2U}{V}$$

9. $U = \frac{1}{2} k \mathbf{x}^2$. Solve for $\mathbf{x}$.

Multiply both sides by 2:

$$2U = k \mathbf{x}^2$$

Divide both sides by k:

$$\frac{2U}{k} = \mathbf{x}^2$$

Take the square root of both sides:

$$\sqrt{\frac{2U}{k}} = \mathbf{x}$$

$$\mathbf{x} = \sqrt{\frac{2U}{k}}$$

10. $T = 2\pi\sqrt{\dfrac{\mathbf{L}}{g}}$. Solve for $\mathbf{L}$.

Divide both sides by 2π:

$$\frac{T}{2\pi} = \sqrt{\frac{\mathbf{L}}{g}}$$

Square both sides:

$$\frac{T^2}{4\pi^2} = \frac{\mathbf{L}}{g}$$

Multiply both sides by g:

$$\frac{gT^2}{4\pi^2} = \mathbf{L}$$

$$\mathbf{L} = \frac{gT^2}{4\pi^2}$$

11. $F = \dfrac{Gm_1 m_2}{\mathbf{r}^2}$. Solve for $\mathbf{r}$.

Multiply both sides by $\mathbf{r}^2$:

$$F\mathbf{r}^2 = Gm_1 m_2$$

Divide both sides by F:

$$\mathbf{r}^2 = \frac{Gm_1 m_2}{F}$$

Take the square root of both sides:

$$\mathbf{r} = \sqrt{\frac{Gm_1 m_2}{F}}$$

12. $\dfrac{h_i}{h_o} = -\dfrac{s_i}{s_o}$. Solve for s_o.

Cross-multiply:

$$h_i s_o = -h_o s_i$$

Divide both sides by h_i:

$$s_0 = -\dfrac{h_o s_i}{h_i}$$

13. $\dfrac{1}{R_{EQ}} = \dfrac{1}{R_1} + \dfrac{1}{R_2} + \dfrac{1}{R_3}$. Solve for R_3.

Subtract $\dfrac{1}{R_1} + \dfrac{1}{R_2}$ from both sides:

$$\dfrac{1}{R_{EQ}} - \dfrac{1}{R_1} - \dfrac{1}{R_2} = \dfrac{1}{R_3}$$

Take the reciprocal of both sides:

$$\dfrac{1}{\dfrac{1}{R_{EQ}} - \dfrac{1}{R_1} - \dfrac{1}{R_2}} = R_3$$

$$R_3 = \dfrac{1}{\dfrac{1}{R_{EQ}} - \dfrac{1}{R_1} - \dfrac{1}{R_2}}$$

This equation could be solved further with several more algebraic steps.

14. $F = qvB \sin\theta$. Solve for θ.

Divide both sides by qvB:

$$\dfrac{F}{qvB} = \sin\theta$$

Take the inverse sine of both sides:

$$\theta = \sin^{-1}\left[\dfrac{F}{qvB}\right]$$

15. $\mu mg \cos\theta = mg \sin\theta$. Solve for μ.
    Divide both sides by $mg\cos\theta$:

$$\mu = \frac{mg \sin\theta}{mg \cos\theta} = \frac{\sin\theta}{\cos\theta} = \tan\theta$$

Name _____

Period _____

Literal Equations
Manipulating Variables and Constants

EXERCISES

Directions: For each of the following equations, solve for the variable in **bold** print. Be sure to show each step you take to solve the equation for the **bold** variable.

1. $v = \mathbf{a}t$

2. $P = \dfrac{F}{\mathbf{A}}$

3. $\lambda = \dfrac{\mathbf{h}}{p}$

4. $F(\mathbf{\Delta t}) = m\Delta v$

5. $U = \dfrac{G\mathbf{m_1}m_2}{r}$

6. $C = \dfrac{5}{9}(\mathbf{F} - 32)$

7. $v^2 = v_0^{\,2} + 2\mathbf{a}\Delta x$

8. $K_{avg} = \dfrac{3}{2}k_B\mathbf{T}$

9. $K = \dfrac{1}{2}m\mathbf{v}^2$

10. $v_{rms} = \sqrt{\dfrac{3RT}{\mathbf{M}}}$

11. $v_{rms} = \sqrt{\dfrac{3\mathbf{k}_B T}{\mu}}$

12. $F = \dfrac{1}{4\pi\varepsilon_0}\dfrac{Kq_1 q_2}{\mathbf{r}^2}$

13. $\dfrac{1}{s_i} + \dfrac{1}{s_o} = \dfrac{1}{\mathbf{f}}$

14. $\dfrac{1}{C_{EQ}} = \dfrac{1}{\mathbf{C}_1} + \dfrac{1}{C_2}$

15. $V = \dfrac{4}{3}\pi\mathbf{r}^3$

16. $P + \mathbf{D}gy + \dfrac{1}{2}\mathbf{D}v^2 = C$

17. $P + Dgy + \dfrac{1}{2}D\mathbf{v}^2 = C$

18. $x = x_0 + v_0 t + \dfrac{1}{2}\mathbf{a}t^2$

19. $n_1 \sin\theta_1 = n_2 \sin\boldsymbol{\theta_2}$

20. $mg\sin\theta = \mu mg\cos\boldsymbol{\theta}\left(\dfrac{M+m}{m}\right)$

Graphing Skills
Reading, Constructing and Analyzing Graphs

Unit Overview

OBJECTIVE

The purpose of this lesson is to provide the teacher with basic graphing skill lessons to be used throughout the science course of study. There are many different kinds of graphs and each has a fairly specific use. By teaching graphing at all grade levels students should be able to choose the best type of graph to represent their data. Students should also become familiar with ways to analyze their graphed information realizing that there is meaning to their graph.

LEVEL
All

NATIONAL STANDARDS
UCP.1, UCP.2, G.2

TEKS
6.2(E)
7.2(E)
8.2(E)
IPC: 2(C)
Biology: 2(C)
Chemistry: 2(D)
Physics: 2(C), 2(E)

CONNECTIONS TO AP
Each of the AP Science courses requires that students are able to read, construct and analyze graphs.

TIME FRAME
45 minutes

MATERIALS
> graph paper
> pencil
> data

TEACHER NOTES

Graphing is a skill that should be introduced at each grade level and reinforced throughout the year whenever data is available. The analysis level of the graph will vary depending on the mathematical ability of your students. The lessons presented here may be used as stand alone lessons or may be combined as a general review of graphing skills.

Graphing is an essential tool in science. Graphs enable us to visually communicate information. The lessons that follow will focus on bar graphs, pie charts and line graphs. Goals for this series of lessons on graphing include:

- Choosing an appropriate display for data (which type of graph to construct)
- Identifying data to be displayed on the x and y axes
- Scaling a graph properly
- Labeling a graph with axes labels, title, units, and legend or key if necessary
- Extrapolating and interpolating data points
- Understanding relevant relationships such as slope and area under the curve

Graphing Skills
Reading, Constructing and Analyzing Graphs

Bar Graphs and Histograms

OBJECTIVE
Students will become familiar with basic bar graphing skills to be used throughout their science course of study.

LEVEL
All

NATIONAL STANDARDS
UCP.1, UCP.2, G.2

TEKS
6.2(E)
7.2(E)
8.2(E)
IPC 2(C)
Biology 2(C)
Chemistry 2(D)
Physics 2(C), 2(E)

CONNECTIONS TO AP
Each of the AP Science courses requires that students are able to read, construct and analyze graphs.

TIME FRAME
45 minutes

MATERIALS
(For each student working individually)

> 4 sheets of quadrille graph paper
> pencil
> data

TEACHER NOTES

Graphing is a skill that should be introduced at each grade level and reinforced throughout the year whenever data is available. Before allowing students to begin this exercise, you should do the following:

- Distinguish between the four types of graphs represented in the student introduction. Key points to be made are as follows:
  - *Simple Bar Graph*: The width of bars must be the same.
  - *Grouped Bar Graph*: The width of bars must be the same. Each bar within a group needs some distinguishing mark—different colors, different markings, etc. The student must provide a legend so that the graph may easily be interpreted.
  - *Composite Bar Graph*: The width of bars must be the same. Each different component of the bar must have some distinguishing mark—different colors, different markings, etc.
  - *Histogram:* The width of bars must be the same. Clearly make the distinction here for the students that in a histogram the bar itself represents a range of independent variables rather than a single value.
- Encourage students to decide which sets of data belong on the *x*-axis and the *y*-axis. The key here is to place the independent variable on the *x*-axis and the dependent variable on the *y*-axis.
- Show the students how to properly scale a graph. The scale represents the range of frequency values shown on the graph. Visually show students how to properly accomplish this task by making an overhead transparency of a piece of the graph paper. Write down the range, count the squares on the graph paper and decide the scale by spacing appropriately. Emphasize the importance of using the entire length and width of the paper when creating the axes.
- Demonstrate proper labeling of a graph with a title, axes labels and units, and keys if necessary. Titles are usually given at the top of each graph and provide an overview of the information that is given in the graph. The axes labels should provide specific information as to what is represented. It is customary to label a graph in the format *y* vs *x*. For example if a graph is described as "Temperature vs Time" then temperature should be on the *y* axis and time should be on the *x*-axis. Depending on the level of your students you may wish to require this format.
- In analyzing a graph, show the students how to read the graph and interpret their graph by using interpolation and extrapolation. Point out that it is difficult to identify specific interpolation (between) points on bar graphs but that rough comparisons are easy. Extrapolation (beyond) points is also difficult (if not impossible) to determine on bar graphs.
- As students work on this activity monitor them closely to ensure that they have correct labels and scales on the graphs they are constructing.

SAMPLE DATA
Student graphs should look similar to the samples found below.

PART I: SIMPLE BAR GRAPH

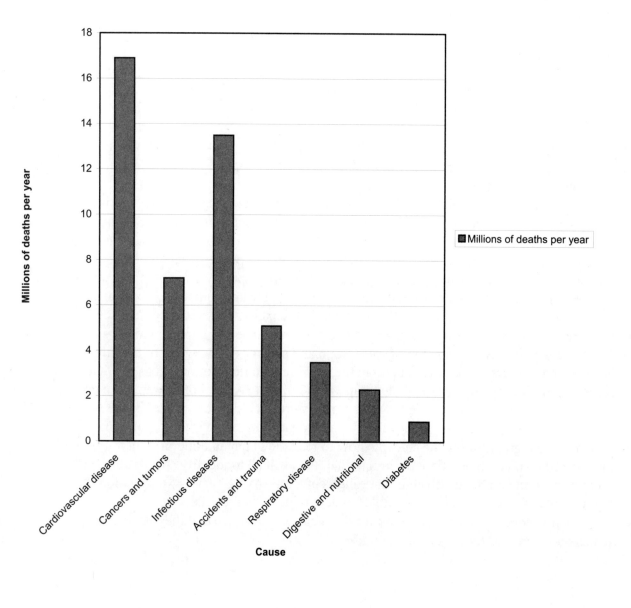

PART II: GROUPED BAR GRAPH

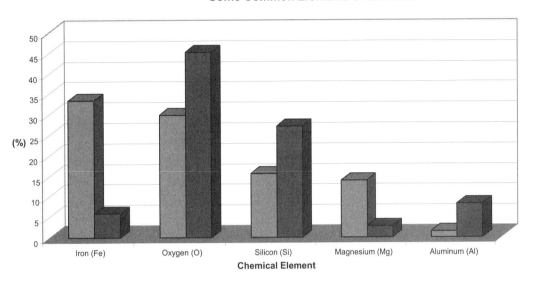

PART III: COMPOSITE BAR GRAPH

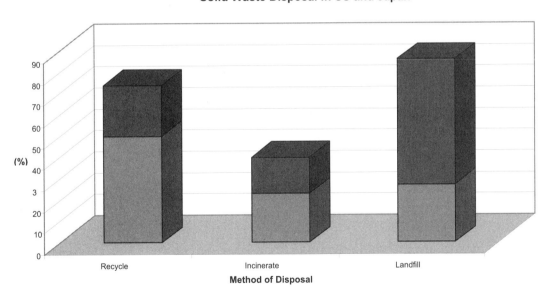

PART IV: HISTOGRAM

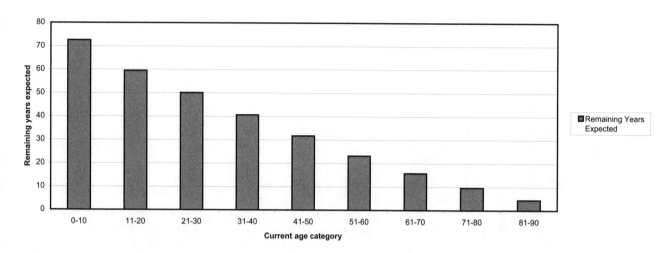

Life Expectancies in the US

POSSIBLE ANSWERS TO THE CONCLUSION QUESTIONS

Using the graphs that you constructed, answer the following questions.

PART I: SIMPLE BAR GRAPH
1. How many deaths due to accidents and trauma occur per year?
   - 5.1 million deaths

2. Can you predict the number of deaths due to cancers and tumors for the next ten years? Explain.
   - No. This type of extrapolation is a weakness of bar graphs.

PART II: GROUPED BAR GRAPH
1. What element is the most abundant in the Earth's crust?
   - Oxygen — about 45%.

2. What element is the most abundant within the Earth?
   - Iron — about 33%.

PART III: COMPOSITE BAR GRAPH

1. What is the favored method of waste disposal for the Japanese population? Cite possible reasons for this.
   - Recycling — about 50% of their waste is disposed of this way.
   - Possible reasons might include: minimum available undeveloped land; overcrowded population.

2. What is the favored method of waste disposal for the US population? Cite possible reasons for this.
   - Landfill — about 60% of waste is disposed of this way.
   - Possible reasons might include: plenty of undeveloped land; expense associated with recycling.

PART IV: HISTOGRAM

1. Make a prediction about the remaining years of life that would be expected for someone in the current age category of 91-100.
   - Some students will attempt to find a mathematical trend in the data by looking at how the years change over each interval. Students should answer between 0 and 2 years if using this type of pattern.

2. Is the answer to question 7 an accurate number? Why or why not? Cite specific reasons.
   - No, the answer in 7 is not an accurate number.
   - This is very difficult to predict from a graph of this type.

3. What type of data is easily represented by a bar graph?
   - Data that compares amounts or frequency of occurrence.

4. Why is a legend (or key) necessary in the grouped and composite bar graphs?
   - Legends are necessary to distinguish between the bars.

5. Explain why it is difficult to make direct comparisons between recycling in Japan and the US using the composite bar graph that you drew.
   - It is difficult for comparison since the bars are merged together. In order to compare accurately, you would have to accurately measure the length of each piece separately.

6. What is the importance of scaling?
   - Scaling is important for making accurate comparisons.

TEACHER PAGES

7. Distinguish between the dependent and the independent variable for each of the graphs that were constructed. On which axis should the independent variable be placed?

	Dependent Variable	**Independent Variable**
Simple Bar Graph	Millions of deaths per year	Cause of death
Grouped Bar Graph	Percentages in Earth and crust	Chemical elements
Composite Bar Graph	Percentage of waste	Method of disposal
Histogram	Remaining years expected	Current age

- The *x*-axis is usually used for the independent variable.

Graphing Skills
Reading, Constructing and Analyzing Graphs

Pie Charts

OBJECTIVE
Students will become familiar with basic pie chart graphing skills to be used throughout their science course of study.

LEVEL
All

NATIONAL STANDARDS
UCP.1, UCP.2, G.2

TEKS
6.2(E)
7.2(E)
8.2(E)
IPC 2(C)
Biology 2(C)
Chemistry 2(D)
Physics 2(C), 2(E)

CONNECTIONS TO AP
Each of the AP Science courses requires that students are able to read, construct and analyze graphs.

TIME FRAME
45 minutes

MATERIALS
(For each student working individually)

2 sheets of blank paper	data
pencil	protractor
compass	colored pencils

TEACHER NOTES

Graphing is a skill that should be introduced at each grade level and reinforced throughout the year whenever data is available. Before allowing students to begin this exercise, you should do the following:

- Emphasize the usefulness of pie charts for showing percentages.
- Remind students that each pie chart should always total 100%. They will be asked to calculate the percentage for the second set of data. To find the percentage:

$$\% = \frac{\text{specific sample of data}}{\text{total data collected}} \times 100$$

- Point out that a legend or key will be necessary for this type of display.
- Reinforce that labels for each wedge and a title are always necessary.
- Students may need to be instructed on proper use of a compass and protractor. An easy way to construct a pie chart is to have students draw a circle with the compass. Using the protractor, make four marks on the outside of the circle in 90° intervals. Next, have the students divide each quadrant into five equal sections. Each section will represent 5%. This makes estimation of points quite simple.

SAMPLE DATA

Student graphs should look similar to the samples found below.

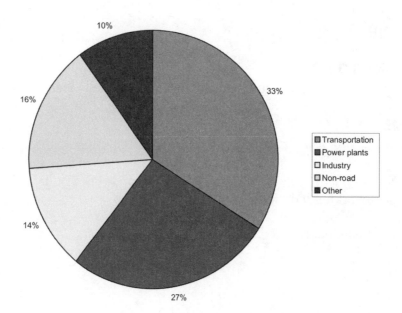

Sources of Nitrogen Oxide Pollutants

10%
33%
16%
14%
27%

Legend:
- Transportation
- Power plants
- Industry
- Non-road
- Other

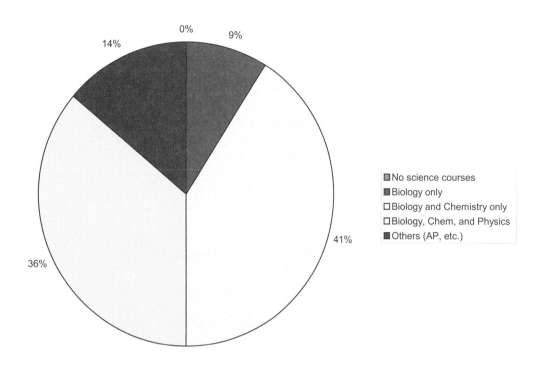

Science Courses Completed by Adults Ages 25-30

Legend:
- No science courses
- Biology only
- Biology and Chemistry only
- Biology, Chem, and Physics
- Others (AP, etc.)

POSSIBLE ANSWERS TO THE CONCLUSION QUESTIONS

Using the graphs that you constructed, answer the following questions:

Sample Data Set 1: Sources of Nitrogen Oxide Air Pollution

1. How much nitrogen oxide air pollution is due to transportation?
   - 33%

2. Not taking into account the set of data labeled "other", what category contributes the least nitrogen oxide air pollutants into our environment?
   - Industry

3. Can you predict from a graph of this type the amount of nitrogen oxide air pollution that will be contributed by industry in the next ten years? Explain your reasoning.
   - No
   - The pie graph has no time period associated with it.

Sample Data Set 2:

1. Why must a percentage calculation be performed on the data before making the graph?
   - Pie graphs show parts of a whole, or 100%. Since there were 500 people polled, the data had to be calculated as if the sample size were 100.

2. Describe the trend displayed by this pie chart. Be specific.
   - Everyone polled had taken at least one science course. Most of those polled had taken at least two courses and many had taken three courses. The "other" category represents a small population of students who continued their science course of study or took a different science elective that was not listed.

3. Most adults polled between the ages of 25-30 years of age completed which science courses during their high school career?
   - Biology and chemistry

4. Describe the type of data that can be displayed using pie charts. List three specific places where you might see pie charts printed.
   - Pie charts visually display data that can easily be categorized into percentages.
   - Newspapers, science textbooks, and consumer labeling.

Graphing Skills
Reading, Constructing and Analyzing Graphs

Line Graphs

OBJECTIVE

Students will become familiar with basic line graphing skills to be used throughout their science course of study.

LEVEL

All

NATIONAL STANDARDS

UCP.1, UCP.2, G.2

TEKS

6.2(E)
7.2(E)
8.2(E)
IPC 2(C)
Biology 2(C)
Chemistry 2(D)
Physics 2(C), 2(E)

CONNECTIONS TO AP

Each of the AP Science courses requires that students are able to read, construct and analyze graphs.

TIME FRAME

45 minutes

MATERIALS

(For each student working individually)

4 sheets of quadrille graph paper	data
pencil	ruler

TEACHER NOTES

Graphing is a skill that should be introduced at each grade level and reinforced throughout the year whenever data is available. In this exercise students will construct line graphs using paper and pencil. Line graphs will be constructed and analyzed more than any of the other types of graphs throughout the science course of study so you should make sure that all students are comfortable with this activity.

Upon completion of the graphing exercise, analysis of the graph follows. Students should understand basic paper/pencil construction methods of graphing before proceeding to the Foundation Lessons which use Microsoft Excel and Graphical Analysis. Before allowing students to begin this exercise, you should have students make a practice graph at their desks while you make one at the overhead or chalkboard. In your example discuss the following points:

- Ask students if they can identify the *x* and *y*- axes. Label these on your sample.
- Ask students about the terms independent and dependent variable. Do they know where to place each on their graph? Below the *x*-axis write the label "independent" and beside the *y*-axis write the label "dependent". Remind the students that each of the variables should have some type of unit associated with them. Write the word "unit" in parentheses after the word "independent" and "dependent" variable on your example.
- Ask students to identify the last missing component of the graph. [Title] Write the term "title" at the top of the graph. Point out that it is common to title a graph using the dependent variable (*y*-axis vs. the independent variable (*x*-axis) format. Encourage students to give descriptive titles to their graph and not just re-name the axes. The graphs that the students construct should have "good" titles. Titles that tell exactly what information the author is trying to represent with the graph. The title should be concise, clear and complete.
- Using any set of generic data points, illustrate to students how to determine the range for each variable and how to determine the scale for each axis. Point out that graph paper must always be used to construct a graph. Each square along an axis must represent the same increment. Encourage students to use as much space as available to construct their graph. Label each axis on the graph with the proper numbers according to data given.
- Illustrate how to plot points on the graph. Emphasize that each point represents both an *x* and a *y* component. Remind students that the plotting of points in science is the same as plotting an ordered pair in the math class. (in math the points are always given as (*x,y*). Encourage the use of pencil to plot the data set.
- Discuss with students the importance of using a graph to understand relationships. In the science classroom graphs are not generally connect-the-dot graphs. It is common practice to draw the best smooth curve or the line of best fit that relates the data. Illustrate how this is done with the data previously plotted. If more than one set of data is to be displayed on any one graph, remind students that a legend or key will be necessary to identify each line.
- Point out to students that as their mathematical and scientific skills increase, the usefulness and meaning of their graphs will also increase.
- Illustrate the following analysis techniques:
  - Interpolation of data — find a value that lies on the smooth curve or line *between* two actual data points.
  - Extrapolation of data — find a value that lies on the smooth curve or line beyond the actual plotted points. Data can be extrapolated both on the front and back end of the line/curve.

o   Linear regression of data — If the plotted relationship generates a straight line, have students write the equation for the straight line in slope-intercept form [y = mx + b]. To calculate the slope of the line use the equation:

$$slope = \frac{rise}{run} = \frac{\Delta y}{\Delta x} = \frac{y_2 - y_1}{x_2 - x_1}$$

The y-intercept can be found by extending the line of best fit backwards until it crosses the y-axis

POSSIBLE ANSWERS TO THE CONCLUSION QUESTIONS AND SAMPLE DATA

Sample Data Set 1: The following set of data was collected while experimenting with position and time of a miniature motorized car traveling on a straight track.

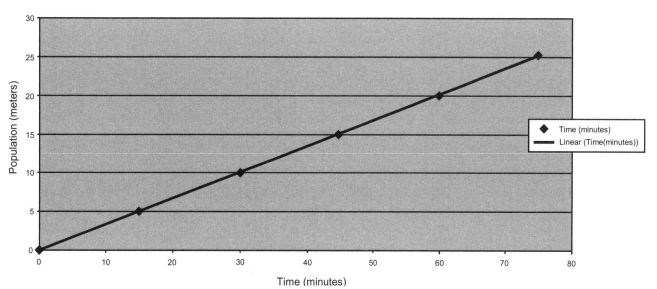

Motion: Position vs. Time

1. What is the independent variable for this graph? Explain.
   - Time is the independent variable. It is the property that is controlled by the experimenter. It is plotted on the x-axis.

2. What would be the position of the car after 25 minutes?
   - About 7.5 to 8.0 meters. Arriving at this answer requires that the student use knowledge of interpolation.

3. If the experiment were carried out for 80 minutes, what would be the position of the car?
   - About 28 meters. Arriving at this answer requires that the student use knowledge of extrapolation.

4. Calculate the slope of the line drawn. What does the slope of this line represent? Explain.

- slope $= \dfrac{\text{rise}}{\text{run}} = \dfrac{\Delta y}{\Delta x} = \dfrac{y_2 - y_1}{x_2 - x_1}$

- Students may choose any set of points to solve. Answers may vary slightly. Possible solution might be: slope $= \dfrac{\text{rise}}{\text{run}} = \dfrac{\Delta y}{\Delta x} = \dfrac{10m - 5m}{30\text{min} - 15\text{min}} = \dfrac{5m}{15\text{min}} = .33\,\text{m/min}$

- The slope represents velocity — distance per time.

5. Write the equation for a straight line including the value that was determined for slope.

- y   =   m   x   + b

  position (m)   = (.33 m/min) (time) + 0

Sample Data Set 2: The following set of data was collected during an experiment to find the density for an unknown metal.

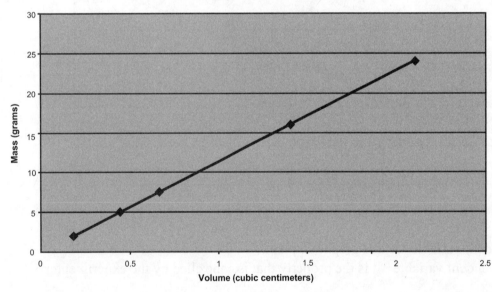

1. What values were considered when creating the scale for each axis in this experiment?
   - Mass: the range of the numbers was from 2.00 to 24.00 grams. The scale was chosen so that each line represents 5 grams. (student answers may vary as chosen scales may vary)
   - Volume: the range was from 0.18 to 2.11 cm^3. The scale was chosen so that each mark represented 0.50 cm^3.

2. What does a data point on this graph actually represent?
   - Each data point represents a mass and a volume measurement. The relationship between the two is the density which is the slope of the best-fit line.

3. What volume would a 20.00 gram sample of this substance occupy?
   - About 1.75 cm^3. Arriving at this answer requires that the student use knowledge of interpolation.

4. Calculate the density of the substance. (HINT: calculate the slope of the line.)
   - $slope = \dfrac{rise}{run} = \dfrac{\Delta y}{\Delta x} = \dfrac{y_2 - y_1}{x_2 - x_1}$ Students may choose any set of points to solve. Answers may vary

     slightly. Possible solution might be: $slope = \dfrac{rise}{run} = \dfrac{\Delta y}{\Delta x} = \dfrac{15g - 5g}{1.44cm^3 - .44cm^3} = \dfrac{10}{1.00} = 10.0\,g/cm^3$

   - The slope represents density — mass per unit volume.

5. Write the equation for a straight line including the value that was determined for slope.
   - y $\quad=\quad$ m $\qquad$ x $\qquad$ + $\quad$ b

     mass (g) $\quad=\quad$ (10.0 g/cm^3 (volume(cm^3)) + 0

6. Use the equation and find the mass when the volume is 5.00 cm^3.

   - y $\quad=\quad$ m $\qquad$ x $\qquad$ + $\quad$ b

     mass (g) $\quad=\quad$ (10.0 g/cm^3) (volume (cm^3)) + 0

     mass (g) $\quad=\quad$ (10.0 g/cm^3) (5.00 cm^3) + 0

     mass (g) $\quad=\quad$ 50.0 grams

Sample Data Set 3: The following set of data was collected during an experiment studying the effect of light intensity on rate of photosynthesis.

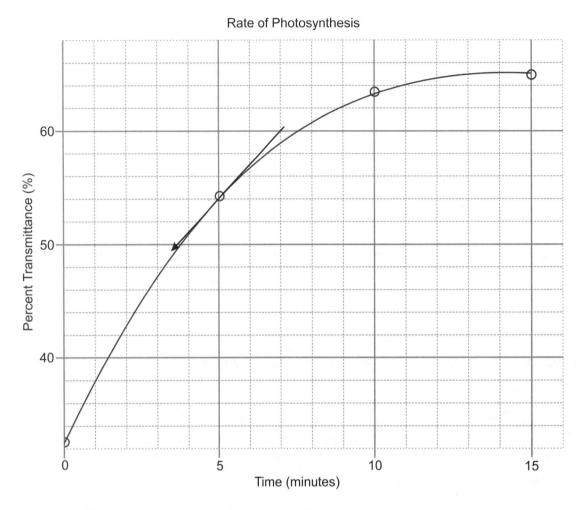

1. Does this graph represent a linear relationship? Why or why not?
   - This graph is not a linear relationship. The slope of the line changes with time.

2. What is the dependent variable in this graph? Explain.
   - The dependent variable is the percent transmittance. Time is the independent variable, the one that the person performing the experiment has control over. At each time interval, the transmittance is measured.

3. If the experiment were continued for 30 minutes, what trend in percent transmittance could be expected?
   - The graph begins to level somewhat between 10 and 15 minutes. It seems logical that the percent transmittance would soon level off or begin to fall.

Note: Middle school teachers may wish to omit the remainder of this lesson.

4. Calculate the slope of the line at 5 minutes. What does this represent?
   - A tangent line must be drawn at 5 minutes.
   - $\text{slope} = \dfrac{\text{rise}}{\text{run}} = \dfrac{\Delta y}{\Delta x} = \dfrac{60\% - 50\%}{7\,\text{min} - 4.5\,\text{min}} = \dfrac{10}{2.5} = 4.09\%/\text{min}$
   - The slope represents the instantaneous rate. Notice that the slope of the line is different at 10 minutes.

Sample Data Set 4: The following set of data was collected during a titration experiment of a diprotic acid and sodium hydroxide.

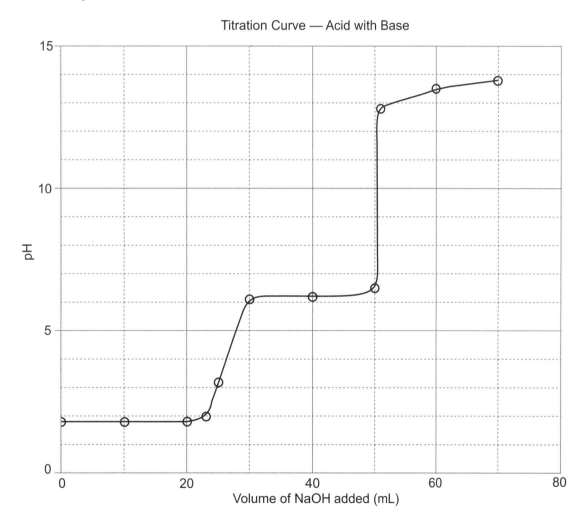

Titration Curve — Acid with Base

1. What is the pH of the solution after 20.0 mL of NaOH are added? After 30.0 mL are added? Would it have been easy to predict this answer?
   - The pH at 20.0 mL = 1.80; the pH at 30.0 mL = 6.10
   - The graph is not a direct relationship and thus, would not have made this easy to predict such a large jump in pH.

2. Graphs often help us to understand the progress of a chemical reaction. In the titration graph for this set of data, there are two relatively sharp, upward curves. The middle of these steep rising portions represent equivalence points (point at which the moles of acid and base are equal). Identify the volume of NaOH needed to reach each of the equivalence points.

- The first equivalence point occurs around 28 mL. (the midpoint of the sharp, upward curve must be taken)
- The second equivalence point occurs around 50 mL.

3. What is the pH at 65 mL. What is the pH expected to do beyond this point with greater additions of the base NaOH? Explain.

- The pH at 65 mL is approximately 13.50.
- Beyond 65 mL the pH will gradually rise and level off. Since NaOH is a strong base it will eventually reach close to a pH of 14.00.
- The curve begins leveling off after the second equivalence point, therefore, the pH will not drastically change with more NaOH added. The pH after the equivalence point is only due to the amount of NaOH that is in excess of the acid.

Graphing Skills
Reading, Constructing and Analyzing Graphs

Bar Graphs and Histograms

Bar graphs are very common types of graphs. They are found in almost all science books, magazines, and newspapers. They can be useful tools in scientific study by allowing us to visually compare amounts or frequency of occurrences between different data sets. Bar graphs can be used to show how something changes over time or to compare items with one another. When reading or constructing this type of graph you should pay close attention to the title, the label on the axes, the unit or scale of the axes, and the bars.

In a simple bar graph the specific group or experimental subject is assigned the x-axis (horizontal) and the y-axis (vertical) is known as the frequency axis. In general, the x-axis will be divided into time periods or measurements while the y-axis is designated for the frequency of occurrences. When data is grouped, the x-axis always represents the grouped data while the y-axis shows the frequency data. A composite bar graph is often useful when displaying the sum of various dependent variables when the values are a fraction of the whole. Histograms are very similar to simple bar graphs with one exception — the bar represents a range of values rather than one single value and the intervals must all be of equal magnitude. Study the sample graphs below before completing this exercise.

Simple Bar Graph

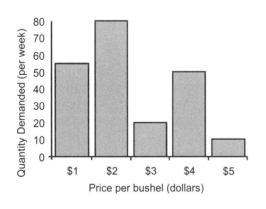

Price of Wheat vs. Quantity Demanded

Grouped Bar Graph

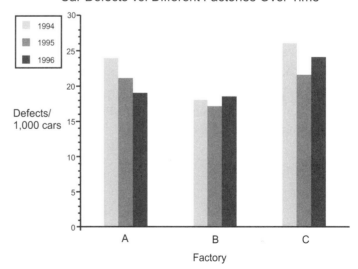

Car Defects vs. Different Factories Over Time

Composite Bar Graph

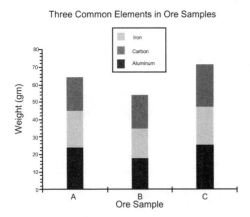

Histogram

Population Distribution

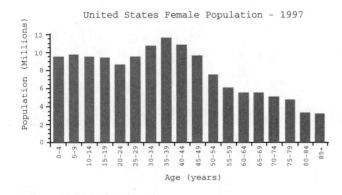

PURPOSE

In this exercise you will create simple bar graphs, grouped bar graphs, composite bar graphs and histograms. You will be expected to properly label each of your graphs and analyze each one by making statements about trends in the data.

MATERIALS

4 sheets of graph paper   data
pencils   straight edge

PROCEDURE

PART I: SIMPLE BAR GRAPH

1. Obtain one piece of graph paper and a pencil.

2. Study the data table below.

Leading Causes Of Death Worldwide	
Cause	Deaths Per Year (millions)
Cardiovascular disease	16.9
Cancers and tumors	7.2
Infectious diseases (includes AIDS, malaria, etc.)	13.5
Accidents and trauma	5.1
Respiratory disease	3.5
Digestive and nutritional	2.3
Diabetes	0.9

3. Choose the data to be graphed on the *x*-axis and the *y*-axis.

4. Survey the data and determine an appropriate scale for each axis. Be sure to utilize as much of the graph paper as possible o display your data. Use your pencil to lightly mark the scale of your *x* and *y*-axes. Have your teacher check your scale before proceeding any further. When making a bar graph, the individual bars should be constructed with the same width. You may decide the width of your bars.

5. When your teacher approves, construct your simple bar graph. Be sure to label each axis with units and give your graph a title.

PART II: GROUPED BAR GRAPH

1. Study the following data and follow the same procedure as Part I with a clean sheet of graph paper. This data should be graphed as a grouped bar graph and include a legend or key to indicate what each bar represents.

Some of the Most Common Chemical Elements of the Earth		
Chemical Element	Whole Earth (%)	Crust (%)
Iron (Fe)	33.3	5.8
Oxygen (O)	29.8	45.2
Silicon (Si)	15.6	27.2
Magnesium (Mg)	13.9	2.8
Aluminum (Al)	1.5	8.2

PART III: COMPOSITE BAR GRAPH

1. Study the following data and follow the same graphing procedure with a clean sheet of graph paper. This data should be graphed as a composite bar graph. You will need to include a legend and be sure to place the waste disposal methods in the same order for each bar drawn.

Solid Waste Recycled, Incerated, and Landfilled in US and Japan		
	Japan	US
Recycle (%)	50	24
Incinerate (%)	23	17
Landfill (%)	27	59

PART IV: HISTOGRAM

1. Study the following data and follow the same graphing procedure. This data should be graphed as a histogram. It is important that histograms have the same interval and width for each bar. For example, each bar might represent 10 years in the data table below.

Life Expectancies in the US	
Current Age	Remaining Years Expected
0-10	72.6
11-20	59.5
21-30	50.1
31-40	40.7
41-50	31.7
51-60	23.2
61-70	15.8
71-80	9.7
81-90	4.5

Graphing Skills
Reading, Constructing and Analyzing Graphs

Bar Graphs and Histograms

CONCLUSION QUESTIONS

Using the graphs that you constructed, answer the following questions.

PART I: SIMPLE BAR GRAPH

1. How many deaths due to accidents and trauma occur per year?

2. Can you predict the number of deaths due to cancers and tumors for the next ten years? Explain.

PART II: GROUPED BAR GRAPH

1. What element is the most abundant in the earth's crust?

2. What element is the most abundant within the earth?

PART III: COMPOSITE BAR GRAPH

1. What is the favored method of waste disposal for the Japanese population? Cite possible reasons for this.

2. What is the favored method of waste disposal for the US population? Cite possible reasons for this.

PART IV: HISTOGRAM

1. Make a prediction about the remaining years of life that would be expected for someone in the current age category of 91-100.

2. Is the answer to question 7 an accurate number? Why or why not? Cite specific reasons.

3. What type of data is easily represented by a bar graph?

4. Why is a legend (or key) necessary in the grouped and composite bar graphs?

5. Explain why it is difficult to make direct comparisons between recycling in Japan and the US using the composite bar graph that you drew.

6. What is the importance of scaling?

7. Distinguish between the dependent and the independent variable for each of the graphs that were constructed. On which axis should the independent variable be placed?

	Dependent Variable	Independent Variable
Simple Bar Graph		
Grouped Bar Graph		
Composite Bar Graph		
Histogram		

Graphing Skills
Reading, Constructing and Analyzing Graphs

Pie Charts

Pie charts are very commonly found in newspapers, magazines and textbooks. A pie chart is a very good way to represent percentages.

PURPOSE
In this activity you will practice constructing and analyzing basic pie charts.

MATERIALS

blank paper   data
pencil   protractor
compass   colored pencils

Safety Alert
1. The sharp point on the compass should only be placed on paper.

PROCEDURE

1. After observing your teacher demonstrate the use of the compass and protractor, use Sample Data Set 1 to construct a pie chart on a piece of blank paper. Use different colors to represent different sections of your graph.

2. Be sure to label your chart with an appropriate title and be sure to provide a legend or key that distinguishes each component.

Sample Data Set 1: Sources of Nitrogen Oxide Air Pollution

Sources of Nitrogen Oxides	Percentages (%)
Power plants	53%
Transportation	68%
Industry	27%
Non-road	32%
Other	20%

3. Use Sample Data Set 2 to construct a second pie chart on another blank piece of paper. Be sure to label appropriately. *Note*: Before beginning construction of this graph, you must calculate component percentages. Show your work on the student answer page.

Sample Data Set 2: 500 adults between the ages of 25-30 were polled as to which science courses they completed in their high school years. The following data was collected.

Science Courses Completed	Number of Adults
No science courses	0
Biology only	45
Biology and Chemistry only	205
Biology, Chemistry and Physics only	180
Other classes not listed (AP, etc.)	70

Name _____

Period _____

Graphing Skills
Reading, Constructing and Analyzing Graphs

Pie Charts

ANALYSIS

1. Staple your two graphs behind this answer page.

2. Show your work here for Sample Data 2 - percentage calculations.

CONCLUSION QUESTIONS

Using the graphs that you constructed, answer the following questions:

Sample Data Set 1: Sources of Nitrogen Oxide Air Pollution

1. How much nitrogen oxide air pollution is due to transportation?

2. Not taking into account the set of data labeled "other", what category contributes the least nitrogen oxide air pollutants into our environment?

3. Can you predict from a graph of this type the amount of nitrogen oxide air pollution that will be contributed by industry in the next ten years? Explain your reasoning.

Sample Data Set 2:

1. Why must a percentage calculation be performed on the data before making the graph?

2. Describe the trend displayed by this pie chart. Be specific.

3. Most adults polled between the ages of 25-30 years of age completed which science courses during their high school career?

4. Describe the type of data that can be displayed using pie charts. List three specific places where you might see pie charts printed.

Name _____

Period _____

Graphing Skills
Reading, Constructing and Analyzing Graphs

Line Graphs

There are all kinds of charts and graphs used in the science classroom. Graphs are useful tools in science. Trends in data are easy to visualize when represented graphically. A line graph is beneficial in the classroom for many different types of data. Line graphs are probably the most widely used scientific graph. They can be used to show how something changes over time, the relationship of two quantities, and can be readily used to *interpolate* (predict between measured points on the graph) and *extrapolate* (predict beyond the measured points along the same slope) data points that were not actually measured in the lab setting. The analysis of these graphs provides very valuable information.

PURPOSE
In this activity you will learn the basic procedure for constructing and analyzing line graphs.

MATERIALS

4 sheets of graph paper	data
pencil	ruler

PROCEDURE

1. Follow along with your teacher as a sample line graph is constructed. Label a blank piece of graph paper as your teacher explains the important components of a line graph.

2. Use the sample sets of data below to construct line graphs. Place only one graph on each sheet of graph paper and use as much of the graph as possible to display your points. ***Do not connect the dots!*** Draw the best smooth curve or line of best fit as your teacher demonstrated.

3. Following the steps below will help ensure that all components of the graph are correctly displayed.
   a. **Identify the variables.** Independent on the x-axis and dependent on the y-axis.
   b. **Determine the range.** Subtract the lowest value data point from the highest value data point—for each axis separately.
   c. **Select the scale units.** Divide each axis uniformly into appropriate units using the maximum amount of space available. (Remember that the axes may be divided differently but each square along the same axis must represent the same interval.)
   d. **Number and label each axis.** Be sure to include units where appropriate as part of the axis label.
   e. **Plot the data points as ordered pairs.** (x,y)

   f. **Draw the best straight line or best smooth curve**. Use a straight edge to draw your line in such a way that equal numbers of points lie above and below the line.

   g. **Title the graph**. The title should clearly describe the information contained in the graph. It is common to mention the dependent variable first followed by the independent variable.

4. After creating graphs for the 4 data sets below, use the graphs to answer the conclusion questions on your student answer page.

Sample Data Set 1: The following set of data was collected while experimenting with position and time of a miniature motorized car traveling on a straight track.

Position (meters)	Time (minutes)
0	0
15	5
30	10
45	15
60	20
75	25

Sample Data Set 2: The following set of data was collected during an experiment to find the density for an unknown metal.

Mass (g)	Volume (cm^3)
2.00	0.18
5.00	0.44
7.50	0.66
16.00	1.41
24.00	2.11

Sample Data Set 3: The following set of data was collected during an experiment studying the effect of light intensity on rate of photosynthesis.

Percent Transmittance (%)	Time (minutes)
32.5	0
54.3	5
63.5	10
65.0	15

Sample Data Set 4: The following set of data was collected during an acid-base titration experiment.

pH	Volume of NaOH (mL)
1.80	0.00
1.80	10.00
1.82	20.00
2.00	23.00
3.20	25.00
6.10	30.00
6.20	40.00
6.50	50.00
12.80	51.00
13.50	60.00
13.80	70.00

Name _____

Period _____

Graphing Skills
Reading, Constructing and Analyzing Graphs

Line Graphs

DATA AND OBSERVATIONS

Staple your completed graphs behind this answer page.

CONCLUSION QUESTIONS

Using the graphs that you constructed, answer the following questions:

Sample Data Set 1:

1. What is the independent variable for this graph? Explain.

2. What would be the position of the car after 25 minutes?

3. If the experiment were carried out for 80 minutes, what would be the position of the car?

4. Calculate the slope of the line drawn. What does the slope of this line represent? Explain.

5. Write the equation for a straight line including the value that was determined for slope.

Sample Data Set 2:

1. What values were considered when creating the scale for each axis in this experiment?

2. What does a data point on this graph actually represent?

3. What volume would a 20.00 gram sample of this substance occupy?

4. Calculate the density of the substance. (HINT: calculate the slope of the line.)

5. Write the equation for a straight line including the value that was determined for slope.

6. Use the equation and find the mass when the volume is 5.00 cm^3.

Sample Data Set 3:

1. Does this graph represent a linear relationship? Why or why not?

2. What is the dependent variable in this graph? Explain.

3. If the experiment were continued for 30 minutes, what trend in percent transmittance could be expected?

4. Calculate the slope of the line at 5 minutes. What does this represent?

Sample Data Set 4:

1. What is the pH of the solution after 20.0 mL of NaOH are added? After 30.0 mL are added? Would it have been easy to predict this answer?

2. Graphs often help us to understand the progress of a chemical reaction. In the titration graph for this set of data, there are two relatively sharp, upward curves. The middle of these steep rising portions represent equivalence points (point at which the moles of acid and base are equal). Identify the volume of NaOH needed to reach each of the equivalence points. .

3. What is the pH at 65 mL. What is the pH expected to do beyond this point with greater additions of the base NaOH? Explain.

Microsoft Excel
Using Excel in the Science Classroom

OBJECTIVE
Students will take data and use an Excel spreadsheet to manipulate the information. This will include creating graphs, manipulating data, finding averages and calculating standard deviation.

LEVEL
All

NATIONAL STANDARDS
UCP.1, UCP.2, A.1, A.2, E.1, E.2, G.2

TEKS
6.2 (E), 6.4(A)
7.2(E), 7.4(A)
8.2 (E), 8.4(A), 8.4(B)
IPC: 2(C)
Biology: 2(C)
Chemistry: 2(D)
Physics: 2(C), 2(E)

CONNECTIONS TO AP
Graphing skills, data management, using technology

TIME FRAME
30 minutes (for each lesson)

MATERIALS
Computers with Microsoft Excel software

TEACHER NOTES
This foundation lesson contains four sub-lessons: bar graphs, line graphs, scatter plots with linear regressions, and data management. You may want to teach each lesson as a stand alone, or as they are relevant to a current lab. The graphing and data lessons can be completely independent of one another.

Sample data has been provided for you to use if you would like to teach these as a stand-alone lesson. It is probably best used as a follow up to a data collection lab so that students can use real data.

Microsoft Excel
Using Excel in the Science Classroom

Part I: How to Make a Bar Graph

PURPOSE

To use the software program Microsoft Excel to generate a bar graph.

MATERIALS

data from this handout
computer
Microsoft Excel software

PROCEDURE

In science class you have collected data to see how much the density of water changes as you add grams of salt. Your teacher wants you to take the data and produce and a bar graph using Excel. The data is as follows:

Grams of Salt	Density
0	1.00
5	1.03
10	1.07
15	1.11
20	1.14

1. Open the Excel program on your computer. A blank workbook will appear. Notice that the columns are identified with letters and the rows are identified by numbers.

2. In the box "A1", type Grams of Salt.

3. In the box "B1", type Density. If you need to make a box larger, take your cursor to the top of the column and place it between two boxes. A double arrow should appear and you can stretch the column to the size you need.

4. Enter the data in the boxes below each section. Be careful to enter the coordinating data in the correct row.

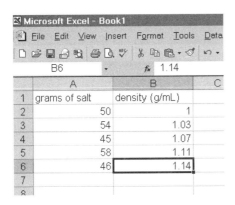

5. On your toolbar there is a very small, colorful bar graph icon. This is called the Chart Wizard. Click on the Chart Wizard icon.

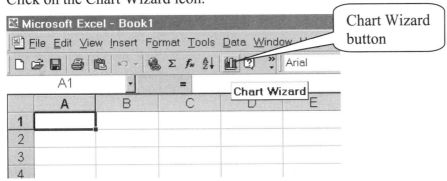

6. After clicking the Chart Wizard button, the first window that opens identifies chart type. Choose "Column" on the left-hand side, and under the chart sub-type on the right side click on the first choice available.

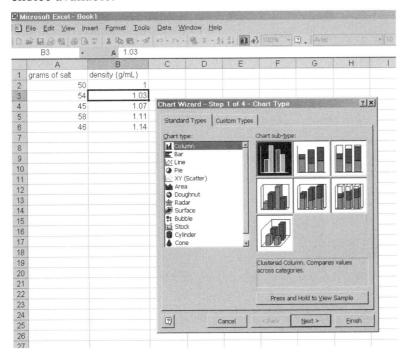

7. Click Next.

8. The next window that appears has two tabs: Data Range and Series. Click on Series and **remove all existing data sets from the series box**.

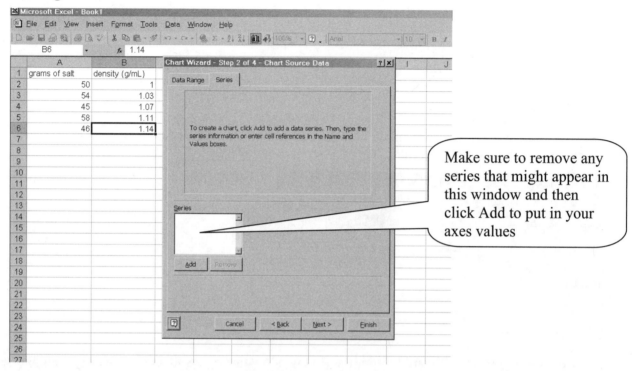

Make sure to remove any series that might appear in this window and then click Add to put in your axes values

9. Now click **Add** to add your data series. On the bottom of the window is "Category (X) Axis Labels". In the right corner of the Category (X) axis labels is a small button with a tiny graph containing a red arrow. Click on this button.

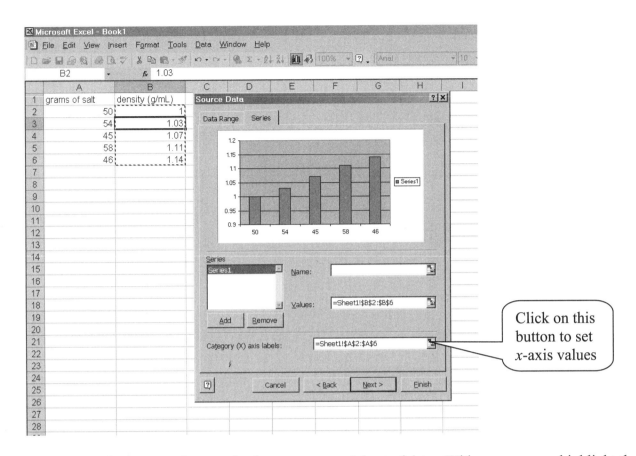

10. Clicking on the button takes you back to your spreadsheet of data. With your mouse, highlight the data you want on the *x*-axis, in this case, the grams of salt, or boxes B1-F1. Press Enter after highlighting.

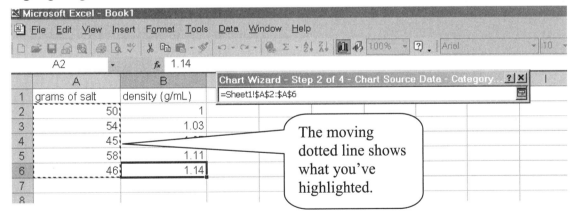

11. The chart wizard screen should now reappear. Click on the small graph button next to the spot labeled Values. This is your *y*-axis label.

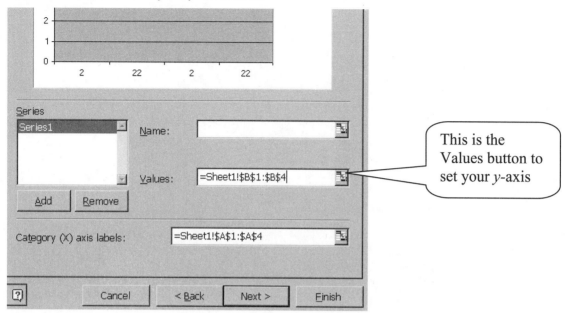

This is the Values button to set your *y*-axis

12. After clicking on the Values button, the computer takes you back to your spreadsheet and now you want to highlight your *y*-axis values, in this case density, B2-F2. Press Enter after highlighting.

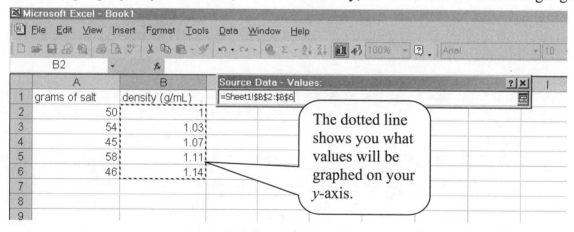

The dotted line shows you what values will be graphed on your *y*-axis.

13. The Chart Wizard screen will reappear. Click next on the bottom of the screen.

14. The new screen allows you to name your graph and label your axis. Fill in the blanks with the appropriate information and click Finish.

15. You now have finished your bar graph and Excel will ask you if you want the graph to appear on your spreadsheet, or on a separate page. Choose whichever you need. Below is a copy of the graph inserted into the spreadsheet page.

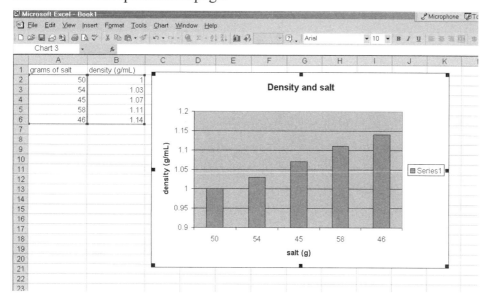

16. You may now print your completed graph by selecting print from the file menu on the task bar.

Microsoft Excel
Using Excel in the Science Classroom

Part II: How to Make a Line Graph

PURPOSE

To use the software program Microsoft Excel to create a line graph.

MATERIALS

data from this handout
computer
Microsoft Excel software

PROCEDURE

In science class you have collected data to see how much the density of water changes as you add grams of salt. Your teacher wants you to take the data and produce and a line graph using Excel. The data is as follows:

Grams of Salt	Density
0	1.00
5	1.03
10	1.07
15	1.11
20	1.14

1. Open an Excel Workbook. Notice that the columns are identified with letters and the rows are identified by numbers.

2. In the box "A1", type Grams of Salt.

3. In the box "B1", type Density. If you need to make a box larger, take your cursor to the top of the column and place it between two boxes until a double arrow appears. Now stretch the column to the size you need.

4. Enter the data in the boxes below each section. Be careful to enter the coordinating data in the correct row.

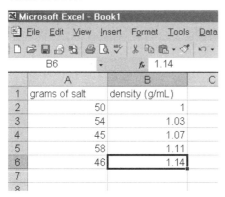

5. On your toolbar there is a very small, colorful bar graph icon. This is called the Chart Wizard. Click on the Chart Wizard icon.

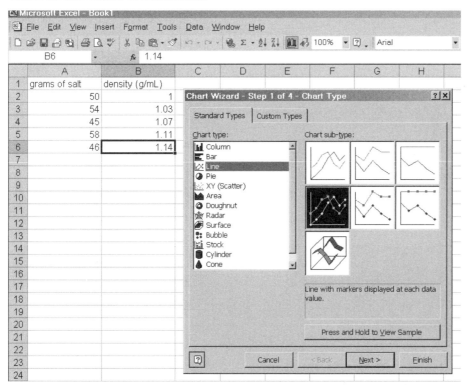

6. The first window to open is to identify chart type. Choose "Line" on the left hand side, and under the chart sub-type on the right side click on the first choice on the second line. Click Next.

7. The next window that appears has two tabs: Data Range and Series. Click on Series and **remove all existing data sets from the series box**.

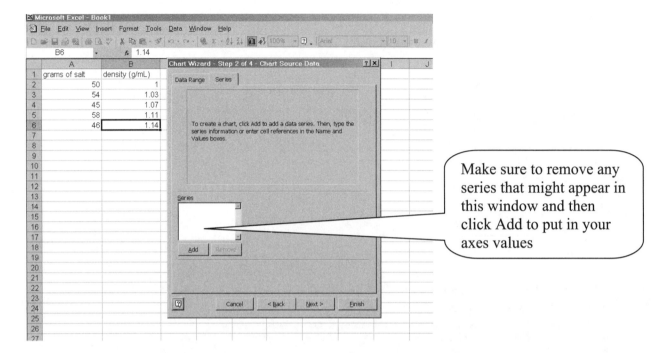

Make sure to remove any series that might appear in this window and then click Add to put in your axes values

8. Now click **Add** to add your data series. On the bottom of the window is "Category (X) Axis Labels". In the right corner of the Category (X) axis labels is a small button with a tiny graph containing a red arrow. Click on this button.

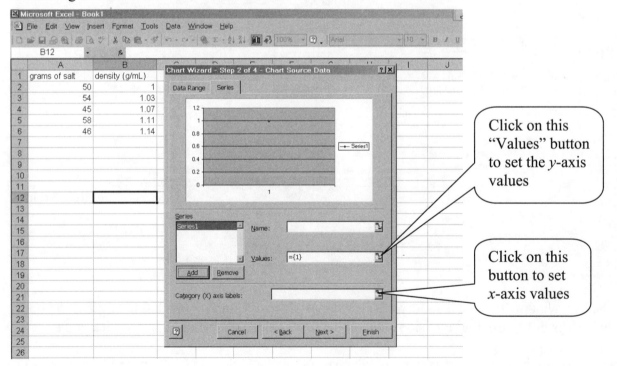

Click on this "Values" button to set the *y*-axis values

Click on this button to set *x*-axis values

9. Clicking on the button takes you back to your spreadsheet of data. With your mouse, highlight the data you want on the *x*-axis, in this case the grams of salt, or boxes B1-F1. Press Enter after highlighting.

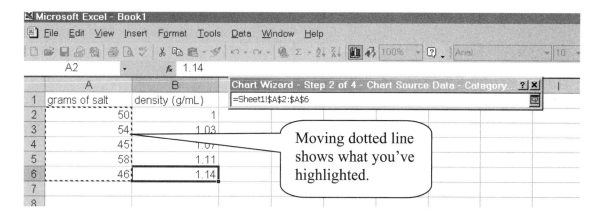

10. The Chart Wizard screen should now reappear. Click on the small graph button next to the spot labeled Values. This is how you add data to your *y*-axis.

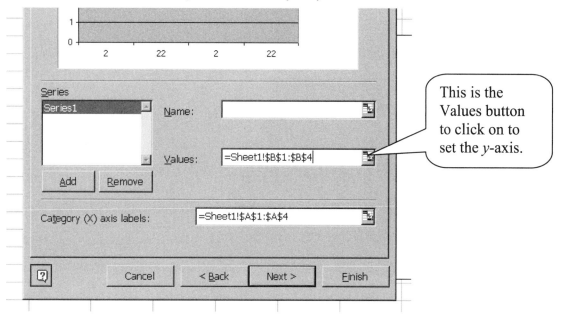

11. Clicking on the button takes you back to your spreadsheet and you now want to highlight your *y*-axis values, in this case density, B2-F2. Press Enter after highlighting.

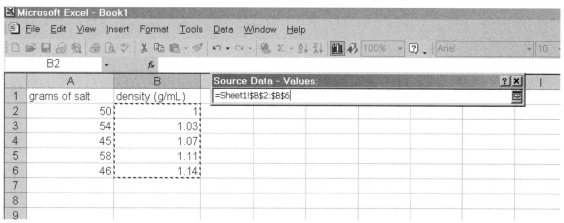

12. The Chart Wizard screen should reappear. Click Next on the bottom of the screen.

13. The new screen allows you to name your graph and label your axis. Fill in the blanks with the appropriate information and click Finish.

14. You now have finished your graph and Excel will ask you if you want the graph to appear on your spreadsheet, or on a separate page. Choose the one you need. Below is a copy of the graph inserted into the spreadsheet page.

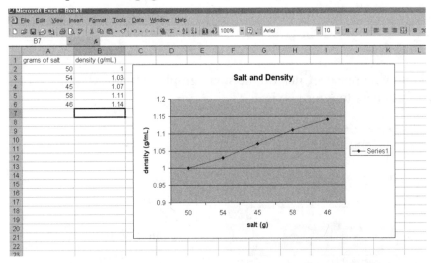

15. You may now print your completed graph by selecting print from the file menu on the task bar.

Microsoft Excel
Using Excel in the Science Classroom

Part III: How to Create a Scatter Plot and Linear Regression Equation

PURPOSE

To use the software program Microsoft Excel to create a line graph.

MATERIALS

data from this handout
computer
Microsoft Excel software

PROCEDURE

In science class you have collected data to see how much the density of water changes as you add grams of salt. Your teacher wants you to take the data and produce and a line graph using Excel. The data is as follows:

Grams of Salt	Density
0	1.00
5	1.03
10	1.07
15	1.11
20	1.14

1. Open an Excel Workbook. Notice that the columns are identified with letters and the rows are identified by numbers.

2. In the box "A1", type Grams of Salt.

3. In the box "B1", type Density. If you need to make a box larger, take your cursor to the top of the column and place it between two boxes until a double arrow appears. Now stretch the column to the size you need.

4. Enter the data in the boxes below each section. Be careful to enter the coordinating data in the correct row.

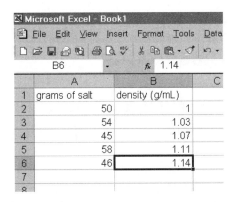

5. To create a scatter plot graph, click on the small, colorful bar graph icon. This is called the Chart Wizard.

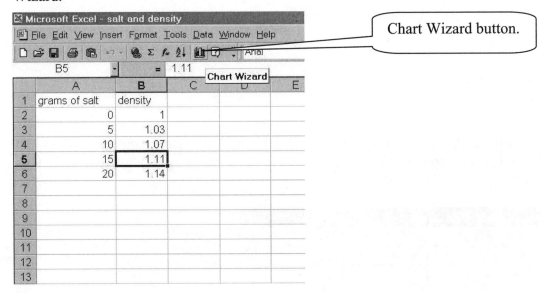

Chart Wizard button.

6. After clicking on the chart wizard button a dialogue screen will appear that allows you to choose Chart type. Choose XY SCATTER as the type of chart. *Do not choose a subtype with any lines connecting the dots*. Click Next.

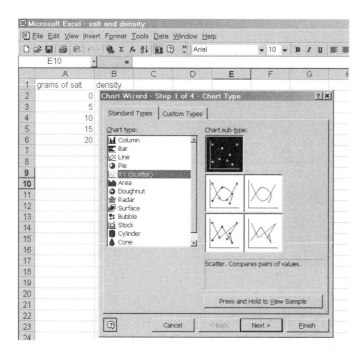

7. At the next dialogue box you will see a preview of your graph. Click the Series tab at the top of the box.

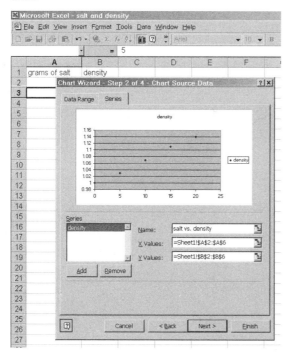

8. On the bottom left of this box it lists the series of data being plotted. To the right there is a blank cell where you can name this series. Name the series in terms of what variables are being graphed in the *y* vs. *x* format (i.e. density vs. salt).

9. Under the Name cell there are two cells that are labeled <u>X</u> values and <u>Y</u> values. The letters in these boxes correspond to the columns in the worksheet. Make sure the data is plotted on the correct axis. If they are not where you want them, click on the small button next to the <u>X</u> values button and it will take you back to your data table. Highlight the column you want to be plotted on your *x*-axis. Do the same for the *y*-axis.

10. When you are satisfied that the correct columns are being plotted and you have named your series, click Next. The next dialogue box, Chart Options, gives you the opportunity to label your axes (include units!). Click Next when you are finished.

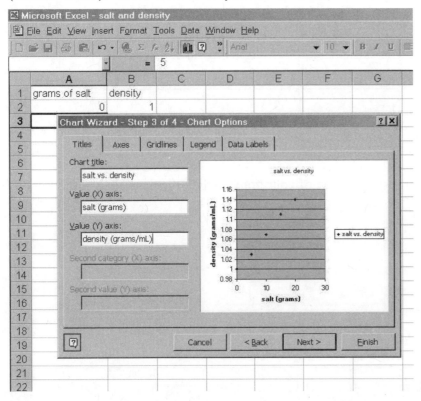

11. The final dialogue box will ask you if you want the graph to appear on your spreadsheet, or on a separate page. Click Finish when you are done.

12. To add a mathematically calculated regression line or best fit curve, choose Add Trendline from the Chart pull-down menu on your toolbar.

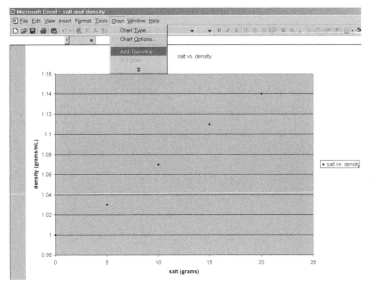

13. The next dialogue box allows you to choose the type of regression you desire.

14. The Options tab allows you to see the mathematical equation and correlation constant (R^2) if the boxes are checked for these options.

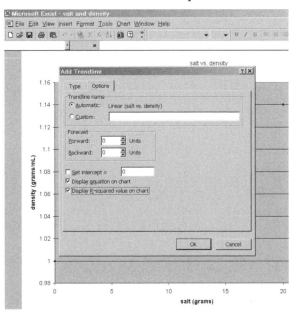

15. If you need to extrapolate data beyond the range of data you have calculated, increase the numbers in the forecast box.

16. If the preset *y*-intercept of 0 causes your graph axis and area to shift too much, set your *y*-intercept more within your data range.

17. You can also title this regression line something that tells about its origin (i.e. linear regression, power regression, etc.)

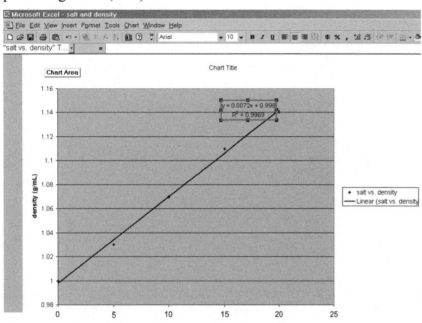

Microsoft Excel
Using Excel in the Science Classroom

Part IV: How Excel Can Manage Data

PURPOSE

To use the software program Microsoft Excel for manipulating data and determining statistical information.

MATERIALS

data from this handout
computer
Microsoft Excel software

PROCEDURE

The table below contains data collected to see how the circumference of the human head relates to the length of the face. For 5 students the data is as follows:

Circumference of Head (cm)	Length of face (cm)
50	11
54	13
45	10
58	14
46	9

1. Open an Excel Workbook. Notice that the columns are identified with letters and the rows are identified by numbers.

2. In the box "A1", type Circumference of Head (cm).

3. In the box "B1", type Length of Face (cm). If you need to make a box larger, take your cursor to the top of the column and place it between two boxes until the double arrows appear. Click and stretch the column to the size you need.

4. Enter the data in the boxes below each section. Be careful to enter the coordinating data in the same row.

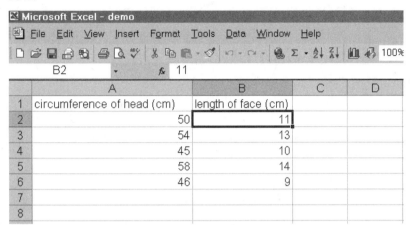

5. You are now going to have the computer calculate an index value for each person by dividing the length of the face by the circumference of the head. Label the new column in C1, skull index.

6. Click in box C2. Notice on the lower tool bar there is an empty box next to a small *fx*. Put your cursor in the box and type an equal sign (=).

7. Following the equal sign enter B2/A2. Press Enter.

8. Your spreadsheet will now reappear and you will see an index number in box C2. Right click your mouse on C2 and choose copy and then drag your mouse down column C for as far as there is data. This will apply the same formula to all of these cells.

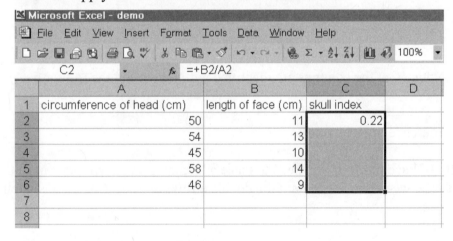

9. Press enter. Excel will calculate and fill in all the indices.

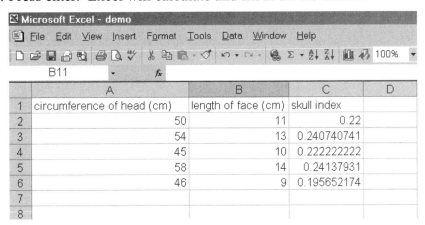

10. To reduce the numbers to two significant figures, right click on the number in cell C3 and select format cell. Click in number and then choose 2 decimal places.

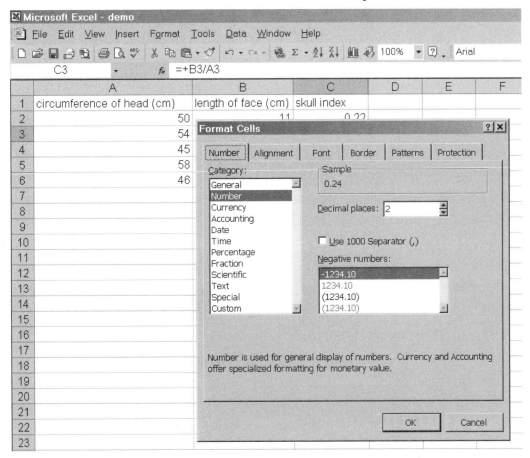

11. Click OK to exit this dialogue box and notice that one cell has changed to two decimal points. Right click on cell C3, select copy, and drag down the rest of your column. Press Enter and all numbers should change to two decimals.

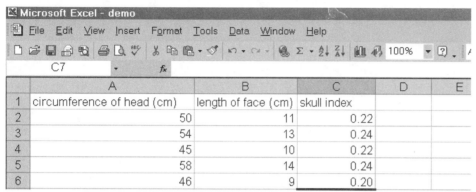

12. To calculate the average of the skull index, click in box C7, below your last index value.

13. Click on *fx* and choose average from the select a function box.

14. The next dialogue box asks you to identify what you want averaged. Highlight the five index values and then press Enter.

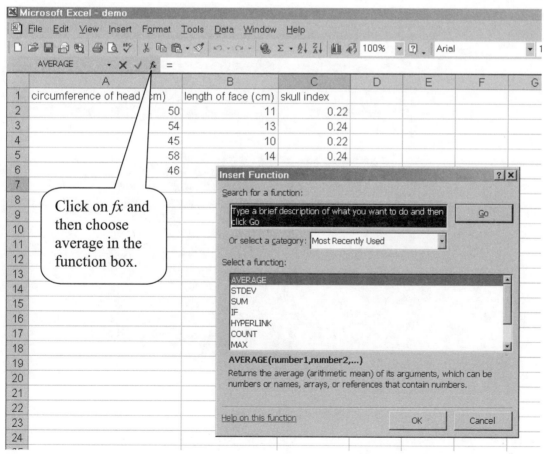

15. The next dialogue box asks you to identify what you want averaged. Highlight the five index values and then press enter. The average value will appear in cell C7.

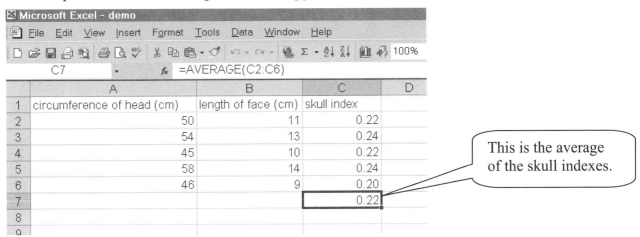

This is the average of the skull indexes.

16. To calculate the sum, standard deviation, maximum or minimum you would follow the same procedure except in step #13 choose the appropriate function.

17. To graph your data follow the procedure outlined in Microsoft Excel Part I or II.

Graphing Calculator
Using the TI-83+ in the Science Classroom

OBJECTIVE
Students will follow the teacher's oral instructions through a series of calculator menus and settings. Students will then enter sets of data to generate graphs, regression equations, and interpolations.

LEVEL
All

NATIONAL STANDARDS
UCP.1, UCP.2, A.1, A.2, E.1, E.2, G.2

TEKS
6.2 (E), 6.4(A)
7.2(E), 7.4(A)
8.2 (E), 8.4(A), 8.4(B)
IPC: 2(C)
Biology: 2(C)
Chemistry: 2(D)
Physics: 2(C), 2(E)

CONNECTIONS TO AP
All AP science courses use graphing skills to analyze data. The calculator is a tool to assist in manipulating data, seeing relationships, and drawing conclusions.

TIME FRAME
90 minutes or two class periods

MATERIALS

graphing calculator
teaching calculator

link cords
view screen and overhead projector
OR TI Presenter

TEACHER NOTES

During the past decade, the graphing calculator, along with its companion data collection devices and probes, have swept the educational field in both mathematics and science. More than any other technology, it has changed the way science is taught. Because we are able to obtain excellent data very rapidly, we may now spend our time in analysis and synthesis of the data. This tool has been found to be extremely effective helping students at all levels understand complex scientific systems.

Unfortunately, many teachers were in school in a time BC (before calculator) and the graphing calculator may seem intimidating. These notes are intended to be a quick guide to navigation of the graphing calculator. The notes are based on a TI-83+, however, other calculators have similar functions.

This document is meant to be a guide to start students on the graphing calculator. It is also designed for you to use as a reference. You should take the students through the basic steps of using the calculator. If a teaching calculator and view screen is available, it facilitates the presentation.

Suggested Teaching Procedure:

TIP #1: You cannot hurt your calculator.

1. *Layout and Important buttons:*

    There are several important keys. Use the diagram to locate each of them.

    ENTER   Used to enter data and execute functions.

    2nd   Accesses the yellow function above each of the keys

    ALPHA   Accesses the green letters above the keys and allows for typing in text.

    ◄ ▲ ► ▼   Operate like a mouse or computer arrows.

    APPS   Store applications which drive data collection devices or other study aids.

    Y= WINDOW ZOOM TRACE   Graphing keys located across top that allow for graphing. They also serve as function keys for some applications.

2. *Adjusting the brightness of the screen*

Turn your calculator on. Use [2nd] [▲] or [2nd] [▼] arrow to increase or decrease the darkness of the printing on the screen. A number at the top right indicates your battery setting. If you are using rechargeable batteries, recharge the batteries when the setting level is at 6. You will get a warning if your batteries are running low.

3. *Setting mode*

Select [MODE] to see a selection menu, which looks like the one below. Use [◄][▲][►][▼] to move around on the screen. We will move through each line of this menu to see which of these functions you will most likely be using. Because many of the programs that drive the probes set the decimal at 2 or 3 decimal places, it is often necessary to re-set this function back to Float.

4. *Using the CATALOG*

The CATALOG has all of the calculator's functions listed in it. You will find the CATALOG by pressing [2nd][0]. We will use the catalog to turn on statistical diagnostics. When you turn on diagnostics, you will enable your calculator to provide you with a correlation of regression when you attempt to curve fit. The closer the value of the "r" is to 1.000 or –1.000, the better the data fit the function which is analyzed. You will notice when you open the catalog it is locked into [ALPHA] mode. You can tell this by the 🄰 in the middle of the curser. You can go to any letter in the alphabet by hitting that key. Press "D" which is above the [x⁻¹] button. Arrow down, [▼], to **Diagnostics On**. Select it by pressing [ENTER]. Execute it by pressing [ENTER] again. It will say, "DONE".

Tip #1: If you are selecting a function from a menu of functions, pressing [ENTER] once selects the function. Pressing [ENTER] a second time actually executes the function.

Tip #2: If you are having difficulty in navigation, read the entire screen!

5. *Numbers in Scientific Notation*

The EE function must be used. This is the [2nd] [,]. The [,] is located above the 7 key. It is very important that you not use the x 10[^] method of putting numbers in scientific notation. Operations that are commutative will work regardless of how the number is entered. Non-commutative operations will not be calculated properly.

- **Exercise:** Note how entering the number incorrectly affects the value of the answer: The calculator interprets 2.5E-2 as a single number and divides the numerator by that number. In the second, and incorrect example, the calculator interprets 2.5 as one factor and 10^{-2} as another factor, so the numerator is divided by 2.5 and multiplied by 10^{-2}.

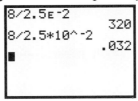

6. *Managing memory in TI-83+*

Enter [2nd][+] to get the following screen. This menu allows you to do several important operations. The operations you will be using most include the following:

- **1: About** will tell you about your operating system. Because the 83+ has a flash memory, it is possible to constantly update your operating system as new advances come out. To do this, you will need a GraphLink cable and a program called TI Connect. You can download this program *free* from the Internet.

- **2: Mem Mgmt/Del** allows you to delete programs or applications you may not be using.

- **4: ClrAllLists** is the easiest way to remove data from multiple lists.

- **7: Reset** will reset all of your memory. At this point you will have the choice of resetting your RAM or just your defaults.

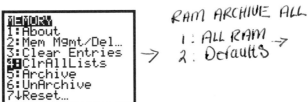

One of the advantages to the TI-83+ is that it allows you to place programs into an archived file to free up memory that is needed to do other tasks. Programs that are in archived files have an * in front of them. To take them out of archives, arrow to 2:Mem Mgmt/Del. Arrow to 7:Prgm. Scroll to the program you wish to take out of archives and press [ENTER]. This feature toggles on and off.

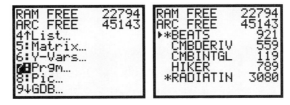

7. *Linking calculators and transferring programs*

Firmly press the link cord into both calculators. Press [2nd] [X,T,Θ,*n*] for the calculator receiving information.

Arrow ▶ to receive and press ENTER. The screen should look like the one below.

Press ENTER. The receiving calculator should now display "Waiting".

Now set up the transmitting calculator by pressing 2nd X,T,Θ,*n*. This window displays the various information or items available for transfer. Use the down arrow, ▼, to select the desired information or item to be transferred. Press ENTER. A new window appears displaying more specific items. Use the down arrow, ▼, to select the item to be transferred and press ENTER. Additional items may be selected for transfer using this same method.

When you have finished selecting your items, arrow, ▶, to TRANSMIT and press ENTER. If a selected item is already present on the receiving calculator, such as a List, a dialog box will appear allowing the user to overwrite that particular item.

8. ***Entering numbers in a list***

Numbers to be analyzed or graphed are stored in the calculator's lists. The lists are handled by the statistical function. Press the STAT key.

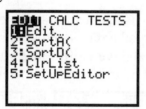

The operations you can perform from this menu are:

    1: Edit: This is where you will go to enter or change numbers in the list

    2: Sort A(: This will sort the numbers from smallest to largest

    3: Sort D(: This will sort the numbers from largest to smallest

    4: ClrList: This will allow you to clear a specific list, however you must enter an argument specifying which lists to clear.

    5: SetUpEditor: Run this option if you need to reestablish lists that have been deleted or altered

9. *Managing Lists*

Press [STAT] then select EDIT, press [ENTER]. You have 6 lists and may enter up to 999 data points in each list. You may also create additional lists with the LIST function. However, that is rarely needed for ordinary classroom work.

Clearing lists: There are three major ways to clear lists.

     (a) Use the arrow key to move to the top of the list where it says L1. Press clear, press [ENTER]. That specific list will be cleared.

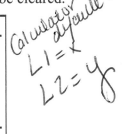

     (b) Go to [2nd][+] to get to (mem)ory. Go to 4:ClearAllLists. Press [ENTER][ENTER].

     (c) Go to [STAT]. Arrow to 4:ClrList. Press [ENTER]. Clear Lists 1 and 2 by pressing [2nd][1][,][2nd][2]). This method will only clear the list(s) that you tell it to clear. If you want to clear multiple lists, separate them by commas.

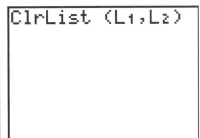

10. *Working with sample data:*

This data is similar to the Middle Grades Chemistry lesson titled *"What's That Liquid?"*. It involves a simple density problem.

A Physical Science class took the following data. Students poured a liquid into a graduated cylinder, and took the mass of several pre-determined volumes.

Enter the volume in L_1. Enter the mass in L_2.

1 = perfect line

Volume (mL)	Mass (g)
2	52.4
5	56.0
8	59.6
15	68.0
20	74.0

11. *Clearing [Y=] functions*

Make sure that all equations from any previous user are removed from [Y=]. To do this, press [Y=] and position your cursor on the = sign and press [CLEAR].

12. *Setting up STAT PLOT*

Go to [2nd] [Y=] to get to STAT PLOT.

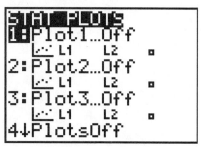

Press [ENTER] to get to Plot 1. Use the down arrow, [▾], and [ENTER] to turn on Plot 1. Move down a row. Select the first choice, a scatterplot. Be sure that XList is L_1 and YList is L_2. Arrow to the next row and choose a point protector style.

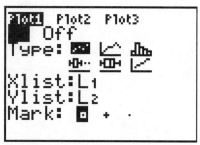

13. *Graphing Statistical Data*

Statistical data are most easily graphed by using the ZOOMSTAT. Press [ZOOM] and arrow down to ZOOMSTAT. Since this is also the 9th choice, it can be accessed by [ZOOM] [9] as well. Your graph should now be shown on the screen.

 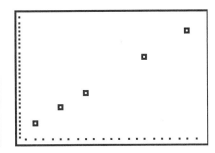

Tip #3: If you THINK you should be seeing a graph but you are not, try using ZOOM 9 to re-set your window to the range that includes your data.

14. **Regressions**

If the data in your graph looks as though it might be linear, you should run a linear regression to find out the function. If students have done a considerable amount of graphing, they should understand the meaning of "line of best fit".A linear regression is simply a mathematical "line of best fit". Press STAT and side arrow ▸ over to CALC. Arrow down to 4:LinReg (ax + b). You should see the following screen.

```
EDIT CALC TESTS
1:1-Var Stats
2:2-Var Stats
3:Med-Med
4:LinReg(ax+b)
5:QuadReg
6:CubicReg
7↓QuartReg
```

Press ENTER ENTER to get to the screen shown below.

```
LinReg
 y=ax+b
 a=1.2
 b=50
 r²=1
 r=1
```

This will give a linear equation in the form of $y = ax + b$. In this case, the function is $y = 1.2x + 50$. The correlation coefficient or correlation of regression is given by "r" and in this case is 1. That would indicate that this is a very nice linear function. In a real-life laboratory situation, the correlation coefficient would not usually be 1.00, but rather more like 0.999 or 0.998. Again, the closer the correlation coefficient is to 1.000 or -1.000, the more likely the data fit the function chosen.

15. ***Pasting the function into*** Y=

This is one of the more difficult sets of keystrokes to make. Open Y= Press VARS. Since we are dealing with statistical data, arrow down ▾ to Statistics. ENTER. Arrow over ▸ to EQ and ENTER. The equation for the line in the form of $y = ax + b$ will be pasted into your Y= menu. Now press ZOOM 9 to see both the scatterplot and the regression line.

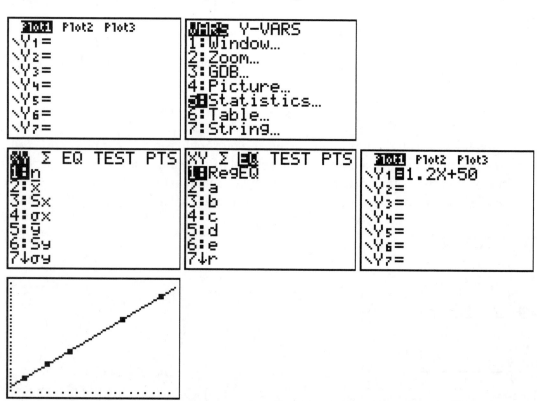

16. ***A second method to paste a function into*** Y=

If you are reasonably certain of the function that the data expresses, it is possible to run the regression and paste it into Y= in one step. Follow the screens to perform this set of operations. Start with STAT Calc 4:LinReg ENTER. Then go to VARS and side arrow ▸ to Y-VARS. Press ENTER ENTER ENTER. You can see the regression equation and it will be pasted into Y= at the same time. View your graph by pressing ZOOM 9.

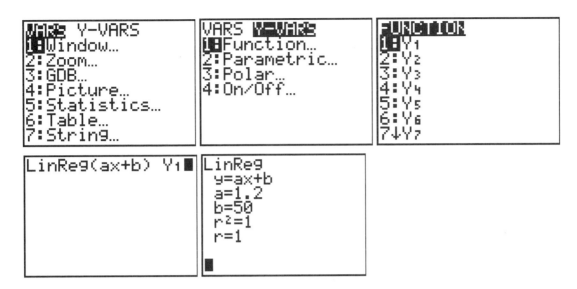

17. **Tracing on the graph**

Go to TRACE. In the upper left corner you should see a P1:L_1,L_2. This means that you are in a statistical plot of your data. Use the sideways arrows to move from data point to data point. The X and Y values are shown at the bottom of the screen. Now press the up arrow, ▲. Notice that the designation in the upper right hand corner has changed, and you should have your regression equation on the screen. Your cursor is now on the regression line so that you may interpolate values. Again the X and Y values are shown at the bottom of the screen.

18. **Helping students interpret the graph**

This is a very good time to help students realize that a graph is simply a mathematical picture of a real world situation. You can talk about independent and dependent variables, rewrite the equation in terms of words, and discuss the meaning of the value for b. For the example that we were using, Y is the mass of the liquid and the graduated cylinder; X is the volume of the liquid. The slope of the line, rise over run, is mass/volume and thus is the density. The y-intercept is the mass of the empty graduated cylinder. The equation can then be expressed in words: Total Mass = Density x Volume + Mass of graduated cylinder.

19. **Calculating a value**

We would like to be able to use this equation that has been developed. Students can be asked to apply their algebraic knowledge to solve certain problems about the situation. They can also use the graph itself to interpret information and make predictions. Let us say that we would like to know the total mass when there is 17-mL of liquid in the graduated cylinder. Press 2nd TRACE to get to CALC. 1: VALUE. Press ENTER You will see a prompt at the bottom of your graph that says X=. Enter 17 and press ENTER. The mass of the cylinder and liquid will be displayed at the bottom.

 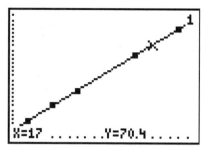

POSSIBLE ANSWERS TO THE CONCLUSION QUESTIONS

1. Enter the following data in your calculator and answer the questions.

Pressure of a gas at various temperatures

L_1 Temperature (Kelvins)	L_2 Pressure (torr)
200	600
250	750
300	900
350	1050
400	1200

a. What is the equation for this function?

- $y=3x$

b. What is the pressure when the temperature is 0 Kelvin?

- 0 torr

2. Enter this as Pressure vs. Distance below the surface of water. Remember that the title of a graph is always the dependent variable (*y*-axis) vs. independent variable (*x*-axis). It is easiest if the *x* value is placed in L_1 and the y value is placed in L_2.

Water pressure is measured at various depths in the ocean.

Distance below surface of water (feet)	Pressure (lb/in^2)
5	16.93
20	23.61
33	29.40
50	36.97

a. What is the pressure at the surface of the water?

- 14.7 lb/in^2

b. What does this represent?

- The atmospheric pressure at the surface.

TEACHER PAGES

3. On a distant planet, far, far away, the atmosphere is different from that of Earth. The speed of sound is not the same. The following data were collected:

Independent variable Temperature (Degrees Celsius)	Dependent variable Speed of sound (m/s)
0	276
10	289
20	302
30	315
50	341

a. What is the slope of this line?

- 1.3 m/s $^{\circ}$C

b. What is the temperature when the speed of sound is 295?

- 14.6 $^{\circ}$C

4. A small toy car is given a push to start it rolling down an inclined plane. The distance verses time is plotted as follows.

Time (seconds)	Distance (meters)
1	15.00
2	50.00
3	105.0
4	180.0
5	275.0
10	1050

a. Does this look like a linear function?

- No, it is curved.

b. Can you describe the motion of the car in words?

- The car is going faster and faster.

Challenge: What is the mathematical function that describes this data?

- $y = 10x^2 + 5x + 0$

What does the "b" in this particular equation represent?

- The initial speed before the car started down the inclined plane.

Graphing Calculator
Using the TI-83+ in the Science Classroom

The graphing calculator is an extremely valuable and powerful tool in both math and science classes. You will be using it in many of your classes during your educational experience. The document provided is intended to be a quick reference for many of the skills you will be using in science. Your teacher will walk you through the basic maneuvers and skills you are expected to know.

PURPOSE
In this activity you will practice using the graphing calculator to analyze data.

> graphing calculator
> pencil

PROCEDURE
Your teacher will take you through the basic functions found on your graphing calculator. You should follow along and do each of the steps on your own calculator. When you get home, go through the steps again on your own, using this guide.

1. *Layout and important buttons*:

   There are several important keys. Use the diagram to locate each of them.

   [ENTER] Used to enter data and execute functions.

   [2nd] Accesses the yellow function above each of the keys

   [ALPHA] Accesses the green letters above the keys and allows for typing in text.

   [◄][▲][►][▼] Operate like a mouse or computer arrows.

   APPS Store applications which drive data collection devices or other study aids.

   [Y=][WINDOW][ZOOM][TRACE] Graphing keys located across top that allow for graphing. They also serve as function keys for some applications.

2. *Adjusting the brightness of the screen*

   Turn your calculator on. Use [2nd] [▲] or [2nd] [▼] arrow to increase or decrease the darkness of the printing on the screen. A number at the top right indicates your battery setting. If you are using rechargeable batteries, recharge the batteries when the setting level is at 6. You will get a warning if your batteries are running low.

3. *Setting mode*

Select MODE to see a selection menu, which looks like the one below. Use the ◀▲▼▶ arrows to move around on the screen. We will move through each line of this menu to see which of these functions you will most likely be using. Because many of the programs that drive the probes set the decimal at 2 or 3 decimal places, it is often necessary to re-set this function back to Float.

4. *Using the CATALOG*

The CATALOG has all of the calculator's functions listed it. You will find the CATALOG by pressing 2nd 0. We will use the catalog to turn on statistical diagnostics. When you turn on diagnostics, you will enable your calculator to provide you with a correlation of regression when you attempt to curve fit. The closer the value of the "r" is to 1.000 or –1.000, the better the data fit the function which is being analyzed. You will notice when you open the catalog it is locked into ALPHA mode. You can tell this by the ▣ in the middle of the curser. You can go to any letter in the alphabet by hitting that key. Press "D" which is above the x^{-1} button. Arrow down, ▼, to Diagnostics On. Select it by pressing ENTER. Execute it by pressing ENTER again. It will say, "DONE".

Tip #1: If you are selecting a function from a menu of functions, pressing ENTER once selects the function. Pressing ENTER a second time actually executes the function.

Tip #2: If you are having difficulty in navigation, read the entire screen!

5. *Numbers in Scientific Notation*

The EE function must be used. This is the [2nd] [,]. The [,] is located above the 7. It is very important that you not use the x 10[^] method of putting numbers in scientific notation. Operations that are commutative will work regardless of how the number is entered. Non-commutative operations will not be calculated properly.

- **Exercise:** Note how entering the number incorrectly affects the value of the answer: The calculator interprets 2.5E-2 as a single number and divides the numerator by that number. In the second, and incorrect example, the calculator interprets 2.5 as one factor and 10^{-2} as another factor, so the numerator is divided by 2.5 and multiplied by 10^{-2}.

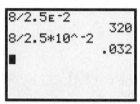

6. *Managing memory in TI-83+*

Enter [2nd][+] to get the following screen. This menu allows you to do several important operations. The operations you will be using most include the following:

- **1: About** will tell you about your operating system. Because the 83+ has a flash memory, it is possible to constantly update your operating system as new advances come out. To do this, you will need a GraphLink cable and a program called TI Connect. You can download this free program from the Internet.
- **2: Mem Mgmt/Del** allows you to delete programs or applications you may not be using.
- **4: ClrAllLists** is the easiest way to remove data from multiple lists.
- **7: Reset** will reset all of your memory or your defaults.

One of the advantages to the TI-83+ is that it allows you to place programs into an archived file to free up memory that is needed to do other tasks. Programs that are in archived files have an * in front of them. To take them out of archives, arrow to 2:Mem Mgmt/Del. Arrow to 7:Prgm. Scroll to the program you wish to take out of archives and press [ENTER]. This feature toggles on and off.

 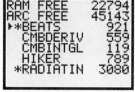

7. *Linking calculators and transferring programs*

Firmly press in the link cord into both calculators. Press [2nd] [X,T,Θ,*n*] for the calculator receiving information.

Arrow [▶] to receive and press [ENTER]. The screen should look like the one below.

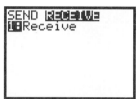

Press [ENTER]. The receiving calculator should now display "Waiting".

Now set up the transmitting calculator by pressing [2nd] [X,T,Θ,*n*]. This window displays the various information or items available for transfer. Use the down arrow, [▾], to select the desired information or item to be transferred. Press [ENTER]. A new window appears displaying more specific items. Use the down arrow, [▾], to select the item to be transferred and press [ENTER]. Additional items may be selected for transfer using this same method.

When you have finished selecting your items, arrow, [▶], to TRANSMIT and press [ENTER]. If a selected item is already present on the receiving calculator, such as a List, a dialog box will appear allowing the user to overwrite that particular item.

8. *Entering numbers in a list*

Numbers to be analyzed or graphed are stored in the calculator's lists. The lists are handled by the statistical function. Press the [STAT] key.

The operations you can perform from this menu are:

1: Edit: This is where you will go to enter or change numbers in the list

2: Sort A(: This will sort the numbers from smallest to largest

3: Sort D(: This will sort the numbers from largest to smallest

4: ClrList: This will allow you to clear a specific list, however you must enter an argument specifying which list(s) to clear. If you want to clear multiple lists, separate them by commas.

5: SetUpEditor: Run this option if you need to reestablish lists that have been deleted or altered

9. *Managing Lists*

From the STAT EDIT screen, press ENTER. You have 6 lists and may enter up to 999 data points in each list. You may also create additional lists with the LIST function. However, that is rarely needed for ordinary classroom work.

Clearing lists: There are three major ways to clear lists.

(a) Use the arrow key to move to the top of the list where it says L1. Press clear, press ENTER. That specific list will be cleared.

(b) Go to 2nd + to get to (mem)ory. Go to 4:ClearAllLists. Press ENTER ENTER.

(c) Go to STAT. Arrow to 4:ClrList. Press ENTER. Clear List 1 and 2 by pressing 2nd 1 , 2nd 2 and pressing ENTER. This method will only clear the lists that you tell it to clear.

10. ***Working with sample data:***

Sample Data: Physical Science students poured liquid into a graduated cylinder and measured the mass of several pre-determined volumes.

Enter the volume in L_1. Enter the mass in L_2.

Volume (mL)	Mass (g)
2	52.4
5	56.0
8	59.6
15	68.0
20	74.0

11. ***Clearing [Y=] functions***

Make sure that all equations from any previous user are removed from [Y=]. To do this, press [Y=] and position your cursor on the = sign and press [CLEAR].

12. ***Setting up STAT PLOT***

Go to [2nd] [Y=] to get to STAT PLOT.

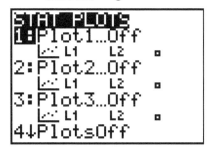

Press [ENTER] to get to Plot 1. Use the down arrow, [▼], and [ENTER] to turn on Plot 1. Move down a row. Select the first choice, a scatterplot. Be sure that XList is L_1 and YList is L_2. Arrow to the next row and choose a point protector style.

13. **Graphing Statistical Data**

Statistical data are most easily graphed by using the ZOOMSTAT. Press ZOOM and arrow down to ZOOMSTAT. Since this is also the 9th choice, it can be accessed by ZOOM 9 as well. Your graph should now be shown on the screen.

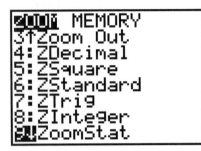 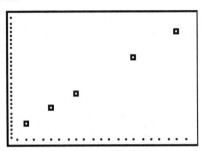

Tip #3: If you think you should be seeing a graph but you are not, try using ZOOM 9 to re-set your window to the range that includes your data.

14. **Regressions**

If the data in your graph looks as though it might be linear, you should run a linear regression to find out the function. If students have done a considerable amount of graphing, they should understand the meaning of "line of best fit". A linear regression is simply a mathematical "line of best fit". Press STAT and side arrow ▶ over to CALC. Arrow down to 4:LinReg (ax + b). You should see the following screen.

```
EDIT CALC TESTS
1:1-Var Stats
2:2-Var Stats
3:Med-Med
4 LinReg(ax+b)
5:QuadReg
6:CubicReg
7↓QuartReg
```

Press ENTER ENTER to get to the screen shown below.

```
LinReg
 y=ax+b
 a=1.2
 b=50
 r²=1
 r=1
```

This will give a linear equation in the form of $y = \mathbf{a}x + \mathbf{b}$. In this case, the function is $y = 1.2x + 50$. The correlation coefficient or correlation of regression is given by "r" and in this case is 1. That would indicate that this is a very nice linear function. In a real-life laboratory situation, the correlation coefficient would not usually be 1.00, but rather more like 0.999 or 0.998. Again, the closer the correlation coefficient is to 1.000 or –1.000, the more likely the data fit the function chosen.

15. ***Pasting the function into*** [Y=]

This is one of the more difficult sets of keystrokes to make. Open [Y=] Press [VARS]. Since we are dealing with statistical data, arrow down [▾] to Statistics. Press [ENTER]. Arrow over [▸] to EQ and press [ENTER]. The equation for the line in the form of $y= ax + b$ will be pasted into your [Y=] menu. Now press [ZOOM] [9] to see both the scatterplot and the regression line.

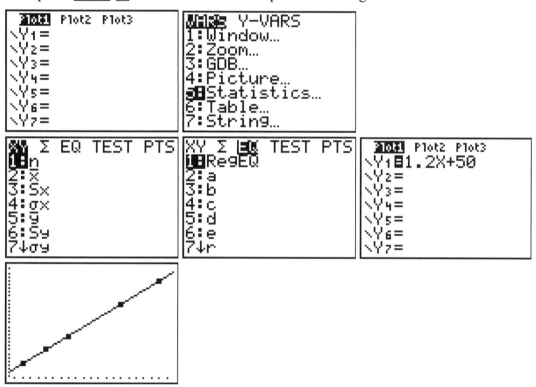

16. *A second method to paste a function into* $\boxed{Y=}$

If you are reasonably certain of the function that the data expresses, it is possible to run the regression and paste it into $\boxed{Y=}$ in one step. Follow the screens to perform this set of operations. Start with $\boxed{STAT}$ Calc 4:LinReg $\boxed{ENTER}$. Then go to $\boxed{VARS}$ and side arrow $\boxed{\blacktriangleright}$ to Y-VARS. Press $\boxed{ENTER}$$\boxed{ENTER}$$\boxed{ENTER}$. You can see the regression equation and it will be pasted into $\boxed{Y=}$ at the same time. View your graph by pressing $\boxed{ZOOM}$$\boxed{9}$.

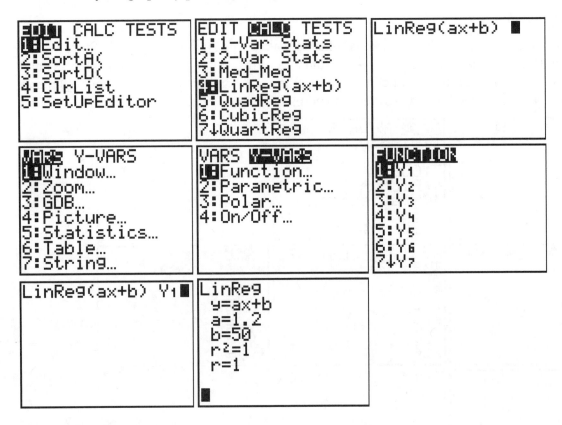

17. *Tracing on the graph*

Press $\boxed{TRACE}$. In the upper left corner you should see a P1:L_1,L_2. This means that you are in a statistical plot of your data. Use the sideways arrows to move from data point to data point. The X and Y values are shown at the bottom of the screen. Now press the up arrow, $\boxed{\blacktriangle}$. Notice that the designation in the upper right hand corner has changed, and you should have your regression equation on the screen. Your cursor is now on the regression line so that you may interpolate values. Again the X and Y values are shown at the bottom of the screen.

18. ***Calculating a value***

We would like to be able to use this equation that has been developed. Let us say that we would like to know the total mass when there is 17 mL of liquid in the graduated cylinder. Press 2nd TRACE to get to CALC. 1: VALUE. Press ENTER. You will see a prompt at the bottom of your graph that says X=. Enter 17 ENTER. The mass of the cylinder and liquid will be displayed at the bottom.

 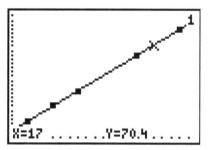

Name _____

Period _____

Using the Graphing Calculator
Analysis of Data Sets

EXERCISES

1. Enter the following data in your calculator and answer the questions.

   Pressure of a gas at various temperatures

L_1 Temperature (Kelvins)	L_2 Pressure (torr)
200	600
250	750
300	900
350	1050
400	1200

   (handwritten) $y = ax + b$ $a = 3$ $b = 6$

   a. What is the equation for this function?

   (handwritten) $y = 3x + 0$

   b. What is the pressure when the temperature is 0 Kelvin?

   (handwritten) 0 $\dfrac{150}{4}$ $600.$

2. Enter this as Pressure vs. Distance below the surface of water. Remember that the title of a graph is always the dependent variable (*y*-axis) vs. independent variable (*x*-axis). Analysis is easiest if the *x* value is placed in L_1 and the *y* value is placed in L_2.

   Water pressure is measured at various depths in the ocean.

Distance below surface of water (feet)	Pressure (lb/in^2)
5	16.93
20	23.61
33	29.40
50	36.97

$$y = .476x + 14.70$$

a. What is the pressure at the surface of the water?

14.70

b. What does this represent?

the deeper we go the greater the pressure.

3. On a distant planet, far, far away, the atmosphere is different from that of Earth. The speed of sound is not the same. The following data were collected:

Independent variable Temperature (Degrees Celsius)	Dependent variable Speed of sound (m/s)
0	276
10	289
20	302
30	315
50	341

a. What is the slope of this line?

$$y = 1.3x + 276$$

b. What is the temperature when the speed of sound is 295?

$$295 = 1.3x + 276$$
$$-276 \qquad -276$$

$$19 = 1.3x$$

$$\frac{19}{1.3} = \frac{1.3x}{1.3}$$

14.615

≈ 15

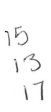

.29533
.84709
1.122.

4. A small toy car is given a push to start it rolling down an inclined plane. The distance verses time is collected as follows.

Time (seconds)	Distance (meters)
1	15.00
2	50.00
3	105.0
4	180.0
5	275.0
10	1050

a. Does this look like a linear function?

b. Can you describe the motion of the car in words?

Challenge: What is the mathematical function that describes this data?

What does the "b" in this particular equation represent?

Data Collection Devices
Determining the Amount of Energy Found in Food

OBJECTIVE
Students will determine the amount of energy in a sample of peanut and walnut. The students will calculate the percent yield and percent error in the experiment and evaluate the sources of error.

LEVEL
All

NATIONAL STANDARDS
UCP.2, UCP.3, A.1, B.5, C.5, E.1, E.2, G.2

TEKS
6.1(A), 6.2(A), 6.2(B), 6.2(C), 6.2(D), 6.2(E), 6.4(A), 6.9(A)
7.1(A), 7.2(A), 7.2(B), 7.2(C), 7.2(D), 7.2(E), 7.4(A)
8.1(A), 8.2(A), 8.2(B), 8.2(C), 8.2(D), 8.2(E), 8.4(A), 8.4(B), 8.10(A), 8.10(C)
IPC: 1(A), 2(A), 2(B), 2(C), 2(D), 8(A), 8(B)
Biology: 1(A), 2(A), 2(B), 2(C), 2(D), 9(A)
Chemistry: 1(A), 2(A), 2(B), 2(C), 2(D), 2(E), 5(A), 5(B), 5(C)
Physics: 1(A), 2(A), 2(B), 2(C), 2(D), 2(E), 2(F), 7(A), 7(B)

CONNECTIONS TO AP
Use of probeware enhances the AP student to accurately and quickly obtain data for a variety of experiments. Energy is a common theme studied in all AP science courses.

TIME FRAME
50 minutes

MATERIALS
(For a class of 28 working in pairs)

14 TI-83 or TI-83 + graphing calculators
(used for graphing calculator lab)
14 computers
(used for computer lab)
14 10 cc syringes
(used for graphing calculator lab)
14 sheets heavy aluminum foil
14 temperature probes
14 30 mL test tubes
1 package of peanuts
28 goggles
28 aprons

14 Lab Pro's or CBL's
(used for graphing calculator lab)
14 Lab Pro's or Serial Interface Device
(used for computer lab)
14 10 mL graduated cylinder
(used for computer lab)
14 packages of matches
14 calorimeters
1 balance
1 package of walnuts
14 test tube clamps
14 corks with inserted needle

TEACHER NOTES

The calorimeter can be made by using a tin vegetable can and drilling a center hole large enough to fit a 30 mL test tube and many small holes to vent. See diagram in step 11 of student procedures.

1. Calculator Information

Students will need a graphing calculator and either the CBL 2 (Calculator-Based Laboratory System) or LabPro® interface. TI 83, 83+, 85, 86, 89 or 92 calculators may be used. The temperature probe that is needed for this lab comes from Vernier. The calculators and interfaces can also be purchased from Vernier. The address is:

> Vernier Software Company
> 13979 SW Millikan Way
> Beaverton, OR 97005-2886
> (503)-277-2299
> www.vernier.com

2. Calculator Software Information

This activity uses DATAMATE as the data collection program. This program must be updated periodically and can be downloaded free from Vernier. A TI-GraphLink cable is necessary for connecting the CBL 2 or LabPro to a computer for download or upload processes. Once updated, it can be installed on the graphing calculators directly from the CBL 2 or LabPro.

3. Computer Hardware Information

Macintosh System Requirements
> Power Macintosh or newer computer
> At least 32 MB of RAM
> Operating System 7.5 or newer

Windows System Requirements
> Windows 98/NT/ME/2000/XP
> At least 32 MB of RAM
> Pentium processor-based or compatible PC

4. Computer Software Information

The software that is used in conjunction with the temperature probe is **Logger Pro 3® Software**. It can be purchased for the PC or Mac. Start the program. Under file there will be a number of folders, open "Probes & Sensors", then open "Temperature Probes", then open, "Stainless Steel Temp Probe.MBL" or "Direct Connect Temp Probe.MBL" for the Direct Connect Temperature probe.

One nice feature of Logger Pro Software is that there are templates for the probes already installed. They can be modified to fit your own experiment and then saved under your own title for the students. The following steps work nicely for this experiment:

- It is important that you try this lab and software settings before working with students so that you can modify the templates or instructions to meet the particular needs of your own experiment.

- Be sure you have the experimental apparatus set up for this experiment by connecting the interface to the computer through the modem port and the temperature probe to Port 1 or Ch 1on the interface.

- Pull down the **Set Up Menu** and highlight **Data Collection**. Then click on **Mode**. Make sure that **Real Time Collect** is selected. Next, click on **Sampling** and select **minutes** and key in **5**. Below that, move the slide bar so that 10 pts/minute is selected. Press **OK**.

- The axes of the graph are modified by first highlighting the last number on the *x*-axis, Time (minutes). Now type in the amount of time that you wish to run the experiment. Type in 5. This means that the experiment will run for five minutes. The *y*-axis should have a range of 15-105. If it does not have this range, highlight the top number, type in 105 and then highlight the lower number and type in 15. The temperature of the water should increase 20-30 degrees Celsius.

- Once you have accomplished this, you can save your experiment so that you do not have to set up the software again. Pull down the **File Menu** and highlight **Save Experiment As**. A dialog box will appear. Type in the name you would like to use for your lab. Decide where you want to save this on the hard drive or a disk. Highlight **Save**. Another dialog box will appear and ask you if you would like to save the calibration for this experiment. Click **Yes**. Now the experiment can be used by your students without having to set up the software. As long as you use that same probe with the same computer, the calibration should be good for a long while.

ANSWERS TO PRE-LAB QUESTIONS

1. After reading the background information, define the terms autotroph and heterotroph.
   - An autotroph can make its own energy rich organic compounds from inorganic compounds. Heterotrophs can not make their own energy rich organic compounds and must obtain them from the environment.

2. What are two purposes of food?
   - Food is providing energy to the organism and comprises the building blocks for synthesizing their own polymers.

3. What is free energy?
   - Free energy is the amount of energy available to do work.

4. What is entropy?
   - Entropy is a measurement of disorder in the system.

5. What is a calorie, a Calorie and a kilocalorie?
   - A calorie is the amount energy it takes to raise the temperature of one gram of water, one degree Celsius. One thousand calories is equal to one Calorie or one kilocalorie.

6. Write the equations to be used to determine the number of calories, percent yield, and percent error.

$$\text{calories} = (\text{Final Temperature} - \text{Beginning Temperature}) * 10 \text{ g}$$

$$\text{Kilocalories or Calories} = \frac{\text{calories}}{1000}$$

$$\text{calories} = 10 \text{ g } H_2O \times (\text{final temperature} - \text{initial temperature}) \times 1\frac{\text{calorie}}{g \cdot {}^\circ C}$$

$$\text{Kilocalories or Calories} = \frac{\text{calories}}{1000}$$

$$\frac{\text{kcal}}{\text{g of food}} = \frac{\text{kcal}}{\text{mass of sample}}$$

$$\text{Percent Error} = \frac{(\text{Experimental Value} - \text{Actual Value})}{\text{Actual Value}} \times 100$$

$$\text{Percent Yield} = \frac{\text{Experimental Value}}{\text{Actual Value}} \times 100$$

$$\text{Percent Error} = \frac{(\text{Experimental Value} - \text{Actual Value}) * 100}{\text{Actual Value}}$$

POSSIBLE ANSWERS TO THE CONCLUSION QUESTIONS AND SAMPLE DATA

						Data Table
Experiment	**Mass of Sample**	**Temperature Before Burning**	**Temperature After Burning**	**# of calories**	**# of kilocalories**	**# of kilocalories/gram**
Peanut # 1	0.23	22	62	400	0.4	1.74
Peanut # 2	0.2	21.5	65	435	0.435	2.18
Peanut # 3	0.24	22	61	390	0.39	1.63
Average						1.85
Walnut # 1	0.2	22	67	450	0.45	2.25
Walnut # 2	0.18	22	62	400	0.4	2.22
Walnut # 3	0.21	21.5	39	475	0.475	2.26
Average						2.24

1. Which nut contained the greatest number of kilocalories per gram?
   - The walnut has the greatest number of kilocalories per gram 2.24 kc/g

2. What was the shape of the graph displayed on your screen? What relationship did it establish?
   - The relationship is not linear. At first it increases and then slows down.

3. The nutritional labels show that peanuts are 6.00 kcal/gram and walnuts are 7.02 kcal/gram. What was your percent yield and was it greater than or less than the actual value?
   - Walnut = (2.24 / 7.02) * 100 = 31.9%
   - Peanut = (1.85 / 6.00) * 100 = 30.1%

4. Give two reasons why your answer varied from the actual values.
   a. One reason the percent yield is less is that some of the nut was not burned.
   b. A second is a large percent of the heat escaped into the environment and did not go into the test tube.

5. What is your percent error? Do you think that it is acceptable? Why or why not?
   - The percent error is 68% and 69% respectively

6. Nuts are high in lipids. What foods might you investigate to determine their calorie content for carbohydrates? Proteins?
   - Answers will vary.

7. A student fails to replace the 10 grams of water in the test tube between the second and third trials for the walnut. How will this error affect the calculated energy result? Your answer should clearly state that the calculated value will increase, decrease or remain the same. Mathematically justify your answer.

- The calculated energy value will decrease. The water was extremely hot as a result of trial 2. There is not sufficient time between trials for the very hot water to cool back down to room temperature. This will cause the change in temperature (ΔT) to be reported as too low.

    Since $\text{Energy} = \text{mass} \times \Delta T \times 1\dfrac{\text{calorie}}{\text{g} \cdot {}^\circ\text{C}}$ the quantity of energy will be reported as too low.

REFERENCE

Milani, Jean P., Revision Coordinator. *Biological Science, A Molecular Approach*. Dubuque: Kendall Hunt Publishing Company, 1990. pp. 628-630

TEACHER PAGES

Data Collection Devices
Determining the Amount of Energy Found in Food Using a Graphing Calculator

All living systems require energy. In the case of autotrophs such as plants and related organisms, they have the ability to take inorganic compounds from the environment and make high energy compounds required for their survival. Heterotrophs, however, are unable to synthesize high energy, organic nutrients from inorganic compounds and therefore must obtain these high-energy nutrients from the environment. From a human point of view, their nutrition translates into our food. All food contains energy. The amount it contains depends on the type of food it is and the organic compounds it contains.

Food is a source of both energy and the basic building blocks needed for the synthesis of macromolecules. Living systems follow the 1st and 2nd Laws of Thermodynamics. The first law states that energy cannot be created or destroyed but only converted from one form to another. This energy conversion is never 100%. Energy is always lost in the form of heat. For living systems, this is especially true. The energy content in food is converted from chemical energy to mechanical energy as muscles contract. In fireflies, the chemical energy found in food is converted to electromagnetic energy as they light up the summer skies with their bioluminescence. It is also possible to convert the chemical energy in a candy bar into another form of chemical energy. The sugar in the candy converts to fat. The fat will be later stored in our adipose tissue. As stated earlier, heat energy always accompanies any energy transfer. This type of energy is never found by itself but is always associated with one of the other forms of energy. In each of the previous examples, as energy converts from one form to another a certain amount of heat energy was released.

Free energy is the amount of energy available to do work. Heat energy that is released during an energy conversion is considered to be wasted energy and is not available to do work. This explains why energy conversions are not 100% efficient. Suppose that a candy bar had 180 kilocalories of chemical energy. Suppose you ate the candy bar in order to have mechanical energy available to play a volleyball game. At best, your body would only use 72 kilocalories of the 180 available and the rest would be released as heat energy. That's over 108 kilocalories lost!

180 kcal Total Energy				180 kcal Total Energy
Candy Bar	—— ENERGY CONVERSION →	Muscle Movement	+	Heat Energy
180 kcal	————————→	72 kcal	+	108 kcal
Free Energy	————————→	Free Energy	+	Wasted Energy

This brings us to the Second Law of Thermodynamics. Suppose there is a closed system which means energy can not enter or leave the system. This would mean that its *total* energy content could not increase or decease, it must therefore remain constant. As time passes, energy conversions occur, and more and more of the free energy available to do work converts to unusable heat energy. The unusable heat energy is unavailable to do work. If this process continued indefinitely, then eventually there would be no free energy and the system would decease in its order, becoming more chaotic. The Second

Law of Thermodynamics states that in a closed system, one in which energy can not enter or leave, there will be an increase in the system's disorder. Entropy is a measurement of the disorder of a system.

In biology, energy is often measured in calories. A calorie is the amount of heat energy needed to raise the temperature of one gram of water one degree Celsius. The amount of energy found in food is measured in kilocalories (symbolized by C or called Calories). So if a candy bar says it has 180 Calories, it really means that it has 180,000 calories. A calorimeter is used to determine the amount of calories found in food.

In this investigation you will burn a sample of food with known mass. Above the burning food sample is a test tube of water that contains 10 mL of water or 10 g of water. The temperature of the water is recorded before and after the sample is burned. The change in temperature, ΔT is calculated.

By using the equation $Energy = mass \times \Delta T \times 1 \dfrac{calorie}{g \cdot °C}$ the change in temperature, ΔT, is multiplied by the number of grams of water used (10 g) and the specific heat of the water. This allows you to calculate the amount of heat energy absorbed by the water. Since the energy absorbed by the water was the result of the burning the food, it also represents the amount of energy contained in the food sample. Since the amount of energy you calculated is in calories, it needs to be converted to kilocalories using the relationship, 1000 calories equals one kilocalories or Calorie. To make the energy relative to the particular sample you used, the energy value should be divided by the mass of your sample. This will allow the energy per gram of your sample to be compared to the energy per gram of other samples.

$$calories = 10 \text{ g } H_2O \times (\text{final temperature - initial temperature}) \times 1 \dfrac{calorie}{g \cdot °C}$$

$$\text{Kilocalories or Calories} = \dfrac{calories}{1000}$$

$$\dfrac{kcal}{g \text{ of food}} = \dfrac{kcal}{\text{mass of sample}}$$

This lab will be repeated three times to obtain an average energy value and increase the validity of your experiment. Your teacher will give you the actual number of the Calories found in the food sample used. You are to determine the percent yield.

$$\text{Percent Yield} = \dfrac{\text{Experimental Value}}{\text{Actual Value}} \times 100$$

You can also determine the percent error by using the formula below:

$$\text{Percent Error} = \dfrac{(\text{Experimental Value - Actual Value})}{\text{Actual Value}} \times 100$$

PURPOSE

In this activity you will determine the amount of energy found in a walnut and a peanut by using a calorimeter and compare it to the actual value.

MATERIALS

temperature probe
interface
matches
test tube clamps
calorimeter
balance
10 cc syringe

graphing calculator
aprons
test tube
heavy aluminum foil
goggles
samples of walnut and peanut

Safety Alert
1. CAUTION: Needles are sharp. Exercise care.
2. CAUTION: Be sure to wear goggles anytime a flame is present in the room.
3. CAUTION: If you burn yourself, immediately place the burned area under cold running water and notify your teacher.

PROCEDURE

1. Answer the pre-lab questions on your student answer page.

2. In the space marked HYPOTHESIS on your student answer page, identify whether the peanut or walnut contains the most energy.

3. Measure out three pieces of peanut and three individual pieces of walnut, approximately 0.2 g each. Determine the exact mass of each piece and record this information on your data table. Do not get the pieces mixed. Be able to identify which piece has what mass.

4. Slide the calculator and CBL2 or LabPro Interface in the bottom part of the cradle and it will click into place. Snap the calculator into the top portion of the cradle.

$$q = mc \Delta T$$

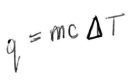

$$Time = X$$
$$Temp) = y)$$

$$Max - Min =$$

5. Plug the short black link cable into the link port on the bottom of the TI Graphing Calculator and the interface. Follow the instructions below to activate the **DATAMATE** program for data collection.

a. TI-83/TI-73 ~~APPS~~ *APPS*

- Press **PRGM**, then press the number key that precedes the **DATAMATE** program. Press **ENTER**, then press **CLEAR** when you reach the Main screen.

b. TI-83+

- Press **APPS**, and then press the calculator key for the number that precedes the **DATAMATE** program. Press **ENTER**, then press **CLEAR** when you reach the Main screen.

c. TI-86

- Press **PRGM**, then **F1** to select <**NAMES**>, and press a menu key to select <**DATAM**> (usually F1). Press **ENTER** and then press **CLEAR** when you reach the Main screen.

d. TI-89/TI-92

- Press **2nd**, **—** [VAR LINK]. Use **▼** or cursor pad to scroll down to "Datamate", then press **ENTER**. Press **)** to complete the open parenthesis that follows "Datamate" on the entry line and press **ENTER**. When you reach the Main screen, press **CLEAR**.

6. Plug the temperature probe into channel 1 of the interface. At this time the interface should have automatically identified your temperature probe and the correct temperature should be displayed in the upper right hand corner. If the correct temperature is not displayed, do the following:

- Select **SETUP** from the MENU by pressing **1**
- Select "CH1" from the MENU and press **ENTER**. , Stainless steel probe
- Select TEMPERATURE from the SELECT SENSOR MENU and press **ENTER**.
- Select the type of temperature probe that is connected to the interface.
- Press **1** to indicate o.k.

7. Select **SET UP** from the MENU by pressing **1**.

- Select MODE by pressing the **▲** key and then **ENTER**.
- Select TIME GRAPH by pressing **2** and then **2** again to change the time settings.

8. Set up the calculator and the interface for data collection.

- Enter "5" as the time between samples, in seconds. Press **ENTER**.
- Enter "60" as the number of samples (the interfaced will collect data for 5 minutes), press **ENTER**.

9. Another window will appear with the summary of the probes and the length of the experiment. Press [1] to indicate OK. Press [1] again to return to the main menu.

10. A new window will appear, and the calculator is now ready to start the experiment. DO NOT press [2] until you are ready to run the experiment.

11. Obtain a calorimeter. Place one piece of the walnut on the needle anchored in the cork.
    - CAUTION: Needles are sharp. Exercise care when doing this step.
    - CAUTION: Be sure to wear goggles anytime a flame is lit in the room.

12. Place the nut, needle, and cork setup on a folded piece of heavy-duty aluminum foil and set the calorimeter over the setup. Put a rolled piece of masking tape around the test tube to prevent it from sliding through the hole. Slide the test tube into the hole in the top of the calorimeter. Adjust the test tube so that it is about 2 cm above the nut. See the diagram below:

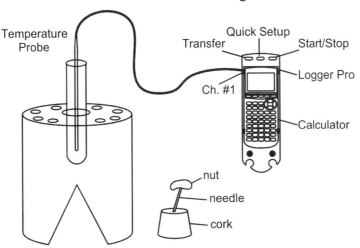

Calorimeter set-up with probe

13. Remove the calorimeter with the test tube in it from over the nut, being careful not to change the position of the test tube. Measure 15 cc of water and pour it in the test tube.

14. Put the temperature probe into the test tube. After about 15 seconds, the temperature should stabilize. Press [2] to begin data collection. The experiment will run for 5 minutes. There should be 4 short beeps and the quick setup light will flash. You should notice that the temperature is being graphed as the data is being collected.

15. When the experiment is complete, four short beeps will sound and the quick set up light will flash. Now a labeled, fitted graph will be displayed.

16. Using a match, set fire to the nut. Quickly and carefully, position the calorimeter over the burning nut.
    - CAUTION: If you burn yourself, immediately, place the burned area under cold running water. Notify your teacher.

17. You should notice that the temperature is being graphed as the data is being collected. It will continue taking data for five minutes. When the experiment is complete, 4 short beeps will sound and the quick setup light will flash. Upon completion of the data collection a labeled, fitted graph will be displayed.

18. Burn the nut completely. If your nut burns out before it is completely burned, quickly relight it.

    Use the ⬅ and ➡ keys to move the cursor. View the data points displayed at the bottom of the graph. Use these keys to determine the minimum and maximum temperature values of the water and record these in your data table on your student answer page.

19. Press **ENTER** on the calculator and you will return to the main menu. The parameters that you set for this experiment are still in the calculator. Another run of the experiment can be done without resetting the calculator.

20. Repeat this experiment two more times with the other pieces of walnut and three more times with the pieces of peanut. It is important not to mix the pieces of nuts so that correct energy per gram values will be obtained.
    - Caution: Be sure to get 10 cc of new water and take care in the handling of the test tube, as it is extremely hot.

21. Clean up your area and return your equipment to its original condition.

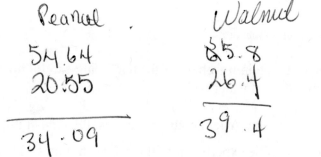

Peanut .
54.64
20.55
——————
34.09

Walnut .
65.8
26.4
——————
39.4

Data Collection Devices
Determining the Amount of Energy Found in Food
Using a Computer

All living systems require energy. In the case of autotrophs such as plants and related organisms, they have the ability to take inorganic compounds from the environment and make high energy compounds required for their survival. Heterotrophs, however, are unable to synthesize high energy, organic nutrients from inorganic compounds and therefore must obtain these high-energy nutrients from the environment. From a human point of view, their nutrition translates into our food. All food contains energy. The amount it contains depends on the type of food it is and the organic compounds it contains.

Food is a source of both energy and the basic building blocks needed for the synthesis of macromolecules. Living systems follow the 1st and 2nd Laws of Thermodynamics. The first law states that energy cannot be created or destroyed but only converted from one form to another. This energy conversion is never 100%. Energy is always lost in the form of heat. For living systems, this is especially true. The energy content in food is converted from chemical energy to mechanical energy as muscles contract. In fireflies, the chemical energy found in food is converted to electromagnetic energy as they light up the summer skies with their bioluminescence. It is also possible to convert the chemical energy in a candy bar into another form of chemical energy. The sugar in the candy converts to fat. The fat will be later stored in our adipose tissue. As stated earlier, heat energy always accompanies any energy transfer. This type of energy is never found by itself but is always associated with one of the other forms of energy. In each of the previous examples, as energy converts from one form to another a certain amount of heat energy was released.

Free energy is the amount of energy available to do work. Heat energy that is released during an energy conversion is considered to be wasted energy and is not available to do work. This explains why energy conversions are not 100% efficient. Suppose that a candy bar had 180 kilocalories of chemical energy. Suppose you ate the candy bar in order to have mechanical energy available to play a volleyball game. At best, your body would only use 72 kilocalories of the 180 available and the rest would be released as heat energy. That's over 108 kilocalories lost!

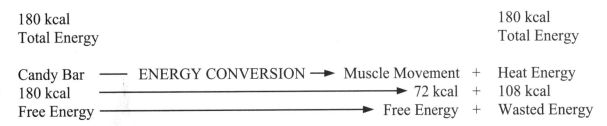

This brings us to the Second Law of Thermodynamics. Suppose there is a closed system which means energy can not enter or leave the system. This would mean that its *total* energy content could not increase or decease, it must therefore remain constant. As time passes, energy conversions occur, and more and more of the free energy available to do work converts to unusable heat energy. The unusable heat energy is unavailable to do work. If this process continued indefinitely, then eventually there would be no free energy and the system would decease in its order, becoming more chaotic. The Second

Law of Thermodynamics states that in a closed system, one in which energy can not enter or leave, there will be an increase in the system's disorder. Entropy is a measurement of the disorder of a system.

In biology, energy is often measured in calories. A calorie is the amount of heat energy needed to raise the temperature of one gram of water one degree Celsius. The amount of energy found in food is measured in kilocalories (symbolized by C or called Calories). So if a candy bar says it has 180 Calories, it really means that it has 180,000 calories. A calorimeter is used to determine the amount of calories found in food.

In this investigation you will burn a sample of food with known mass. Above the burning food sample is a test tube of water that contains 10 mL of water or 10 g of water. The temperature of the water is recorded before and after the sample is burned. The change in temperature, ΔT is calculated.

By using the equation $\text{Energy} = \text{mass} \times \Delta T \times 1 \dfrac{\text{calorie}}{\text{g} \cdot {}^\circ\text{C}}$ the change in temperature, ΔT, is multiplied by the number of grams of water used (10 g) and the specific heat of the water. This allows you to calculate the amount of heat energy absorbed by the water. Since the energy absorbed by the water was the result of the burning the food, it also represents the amount of energy contained in the food sample. Since the amount of energy you calculated is in calories, it needs to be converted to kilocalories using the relationship, 1000 calories equals one kilocalories or Calorie. To make the energy relative to the particular sample you used, the energy value should be divided by the mass of your sample. This will allow the energy per gram of your sample to be compared to the energy per gram of other samples.

$$\text{calories} = 10 \text{ g H}_2\text{O} \times (\text{final temperature} - \text{initial temperature}) \times 1 \frac{\text{calorie}}{\text{g} \cdot {}^\circ\text{C}}$$

$$\text{Kilocalories or Calories} = \frac{\text{calories}}{1000}$$

$$\frac{\text{kcal}}{\text{g of food}} = \frac{\text{kcal}}{\text{mass of sample}}$$

This lab will be repeated three times to obtain an average energy value and increase the validity of your experiment. Your teacher will give you the actual number of the Calories found in the food sample used. You are to determine the percent yield.

$$\text{Percent Yield} = \frac{\text{Experimental Value}}{\text{Actual Value}} \times 100$$

You can also determine the percent error by using the formula below:

$$\text{Percent Error} = \frac{(\text{Experimental Value} - \text{Actual Value})}{\text{Actual Value}} \times 100$$

PURPOSE

In this activity you will determine the amount of energy found in a walnut and a peanut by using a calorimeter and compare it to the actual value.

MATERIALS

temperature probe
matches
test tube clamps
calorimeter
balance
10 mL graduated cylinder
computer with Logger Pro® installed

aprons
test tube
heavy aluminum foil
goggles
samples of walnut and peanut
Lab Pro® interface box

Safety Alert

1. CAUTION: Needles are sharp. Exercise care.
2. CAUTION: Be sure to wear goggles anytime a flame is present in the room.
3. CAUTION: Matches are flammable. If you burn yourself, immediately place the burned area under cold running water. Notify your teacher.

PROCEDURE

1. Answer the pre-lab questions on your student answer page.

2. In the space marked HYPOTHESIS on your student answer page, identify whether the peanut or walnut contains the most energy.

3. Plug the temperature probe into Port 1 or CH 1 of the interface box.

4. Mass three individual pieces of peanut and three individual pieces of walnut, approximately 0.2 g each. Record the exact mass of each piece in your data table. Store each nut piece on a small square of paper labeled with its mass. It is very important not to mix up the pieces.

5. Turn on the computer. Click on the folder called **Experiment Templates**, and open **Calorimeterss** (for stainless steel probe) or **Calorimeterdc** (for direct connect temperature probe). Logger Pro® should open.
   - CAUTION: Electricity is being used. Take care not to spill any liquids on any of the computer equipment or electrical outlets.

6. Obtain a calorimeter.

7. Place one of the walnuts on the needle anchored in the cork.
   - CAUTION: Needles are sharp. Exercise care.

8. Place nut, needle, and cork setup on a folded piece of heavy-duty aluminum foil and set the colorimeter over the setup. Put a rolled piece of masking tape around the test tube to prevent it from sliding through the hole. Slide the test tube into the hole in the top of the calorimeter. Adjust the test tube so that it is about 2 cm above the nut. See the diagram below:

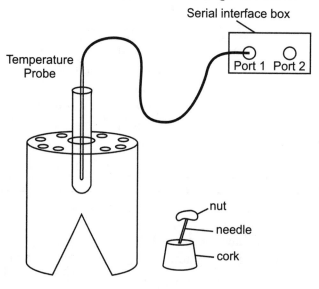

Calorimeter set-up with probe

9. Remove the calorimeter with the test tube in it from over the nut, being careful not to change the adjust position of the test tube. Measure 10 mL of water and pour it in the test tube.

10. Put the temperature probe into the test tube.

11. Using a match, set fire to the nut. Quickly and carefully position the calorimeter over the burning nut.
    • CAUTION: If you burn yourself, immediately, place the burned area under cold running water. Notify your teacher.

12. Click **Collect** and the temperature probe should start taking temperature readings immediately. Record the starting temperature.

13. Burn the nut completely. If your nut burns out before it is completely burned out, quickly relight it. When it burns out, look at the computer screen and record the final temperature. Note the shape of the graph and its axes.

14. Pull down the **Analyze Menu** and highlight **Statistics**. A statistics box will appear. It will list the maximum and minimum temperature. Record this information in your data table on the student answer page. Click the very small, gray square in the corner of the statistics window to close it.

15. Repeat this experiment two more times with the other pieces of walnut and three more times with the pieces of peanut. If you would like to keep the data that is being collected so that you can compare the various runs of the experiment, pull down the **Data Menu** and click **Store Latest Run**. This will cause the line on the graph to become a lighter red. If you do not wish to save your data you can **Delete the Latest Run** when you pull down the **Data Menu**. If you do neither of these, the Logger Pro® software will ask you if you would like to erase the previous data, you must click **Yes** to continue. It is important not to mix the pieces of nuts so that correct energy per gram values will be obtained.

16. If your teacher would like you to print a graph for this lab, pull down the **File Menu** and highlight **Print**. A dialog box will appear. Type in the number of desired copies and click **OK** or press Enter on the keyboard.

17. After you have finished close all windows on the computer. Clean up your area and return your equipment to its original condition.

Name _____

Period _____

Data Collection Devices
Determining the Amount of Energy Found in Food

HYPOTHESIS

DATA AND OBSERVATIONS

Data Table						
Experiment	Mass of Sample	Temperature Before Burning	Temperature After Burning	# of calories	# of kilocalories	# of kilocalories/gram
Peanut # 1						
Peanut # 2						
Peanut # 3						
Average						
Walnut # 1						
Walnut # 2						
Walnut # 3						
Average						

PRE-LAB QUESTIONS

1. After reading the background information, define the terms autotroph and heterotroph.

2. What are two purposes of food?

3. What is free energy?

4. What is entropy?

5. What is a calorie, a Calorie and a kilocalorie?

6. Write the equations to be used to determine the number of calories, percent yield, and percent error.

CONCLUSION QUESTIONS

1. Which nut contained the greatest number of kilocalories per gram?

2. What was the shape of the graph that screen displayed for you (if using a graphing calculator) or that Logger Pro® displayed for you (if using a computer) and what relationship did it establish?

3. The nutritional labels show that peanuts are 6.00 kcal/gram and walnuts are 7.02 kcal/gram. What was your percent yield and was it greater than or less than the actual value?

4. Give two reasons why your answer varied from the actual values.

5. What is your percent error? Do you think that it is acceptable? Why or why not?

6. Nuts are high in lipids. What foods might you investigate to determine their calorie content for carbohydrates? Proteins?

7. A student fails to replace the 10 grams of water in the test tube between the second and third trials for the walnut. How will this error affect the calculated energy result? You answer should clearly state that the calculated value will increase, decrease or remain the same. Mathematically justify your answer.

Computer Graphing Software
Using Graphical Analysis® 3 or Logger Pro® 3

OBJECTIVE
Students will learn to use Graphical Analysis® 3 or Logger Pro® 3 computer graphing software to graph and analyze data as well as generate a lab report by importing their data and graphs into a word processing document.

LEVEL
All

NATIONAL STANDARDS
UCP.1, UCP.2, A.1, A.2, E.1, E.2, G.2

TEKS
6.2 (E), 6.4(A)
7.2(E), 7.4(A)
8.2 (E), 8.4(A), 8.4(B)
IPC: 2(C)
Biology: 2(C)
Chemistry: 2(D)
Physics: 2(C), 2(E)

CONNECTIONS TO AP
All AP science exams ask students to read and interpret graphs.

TIME FRAME
90+ minutes

MATERIALS
> either Graphical Analysis® 3 software or Logger Pro® 3 software
> computer

TEACHER NOTES

Ideally, this lesson would follow a lesson on graphing calculators and the use of data collection devices. The data from the calculator can be imported into either of these pieces of software for further analysis. It is important to load the TI Connect software *first*, followed by the graphing software you have chosen. *Foundation Lesson VII: Data Collection Devices*, addresses the collection of data using Logger Pro. It is the intent of this lesson to simply use these computer graphing programs to analyze a set of sample data to learn the features of the program and then to generate a lab report by cutting and pasting the data and graphs into a word processing document. The students are directed to skip two of the more mathematically advanced tutorials and told they may do them count for extra credit.

Both of these graphing software programs are available for purchase from Vernier Software & Technology, 13979 SW Millikan Way, Beaverton, Oregon 97005-2886. You can also purchase on-line at www.vernier.com. At the time of this printing, Graphical Analysis® 3 can be purchased at a cost of $80 and Logger Pro® 3 at a cost of $100. This gives your campus a site license that allows you to load the program on any and every computer on campus and give a copy of the program to each of your students.

What is the difference between these two programs? Graphical Analysis® 3, GA3, is simply a user friendly graphing software program that can be used alone or with TI graphing calculator, CBL 2, or LabPro compatibility. Data can be imported from a TI graphing calculator, CBL 2 or LabPro into GA3 in a matter of seconds with the use of a TI-Graph Link cable. Logger Pro® 3, LP3, has GA3 embedded within it, but is also capable of communicating directly with a LabPro and probes to collect data bypassing the need for a calculator. Logger Pro® 3 houses all of the experiment files found in the TI and Vernier Lab Manuals and has more capability than GA3. Logger Pro also contains some sample movies that graph data as they play so that the data can be analyzed. You may also add your own movies and synchronize them with data collection for further analysis. Logger Pro® 3 is the software to purchase if you have access to several computers for student use in your laboratory.

The complete user's manual to either piece of software is available at www.vernier.com.

**IF you are using a USB TI Graph-Link, you need to first load TI Connect from the TI website, http://education.ti.com/us/product/accessory/connectivity/down/download.html.

THEN load the graphing software. TI Connect contains the driver for the USB Graph-Link. Neither Graphical Analysis nor Logger Pro contains this driver. **You will not be able to import data from the calculator if you load the graphing software first. If you are using the older gray or black Graph-links, you do not need TI Connect software at all.

Graphical Analysis® 3 Computer Requirements

Windows requirements:

- Windows 95, Windows 98, Windows 2000, Windows NT 4.x, Windows ME, and Windows XP.
- 133 MHz Pentium processor or better.
- 16 MB physical RAM plus free hard disk space (for virtual memory).
- Color monitor (>=256 colors)

Macintosh requirements:

- Mac OS 8.x, MacOS 9.x, MacOS X.
- 66 MHz PowerPC processor or better.
- 16 MB machine RAM, 8MB for the application partition.

Logger Pro® 3 Computer Requirements

- Windows 98®, 2000, ME, NT, or XP on a Pentium Processor or equivalent, 133 MHz, 32 MB RAM, 25 MB of hard disk space, for a minimum installation.
- Mac OS® 9.2, or Mac OS X (10.1 or newer), with 25 MB of hard disk space for a minimum installation.
- Using the movie feature of Logger Pro will require a faster processor and an additional 100 MB of hard disk space.

Note: Logger Pro cannot be used with the ULI or Serial Box interface.

Loading Graphical Analysis or Logger Pro onto Your Hard Drive

(Remember to load the TI Connect software first, before the graphing software)

To install Logger Pro or Graphical Analysis on a computer running Windows 98/2000/ME/NT/XP, follow these steps: [note that GA3 will run on Windows 95, but LP3 will NOT]

1. Place the software CD in the CD-ROM drive of your computer.

2. If you have Autorun enabled, the installation will launch automatically; otherwise choose Settings→Control Panel from the Start menu. Double click on Add/Remove Programs. Click on the Install button in the resulting dialog box.

3. The software installer will launch, and a series of dialog boxes will step you through the installation. You will be given the opportunity to either accept the default directory [recommended] or enter a different directory.

To install Graphical Analysis or Logger Pro on a computer running MacOS 8.x, MacOS 9.x, MacOS X, follow these steps:

1. Place the Graphical Analysis CD in the CD-ROM drive of your computer.

2. Double-click on the Install Graphical Analysis icon and follow the directions.

Setting Preferences Within the Software

There is one aspect of both programs that needs to be changed. The setting in the start-up file for both programs has the "Connect Lines" or "Connected Points" feature as the default. One of your primary goals is to have Pre-AP students mathematically model data rather than draw a dot-to-dot picture of the data. You will spend a great deal of time encouraging students to break this habit. You want students to mathematically model their data using curve-fitting so that they can use the equation of the curve to interpolate and extrapolate. This setting is easy to change, but must be done on each computer or on the master *copy* of the software that you issue to students. Follow these steps:

1. Start the program and double click on the blank graph. This dialogue box appears:

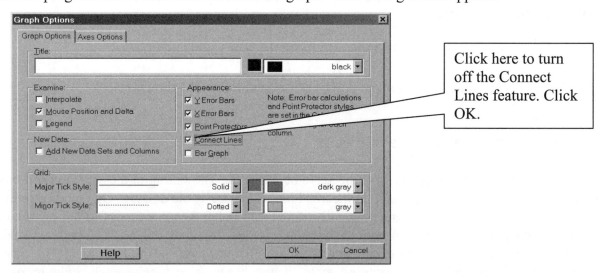

Click here to turn off the Connect Lines feature. Click OK.

2. Go to File and select preferences and make sure the Start up file box is checked. You can also change other features such as removing the automatic curve fitting option for students if you wish.

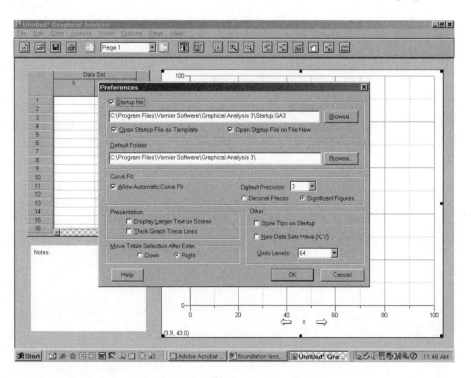

3. Go to File and select Save As. If using GA3, select the file titled startup.ga3. If using LP3, select the file titled startup.xmbl.

4. Right click on the file and scroll to the bottom of the pop up box and select Properties. The Properties dialogue box will appear. Click to remove the Read Only status of the startup file. Click OK. Click SAVE. The program will inform you that the file already exists and ask you if you wish to replace it, click YES.

5. Failure to disable the read only status will result in this error message:

Click OK and simply repeat step 4.

6. If you did not get the error message, you were successful. However, you must restore your new and improved startup file to Read Only status to protect it from corruption. Go to File, Save As and select the startup file again. Right click and select properties at the bottom of the pop up box. Click to replace the Read Only checkmark. Click OK. Click Cancel.

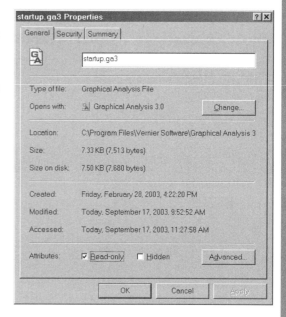

7. Repeat these steps on each of your student computers.

8. Since you may make copies of the CD for distribution to students you may wish to do these steps **once** on your own computer. When you are setting up the files to burn your master *copy* for distribution, simply delete the factory startup file before you burn the disk and replace it with the one from the hard drive on the computer you are working from once you have completed steps 1-7. The startup file is found in C:\Program Files\Vernier Software\Graphical Analysis 3 if using GA3. The startup file is found in C:\Program Files\Vernier Software\Logger Pro 3 if using LP3. Performing this step *before* burning your master *copy* CD for student distribution will save you a great deal of time and make the student's home computer begin the program just as it does at school.

NOTE: Since the tutorials in this lesson do NOT use the startup file, do not be surprised if this change is not apparent. It will take effect anytime a new file is created.

STUDENT SAMPLE

This sample can serve as your grading key. Since all students are following the same set of tutorials, their reports should be nearly identical and thus easy to grade.

Learning to Use Graphical Analysis [or Logger Pro]
Ima Student
8th Grade Science, Period 4

Basic Operations Tutorial Results

How much garbage was generated per person in 1994?
• 2511 lb/person/yr

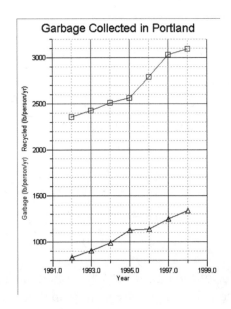

Customization Tutorial Results

	Data Set	
	X	**Y**
1	1.00	2.00
2	2.00	4.00
3	3.00	5.80
4	4.00	6.70
5	5.00	8.20
6	6.00	10.0
7	7.00	12.0
8	8.00	14.3
9	9.00	15.6
10	10.0	17.0
11		
12		
13		
14		
15		
16		
17		
18		

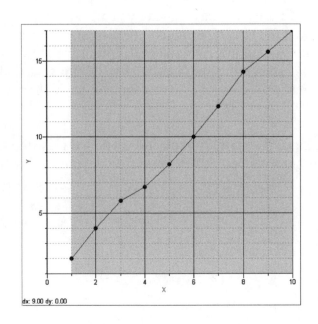

Viewing Graphs Tutorial Results

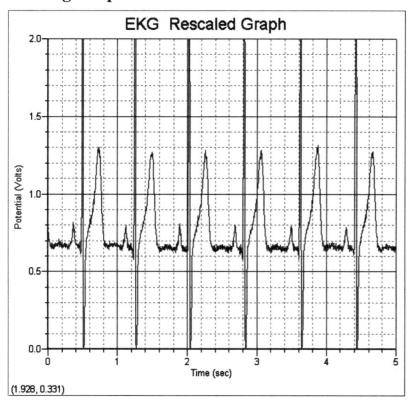

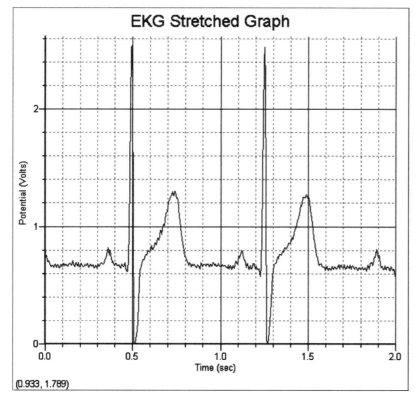

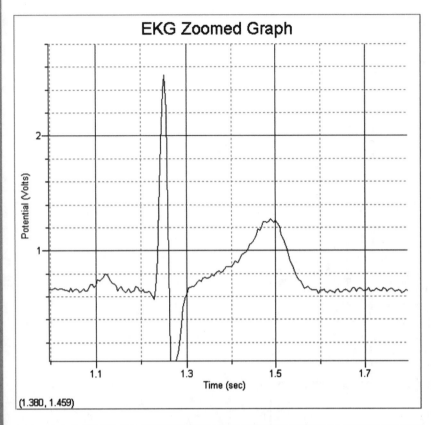

(1.380, 1.459)

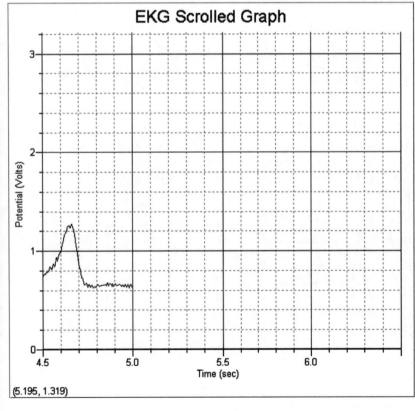

(5.195, 1.319)

Linearization Part 1 Tutorial Results

Is Population proportional to Time Interval ^2?

- Yes since the graph of Population vs. Time Interval squared is linear.

The top graph on the right shows a different kind of relationship. What is it?

- It is an inverse relationship since y decreases as x increases.

Linearization Part 2 Tutorial Results

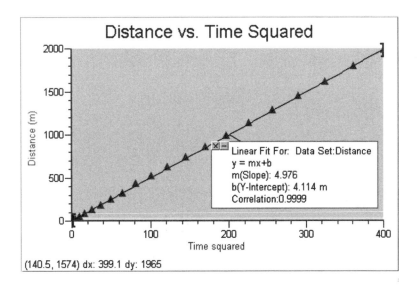

Is the graph now a straight line? If so, then you've found the relationship.

- Yes it is. The equation for the straght line is $y = 4.976x + 4.11$ with a correlation of 0.999. This means that Distance = 4.976 (time^2) + 4.11 m.

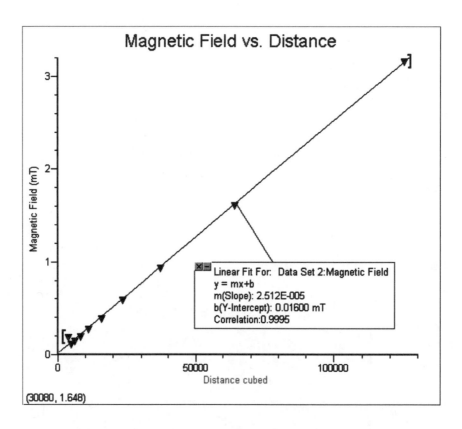

Computer Graphing Software
Using Graphical Analysis® 3

You will be asked to graph and analyze a vast amount of data in this course as well as future science courses. In the past, this was a daunting and time consuming task. Your school has purchased either Graphical Analysis® 3 (GA3) to greatly simplify analyzing and communicating the data you have collected during your laboratory exercises. The program also allows you to cut and paste your data tables and graphs into a word processing program such as Microsoft Word. This will be very helpful in generating lab reports. The site license your school has purchased allows you to load a copy of the software on any computer in your school or at home.

PURPOSE
In this activity you will learn the basic features of Graphical Analysis or Logger Pro computer graphing software so that you may use this valuable tool throughout this course and future science courses.

MATERIALS
Computer graphing software, Graphical Analysis® 3

PROCEDURE FOR GRAPHICAL ANALYSIS® 3 SOFTWARE

1. Start the program by double clicking either the Graphical Analysis icon.

2. The program may greet you with a Tip of the Day. Read the tip, close it and go to File and select Open or simply click on the second icon from the right to open files. Open the Tutorials folder and this dialogue box will appear:

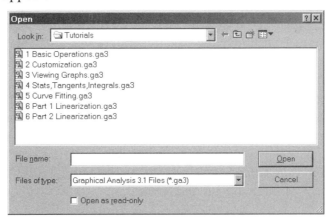

3. Double click on "1 Basic Operations.ga3" to open the file.

4. Without closing GA3, open a word processing program such as Microsoft Word®. Start a new file and title this document "Learning to Use Graphical Analysis". Your title should be centered with 18 point bold font. Type your name under your title making it centered with 16 point bold font. Type your course name and class period together under your name making it centered with 14 point bold font. Press the Enter key three times and type "Basic Operations Tutorial Results". Subtitles such as this one should be left justified, bold 14 point font. Press Enter twice, change the font to 12 point, and remove the bold. This will be the formatting for all of the body text in today's lab report. SAVE this document. Do not close the program; just minimize it since you will be going back and forth between GA3 and your document.

5. Go back to GA3 by clicking on it at the bottom of your screen. Read and follow all instructions contained within the tutorial. As you proceed through this tutorial, the pop up dialogue boxes may obscure the directions. You can move the dialogue boxes by clicking on the blue title bar of the box and dragging them to another part of the screen, out of your way. Don't forget that you will get tool tips if you let your mouse pointer hover over a button, these are very useful.
   a. When you reach page 5, you will be instructed to title your graph "Garbage vs. Year". This is not a title, but rather a restatement of the axes. Title your graph "Garbage Collected in Portland" instead.
   b. You will be asked questions throughout these tutorials about the data presented. In all cases, cut and paste the questions from GA3 by clicking on the box containing the question, highlighting the question, and pressing Ctrl+C to copy the question to the clipboard. Go to your word processing document and place your cursor under the appropriate subtitle heading and press Ctrl+V to paste the question into your document. Type your answer to the question. Pay special attention to the units of the answer. No naked numbers!

6. Continue through the tutorial until you reach the "Congratulations" page. Make sure you have cut and pasted (as well as answered) all questions throughout the tutorial. Once you are finished, go back 2 pages to your graph. Check to make sure it has a correct title, then click on the graph to select it and copy it onto the clipboard using Ctrl+C. Return to your document and press Enter twice to separate the graph from the questions asked. Use Ctrl+V to paste your completed graph from the first tutorial into this document. Resize the graph in your document so it takes up less space. Reduce it by a factor of *about* one half. Resize all of your graphs by a factor of *about* one half throughout the rest of your document.

7. Go back to GA3 and go to File, Open and select "2 Customization". **Do NOT save your changes to the first tutorial.** Proceed through this second tutorial.

8. Continue until you reach the "Congratulations" page. Once you are finished, go back 2 pages to your graph. Click on the graph to select it and copy it onto the clipboard using Ctrl+C.

9. Go back to your minimized word processing document. Press Enter twice and type the subtitle "Customization Tutorial Results". Press Enter once followed by Ctrl+V to paste your customized graph into this document. Resize the graph in your document. SAVE. Do not close the program, just minimize it.

10. Go back to GA3 and page back to page 3. Cut and paste the improved data table into your word processing document. Resize the data table in your document.

11. Go back to Graphical Analysis and go to File, Open and select "3 Viewing Graphs". **Do NOT save your changes to the second tutorial.** Proceed through this third tutorial.

12. Once you have rescaled the graph at the end of the first page, change its title to "EKG Rescaled Graph" and copy it to the clipboard.

13. Go back to your minimized word processing document. Press Enter once followed by Ctrl+V to paste your rescaled graph into this document. Resize the graph in your document. SAVE. Do not close the program, just minimize it.

14. Go back to Graphical Analysis and continue the tutorial. Once you have stretched the graph on page 2 of the tutorial, change its title to "EKG Stretched Graph" and copy it to the clipboard.

15. Go back to your minimized word processing document. Press Enter twice and type the subtitle "Viewing Graphs Tutorial Results". Press Enter once followed by Ctrl+V to paste your zoomed graph into this document. Resize the graph in your document. SAVE. Do not close the program, just minimize it.

16. Go back to Graphical Analysis and continue the tutorial. Once you have zoomed in on the graph on page 3 of the tutorial, change its title to "EKG Zoomed Graph" and copy it to the clipboard.

17. Go back to your minimized word processing document. Press Enter twice and type the subtitle "Viewing Graphs Tutorial Results". Press Enter once followed by Ctrl+V to paste your zoomed graph into this document. Resize the graph in your document. SAVE. Do not close the program, just minimize it.

18. Go back to Graphical Analysis and continue the tutorial. Continue until you reach the "Congratulations" page. Once you are finished, go back 1 page to your graph. Click on the graph to select it, change its title to "EKG Scrolled Graph" and copy it onto the clipboard using Ctrl+C.

19. Go back to your minimized word processing document. Press Enter once followed by Ctrl+V to paste your scrolled graph into this document. Resize the graph in your document. Resize the graph in your document. SAVE. Do not close the program, just minimize it.

20. Go back to Graphical Analysis and go to File, Open and select "6 Linearization Part 1" [Yes, we are skipping tutorials 4 & 5, they will be extra credit!]. **Do not save your changes to the third tutorial**. Proceed through this next tutorial. Be sure to cut and paste the two questions into your report under the subtitle "Linearization Part 1 Tutorial Results". There are no graphs to paste into your report, but you will need to use the skills presented in this tutorial to be successful in Part 2. You may want to print the last page of this tutorial for future reference.

21. Go to File, Open and select "6 Linearization Part 2". Follow all of the directions on page one and paste your data tables and your graphs into your report under the subtitle "Linearization Part 2 Tutorial Results". Be sure and run the linear regression on your data so that your graph shows the statistics box. Do this by highlighting the data on your graph and pressing the [R=] button.

22. SAVE. Continue to the end of the tutorial and paste your second graph into your report. [Hint: The shape of the second graph tells you it is an inverse function.] When you calculate your column you will need to use the reciprocal of one of the variables. Be sure and include the algebraic equation for your line of best fit once you have pasted the graphs into your document.

23. You may add the appropriate graphs from tutorials 4 & 5 to your report for extra credit. Be sure and include the proper subtitle headings and continue the proper format for the report.

Computer Graphing Software
Using Graphical Logger Pro® 3

You will be asked to graph and analyze a vast amount of data in this course as well as future science courses. In the past, this was a daunting and time consuming task. Your school has purchased either Logger Pro® 3 (LP3) to greatly simplify analyzing and communicating the data you have collected during your laboratory exercises. The program also allows you to cut and paste your data tables and graphs into a word processing program such as Microsoft Word. This will be very helpful in generating lab reports. The site license your school has purchased allows you to load a copy of the software on any computer in your school or at home.

PURPOSE
In this activity you will learn the basic features of Graphical Analysis or Logger Pro computer graphing software so that you may use this valuable tool throughout this course and future science courses.

MATERIALS
Computer graphing software, Logger Pro® 3

PROCEDURE FOR LOGGER PRO® 3 SOFTWARE

1. Start the program by double clicking the Logger Pro icon.

2. The program may greet you with a Tip of the Day. Read the tip then close it. You may also be asked about continuing without an interface attached. Choose to continue without an interface and click OK. Go to File and select Open or simply click on the second icon from the right to open files. Open the Tutorials folder and this dialogue box will appear:

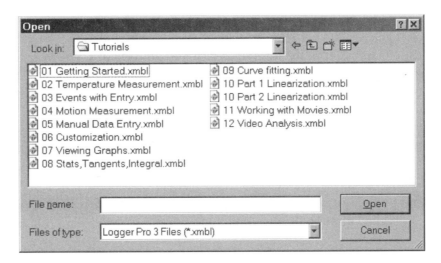

3. Double click "01 Getting Started.xmbl" to open the file.

4. Without closing LP3, open a word processing program such as Microsoft Word®. Start a new file and title this document "Learning to Use Logger Pro". Your title should be centered with 18 point bold font. Type your name under your title making it centered with 16 point bold font. Type your course name and class period together under your name making it centered with 14 point bold font. Press the Enter key three times and type "Manual Data Entry Tutorial Results". Subtitles such as this one should be left justified, bold 14 point font. Press Enter twice, change the font to 12 point, and remove the bold. This will be the formatting for all of the body text in today's lab report. SAVE this document. Do not close the program; just minimize it since you will be going back and forth between LP3 and your document.

5. Go back to LP3 by clicking on it at the bottom of your screen. Read and follow all instructions contained within the tutorial. Continue through this tutorial until you reach the "Congratulations" page.

6. Go to File, Open and select "05 Manual Data Entry". **Do NOT save your changes to the first tutorial.** As you proceed through this tutorial, the pop up dialogue boxes may obscure the directions. You can move the dialogue boxes by clicking on the blue title bar of the box and dragging them to another part of the screen, out of your way. Don't forget that you will get tool tips if you let your mouse pointer hover over a button, these are very useful.
   a. When you reach page 5, you will be instructed to title your graph "Garbage vs. Year". This is not a title, but rather a restatement of the axes. Title your graph "Garbage Collected in Portland" instead.
   b. You will be asked questions throughout these tutorials about the data presented. In all cases, cut and paste the questions from LP3 by clicking on the box containing the question, highlighting the question, and pressing Ctrl+C to copy the question to the clipboard. Go to your word processing document and place your cursor under the appropriate subtitle heading and press Ctrl+V to paste the question into your document. Type your answer to the question. Pay special attention to the units of the answer. NO naked numbers!

7. Continue until you reach the "Congratulations" page. Make sure you have cut and pasted (as well as answered) all questions throughout the tutorial. Once you are finished, go back 2 pages to your graph. Check to make sure it has a correct title, then click on the graph to select it and copy it onto the clipboard using Ctrl+C. Return to your document and press Enter twice to separate the graph from the questions asked. Press Ctrl+V to paste your completed graph from the first tutorial into this document. Resize the graph in your document so it takes up less space. Reduce it by a factor of *about* one half. Resize all of your graphs by a factor of *about* one half throughout the rest of the document.

8. Go back to LP3 and go to File, Open and select "6 Customization". **Do NOT save your changes to the first tutorial.** Proceed through the customization tutorial.

9. Continue until you reach the "Congratulations" page. Once you are finished, go back 2 pages to your graph. Click on the graph to select it and copy it onto the clipboard using Ctrl+C.

10. Go back to your minimized word processing document. Press Enter twice and type the subtitle "Customization Tutorial Results". Press Enter once followed by Ctrl+V to paste your customized graph into this document. Resize the graph in your document. SAVE. Do not close the program, just minimize it.

11. Go back to LP3 and page back to page 3. Cut and paste the improved data table into your word processing document. Resize the data table in you document.

12. Go back to Logger Pro and go to File, Open and select "07 Viewing Graphs". **Do NOT save your changes to the customization tutorial.** Proceed through the viewing graphs tutorial.

13. Once you have rescaled the graph at the end of the first page, change its title to "EKG Rescaled Graph" and copy it to the clipboard.

14. Go back to your minimized word processing document. Press Enter once followed by Ctrl +V to paste your rescaled graph into this document. Resize the graph in your document. SAVE. Do not close the program, just minimize it.

15. Go back to LP3 and continue the tutorial. Once you have stretched the graph on page 2 of the tutorial, change its title to "EKG Stretched Graph" and copy it to the clipboard.

16. Go back to your minimized word processing document. Press Enter twice and type the subtitle "Viewing Graphs Tutorial Results". Press Enter once followed by Ctrl+V to paste your zoomed graph into this document. Resize the graph in your document. SAVE. Do not close the program, just minimize it.

17. Go back to LP3 and continue the tutorial. Once you have zoomed in on the graph on page 3 of the tutorial, change its title to "EKG Zoomed Graph" and copy it to the clipboard.

18. Go back to your minimized word processing document. Press Enter twice and type the subtitle "Viewing Graphs Tutorial Results". Press Enter once followed by Ctrl+V to paste your zoomed graph into this document. Resize the graph in your document. SAVE. Do not close the program, just minimize it.

19. Go back to LP3 and continue the tutorial. Continue until you reach the "Congratulations" page. Once you are finished, go back 1 page to your graph. Click on the graph to select it, change its title to "EKG Scrolled Graph" and copy it onto the clipboard using Ctrl+C.

20. Go back to your minimized word processing document. Press Enter once followed by Ctrl+V to paste your scrolled graph into this document. Resize the graph in your document. Resize the graph in your document. SAVE. Do not close the program, just minimize it.

21. Go back to LP3 and go to File, Open and select "10 Linearization Part 1" [Yes, we are skipping tutorials 08 and 09, they will be extra credit!]. Do not save your changes to the tutorial. Proceed through this next tutorial. Be sure to cut and paste the two questions into your report under the subtitle "Linearization Part 1 Tutorial Results". There are no graphs to paste into your report, but you will need to use the skills presented in this tutorial to be successful in Part 2. You may want to print the last page of this tutorial for future reference.

22. Go to File, Open and select "10 Linearization Part 2". Follow all of the directions on page one and paste your data tables and your graphs into your report with the subtitle "Linearization Part 2 Tutorial Results". Be sure and run the linear regression on your data so that your graph shows the statistics box. Do this by highlighting the data on your graph and pressing the [icon] button.

23. SAVE. Continue to the end of the tutorial and paste your second graph into your report. [Hint: The shape of the second graph tells you it is an inverse function.] When you calculate your column you will need to use the reciprocal of one of the variables. Be sure and include the algebraic equation for your line of best fit once you have pasted the graphs into your document.

24. You may add the appropriate graphs from tutorials 4 & 5 to your report for extra credit. Be sure and include the proper subtitle headings and continue the proper format for the report.

Essay Writing Skills
Developing a Free Response

OBJECTIVE

This lesson is designed to introduce the students to the skill of planning appropriate free response essays.

LEVEL

All

NATIONAL STANDARDS

UCP.1, UCP.2, G.1, G.2

TEKS

6.2(C), 6.2(D)
7.2(D)
8.2(D)
IPC: 2(D)
Biology: 2(D)
Chemistry: 2(E)
Physics: 2(D)

CONNECTIONS TO AP

Writing appropriate free response answers is a fundamental skill needed in both AP Biology and AP Environmental Science.

TIME FRAME

45 minutes

MATERIALS

transparencies of Practice Essay #1-4
student copies of the practice pages

transparency of *Writing a Free Response*
strategies page

TEACHER NOTES

An appropriately written Free Response essay for an AP Biology or AP Environmental Science exam is markedly different from the type of essay that students are typically asked to write in English courses. For this reason, the skill of writing an essay in AP Biology and AP Environmental Science must be explicitly addressed. This activity is designed to help students understand how to set up a mechanical outline or plan to use when writing free response essays in AP science courses. It emphasizes the need for planning a response prior to writing. This planning and pre-thinking approach gives the students a tool to use and enables them to dissect complex prompts into manageable units. This practice activity focuses on the mechanics of dissecting the prompt rather than on writing the specific content. Once students have mastered the skill of planning or outlining a free response prompt, they need to practice writing essays using specific content throughout the school year.

SUGGESTING TEACHING PROCEDURE:

1. Explain the need for knowing how to write appropriate essays in a science class and describe strategies for writing essays using the handout: Tips for Writing Free Response Essays as your guide.

2. Show the transparency of Practice Essay #1 and explain how to use the strategies covered in step 1 with the sample essay. Explain to students how a mechanical outline of a response to this prompt might look (see answer key to student pages). Note, at this point, students may not be able to provide the correct content for the response, so focus on the mechanics of what should be included.

3. Distribute copies of the Essay Writing Outline Practice pages and show the transparency of Practice Essay # 2 as you model how to outline an appropriate response for the students (see answer key for Practice Essay #2). Elaborate on tip 5 by showing the students a possible outline for the question.

4. Focus student attention on Practice Essay #3 and allow students time to work alone to outline the major items that should be included in an appropriate free response. After 2-3 minutes of individual planning, have students compare their outline to that of their partners. Call on two or three volunteer pairs to write their outlines on the board for all students to see.

5. Have students read the prompt for Practice Essay #4 and prepare an outline of the items that should be included in a well designed free response. Restate tip 5 from the strategies list.

6. Show the students the sample student answer to essay # 3. Ask them to look at the response and identify 4 things that this student could do to make this essay better. Use the annotated and revised student response to show the students how the incorrect answer could be written more appropriately.

7. Follow up this activity in future lessons by including the outlining and writing of free response type questions in your daily warm-ups, daily quizzes, homework assignments, and major tests. The following list of grading hints taken from Advanced Placement and TEKS: A Lighthouse Initiative for Texas Science Classrooms may assist you as you approach the inclusion of free response questions in your assessment strategies:
   a. Start early in the year writing a free response (over a simple topic) in class and going over the rubric for it. Sharing the rubric with the students helps them gain insight into what types of information should and could be included.
   b. Create a rubric with positive points when you write the question. Students are more willing to take a chance when writing an essay in which points are collected rather than lost.
   c. Highlight or check off correct parts of free response answers as you grade them to make it easier for you to add up the points.
   d. Grade all free response answers at one sitting to develop a flow/pattern and encourage consistency in your grading.

e. On math problems, a correct answer with correct unit and work shown clearly earns full credit. Give partial credit for
   i. Set up of problem
   ii. Correct labels
   iii. Hint-look for final answer; if correct, just scan that work is present.

f. Encourage students to separate and label each section as this will help them organize their response and allow for easier grading.

g. Go over the rubric with the entire class to eliminate the need for making individual remarks. Alternately, you can use colored highlighter to mark the papers using a code such as blue — this statement scores a point, yellow — this statement included unnecessary or off topic information, and pink — this sentence contains incorrect information.

h. Use the College Board prompts and rubrics whenever possible

Tips for Writing Free Response Essays
AP Biology and AP Environmental Science*

1. Read the question twice.

2. Dissect the question to determine exactly what is being asked. (Highlight or underline)

3. Prepare a skeleton outline of the main components of your response.

4. Begin answering the question in the order it is written and DO NOT restate the question or write an introductory paragraph.

5. If the question says to 'discuss' or 'describe'
   a. Define the topic.
   b. Describe or elaborate on the topic.
   c. State an example of that topic.

6. If the question says to 'compare and contrast'
   a. Clearly state what the items have in common.
   b. Clearly state how items are different.

7. If the question asks for a graph to be made
   a. Label each axis with a name and units.
   b. Title the graph.
   c. Scale and number the axes correctly.
   d. Use the correct type of graphs (line or bar).

8. If the question asks a mathematical problem
   a. Show every single step of all work.
   b. Set up problems so that units cancel out (dimensional analysis).
   c. Write answers with units.
   d. If numbers are very large or very small, use scientific notation.

9. If the question asks for lab design
   a. State a hypothesis in the "if, then" format.
   b. Describe each step of a planned experiment in detail including what will be measured and how often the readings will be taken.
   c. Clearly identify the control(s).
   d. State that the experiment will have multiple trials for validity.
   e. Describe the expected results.

10. For ALL questions
    a. Answer in complete sentences, DO NOT use lists, charts, outlines in your final response.
    b. Label each section of your response as it is labeled in the question.
    c. Diagrams can support your statements but will not be scored.
    d. For every statement you write, ask yourself "why". If there is an answer to that why, keep on writing!
    e. Do not answer more than what is asked for. For example, if the question says to choose 3 out of 5 topics, ONLY answer 3 of the 5; If the question asked about RNA specifically, do not discuss DNA replication.
    f. Remember — this writing is timed. Use your time wisely.

---

*Edited from strategies list provided on page 85 of *A Lighthouse Initiative for Texas Science Classrooms*.

POSSIBLE ANSWERS TO THE CONCLUSION QUESTION AND SAMPLE DATA

1) Practice Essay#1-#4 could be dissected/outlined as follows:

Practice Essay #1

Main Topic: Carbon & Organic Compounds

    A) Characteristics of carbon atom
       * characteristic #1
       * characteristic #2
       * characteristic #3

    B) Structure and function of
       a. Lipids
          i. Structure
          ii. Function
       b. Proteins
          i. Structure
          ii. Function
       c. Nucleic Acids
          i. Structure
          ii. Function

Practice Essay #2

Main Topic: Usefulness of the Scientific Method

Components
    Name & describe

    Name & describe

    Name & describe

    Name & describe

1. How used in biological discovery # 1

2. How used in biological discovery # 2

Practice Essay #3

Main Topic: Changes in rate of Photosynthesis

   A. Descriptions
How low levels of light affect the rate of photosynthesis.

How high temperature will affect the rate.

How low levels of water will affect the rate.

   B. An adaptation to low levels of light
     a. Describe adaptation
     b. Give an example

Practice Essay #4

Main Topic: Growth curve fluctuations

   A. Explain what is happening in phase A

   B. Three factors that might cause changes in phase B
     a. Factor one
     b. Factor two
     c. Factor three

   C. Strategies
     a. Explain (r) strategy
     b. How (r) strategies effect population size
     c. Explain (K) strategies
     d. How (K) strategies effect population size

Annotated and Revised Student Response to Practice Essay #3

Sections labeled to match prompt

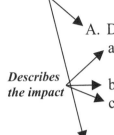

A. Descriptions
   a. Low light levels make it hard for plants to have enough solar energy to go through photosynthesis (so the rate of photosynthesis will go down in low light).

Describes the impact

   b. Low amounts of water will (also reduce the rate of photosynthesis).
   c. When temperatures get really high plants lose too much water when they open their stoma (causing photosynthesis rate to decline) so some plants are adapted to opening their stoma at night in order to take in carbon dioxide to use in photosynthesis.

B. Some plants have special adaptations that allow them to survive in extreme environments (such as low levels of light).
   a. Some plants have broad leaf surfaces to catch any available sun rays and can live in low light levels while some have thin leaves.

Gives an example

Revised Student Answer

A. Low light levels make it hard for plants to have enough solar energy to go through photosynthesis so the rate of photosynthesis will go down in low light. Low amounts of water will also reduce the rate of photosynthesis. hen temperatures get really high plants lose too much water when they open their stoma causing photosynthesis rate to decline so some plants are adapted to opening their stoma at night in order to take in carbon dioxide to use in photosynthesis.

B. Some plants have special adaptations that allow them to survive in extreme environments such as low levels of light. Some plants have broad leaf surfaces to catch any available sun rays and can live in low light levels while some have thin leaves.

REFERENCE
Jones, Kristen, Project Editor. *Advanced Placement and TEKS: A Lighthouse Initiative for Texas Science Classrooms*. Austin: Texas Education Agency, 2003. pg. 87

Essay Writing Skills
Developing a Free Response

An appropriately written free response essay for an AP Biology or AP Environmental Science exam is markedly different from the type of essays that are typically written in English courses. For this reason, you can improve your free response writing skills through the practice of dissecting a prompt and preparing a brief outline of the components that should be included in a quality answer. The free response portion of an AP Exam is a timed exercise and as such requires efficient use of the allotted time. By making an outline prior to actually writing the essay, you will be much more likely to include the important parts of a good response. With practice you will become able to dissect even the most complex prompts into manageable pieces.

PURPOSE
In this activity you will practice the skill of dissecting a free response prompt and preparing a mechanical outline of an appropriate response.

MATERIALS
copy of *Tips for Writing Free Response Essays*
copy of Practice Essay Prompts #1-4

PROCEDURE
1. Read through Tips for Writing Free Response Essays as you teacher explains specific tips mentioned in the document.

2. Read the prompt for Practice Essay #1. Observe and record the sample mechanical outline shown by your teacher of an appropriate response for this prompt.

3. Read the prompt for Practice Essay #2, record the outline of an appropriate response as your teacher goes through this essay with the class.

4. Read the prompt for Practice Essay #3. For 2-3 minutes, work alone to use the tips discussed in Tips for Writing Free Response Essays to prepare a mechanical outline for this prompt. When the individual planning time expires, compare the outline you have designed with that of your partner's. You may be asked to share you outline with the class.

5. Read the prompt for Practice Essay #4 and prepare an outline of the items that should be included in a well designed free response. You will be working alone to prepare your outline. Refer to Tips for Writing Free Response Essays if needed.

6. Read through the sample student response to Practice Essay #4. This response is written incorrectly. In the space below the prompt identify 4 things that this student could have done to make this essay more appropriate.

Name _____

Period _____

Free Response:
Planning for Success

PRACTICE ESSAY # 1

Prompt:
Structure and function are closely related in living systems. For example, the structure of the carbon atom allows it to be the building block of a variety organic compounds.
A. Explain the characteristics of carbon that allow its atoms to provide molecular diversity.
B. Chose three of the following categories of organic compounds and describe each in terms of the compound's structure and function in living organisms.
  1. Carbohydrates
  2. Lipids
  3. Proteins
  4. Nucleic Acids

Main Topic: _____

A. _____
  * _____
  * _____
  * _____

B. _____

  _____

   _____
   _____

   _____

   _____
   _____

   _____

   _____
   _____

PRACTICE ESSAY # 2

Directions: Read the prompt and prepare an outline of the major components that should be included in a response using the lines and bullets as your guide.

Prompt:
The scientific method of problem solving is a useful tool for scientific investigation. Describe the components of the scientific method and cite two examples of how the scientific method has been used to make biological discoveries.

Main Topic: _____

_____

_____
_____
_____
_____
_____

1. _____

2. _____

PRACTICE ESSAY #3

Directions: Read the prompt and prepare a mechanical outline of the major components that should be included in a response

---

Prompt:
The rate of photosynthetic activity may change in various environmental conditions.
a. Describe how each of the following environmental conditions could impact the rate of photosynthesis in a terrestrial plant.
   *low levels of light
   *high temperature
   *low availability of water
b. Select one of the conditions listed above and describe an adaptation that would allow a plant species to photosynthesize effectively in that specific environmental condition.

Main Topic: _____

A _____

   _____

   _____

   _____

B _____

   * _____

   * _____

Essay 4: Ecology question from 2003 exam

Directions: Read the prompt and prepare an outline of the major components that should be included in a response

Many populations exhibit the following growth curve:

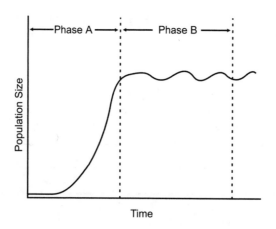

a. <u>Describe</u> what is occurring in the population during phase A.
b. Discuss THREE factors that might cause the fluctuations shown in phase B.
c. Organisms demonstrate exponential (r) or logistic (K) reproductive strategies. <u>Explain</u> these two strategies and <u>discuss</u> how they affect population size over time.

Main Topic: _____

    A. _____

    B. _____

        a. _____

        b. _____

        c. _____

    C. _____

      * _____

      * _____

      * _____

      * _____

Read the following student's response to Essay #3. Identify 4 things that this student could do to make this essay better.

Prompt:
The rate of photosynthetic activity may change in various environmental conditions.
a. Describe how each of the following environmental conditions could impact the rate of photosynthesis in a terrestrial plant.
  *low levels of light
  *high temperature
  *low availability of water
b. Select one of the conditions listed above and describe an adaptation that would allow a plant species to photosynthesize effectively in that specific environmental condition.

Student's Answer

The rate of photosynthetic activity may change in various environmental conditions. Some plants have special adaptations that allow them to survive in extreme environments. Some plants have broad leaf surface to catch any available sun rays and can live in low light levels while some have thin leaves. Low light levels make it hard for plants to have enough solar energy to go through photosynthesis. Low amounts of water will harm plants. When temperatures get really high plants lose too much water when they open their stoma so some plants are adapted to opening their stoma at night in order to take in carbon dioxide to use in photosynthesis.

List 4 things this student could have done differently to improve the quality of this free response:

1.

2.

3.

4.

Chemistry Content
Skills Chart

CONTENT SKILL CHART

AP Connection	Scope	Activities

	PAP
	I. Introductory Concepts
	
	  A. Safety
	
	  B. Scientific measurement
	
	    1. Metric system — Mega, kilo, deci, centi, milli, micro, nano, pico
	      Difference between base units and derived units
	      Significant figures
	      Scientific notation
	      Density
	
	      Temperature conversions
	
	      Dimensional analysis
	
	  C. Graphing and Calculator Skills
	    1. General graphing skills
	
	    2. Calculator graphing skills
	
	    3. Computer graphing skills
	
	  D. Accuracy and precision
	    Matter, change and energy
	
	    The scientific method-hypothesis, theory and law
	    Classification of matter
	    Conservation laws
	    Interpreting a potential energy diagram for exothermic and endothermic reactions
	    Specific heat calculations
	
	    First Law of Thermodynamics
		
		Foundation Lesson: Numbers in Science
		
		Making a Semimicropycnometer
		Foundation Lesson: Literal Equations
		Dimensional Analysis
		
		Foundation Lesson: Graphing Skills
		Foundation Lesson: Graphing Calculator
		Foundation Lesson: Microsoft Excel
		Foundation Lesson: Computer Graphing Software
		
		Physical/chemical change lab
		
		Foundation Lesson: The Scientific Method Chromatography
		
		Endothermic/exothermic reactions lab
		
		Foundation Lesson: Data Collection Devices

AP Connection	Scope	Activities
I. Structure of Matter	II. Structure of Matter	
A. Atomic theory and atomic structure	Atomic theory and atomic structure	
1. Evidence for atomic theory	History of the atom and experiments Atomic models (Dalton, Thomson, Rutherford, Bohr)	**Matter Waves**
2. Atomic masses; determination by chemical and physical means	Atomic mass, average atomic mass	
3. Atomic number and mass number; isotopes	Atomic number and mass number; isotopes	Beanium Lab or similar activity to determine avg. mass
4. Electron energy levels; atomic spectra, quantum numbers, atomic orbitals	Quantum mechanical model	Flame test lab
	Electron configuration of elements and ions; exceptions	**Electron Configuration, Orbital Notation and Quantum Numbers**
	Orbital notation	
	Quantum numbers—assigning values	
	Emission and absorption spectra	**Determining the Wavelength of Laser Light**
	Lewis dot for atoms and ions	
5. Periodic relationships including for example, atomic radii, ionization energies, electron affinities, oxidation states	Chemical periodicity	**Periodic Trends Lesson and graphing**
	Relationship between electron configuration and periodic table	Emphasis on "why" for trends
	Effective nuclear charge	
	Shielding effect	
	Atomic radii	
	Ionic radii	
	Ionization energies	
	Electron affinity	
	Electronegativity	
	Oxidation states	
	Chemical families	

AP Connection	Scope	Activities
C. Nuclear chemistry; nuclear equations, half-lives, and radioactivity; chemical applications	Nuclear chemistry Types of nuclear decay including alpha, beta, gamma Writing and balancing nuclear equations Nuclear stability Half-life calculations Fission and fusion Nuclear reactor Chemical applications: radioisotopes and tracers in nuclear medicine	**Isotopic Pennies** **Red Hot Half-Life**
B. Chemical Bonding 　1. Binding forces	III. Chemical Bonding 　Binding forces	**Bonds—Chemical Bonds**
a. Types: ionic, covalent, metallic, hydrogen bonding, van der Waals (including London Dispersion forces)	Ionic Covalent — bond energies Metallic Intermolecular forces — London, dipole-dipole. Hydrogen bonding, etc.	**Don't Flip Your Lid**
b. Relationships to states, structure, and properties of matter	Relationship to states, structures and properties	
c. Polarity of bonds, electronegativities	Polarity of bonds, electronegativities	
2. Molecular models 　　a. Lewis structures	Bonding models 　Valence bond theory	**Molecular Geometry**
b. Valence bond; hybridization of orbitals, resonance, sigma and pi bonds	Lewis structures for molecules, ions including exceptions to the octet rule 　　Resonance, hybridization	
c. VSEPR	VSEPR/ Molecular geometry 　　Bond angles 　　Sigma and pi bonds	
3. Geometry of molecules and ions, structural isomerism of simple organic molecules and coordination complexes; dipole moments of molecules; relation of properties to structure	Molecular polarity	

AP Connection	Scope	Activities

	IV. Chemical Nomenclature
	    Writing chemical formulas using ionic charges
	    Naming and writing formulas for ionic compounds
	    Naming and writing formulas for binary molecular compounds
	    Naming and writing formulas for acids
	    Naming and writing formulas for organic compounds
	    Naming and writing formulas for complex ions
		Chemical Nomenclature
		Identification of cations and anions in solution

III. Reactions
  A. Reaction Types
	V. Chemical equations
	    Balancing chemical equations
	    Identifying and predicting products of chemical reactions
	    Synthesis, Decomposition, Single displacement, Double replacement, Combustion
	    Net ionic equations
		Net Ionics
		
		Activity series of metals lab
		
		Types of chemical reactions lab

2. Precipitation reactions	Use of solubility rules	**The Eight Solution Problem**

B. Stoichiometry
    1. Ionic and molecular species present in chemical systems; net ionic equations
	VI. Stoichiometry
	    Calculations of molar mass
		Stoichiometry

2. Balancing of equations including those for redox reactions	Mole conversions with molar mass, Avogadro's number and molar volume	
3. Mass and volume relations with emphasis on the mole concept, including empirical formulas and limiting reactants	Percent composition	Empirical formula or formula of hydrate lab

	Empirical formula calculations
	    Molecular formula calculations
	    Mass-mass calculations
	    Mass-volume calculations
	    Volume-volume calculations
	    Percent yield
	    Limiting reactant calculations
	    Thermochemical calculations
		Simple vs. True
		
		Limiting Reactant

AP Connection	Scope	Activities
II. States of Matter	VII. States of Matter	
A. Gases	Gases	
1. Laws of Ideal Gases	Kinetic-molecular theory	
a. Equation of state for an ideal gas	Avogadro's hypothesis—molar volume, gas stoichiometry, and density of gases	Molar volume of a gas lab
b. Partial pressures	Gas laws: Theories and calculations	Other gas labs: Boyle's, Graham's law
2. Kinetic-molecular theory	Boyle's, Charles', Gay-Lussac's, Dalton's, Graham's, Ideal Gas, Henry's	**Charles' Law**
a. Interpretation of ideal gas laws on the basis of this theory		**Airbags**
b. Avogadro's hypothesis and the mole concept		
c. Dependence of kinetic energy of molecules on temperature		
d. Deviations from ideal gas laws		
B. Liquids and Solids	Liquids	
1. Liquids and solids from the kinetic-molecular viewpoint	Equilibrium vapor pressure	
2. Phase diagrams of one-component systems	Phase changes and phase change diagrams	**Heating Curves and Phase Diagrams**
3. Changes of state, including critical points and triple points	Solids	Heat of fusion of ice lab
4. Structure of solids; lattice energies	Amorphous	
	Crystalline solids: ionic, molecular. Metallic, and covalent network solids Analyzing heating curves with calculations	
C. Solutions	VIII. Solutions	
1. Types of solutions and factors affecting solubility	Properties of water	
2. Methods of expressing concentration (The use of normalities is not tested.)	Aqueous solutions and dissolving process	

AP Connection	Scope	Activities
3. Raoult's law and colligative properties (nonvolatile solutes); osmosis	Electrolytes and nonelectrolytes	**Conductivity of Ionic Solutions**

4. Non-ideal behavior (qualitative aspects)	Solution vs. colloids vs. suspensions
	Solubility graphs
	
	Concentration calculations:
	  Molarity, molality, % by mass, % by volume, mole fraction, and parts per million
	Colligative properties:
	  Boiling point elevation, freezing point depression, Raoult's law
	Determination of molar mass using colligative properties
		Effect of temperature on solubility
		
		It's Not Easy Being Green*
		
		Colligative Properties
		Freezing point depression lab and molecular weight

D. Kinetics 1. Concept of rate of reaction	IX. Kinetics Factors that effect rate: temperature, nature of reactants, state of subdivision, concentration, catalysts	
2. Use of experimental data and graphical analysis to determine reactant order; rate constants, and reaction rate laws	Determining rate and order of the reaction from given data	**Chemical Reaction Rates I**
3. Effect of temperature change on rates		**Chemical Reaction Rates II** **The Iodine Clock Reaction**
4. Energy of activation; the role of catalysts	Activation energy and the role of a catalyst	
5. The relationship between the rate-determining step and a mechanism	Using mechanisms to predict order for a reaction	

AP Connection	Scope	Activities
1. Acid-base reactions; concepts of Arrhenius, Bronsted-Lowry, and Lewis; coordination complexes; amphoterism	X. Acids and Bases Properties of acids and bases Theories: Arrhenius, Bronsted-Lowry, Lewis Identify conjugate acid-base pairs Strengths of acids and bases	**Chemistry in Atmosphere**

C. Equilibrium
 	
1. Concept of dynamic equilibrium, physical and chemical; Le Chatelier' principle; equilibrium constants
	Equilibrium
	
	  LeChatelier' principle and predicting shifts in equilibrium
	
	  Law of mass action: Kc and Kp
		General Chemical Equilibrium
		Disturbing Equilibrium

2. Quantitative treatment
  a. Equilibrium constants for gaseous reactions: Kp, Kc
  b. Equilibrium constants for reactions in solution
	Ionization of water
	    pH calculations of strong acids and strong bases
	  Ka, Kb, Ksp calculations
		Acid-Base Equilibrium
		
		How Weak is Your Acid? The Special K

(1) Constants for acids and bases; pK; pH
    (2) Solubility product constants and their application to precipitation and the dissolution of slightly soluble compounds
    (3) Common ion effect; buffers; hydrolysis
	pH of weak acids and weak bases
	Relationship between pH and pOH
	
	Buffer systems
	
	Salt hydrolysis reactions
	Titration curves and calculations
		Neutral or Not?
		Titrations—itrations

E. Thermodynamics
  1. State functions
	XI. Thermodynamics
	  Calculate and interpret the value of ?H using tables of standard values, Hess' law, calorimetry and bond energy
		Thermodynamics
		How Hot is a Candle?
		
		Hess' Law

AP Connection	Scope	Activities
2. First law: change in enthalpy; heat of formation; heat of reaction; Hess's law; heats of vaporization and fusion; calorimetry	Predict the sign of the entropy change in chemical reactions	
3. Second law: entropy; free energy of formation; free energy of reaction; dependence of change in free energy on enthalpy and entropy changes	Calculate and interpret the value of ?S from tables of standard values	
4. Relationship of change in free energy to equilibrium constants and electrode potentials	Calculate and interpret the value of ?G using tables of standard values and Gibb's equation	
3. Oxidation-reduction reactions	XII. Redox reactions	
a. Oxidation number	Assigning oxidation numbers	
b. The role of the electron in oxidation-reduction	Balancing redox reactions using the half-reaction method for acids and bases	**OIL RIG**
	Electrochemistry	**It's Electrifying!**
c. Electrochemistry; electrolytic and galvanic cells; Faraday's laws; standard half-cell potentials; Nernst equation; prediction of the direction of redox reactions	Voltaic cells: Identify components and calculate cell potentials	Voltaic cell lab
	Batteries, fuel cells, and storage batteries	
	Electrochemical cells vs. voltaic cells	Deposition
	XIII. Organic	
IV. Descriptive Chemistry	Draw and name the following:	
A. Chemical reactivity and products of chemical reactions	Alkanes, alkenes, alkynes, alkadienes, cyclic aliphatic, aromatics, halocarbons, alcohols, ethers, aldehydes, ketones, amines, carboxylic acids, esters	Synthesis of an organic compound (soap, polymers, esters)
B. Relationships in the periodic table; horizontal, vertical, and diagonal with examples from alkali metals, alkaline earth metals, halogens, and the first series of transition elements	Isomers	

AP Connection	Scope	Activities
C. Introduction to organic chemistry: hydrocarbons and functional groups (structure, nomenclature, chemical properties)	Draw and recognize and name: structural, geometric and optical isomers	**Long, Long Chains**
	Recognize the importance of biological molecules	**Biochemistry**
		Protein Properties

National
Standards

National Standards

The National Standard codes used in the lessons throughout the Laying the Foundation series are based on the following coding system. We encourage you to read the National Standards for your content area found at http://www.nap.edu/readingroom/books/nses/html.

Unifying Concepts and Processes

UCP.1 Systems, order, and organization

UCP.2 Evidence, models, and explanation

UCP.3 Change, consistency, and measurement

UCP.4 Evolution and equilibrium

UCP.5 Form and function

Science as Inquiry

A.1 Abilities necessary to do scientific inquiry

A.2 Understandings about scientific inquiry

Physical Science

B.1 Structure of atoms

B.2 Structure and properties of matter

B.3 Chemical reactions

B.4 Motions and forces

B.5 Conservation of energy and increase in disorder

B.6 Interactions of energy and matter

Life Science

C.1 The cell

C.2 Molecular basis of heredity

C.3 Biological evolution

C.4 Interdependence of organisms

C.5 Matter, energy, and organization in living systems

C.6 Behavior of organisms

Earth and Space Science

D.1 Energy in the earth system

D.2 Geochemical cycles

D.3 Origin and evolution of the earth system

D.4 Origin and evolution of the universe

Science and Technology

E.1 Abilities of technological design

E.2 Understanding about science and technology

Science in Personal and Social Perspectives

F.1 Personal and community health

F.2 Population growth

F.3 Natural resources

F.4 Environmental quality

F.5 Natural and human-induced hazards

F.6 Science and technology in local, national, and global challenges

History and Nature of Science

G.1 Science as a human endeavor

G.2 Nature of scientific knowledge

G.3 Historical perspective

Syllabi

Developing a Course Syllabus

There are many reasons to develop a course syllabus. The most important reason is to communicate to your students that you have carefully planned your course. It is often thought of as a contract between an instructor and a student. It conveys the message that each of you is committed to meeting the objectives of the course. Research has shown that students who are told what they are supposed to learn and how they are to be evaluated perform better than those who are not so instructed. The syllabus should specify the duties and responsibilities of the student while communicating a commitment from the instructor to respect a timeline. Once the syllabus has been published and distributed, it is important to do your very best not to change it. Students will feel that the rules have been changed in the middle of the game.

Other reasons to develop a course syllabus include:
- Clearly communicating course expectations to administrators and parents.
- Creating an organizational tool for yourself if you teach more than one kind of course.
- Organizing your laboratory time and resources for maximum efficiency.
- Enhancing cooperation and communication with your colleagues so that multiple teachers teaching the same course deliver the same quality of instruction to each student.
- Reducing the amount of time spent deciding what to do each class period; the plan has already been formulated making execution of the plan easier.
- Documenting what has been taught.
- Empowering students to take more responsibility for their education.
- Helping students better manage their time and set priorities while improving their study habits.

Constructing a syllabus is actually quite easy. A suggested procedure follows:
1. Realize that a school year usually consists of 36 weeks and that you are fortunate if you are able to instruct for 32 of those weeks. Factors such as semester exam weeks, TAKS testing, benchmark testing for TAKS, pep-rallies, assemblies and other unscheduled interruptions will consume up to 4 weeks of your instructional time.

2. Examine the course scope found in the front of this guide. Without regard for the fact that you only have 32 weeks, estimate how much time it would take for you to cover each major topic, including laboratory time and testing time.

3. Determine the mid-point of the course, which is the point you wish to reach by the end of the first semester. Tally the ideal number of weeks required for each semester.

4. Obtain an accurate district calendar and count the number of weeks you have for each semester. (The first semester is usually shorter if your district ends the semester before January.)

5. Compare. Do not be alarmed if your tally exceeds 32 weeks. Decide if you can shift the mid-point a bit. If not, it is time to adjust the ideal number of weeks so that they fit into your actual time frame. This may mean radically changing the structure of your course. You may have to change the number of activities that you do in a given unit, change the type of notes students are

expected to take, or other instructional methods to free up time to cover the entire scope of the content.

6. Now that you have determined the number of weeks, break each unit down by day and choose logical starting and stopping points for each class period. Allow a two day cushion at the end of each grading period for unforeseen obstacles such as personal illness, re-teaching, fire drills, etc…

7. If you find it daunting to publish the whole semester on a syllabus, consider publishing it in 3 week intervals. That way the plan stays more current as those unforeseen interruptions occur.

8. Now that you have a plan for student success, stick to the plan. Realize that you will have to teach bell-to-bell in order to successfully complete your plan.

The syllabi that follow were contributed by some of the best and most experienced Pre-AP teachers in the region, representing classes at both public and private schools. Each of these teachers has played an integral part in developing a successful Pre-AP and AP program at their school. Each of them would tell you there is no one right way to teach a Pre-AP science course. It is our hope that you will survey the collection of syllabi presented and formulate your own syllabus to share with your students and colleagues. Understand that your first attempt will require monitoring and adjusting. Keep careful notes about your timeline expectations and what worked and did not work on your initial syllabus. Do not be alarmed if it takes a couple of tries to create a syllabus that works consistently in your classroom year after year.

TEACHER

Rhonda Alexander
Robert E. Lee High School
Tyler, Texas

SCHOOL PROFILE

School Location and Environment: Tyler Texas, an east Texas town in Smith County half way between Dallas and Shreveport. 2500 students attend Robert E. Lee in a community with two public high schools, three private high schools and three institutions offering post-secondary education.

Grades the school contains: 9-12

Type of school: Public

Percentage of minorities: 40%

College Record:
% of graduating seniors attending post-secondary institutions: 20%
% of graduating seniors attending 4-year institutions: 45%

Do you teach the AP course at your school? Yes

If not, is it offered at your school? NA

Number of students taking the PAP science course in your discipline: PreAP Chemistry has an enrollment of approximately 112 students.

Number taking the AP course in your discipline: One section of 19 students

Course sequence of the PAP courses in your school:
Freshman: PAP Biology
Sophomore: PAP Chemistry
Junior: PAP Physics some student take either AP Chemistry or AP Biology and PAP Physics
Senior: can take any combination of the following: Biology AP, Chemistry AP, Physics AP

In what year is the AP course in your discipline taken? AP Chemistry students are split almost equally between juniors and seniors.

SYLLABUS

Pre-AP Chemistry I

Robert E. Lee High School
Tyler, TX

There are between five and six sections of PAP chemistry per year with class size averaging 24 students a section. Some sections of PAP Chemistry are taught on an A/B block schedule and some sections are taught on a 47 minute period which meets everyday. Prerequisites for PAP chemistry are biology I and geometry. The goal of PAP chemistry at Robert E. Lee is to furnish our students a background that will prepare them for the rigor of college level chemistry in a challenging environment where students aspire to learn and develop critical thinking skills.

TOPIC	TIME	LABORATORY
FIRST SEMESTER		
I. Matter and Measurements: A. Lab safety and equipment B. Matter and energy C. Three types changes matter undergoes D. Classification of matter E. Conversion factors and dimensional analysis F. Evaluation and collection of lab data	2 ½ wks	1. Mystery Solutions day 1 lab (Optional) 2. Chemical and Physical Change 3. Endothermic & Exothermic Reactions (Computer-Vernier probes) 4. Accuracy and Precision in Measurement
II. Atomic Structure A. Dalton's Atomic Theory B. Thomson's model C. Rutherford and the nucleus D. Isotopes E. Atomic mass, atomic number, mass number, mole concept F. Bohr's model G. Quantum Theory, quantum numbers, Uncertainty Principle, dual nature H. Electron configuration, core notation orbital diagrams	2 ½ wks	1. Vegium (isotope analogy) 2. Avogadro's Number 3. Flame Test and Spectrum Tubes 4. Quantum Quest
III. Periodic Table and Periodic Properties A. History of the periodic table B. Periodic law C. Organization of the periodic table D. Periodic trends and exceptions	2 ½ wks	1. The Periodic Table Puzzle 2. Determining Moles in Your Signature (Inquiry)

IV. Nomenclature   2 ½ wks
A. Writing formulas
B. Naming binary and nonbinary
   inorganic compounds

V. Calculation of Formulas Predicting   2 wks   1. Empirical Formula
Equations   Determination
A. Molecular weight and the mole   2. Hollow Penny
B. Percent composition
C. Calculating empirical and molecular
   formulas (including problems using
   combustion data)

VI. Predicting Equations   2 ½ wks   Types of Reactions
A. Classification of reactions
B. Identify oxidation reduction reactions
C. Identify oxidizing and reducing
   agents
D. Balancing chemical equations
   including redox
E. Predicting Reactions

VII. Stoichiometry
A. Stoichiometric problems   2 ½ wks   1. A Cycle of Copper Reactions
B. Limiting reagent problems   2. Left Over Al Wire
C. % yield

SECOND SEMESTER
VIII. Bonding and intermolecular   2½ wks   1. Shapes and Polarity of
forces   Molecules-Model lab
A. Description of ionic, covalent and   2. Intermolecular Forces
   metallic bonding   (Computer-Vernier probes)
B. Polar and nonpolar covalent bonds
C. Dot diagrams/ (including those that
   do not follow octet)
D. VSEPR Theory & molecular
   geometry
E. Intermolecular forces and molecular
   polarity
F. Types of solids

IX. Gases   2 ½ wks   Molar Volume of a Gas
A. Behavior of gases and temperature and pressure conversions
B. Boyles, Charles Law and Combine Gas Law
C. Avogadro's Hypothesis and Ideal Gas Law
D. Dalton's Law of Partial Pressure
E. Kinetic Molecular Theory
F. Gas Stoichiometry
G. Ideal vs Real
H. Graham's law

X. Solids and liquid   2 wks
A. Properties of solids and liquids
B. Change of state and phase diagrams
C. Heat calculations for change in temperature and change in state
D. Heating Curve
E. Introduction of the concepts of driving forces and equilibrium

1. Characteristics of Liquids
2. Heat of Fusion of Ice (Computer-Vernier probes)

X1. Solutions and Colligative Properties   2 ½ wks
A. Description of solutions, colloids and suspensions
B. Molarity and molality
C. Colligative properties
D. Solubility rules
E. Net ionic equations

1. A Study of Solutions
2. Using Freezing-Point Depression to Find MW (Computer-Vernier probes)
3. Net Ionic Equations

XI1. Kinetics and Introductory Thermochemistry   2 ½ wks
A. Collision theory and the factors that effect reaction rate
B. Potential energy diagrams
C. Determine rate law from experimental data
C. Endothermic and exothermic
D. Hess's Law
E. Heat of formation

1. Hess' Law (Computer-Vernier probes)
2. Rate of Reactions

XII1. Equilibrium   2 wks   Disturbing Equilibrium
A. Characteristics of Equilibrium
B. Driving Forces
C. Equilibrium expression
D. Solving equilibrium problems
E. Le Chatelier's Principle

XIV. Acid/Base Chemistry   2½ wks   Vinegar Titration
A. Logarithms
B. Acid/Base Theories
C. pH scale
D. Strong and weak pH problems
E. Titration
F. Hydrolysis
G. Common ion problems

XV. Nuclear Chemistry   1 wk
A. Nuclear stability
B. Binding energy and mass defect
C. Radiation
D. Nuclear equations
E. Fission and fusion
F. Nuclear reactors

TEACHER

Carol Brown
Saint Mary's Hall
San Antonio, TX

SCHOOL PROFILE

School Location and Environment: Saint Mary's Hall is a private school located in a suburban area on the near northeast side of San Antonio, Texas. The school is positioned on 60 acres of land and contains 3 separate schools. The school is on a 7 period day with period 8 reserved for tutorials. Periods are 45 minutes long. All AP Science classes receive 1½ credits and meet either 7 or 8 periods each week, allowing for the extra lab time that is necessary. There are 312 students in the Upper School with an average graduating class between 80 to 90 students. All grades are co-educational with an approximately even distribution between males and females.

Grades the school contains: Pre-K to 12

Type of school: Private

Percentage of minorities: Upper School (9-12) is 22% minority

College Record:
% of graduating seniors attending post-secondary institutions: All of our graduates attend college or some type of post-secondary education.
% of graduating seniors attending 4-year institutions: 97%

Do you teach the AP course at your school? Yes

If not, is it offered at your school? NA

Number of students taking the PAP science course in your discipline: Pre-AP Chemistry has an enrollment of approximately 35 students this year.

Number taking the AP course in your discipline: 11 students in AP Chemistry this year, but I expect the enrollment to be higher next year.

Course sequence of the PAP courses in your school:
6th grade: Earth Science
7th grade: Life Science
8th grade: IPC
9th grade: Physics (either conceptual or honors)
10th grade: Chemistry (either ChemCom or honors)
11/12th grade: Two semesters of biological sciences (Cellular Biology, Genetics, Zoology, Human A & P, Field Marine Biology) or the student moves into the AP Sciences. One year of biological science is required for graduation. AP Biology satisfies that requirement. AP Environmental Science satisfies one semester of that requirement.

The majority of the Pre-AP Biology students are freshman. After Pre-AP Biology, the students typically take Pre-AP Chemistry followed by Pre-AP Physics.

In what year is the AP course in your discipline taken? The vast majority of the AP Chemistry students are juniors.

COURSE SYLLABUS

Chemistry I for the Pre-AP Years

Saint Mary's Hall
San Antonio, TX

(3 weeks) Unit One: Introductory Concepts: (Chapters 1-4)
  A. Matter, change, and energy
    1. The scientific method
      a. The place of the hypothesis
      b. The difference between theory and law.
    2. States of matter
    3. Classification of matter
    4. Conservation laws
  B. Scientific measurement
    1. Metric system (SI)
    2. Accuracy and precision
    3. Significant figures
    4. Mass, volume, density, specific gravity
    5. Heat and temperature
  C. Problem solving skills

The student will:
1. Demonstrate in practical situations how the terms experiment, hypothesis, theory, and law fit into the scientific method.
2. Name and characterize the three states of matter.
3. Classify a sample of matter as a substance or a mixture, and classify mixtures as homogeneous or heterogeneous.
4. State the difference between an element and a compound.
5. Classify substances as elements or compounds.
6. Write the symbols of common elements , and, given their symbols, write their names.
7. Distinguish between potential and kinetic energy.
8. Write and apply the mathematical formula for kinetic energy
9. State the law of conservation of matter and energy.
10. Classify changes in matter as physical, chemical, or nuclear.
11. Distinguish between the accuracy and precision of a measurement.
12. Solve problems dealing with error and deviation.
13. Use the rules for significant figures in calculations.
14. Name the metric units of length, mass, and volume.
15. List and define the common metric prefixes.
16. State the difference between the mass and weight of an object.
17. Calculate the density of an object.
18. Convert between Celsius and Kelvin temperature scales.
19. Work problems involving specific heat.
20. Construct conversion factors from equivalent measurements.
21. Use conversion factors in factor-label problem solving.

22. Use dimensional analysis to solve multi-step problems.
23. Correctly name and identify laboratory equipment.

Key Terms:

Chemistry	Homogeneous mixture	Kinetic energy	Mass	Mixture
Phase	Potential energy	Reactant	Solid	Substance
Vapor	Fluid	Accuracy	Celsius temperature	Gram
SI units	Kelvin temperature	Meter	scale	Specific gravity
Temperature	scale	Dimensional	Precision	Compound
Distillation	Weight	analysis	Factor-label	Energy
Gas	Distillate	Filtration	Filtrate	Matter
Observation	Heterogeneous mixture	Hypothesis	Liquid	Scientific law
Solution	Physical change	Product	Reactant	Density
Heat capacity	Theory	Absolute zero	Calorie/ calorie	Metric system
Significant figure or	Specific heat	Joule	Liter	
digit	Volume	Conversion factor		

Laboratory:

Introductory Laboratory skills
Determining specific gravity with a semi micropycnometer
Separation of mixtures—distillation
Determination of density
Measurement and error

(3 ½ weeks) Unit Two: Atomic Structure   (Chapters 5, 13, and 14)

    A. Atomic theory
        1. Atomic number
        2. Atomic mass
        3. Mass number
        4. Isotopes
    B. Electronic structure
        1. Atomic models
            a. Bohr atom
                1) Quantized energy
                2) Emission spectra
            b. Wave mechanical model
                1) Quantum numbers
                2) Electron configuration
                3) Orbital notation
                4) Lewis structures
    C. Chemical Periodicity
        1. Electron configuration and periodicity
        2. Periodic trends
            a. Atomic size
            b. Ionization energy
            c. Electron affinity

    d. Electronegativity
    e. Ionic size
      1) Cations
      2) Anions
  3. Chemical families

The student will:

1. Summarize Dalton's atomic theory.
2. Distinguish between protons, electrons, and neutrons in terms of their relative masses, charges, and position in the atom.
3. Use the atomic number and mass number of an element to find the number of protons, electrons, and neutrons.
4. Define an atomic mass unit.
5. Use the concept of isotopes to explain why the atomic masses of elements are not whole numbers.
6. Calculate the average atomic mass of an element from isotope data.
7. Describe the contributions that Thomson and Rutherford made to the development of atomic theory.
8. Explain how Bohr's model of the atom differed from its predecessors.
9. Compare the quantum mechanical model of the atom with previous models.
10. Describe the general shape of s, p, and d orbitals.
11. Use the Aufbau principle, the Pauli exclusion principle, and Hund's rule to write the electron configurations of the elements.
12. Describes the significance of the four quantum numbers and be able to assign quantum numbers to specific electrons in atoms.
13. Work problems involving wavelength, frequency, and energy of light.
14. Explain the origin of the atomic emission spectrum of an element.
15. Elucidate wave-particle duality of matter.
16. Explain the origin of the periodic table.
17. Distinguish between a period and a group in the periodic table.
18. State the periodic law.
19. Recognize the demarcation of the periodic table into a s block, p block, d block, and f block.
20. Describe variations within a group and a period of atomic radii, ionization energy, electron affinity, and electronegativity.
21. Explain how the shielding effect influences periodic trends.
22. Identify an element as an alkali metal, alkaline earth metal, chalcogen, halogen, or noble gas.

Key Terms:

Atom
Emission spectrum
Quanta
Halogen
d block elements
Spin quantum
number
Nucleus
Pauli exclusion
principle
Group
Atomic mass unit
Neutron
Photon
Family

Atomic number
Electromagnetic
radiation
Lewis dot formula
Transition metal
Orbital
Atomic mass
Absorption
spectrum
Spectrum
Chalcogen
Dalton
Proton
Wavelength
Noble gas

Cathode ray
Electron cloud
Alkali metal
p block elements
Angular momentum
quantum number
Cathode
Electron
configuration
Orbital notation
Inner transition metal
amu
Energy level
Periodic law
Representative
elements

Isotope
Excited state
Electron affinity
f block elements
Magnetic quantum
number
Anode
Frequency
Alkaline earth
metal
Quantum numbers
Carbon-12-exactly-
12 scale
Ground state
Covalent atomic
radius
Tritium

Amplitude
Heisenberg
uncertainty
principle
Period
s block elements
Principal quantum
number
Mass number
Hertz
Electronegativity
Deuterium
Electron
Hund's rule
Ionization energy
Planck's constant

Laboratory:

Physical and Chemical Changes
Determining the atomic weight of beanium
Determining an unknown from its emission spectrum
Wavelength of laser light

(4 weeks) Unit Three: Chemical Bonding (Chapters 15 and 16)
  A. Ionic bonds
  B. Metallic bonds
  C. Covalent bonds
    1. Models of chemical bonding
      a. Valence bond model
      b. Molecular orbital theory
    2. Bond dissociation
    3. Intermolecular attractions
      a. London forces
      b. Dipole-dipole attraction
      c. Hydrogen bonds

The student will:
1. Use the periodic table to find the number of valence electrons in an atom.
2. Draw electron dot formulas of the representative elements.
3. State the octet rule.
4. Describe the formation of a cation or an anion.
5. Give the characteristics of an ionic bond.
6. Explain the electrical conductivity of a melted and aqueous solution of ionic compounds.
7. Use the theory of metallic bonds to explain the physical properties of metals.

8. Describe the formation of a covalent bond between two nonmetallic elements.
9. Describe double and triple covalent bonds.
10. Draw electron dot formulas for simple covalent molecules.
11. Explain the formation of a coordinate covalent bond.
12. Define resonance and draw resonance structures of compounds.
13. Show why some molecules that are exceptions to the octet rule may be paramagnetic.
14. Describe the molecular orbital theory of covalent bonding.
15. Distinguish between bonding and antibonding molecular orbitals.
16. Describes the features of sigma and pi bonds.
17. Use a molecular orbital energy level diagram to interpret stability, bond order, and magnetic properties.
18. Use VSEPR theory to describe the shapes of simple covalently bonded molecules.
19. Describes the shapes of simple molecules using orbital hybridization.
20. Use electronegativity values to determine whether a bond is non polar covalent, polar covalent, or ionic.
21. Show the relationship between polar covalent bonds and polar molecules.
22. Define bond dissociation energy.
23. Name and describe the weak attractive forces that hold molecules together.
24. Identify the characteristics of molecular substances. Laboratory: Covalent bonds and molecular polarity

Key Terms:

Coordination number	Lewis dot structure	Octet rule	Oxidation number	Antibonding orbital
Coordinate covalent bond	Dispersion interaction or force	Hydrogen bond	London force	Nonpolar covalent bond
Diamagnetic	Polar covalent bond	Single covalent bond	Lone pair	Linear molecule
Trigonal bipyramid (TBP)	Sigma bond	σ bond	Ionic bond	Halide ion
Valence	Anion	Bond dissociation energy	Dipole	Double covalent bond
Molecular orbital	Electronegativity		π bond	
Structural formula	van der Waals force	Pi bond	Trigonal planar	Polar molecule
δ charge	Electron dot structure	VSEPR	Valence electron	Octahedral
Bonding orbital	Polar	Metallic bond	Network solid	Cation
Resonance	Tetrahedral angle	Hybridization	Dipole moment	Paramagnetic
Solvent	Solute	Trigonal pyramid	Immiscible	Solution
		Miscible		R_f

Laboratory:
Molecular modeling, bond type, molecule shape and type
Polarity and melting point
Polarity and solubility
Introduction to chromatography

(1½ week) Unit Four: Chemical Nomenclature (Chapter 6)
    A. Writing chemical formulas using ion charges
    B. Naming ionic compounds

C. Naming binary molecular compounds
D. Naming acids

The student will:
1. Define the terms cation and anion and show how they are related to the terms metal and nonmetal.
2. Using experimental data, show that different samples of the same compound obey the law of definite proportions.
3. Differentiate between ionic compounds and molecular compounds, and between formula units and molecules.
4. Using experimental data, show that two different compounds composed of the same two elements obey the law of multiple proportions.
5. Use the periodic table to determine the charge on an ion.
6. Learn the common polyatomic ions and their names.
7. Write the chemical formula of a compound, when given the name of the compound.
8. Name a compound when given the formula.
9. Name an acid when given the formula and vice versa.

Key Terms:

Acid	Anion	Binary compound	Base	Cation
Formula unit	Nonmetal	Molecular	Group	Ion
Periodic Table	Polyatomic ion	compound	Transition metal	Representative
Law of definite	Metalloid	Ionic compound	Law of multiple	element
proportions	(semimetal)	Ternary compound	proportions	

Laboratory:
Identification of cations and anions in solution

(3 weeks) Unit Five: Basic Stoichiometry (Chapters 7, 8 and 9)
A. Measuring matter
1. The mole
a. Molar mass
b. Avogadro's number
c. Molar volume
2. Percent composition
3. Empirical formulas
4. Molecular formulas
B. Chemical reaction
1. Writing and balancing chemical equations
2. Predicting chemical products
a. Combination
b. Decomposition
c. Single displacement
—activity series
d. Double displacement
e. Combustion
C. Stoichiometry

1. Mass-mass problems
2. Mass-volume problems
3. Volume-volume problems
4. Percent yield
5. Limiting reagents

The student will:
1. Identify the mole as the SI unit that measures the "amount of substance."
2. Define a mole as Avogadro's number of representative particles of a substance.
3. Calculate the number of moles in a given number of representative particles of any substance.
4. Distinguish between the terms gram atomic mass, gram molecular mass, gram formula mass, and molar mass.
5. Calculate the mass of one mole of any substance.
6. Calculate the mass of a given number of moles of a substance.
7. Calculate the number of moles in a given mass of a substance.
8. Use the molar volume of a mole of gas (STP) to work mole-volume problems.
9. Calculate the gram molecular mass of a gas from density measurements of gases at STP.
10. Calculate the percentage composition of a substance from its chemical formula or experimental data.
11. Distinguish between an empirical and a molecular formula.
12. Derive empirical and molecular formulas from appropriate experimental data.
13. Identify the reactants and products in a chemical equation.
14. Write a balanced chemical equation when given the names or formulas of all the reactants and products in a chemical reaction.
15. Determine the limiting reagent in a chemical system
16. Classify a reaction as combination, decomposition, single-replacement, double replacement (metathetical), or combustion.
17. Predict the products of simple combination and decomposition reactions.
    18. Interpret a balanced chemical equation in terms of interacting moles, representative particles, masses, and volumes of gases (STP).
19. Perform stoichiometric calculations with balanced equations using moles, mass, representative particles, and volumes of gases (STP).
20. Knowing the limiting reagent in a reaction, calculate the maximum amount of product(s) produced and the amount of any unreacted excess reagent. Laboratory: Percent oxygen in potassium chlorate

Key Terms:

Avogadro's number	Empirical formula	Gram atomic mass or weight	Gram formula mass or weight	Molar mass
Molar volume	Mole			STP
Activity series of elements	Balanced equation	Percent composition	Representative particle	Coefficient
	Synthesis reaction	Catalyst		ΔH
Combination reaction	Net ionic equation	Combustion reaction	Chemical equation	Enthalpy
Double replacement reaction	Endothermic reaction	Single replacement reaction	Decomposition reaction	ΔH_f^o
				Hess's Law
Exothermic reaction	Limiting reagent	Excess reagent	Metathetical reaction	
Standard molar heat of formation	Theoretical yield	Percent yield	Heat of combustion	
		Thermochemical equation	Heat of reaction	
Stoichiometry				

Laboratory:
Percent water in "Light" margarine
Percent water in popcorn
Activity series of metals
Types of chemical reactions

This is the end of the first semester

(1 week) Unit Six: Thermochemistry (Chapter 11)
  A. Heat Changes in chemical reactions
    1. Calorimetry
    2. Phase changes
    3. Problems involving the 1st law of thermodynamics
  B. Hess's Law
    1. Heat of reaction
    2. Standard molar heat of formation
    3. Standard molar heat of combustion
  C. Stoichiometry and heat
  D. Factors which drive reactions
    1. Enthalpy
    2. Entropy
    3. $\Delta G = \Delta H - T\Delta S$

The student will:
1. Classify reactions as exothermic or endothermic.
2. Use the heating curve to solve calorimetry problems.
3. Distinguish between heat capacity and specific heat.
4. Understand the factors that drive chemical reactions: enthalpy and entropy.
5. State the laws of thermodynamics.
6. Use a calorimeter to determine the heat of combustion of a substance.
7. Define the enthalpy of a substance.
8. Use standard heats of formation to calculate the enthalpy change of a reaction.
9. Interpret and/or draw a potential energy diagram for both exothermic and endothermic reactions correctly labeling the axes, the enthalpy of the reaction, and the activation energy.
10. Distinguish between standard molar heat of combustion, molar heat of reaction, and standard molar heat of formation.
11. Solve simple stoichiometric energy problems.
12. Solve Hess's Law problems.

Key Terms:

Exothermic	Endothermic	ΔH	Potential energy	Enthalpy
Entropy	ΔH°_f	Heat of reaction	diagram	Standard molar heat of
Heat of combustion	Thermochemical	Free Energy	Hess's Law	formation
$\Delta G = \Delta H - T\Delta S$	equation	ΔS	Exergonic	Endergonic
System	ΔG	Universe	State function	Heat of solution
	Surroundings			

Laboratory:
Heats of combustion of paraffin

(3 weeks) Unit Seven: The Kinetic Molecular Theory (Chapters 10 and 12)
  A. Kinetic molecular theory-Boltzmann Distribution
  B. Avogadro's hypothesis
  C. The nature of liquids
    1. Equilibrium vapor pressure
    2. Phase changes and phase diagrams
  D. The nature of solids
    1. Amorphous solids
    2. Crystalline solids
      a. Ionic solids
      b. Molecular solids
      c. Metallic solids
      d. Covalent network solids
  E. The nature of gases
    1. Boyle's law
    2. Charles' law
    3. Gay-Lusaac's law
    4. Combined gas laws
    5. Ideal gas law
    6. Graham's law

The student will:
1. Use the kinetic molecular theory to explain the motion, pressure, and temperature of a gas.
2. Relate temperature to the average kinetic energy of the particles in a substance.
3. Use the kinetic molecular theory to explain phenomena such as diffusion, compressibility, or pressure of gases.
4. Be introduced to a Boltzman distribution diagram.
5. Explain the meaning of dynamic equilibrium in phase changes.
6. Describe what happens on a particle level at the melting point and boiling point of a substance.
7. Apply Le Chatelier's principle to systems in equilibrium.
8. Relate the physical properties of solids to their type.
9. Identify the basic regions shown in a phase diagram. Locate each of the following on the diagram: normal boiling point, normal freezing point, triple point, critical point.
10. Explain the significance of absolute zero, giving its value in degrees Celsius and Kelvin.
11. Convert between units of pressure—kPa, atm, and torr.
12. State the values of standard temperature and pressure.
13. State and explain the significance of Avogadro's hypothesis.
14. Use the gas laws to solve problems involving volume, temperature, pressure, and moles.
15. State and use Dalton's law of partial pressures.
16. State and use the ideal gas law.
17. Explain why real gases deviate from the gas laws.
18. State and use Graham's law of diffusion.
19. State and explain the major points of the kinetic molecular theory.

Key Terms:

Solid
torr
Boltzman distribution
   curve
Le Chatelier's
   principle
Triple point
Crystal
Molecular solid
Boyle's Law
Avogadro's principle
Hydrogen Bonding

Liquid
Kilopascal (kPa)
Evaporation
Sublimation
Critical point
Amorphous solid
Metallic solid
Charles' Law
Diffusion
Ideal gas

Gas
Atmosphere
Dynamic
   equilibrium
Boiling
Van der Waals
   forces
Glass
Covalent network
   solid
guy Lussac's Law
Effusion
Ideal gas law

Kinetic molecular
   theory
Kinetic energy
Equilibrium vapor
   pressure
Normal boiling point
Dispersion interaction
Super cooled liquid
Allotrope
Dalton's Law of Partial
   Pressures
Barometer
Scuba

Pressure
Kelvin temperature
Relative humidity
Phase diagram
London forces
Ionic solid
Pressure
Graham's Law
Dipole-dipole
   attraction

Laboratory:

Microscale Diffusion Lab
Boyle's Law and mathematical modeling (If not done in Physics)
Charles' Law
Replacement of hydrogen by a metal
Molecular weight of a gas.
Molar Volume of a Gas.

(2 ½ weeks) Unit Eight: Properties of Solutions (Chapters 17 and 18)
   A. Water
      1. Properties of water
      2. Aqueous solutions
         a. electrolytes and non-electrolytes
         b. colloids and suspensions
   B. Solution formation
      1. Solubility
      2. Concentration units
      3. Colligative properties

The student will:
1. Describe the hydrogen bonding that occurs in water based on the structure of the polar water molecule.
2. Use the concept of hydrogen bonding to explain the following properties of water: high surface tension, low vapor pressure, high specific heat, high heat of vaporization, and high boiling point.
3. Define the terms solution, aqueous solution, solute, and solvent and give an example of each.
4. Use the rule that "like dissolves like" to predict the solubility of one substance in another.
5. Calculate the percent of water in a given hydrate.
6. Give the characteristics of colloids and suspensions that distinguish them from solutions.
7. Explain the difference between saturated, unsaturated, and supersaturated solutions.
8. Use Henry's law to solve gas solubility problems.
9. Define and work problems involving the molarity and molality of a solution.

10. Describe how to prepare dilute solutions from concentrated solutions or solids.
11. Explain on a particle basis how the addition of a solute to a pure solvent causes an elevation of the boiling point and a depression of the freezing point of the resultant solution.
12. Calculate the freezing point depression and the boiling point elevation of various solutions.
13. Determine the molecular mass of an unknown from experimental freezing point depression or boiling point elevation measurements Laboratory: Water of hydration of barium chloride.

Key Terms:

Aqueous	Browning motion	Colloid	Deliquescent	Efflorescent
Electrolyte	Emulsion	Hygroscopic	Non-electrolyte	Solute
Solvation	Solvent	Strong	Weak	Surface tension
Surfactant	Suspension	Water of hydration	Water of crystallization	Suspension
Tyndall effect	Boiling point	Normal boiling point	Colligative property	Concentrated
Dilute	Saturated	Unsaturated	Supersaturated	Henry's law
Immiscible	Miscible	Kb	Kf	Molality (m)
Molarity (M)	[ ]	Mole fraction	Mass percent	

Laboratory:

Water of crystallization of barium chloride
Conductivity of electrolytes
Freezing point depression
Alchemy in the chemistry lab—brass pennies

(3 ½ weeks) Unit Nine: Acids and bases (Chapters 20 and 21)

    A. Properties of acids and bases
    B. Theories of acids and bases
        1. Arrhenius
        2. Lowry-Bronsted
        3. Lewis
    C. Strengths of acids and bases
    D. Equilibrium of the ionization of water
        1. pH of strong acids
        2. pH of weak acids
        3. pOH
        4. Buffer systems
        5. Hydrolysis of salts
    E. Solubility product

The student will:
1. List the properties of acids and bases.
2. Write the equation for the self-ionization of water.
3. Calculate the pH of a solution given the {H+} or {OH-}, or vice versa.
4. Define and give examples of Arrhenius, Bronsted-Lowry, and Lewis acids and bases.
5. Use the Bronsted-Lowry theory to classify substances as acids or bases; identify conjugate acid-base pairs in reactions.

6. Use the extent of ionization and the Ka or Kb to distinguish between strong and weak acids and bases. List the strong acids.
7. Calculate an acid dissociation constant from concentration and pH measurements.
8. Complete and balance a neutralization reaction.
9. Perform calculations involving acid-base reactions.
10. Explain the steps of a titration.
11. Calculate the gram equivalent mass of any acid or base.
12. Define and calculate the normality of a solution.
13. Use the concept of hydrolysis to explain why aqueous solutions of some salts are acidic or basic.
14. Define a buffer, and show with equations how a buffer system works.
15. Calculate concentrations of ions of slightly soluble salts.
16. Use Le Chatelier's principle to explain the common ion effect.

Key Terms:

Acid	Amphoteric	Bronsted acid/base	Diprotic acid	Hydronium ion
K_w	Autoprotolysis	Law of mass	Common ion effect	(H_3O^+)
Indicator	Solubility product	action	Arrhenius	Normality (N)
Monoprotic acid	constant K_{sp}	Amphiprotic	acid/base	Base
End point	pK_a	Neutral solution	Self-ionization	Equilibrium
Lewis acid/base	Neutralization reaction	Standard solution	Alkaline solution	Conjugate acid-base
Weak acid/base	Triprotic acid	Hydroxide ion	pH	pair
	Equilibrium expression	(OH^-)	Hydrolysis	pOH
		Equivalence point		Titration

Laboratory:
Acid/Base Anhydrides
Determination of an Ionization Constant
Percentage of acetic acid in vinegar

(3 weeks) Unit Ten: Organic chemistry (Chapters 25, 26, and 27)
   A. Hydrocarbons
      1. Alkanes
      2. Alkenes
      3. Alkynes
      4. Alkadienes
      5. Cyclic aliphatic compounds
      6. Aromatics
   B. Hydrocarbon substitution compounds (Structures, nomenclature, preparations, and reactions)
      1. Halocarbons
      2. Alcohols
      3. Ethers
      4. Aldehydes
      5. Ketones
      6. Amines
      7. Carboxylic acids

8. Esters

C. Introduction to Biological molecules.

Objectives: Upon completion of this unit the student will be able to:
1. Draw structural formulas of simple alkanes.
2. Recognize structural, condensed, and molecular formulas of the continuous-chain hydrocarbons up to ten carbons.
3. Given the structural formula of an alkane, alkene, alkyne, cyclic hydrocarbon, or alkadiene, name it according to IUPAC rules, or vice versa.
4. Draw and name the isomers of C8H18.
5. Distinguish between structural isomers and geometric isomers.
6. Identify the asymmetric carbon in a stereoisomer.
7. Describe the bonding and structure of benzene.
8. Name and draw the structures of simple cyclic and aromatic compounds.
9. Recognize and identify a molecule's functional groups.
10. Characterize substitution and addition reactions.
11. Identify an alcohol or an amine as primary, secondary, or tertiary.
12. Show how alcohols, aldehydes and ketones, and acids are related by oxidation and reduction reactions.
13. Predict the products of the formation or hydrolysis of an ester.
14. Use IUPAC rules to name halocarbons, alcohols, ethers, aldehydes, esters, carboxylic acids, and ketones.
15. Identify the major groups of biomolecules. Recognize their functional groups and structures.
16. Understand the photosynthetic cycle.
17. Understand the concepts of metabolism, catabolism, and anabolism.

Key Terms:

Aliphatic compound	Alkyne	Alkane	Alkene	Arene
Aromatic	Asymmetric carbon	Cis	Trans	Structural isomer
Geometric isomer	Optical isomer	Isomer	Cracking	IUPAC naming system
Hydrocarbon	Homologous series	Stereoisomer	Substituent	Condensed structural
Alkyl group	Paraffin family	Olefin family	Addition reaction	formula
Alcohol	RX	Halocarbon	Alkyl halide	Substitution reaction
RCHO	Aryl halide	Monomer	Polymer	Aldehyde
Ester	Carbohydrate	Carbonyl group	Carboxylic acid	Dimer
Ether	ROR'	RCOOH	RCOOR'	ROH
Dehydrogenation reaction	Denatured alcohol	Absolute alcohol	Ketone	Fermentation
	Formyl group		Nucleic acid	RCOR'
Hydroxyl group	Peptide bond	Carboxyl group	Esterification	Nucleotide
Peptide	Zwitterion	Protein	Catabolism	Saponification
Triglyceride	RNA	Metabolism	Phospholipids	Anabolism
DNA	Monosaccharides	ATP	Polysaccharides	Enzymes
Amino acids		Disaccharides		Photosynthesis

Laboratory:
Model Lab—Hydrocarbons
Model lab—hydrocarbon substitution families
Synthesis of Esters
Saponification
Polymers

(2 weeks) Unit Eleven: Oxidation-Reduction and Electrochemistry (Chapters 20 and 21)
  A. Balancing redox reactions
     1. Oxidation numbers
     2. Half-reactions
  B. Electrochemical cells
     1. Voltaic cells
       a. Cell potentials
       b. Batteries, fuel cells, and storage batteries
     2. Electrochemical cells

The student will:
1. Use the half-reaction method to balance redox equations.
2. Properly use the terms oxidation, reduction, reducing agent, oxidizing agent.
3. Describe the nature of electrochemical processes.
4. Sketch a voltaic cell, labeling the cathode, the anode, and the direction of the flow of the electrons.
5. Given a voltage cell, identify the half-cell in which oxidation occurs and the half-cell in which reduction occurs.
6. Define the cell potential, and describe how it is determined.
7. Define the standard electrode potential and use them to calculate the standard emf of a cell.
8. Distinguish between electrolytic and voltaic cells; give practical uses of each. Laboratory: Reduction potentials

Laboratory:
The great lemon battery challenge
Micro-chemical cells
The copper cycle

(1 week) Unit Twelve: Kinetics, Thermodynamics, and Equilibrium. (Chapter 17)
  A. Kinetics
     1. Factors that effect rate
       a. temperature
       b. nature of reactants
       c. state of subdivision
       d. concentration
       e. catalysts
     2. Rate laws
  B. Thermodynamics
     1. Reaction coordinates
       a. exothermic reactions

   b. endothermic reactions
 2. Hess's Law
C. Equilibrium
   1. Le Chatelier's Principle
   2. Law of mass action

The student will
1. Relate the ideas of activation energy and the activated complex to the rate of a reaction.
2. Understand the factors that influence the rate of a reaction.
3. Recognize a Boltzmann distribution curve, and relate it to reaction rates.
4. Know the form for a rate law equation.
5. Understand and interpret potential energy diagrams for exothermic and endothermic reactions.
6. Use enthalpy, entropy, and free energy in problems.
7. Work Hess's law problems.
8. Understand the importance of the sign of *H, *S, and *G.
9. State and apply Le Chatelier's principle.
10. Write a law of mass action for a reversible reaction.
11. Understand the meaning of the K_{eq}.

Laboratory:
Kinetics of thiosulfate

All chemistry students take a semester final and a year final. Both tests are cumulative with the year final covering both semesters. Generally the NSTA/ACS High School Exam is used as the end-of-year final. Students are given 2 review days and 3 days of finals each semester.

TEACHER

Denise DeMartino
Westlake High School
Austin, Texas

PRE-AP CHEMISTRY SYLLABUS

Email: ddemarti@eanes.k12.tx.us
Website: www.eanes.k12.tx.us/whs/Departments/Science/Chemistry

Course Description: Chemistry Pre-AP is a first year chemistry course designed to meet the needs of the student who plans on continuing on in AP Chemistry or eventually taking a college chemistry class.

Course Content: The following units will be covered in each course. Objectives for each unit will be passed out in calendar form at the start of each unit.

Unit 1 – Math and Measurement
Unit 2 – Matter Classification & Graphing Calculator Techniques
Unit 3 – Atomic Structure
Unit 4 – The Modern Atom
Unit 5 – Periodicity
Unit 6 – Bonding
Unit 7 – Molecular Structure
Unit 8 – Chemical Nomenclature
Unit 9 – Chemical Calculations

Unit 10 – Chemical Equations
Unit 11 – Stoichiometry
Unit 12 – Gas Laws
Unit 13 – Solutions
Unit 14 – Electrochemistry
Unit 15 – General Equilibrium
Unit 16 – Acid/Base Equilibrium
Unit 17 – Thermochemistry
Unit 18 – Kinetics and Nuclear

Textbook: *Matter and Energy, Glencoe Chemistry*

Required Materials: A binder to hold class handouts, quadrille-lined laboratory notebook with carbon copy capability that can be purchased from instructor, and a scientific calculator. (A TI-83 is suggested but not required)

Grading: Your grade will be determined on the scale

Tests	60%
Labs	20%
Quizzes	20%

Homework: Homework will not be graded in this course although it will be assigned regularly. The day homework is due a key will be posted. The homework will be checked for completion and extra credit will be given. A maximum of 50 extra credit points on a quiz can be earned by completing all homework assignments. *If the homework is not done, the quizzes and tests will be impossible!!!!*

Student Conference Hours: 7:15 to 8:00am daily or after school (by appointment only)
Parent Conference Hours: Please call or preferably e-mail to make an appointment

Assignments: A calendar will be issued every unit with all assignments, quizzes, and tests listed. It is the students' responsibility to keep up with the calendar and anything listed on it. Any changes that need to be made to the calendar will be announced in class.

Make-up Work: Student grades will be posted every Monday. If there are any missing assignments, the student has until the next posting of grades to get the work in. Any incomplete work will at that point result in a zero. If there are extenuating circumstances, the student must individually approach the instructor and make other arrangements. Ultimately, it is the students' responsibility to get the makeup work done quickly.

Students will also be responsible to take any quiz or test scheduled on the day they return from an absence if no new material was covered in regards to that quiz/test. Students who actually miss quizzes and tests need to expect to make them up the day they return either in class, during lunch, before school, or after school.

Tardiness: You will be counted tardy if you are not in class when the bell rings. After 5 minutes, you will be counted absent. I am an instructor who takes tardiness seriously and make no exceptions unless student has an admit from the attendance office.

Progress Reports: Progress reports will be sent home for a parent signature at the 3-week mark of each 6-week grading period.

Lab Discipline: This is covered on a separate sheet of rules. Any repeated violation will result in removal from lab with no make-up possibility. Safety in the lab is most important.

Format for Lab Write-Ups: See "The Laboratory Notebook" on back of this syllabus. Read CAREFULLY!

Pre-AP Chemistry Timeline

Instructor: Denise DeMartino

Traditional Schedule – 55-minute period

1st six weeks

Unit 1 – Math and Measurement (8 days)
- Scientific notation
- Significant digits/math operations
- Metric prefixes
- Factor analysis
- Percent calculations

Unit 2 – Matter Classification/Graphing Techniques (7 days)
- Density
- Graphing by hand
- Calculator graphing
- Classification of matter

Unit 3 – Atomic Structure (8 days)
- Nuclear composition
- Isotopes and nuclide symbols
- % Abundance
- Atomic structure experiments
- Bohr atom and Rydberg Equation
- Periodic structure
- Light behavior and equations

2nd six weeks

Unit 4 – The Modern Atom (7 days)
- Major atomic theories
- Quantum numbers
- Orbital shapes
- Pauli's exclusion
- Electron Configuration
- Lewis Dot Diagrams
- Recognizing excited states

Unit 5 – Periodicity (7 days)
- Predicting oxidation state
- Size of atoms and ions
- First and multiple ionization energies
- Electronegativity
- Metallic Character
- Reactivity

Unit 6 – Bonding (6 days)
- Lewis dot diagrams of molecules
- Exceptions to the octet rule
- Carbon bonding
- Resonance
- Basic Organic Nomenclature

3rd six weeks

Unit 7 – Molecular Structure (10 days)
- Molecular geometries
- Bond angles and polarity
- Advanced organic nomenclature
- Isomers

Unit 8 – Chemical Nomenclature (7 days)
- Binary molecular nomenclature
- Binary ionic nomenclature (mono/multivalent)
- Acid nomenclature
- Stock vs. Old system

Unit 9 – Chemical Calculations (9 days)
- Mole concept
- 2 step mole problems
- % Composition
- Empirical formulas
- Molecular formulas
- Hydrates
- Molarity

Unit 10 -- Chemical Equations (10 days)
- Balancing and writing equations
- Recognizing types of reactions
- Recognizing states of matter
- Predicting products
- Special cases of predicting
- Net ionic equations

4th six weeks
Unit 11 – Stoichiometry (8 days)
- Mass-mass problems
- % Yield calculations
- Energy calculations
- Mass-volume problems
- Volume-volume problems
- Limiting reagent problems

Unit 12 – Gas Laws (9 days)
- Kinetic Theory
- 2 and 3 variable gas laws
- Graham's law
- Ideal gas law
- Gas density and molecular mass

Unit 13 – Solutions (9 days)
- Phase diagrams
- Intermolecular forces
- Units of concentration
- Colligative properties
- Determining MM by FPD and BPE

5th six weeks
Unit 14 – Electrochemistry (9 days)
- Oxidation numbers
- ID redox reactions
- Balancing redox in basic and acidic solutions
- Voltaic Cells
- EMF calculations
- Nernst Equation
- Electrolytic Cells
- Faraday's calculations

Unit 15 – General Equilibrium (7days)
- Concept of Equilibrium
- LeChatlier's Principle
- Equilibrium Calculations (RICE)

Unit 16 – Acids, Bases, and their Equilibrium (8 days)
- Review of acid concepts
- Anhydrides
- pH scales and calculations
- Weak acid pH calculations
- Titration and calculations
- Hydrolysis
- Common ion effect

6th six weeks
Unit 17 – Thermochemistry (7 days)
- Heat transfer and calculations
- Enthalpy
- Hess's law
- Entropy
- Gibbs free energy and spontaneity
- Relationships between E, K and free energy

Unit 18 – Kinetics and Nuclear Chemistry (9 days)
- Factors affecting rates
- Method of initial rates
- Reaction pathways
- Time integrated rate laws
- Graphing zero, 1st, and 2nd order
- Nuclear decay reactions
- Mass defect/binding energy
- Half lives
- Fission and fusion

The Laboratory Notebook
Chemistry Pre-AP & Chemistry Pre-AP/GT

A laboratory notebook should be used to explain lab procedures, record all lab data, show how calculations are made, discuss the results of an experiment, and explain the theories involved.

A record of lab work is an important document, which will show the quality of the lab work that you have done. At some point and time, it may be necessary to show your lab book to a university for credit in a chemistry lab course, so keep you book neat and organized since it may be looked at in the future and might be of some great use.

Getting Started:
1. Use a quadrille-lined book with pages numbered and with carbon copy capability.
2. Write you name and class on the front cover and inside the front cover.
3. <u>Always, Always, Always</u> use black or blue ink!
4. Fill in the table of contents provided in the book. This should be kept current as you proceed during the year. In the table of contents, place the title and the page number where the lab report begins for each lab.
5. If you make a mistake DO NOT ERASE! Just draw ONE LINE through the error and continue. Do not scribble out the error or use white-out. It is expected that some errors will occur. You cannot produce and error free notebook. If you mess up an entire page DO NOT rip it out of the book. Simply draw a line through the page corner to corner and go to the next page.
6. You will keep the original copy of the lab in the book and turn in the carbon copy. If your instructor cannot read carbon copies of the lab easily, the student will receive an automatic 50% for a grade until it is turned in legibly. Late points will be deducted.

Laboratory Reports:
A specific format will be given to you for each lab. You must follow that format and label all sections very clearly. The carbon copy of your lab will be due the day after it is completed in the laboratory, unless the next day is an exam and then the due date will be postponed to the day after that. Make sure your carbon copy is legible. A lab quiz will be given over the lab the day it is due, and you are allowed to use your lab notebook and lab handouts ONLY on this quiz. Be sure to bring the lab books to class on the day of the quiz, otherwise you will have to take the quiz without. *Fifteen points will be taken off for the first day a lab report is late and 5 will be taken off for every additional day.* The following are the different sections that the various labs may consist of. (Remember, a more specific format will be given to you for each lab.)

Pre-Lab Work
(To be completed by the day the lab is done. If not completed by start of class on lab day, 10 points will be taken off)
1. *Title*
   The title should be descriptive. "Experiment 5", for instance, it not a descriptive title.
2. *Date*
   This is the date you performed the experiment.
3. *Purpose*
   A statement summarizing the "point" of the lab. What are you trying to do?

4. *Pre-Lab Questions*

You will be given some questions to answer before the lab is done. You do not need to re-write the question, but you do need to write out your answer in complete sentence. This will be of great help to you on lab quizzes.

5. *Data Tables*

You will need to create any data tables or charts necessary for data collection in the lab.

During the Lab

6. *Data*

Record all your data **<u>directly</u>** in you book. Label all data clearly and always include proper units of measurement. Underline, use capital letters, or use any device you choose to help organize this section well. Space things out neatly and clearly.

Post-Lab Work

7. *Calculations and Graphs*

You should show how calculations are carried out. Your instructor needs to be able to follow you calculations and read your graphs easily. Graphs need to be titled, axis need to be labeled, and units need to be shown on axis. **To receive credit for any graphs they must be at least ½ page in size.**

8. *Conclusions*

In this section you need to make a statement telling what can be concluded from the experiment. You need to discuss what theory was demonstrated in this experiment, what your calculations show, what the purpose of the experiment fulfilled, why does the experiment work or not work, and list any sources of error you had. This needs to be a very meaningful section. In the instructor's opinion, it is the conclusion that really tells how much a student learned from the lab and it will be closely graded.

9. *Post-Lab Questions*

Follow the same procedure as for Pre-Lab Questions

TEACHER

Mary Payton
Carroll High School
Southlake, Texas

SCHOOL PROFILE

School Location and Environment: Carroll High School is located in Southlake, Texas near the Dallas-Fort Worth International Airport. The community is considered suburban and upper middle class with high expectations and a high rate of mobility. CHS is an atypical Texas school in that most of its population is not native to Texas.

Grades the school contains: 9-10

Type of school: Public

Percentage of minorities: fewer than 10%

College Record:
% of graduating seniors attending post-secondary institutions: 15%
% of graduating seniors attending 4-year institutions: 78%

Do you teach the AP course at your school? No

If not, is it offered at your school? No, it is a 9-10 school, but the course if offered at the 11-12 campus, Carroll Senior High School.

Number of students taking the PAP science course in your discipline: Pre-AP Chemistry has an enrollment of approximately 200 students.

Number taking the AP course in your discipline: Two sections that average 20 students per section.

Course sequence of the PAP courses in your school: The majority of the Pre-AP Biology students are freshman. After Pre-AP Biology, the students typically take Pre-AP Chemistry followed by Pre-AP Physics.

In what year is the AP course in your discipline taken? The vast majority of the AP Chemistry students are juniors.

SYLLABUS

Before leaving for the summer all students signed up for Pre-AP Chemistry receive a Summer Assignment Packet that has their ion list to memorize, some instructions on simple formula writing and a timeline for their science fair project. They are held accountable for memorizing the ion list by the second day of school. The formula writing is just to give them a head start, and the science fair information lets those who want to get it over with get started. We are on an A/B Block schedule and see the students for 90 minutes every other day.

Unit 1 (3 days) – Getting Started
Guidelines
Safety
Science Fair
Formula Writing/Nomenclature
  Ionics
  Binary molecular
  Acids
CBL Lab #5 – Introduction to graphing

Unit 2 (6 days) - Math and Measurement
Significant Figures
Scientific Notation
Metric Units*
Density
Temperature Conversions
Dimensional Analysis
Lab – Density of Pennies (analysis of slope)

Unit 3 (5 days) Matter and Energy
Classifying Matter
Properties and Changes
Energy/Specific Heat
CBL Lab #1 – Endo/Exothermic Reactions
CBL Lab #16 – Energy content of foods

Unit 4 (6 days) Atomic Structure
Atomic history and experiments
Isotopes
Average Atomic Mass
Quantum mechanical model
Orbital notation
Electron configurations (by periodic table)
Lewis dot notation for atoms/ions
Quantum numbers (values and assigning)
Predicting oxidation states
Exceptions to electron configurations
Lab – Flame Tests

Unit 5 (4 days) Periodic Properties
Atomic radii
Ionic radii
First ionization energy (emphasis on deviations)
Multiple ionization energy to determine ion charge
Electron affinity
Lab – Alkaline Earth Metal Periodicity

Unit 6 (4 days) Nomenclature/Formula Writing
Review: ionics, binary molecular, acids
Organic (alkanes, alkenes, alkynes, alcohols)
Dry Lab-all nomenclatures/formulas

Unit 7 (5 days) Moles and Composition Stoichiometry
Moles
Mole calculations with mole map
Percent composition
Empirical formulas
Molecular formulas
Combustion analysis
Molarity
Lab

Unit 8 (7 days) Equations and Stoichiometry
Balancing equations
Types of reactions (including special cases)
Stoichiometry (McCormick Method)
        Mass-mass, mass-vol,vol-vol
Percent yield
Limiting reactant
Lab – types of reactions
Lab- Deterring Ca^{+2} from TUMS

Second Semester

Unit 9 (6 days) Bonding and Molecular Geometry
Ionic and covalent bonding
Lewis structures for molecules
Exceptions to the octet rule
Molecular geometry
Bond angles
Resonance
Hybrid orbitals
Sigma and Pi bonds
Polarity
Intermolecular forces
[Complex ions]
CBL Lab # 14 – Conductivity and Concentration
CBL Lab # 9 – Evaporation and Intermolecular Forces

Unit 10 (7 days) Gas Laws
Kinetic theory
Boyles, Charles, Gay-Lussac's Laws
Daltons Law of Partial Pressures
Graham's Law
Ideal Gas Law
Gas density and molecular mass
CBL Lab #6 – Boyle's Law
CBL Lab #7 – Pressure and Temperature Relationships
[Lab-Molar Volume of a Gas]

Unit 11 (2 days) Liquids and Solids
*I give a take home unit on this consisting of
chapter review and extension worksheets
Class time is spent going over
Phase diagrams
Change of state calculations
CBL Lab #4 – Heat of fusion for ice
[CBL Lab #10 – Vapor pressure of liquids]

Unit 12 (7 days) Solutions
Dissolving process
Precipitation Reactions
Net ionic equations
Concentration
Colligative properties
Molar mass determination by freezing point depression
Lab – Ice cream
Lab – Seven solution problem

Unit 13 (5 days) Kinetic and Equilibrium
Rates of reactions
Determining Rate Laws from data
Graphical determination of reaction orders
Mechanisms
Equilibrium concept
Calculating Kc and Kp
LeChatlier's Principle
RICE tables
CBL Lab #30 – Rate Law for Crystal Violet Reaction
[CBL # 20 – Equilibrium Constant Kc]

Unit 14 (5 days) Acid/Base Equilibrium
Acid/Base definitions
Calculation of pH
Ka and weak acid calculations
Titration
Hydrolysis reactions
Lab – Acid/Base titration

Unit 15 (2 days) Introduction to Electrochemistry
Assigning oxidation numbers
Balancing redox equations in acid/base
Dry Lab- balancing redox equations

Unit 16 (1 day) Nuclear Chemistry
Fission and fusion
Nuclear decay
Half Lives

[Unit 17 (2 days) Thermodynamics
Enthalpy
Entropy
Gibb's Free Energy]

Unit 18 (4 days) Qualitative Analysis
Last 1 ½ weeks before exams students run qual scheme and identify unknown.

Eliminates the end-of-the-year-beating-your-head-against-the-wall syndrome!

TEACHER

Charlotte Taggart
Abilene High School
Abilene, Texas

SCHOOL PROFILE

School Location and Environment: Abilene, Texas is a small town of about 100,000 residents. Abilene ISD has the typical problems of most West Texas towns with shrinking student enrollment. Abilene High School is a public High School.

Grades the school contains: approximately 2100 students in grades 9-12

Type of school: Public

Percentage of minorities: 42%; 41% are economically disadvantaged.

College Record:
% of graduating seniors attending post-secondary institutions: 28.7%
% of graduating seniors attending 4-year institutions: 63.1%

Do you teach the AP course at your school? No

If not, is it offered at your school? Yes, it is offered at the school.

Number of students taking the PAP science course in your discipline: Pre-AP Chemistry has an enrollment of approximately 150 students.

Number taking the AP course in your discipline: Usually less than 20 students

Course sequence of the PAP courses in your school: The normal sequence of PAP courses is PAP Biology 9th, and PAP Chemistry 10th. From that point, students proceed to AP Physics B in 11th grade rather than PAP Physics, and other AP science courses in 11th and 12th grades.

In what year is the AP course in your discipline taken? AP Chemistry students are juniors or seniors.

PRE-AP CHEMISTRY SYLLABUS

FALL SEMESTER

UNIT 1 – ALL THAT MATTERS: The Nature of Matter (5 weeks)
- o Introduction to Laboratory Work: Safety, Equipment, Laboratory Techniques
- o Properties of Matter
- o The Periodic Table
  - o Symbols and Names of Elements
- o Atomic Structure
- o Electron Configuration
- o Bonding in Brief
  - o Ionic and Covalent Compounds and Properties
- o Lewis Structures in Brief
- o Classification of Matter
  - o Ions and Charges/Oxidation Numbers
  - o Polyatomic Ions
  - o Writing Formulas and Nomenclature
  - o Naming Hydrates
  - o Naming Acids
  - o Naming Complex Ions
  - o Organic Compounds

UNIT 2 – ENOUGH MATTER: Quantifying Matter (4 weeks)
- o Measuring Matter
  - o Use of Measuring Instruments and Significant Figures
  - o Metrics
  - o Scientific Notation
  - o Graphical Analysis
  - o Temperature Scales and Conversions
  - o Dimensional or Factor Analysis
- o Measuring Reacting Amounts of Matter – The Mole
  - o Mole conversions
  - o Percent Composition
  - o Empirical Formula/Molecular Formula
  - o Molarity

UNIT 3 – ENERGY MATTERS: Interactions of Matter and Energy (2 weeks)
- o Energy and States of Matter
- o Energy Associated With Phase Changes
- o Specific Heat
- o Energy Associated With Chemical Reactions
- o Measuring Energy Changes: Units and Conversions

UNIT 4 – MATTER CHANGES: Chemical Reactions (6 weeks)
- o Writing and Balancing Chemical Equations
  - o Writing Molecular, Complete Ionic, Net Ionic Equations
- o Classifying Reactions
- o Reaction Prediction
- o Stoichiometry
- o Percent Yield
- o Limiting Reactant
- o Energy Changes in Chemical Reactions
  - o Heat of Reaction, Enthalpy
  - o Endothermic/Exothermic Reactions

FALL SEMESTER REVIEWS AND EXAMS (1 week)

SPRING SEMESTER

UNIT 4 REVIEW/WRAP-UP (1/2 week)

UNIT 5 – MATTER TO MATTER: Chemical Bonding (2 1/2 weeks)
- o Bond Types
  - o Ionic Bonds
  - o Bonding Representation: Lewis Structures Showing Ionic and Covalent Bonding
  - o Covalent Bonds and Molecular Geometry
  - o Polarity
  - o Sigma and Pi Bonds
  - o Hybrid Orbitals
  - o Resonance
  - o Octet Rule and Exceptions

UNIT 6 – MATTER STATES ITS CASE: Liquids, Solids and Gases (4 weeks)
- o Kinetic Theory
- o Properties of Liquids
- o Properties of Solids
- o Properties of Gases
- o Intermolecular Forces
- o Phase Diagrams
- o Pressure Units and Conversions
- o Gas Laws
  - o Boyle's Law
  - o Charles' Law
  - o Gay-Lussac's Law
  - o Dalton's Law of Partial Pressures
  - o Graham's Law
  - o Ideal Gas Law/Gas Density/Molar Mass of Gases

UNIT 7 – MIXED-UP MATTER: Solutions (2 weeks)
- o The Dissolving Process – Solvents and Solutes
- o Water – Properties of the Universal Solvent
- o Solution Concentration and Calculations
- o Colligative Properties and Calculations
- o Review of Ionic Reactions

UNIT 8 – MATTER IS DRIVEN: Kinetics and Equilibrium (2 weeks)
- o Reaction Rate and Factors Affecting It
  - o The Rate Law: Analyzing Data and Determining Reaction Orders
  - o Reaction Mechanisms
- o Equilibrium and Le Chatlier's Principle
  - o Predicting Equilibrium Shifts
  - o Equilibrium Constant Calculations
  - o Equilibrium Changes – Calculations

UNIT 9 – THE pH of the MATTER: Acids and Bases (2 weeks)
- o Theories of Acids and Bases: Definitions
  - o Writing Hydrolysis Reactions for Acids
- o Weak and Strong Acids and Bases
  - o Calculations Involving pH and pOH
  - o Equilibrium Constants of Weak Acids; Calculations
  - o Acid-Base Reactions and Titration
- o How Buffers Work

UNIT 10 – MATTER GETS CHARGED UP: Electrochemistry (1 1/2 weeks)
- o Oxidation and Reduction
  - o Definitions, and Identifying Processes
  - o Assigning Oxidation Numbers
  - o Writing Half Reactions
  - o Balancing RedOx Reactions
- o Voltaic Cells
  - o How They Work
  - o Cell Potential Calculations

UNIT 11 – UNSTABLE MATTER: Nuclear Chemistry (1/2 week)
- o Nuclear Decay
  - o Half Life Calculations
  - o Fission and Fusion
  - o Writing Nuclear Equations

UNIT 12 – DYNAMIC MATTER: Thermodynamics (1 week)
- o Spontaneous Reactions – Will It or Won't It?
  - o Enthalpy Considerations
  - o Entropy Considerations
  - o All Things Considered: Gibb's Free Energy

UNIT 13 – EXAM MATTERS (2 weeks)
- o Review Materials, Instructions for Final Exam (Written & Lab Portions)
- o Lab Exam
- o Final Exam

LABORATORY INVESTIGATIONS BY UNIT

LABS FOR UNIT 1
- ➤ First day lab: Micro: Identifying an Unknown Solution/Observing Chemical Reactions
- ➤ Lab Orientation: Introduction to Safety in the Laboratory; Using Laboratory Equipment; Common Laboratory Techniques; Working With a Lab Partner Effectively
- ➤ Lab: Chemical and Physical Properties/Changes
- ➤ Lab: Measuring Energy Changes in Electrons (Wavelength)
- ➤ Lab Activity: Identifying Ionic and Covalent Substances in Solution
- ➤ Dry Lab: Formula Writing and Naming

LABS FOR UNIT 2
- ➤ Lab Activity: Recording Measurements Appropriate to Specific Instruments (Intro to Significant Figures)
- ➤ Lab Activity: Introduction to CBL and Graphical Analysis
- ➤ Lab: CBL #5, Find the Relationship (Using TI 83+ Graphing Calculator)
- ➤ Lab: CBL # 1, Endothermic and Exothermic Reactions
- ➤ Lab: SDL #1: Finding Percent Composition and Formula of 3 Hydrates
- ➤ Lab: SDL #7: Finding Empirical Formula

LABS FOR UNIT 3
- ➤ Lab: CBL #4, Heat of Fusion of Ice
- ➤ Lab: SDL #4, Finding Melting Point of Lauric Acid

LABS FOR UNIT 4
- ➤ Lab Activity: Dry Lab, Reaction Prediction
- ➤ Lab: Types of Reactions
- ➤ Lab: Mole Relationships in a Chemical Reaction
- ➤ Lab: SDL #3, Limiting Reactant
- ➤ Lab: CBL #18, Additivity of Heats of Reaction: Hess's Law

LABS FOR UNIT 5
- ➤ Lab Activity: Lewis Structures
- ➤ Lab: Molecular Geometry (Molecular Models)

LABS FOR UNIT 6
- ➤ Lab: CBL#9, Evaporation and Intermolecular Attractions
- ➤ Lab: CBL#10, Vapor Pressure of Liquids
- ➤ Lab: CBL#6, Boyle's Law
- ➤ Lab: CBL#7, Gay-Lussac's Law

> Lab: Molar Volume of a Gas
> Lab: SDL#2, Finding Room Pressure

LABS FOR UNIT 7
> Lab: CBL#12, Effect of Temperature on Solubility of a Salt
> Lab: CBL#11, Finding the concentration of a Solution: Beer's Law
> Lab: CBL#15, Using Freezing Point Depression to Find Molecular Weight

LABS FOR UNIT 8
> Lab: Rate Law Determination of the Crystal Violet Reaction
> Lab: CBL#18, Additivity of Heats of Reaction: Hess's Law
> Lab: Micro Lab, LeChatlier's Principle: Factors Affecting Equilibrium
> Lab: CBL#20, Chemical Equilibrium: Finding a Constant, K_C

LABS FOR UNIT 9
> Lab: CBL#22, Acid Rain
> Lab: CBL#24, Acid-Base Titrations (I use both traditional and CBL methods of finding equivalence points – see teacher note #2, and I also set up double burets)
> Lab: SDL# 5, Making a pH Scale (Using CBL for pH meter)

LABS FOR UNIT 10
> Lab Activity: Dry Lab, Writing Half Reactions and Balancing RedOx Reactions
> Lab: CBL#28 (modified) Establishing a Table of Reduction Potentials: Micro-Voltaic cells

LABS FOR UNIT 11
> Lab: Finding the Half Life of a Radioactive Substance
> Lab: Alpha, Beta, Gamma Radiation

LABS FOR UNIT 12
> Lab: CBL#19, Heat of Combustion: Magnesium

LABS FOR UNIT 13 – LAB FINAL
> Copper to Copper Lab or Qualitative Analysis (Each Involving an Unknown Sample)
> **PROJECT – SECOND SEMESTER – DATES TBA**

NOTE:
> **"SDL" = Student Designed Laboratory (a series of labs with minimal instructions to aid student development in investigative design – see instruction sheet included)**
> **"CBL" Laboratories come from manuals purchased to accompany hardware**
> **"PROJECTS" vary from year to year, but usually involve the production of hands-on chemistry activities presented to elementary science classes in the spring semester**

Lessons

Making a Semimicropycnometer
Determining the Specific Gravity of an Unknown Liquid

OBJECTIVE

Students will make a glass device called a semimicropycnometer and use the pycnometer to determine the specific gravity of an unknown liquid. They will be able to identify the liquid from its specific gravity.

LEVEL

Chemistry

NATIONAL STANDARDS

UCP.3, A.1, A.2, B.2

TEKS

1(A), 1(B), 4(B)

CONNECTIONS TO AP

The concept of specific gravity and its relationship to density is important in all four AP science courses. Measurement skills are also important in all of the AP science courses.

TIME FRAME

60-75 minutes

MATERIALS

(For a class of 28 working individually)

56-9 inch flint glass Pasteur pipets*	28 crucibles
unknown liquids: dichloromethane, ethyl alcohol, hexane, ethyl acetate, 50 mL of each liquid	one water aspirator with side-armed filter flask Instructions below**
1 ring stand	1 three armed clamp to hold filter flask
28 Bunsen burners	7 balances
ice to treat possible burns	28 triangular files

*Flint glass Pasteur pipets may be ordered from Fisher Scientific—catalogue. A/S67051 250 for $9.00 (1-800-955-1177)

The assembly used to remove the liquid from the pycnometer is made by attaching a side-armed filter flask to a water aspirator with a length of rubber tubing. Clamp the flask on the ring stand with the three armed clamp so that it is secure. Place a one-holed rubber stopper in the flask. When the water is turned on, the large end of the pycnometer may be placed in the hole in the stopper. The liquid in the pycnometer is drawn into the flask and may be disposed of later. Since you will have relatively small amounts to dispose of, pour the mixture of organic liquids into a shallow pan or dish and place it under a fume hood. Allow it to evaporate.

TEACHER NOTES

Specific gravity of a substance is an intensive physical property. It is defined as the density of a substance divided by the density of a reference substance. For liquids, the reference substance is water, which has a density of 1.00 g/mL. Numerically, the specific gravity is the same as the density but is a unitless number. It is abbreviated "s.g.".

The pycnometer is a device which has a controlled volume. Usually the openings are capillary tubes to assure that the volume contained within is exactly the same, regardless of the liquid. In that way, the mass of the unknown liquid may be divided by the mass of an equal volume of water. Since the volumes are the same, they cancel.

$$\frac{D_x}{D_{water}} = \frac{\dfrac{Mass_x}{Volume_{pycnometer}}}{\dfrac{Mass_{water}}{Volume_{pycnometer}}}$$

The advantage of using a semimicropycnometer is the ability to determine the specific gravity of much smaller samples of liquid.

POSSIBLE ANSWERS TO THE CONCLUSION QUESTION AND SAMPLE DATA

Mass of empty pycnometer	1.16 g
Mass of pycnometer filled with unknown liquid	1.90 g
Mass of unknown liquid	0.74 g
Mass of pycnometer filled with water	2.11 g
Mass of water	0.95 g
Specific gravity of unknown liquid	0.78 g
Identity of liquid	Ethyl alcohol

1. Why does the specific gravity have no units?
   - Specific gravity has no units because it is defined as the density of the unknown divided by the density of water and the units cancel.

2. Your data indicate that the specific gravity of your liquid is 0.85. This is halfway between ethyl alcohol and ethyl acetate. Describe the steps you would take to determine the identity of the liquid.
   - Possible methods of determining the identity:
   - Repeat the procedure.
   - Use other methods such as odor to determine the identity.

3. When you fill your pycnometer with the unknown liquid, you get a bubble of air trapped in the pycnometer. Would the calculated value of the specific gravity be less than, more than, or the same as the actual value? Explain your answer.
   - If there is a bubble in the pycnometer, the mass of the unknown liquid would be less and therefore the specific gravity would be less.

Making a Semimicropycnometer
Determining the Specific Gravity of an Unknown Liquid

Specific gravity of a substance is an intensive physical property. It is defined as the density of a substance divided by the density of a reference substance. For liquids, the reference substance is water, which has a density of 1.00 g/mL. Numerically, the specific gravity is the same as the density but is a unitless number. It is abbreviated "s.g.".

The pycnometer is a device which has a controlled volume. Usually the openings are capillary tubes to assure that the volume contained within is exactly the same, regardless of the liquid. In that way, the mass of the unknown liquid may be divided by the mass of an equal volume of water. Since the volumes are the same, they cancel.

$$\frac{D_x}{D_{water}} = \frac{\dfrac{Mass_x}{Volume_{pycnometer}}}{\dfrac{Mass_{water}}{Volume_{pycnometer}}}$$

The advantage of using a semimicropycnometer is the ability to determine the specific gravity of much smaller samples of liquid.

PURPOSE
In this activity you will make a glass device called a semimicropycnometer which will be used to determine the specific gravity of a liquid and to determine the identity of the liquid.

MATERIALS

2 nine-inch flint glass Pasteur pipets
unknown liquid
Bunsen burner
triangular file

crucible
one classroom water aspirator fitted with a length
    of rubber tubing and stopper assembly. Your
    teacher will have this set up.
balance

Safety Alert
1. Eye goggles must be worn throughout this entire lab.
2. Hot glass burns. Be careful. Allow glass ample time to cool after heating.
3. Hold glass in toweling to cut.
4. Organic liquids are flammable. Keep far away from flames.

PROCEDURE

1. Use a 9" Pasteur pipet made from flint glass.

Flint Glass 9" Pasteur Pipet

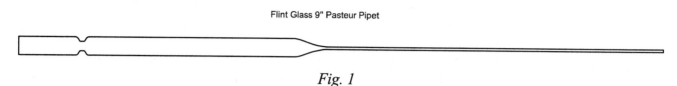

Fig. 1

2. Heat the thick part of the pipet approximately 3 cm from the capillary tube. Pull a second capillary tube about 2 cm in length. After the glass has cooled, cut using a triangular file.

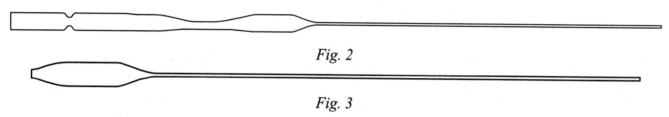

Fig. 2

Fig. 3

3. Close the air ports on the Bunsen burner and turn the flame down until it looks like a small candle flame. Bend the capillary tube so that it is close to the glass tubing.

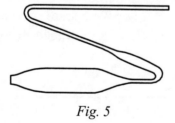

Fig. 4

4. Make a second bend in the capillary tube even with the end of the thick tubing. See figure 5. Cut off any excess capillary tube with a triangular file. Gently make a small scratch on the capillary where you want to break it. It will snap easily at this scratch with a slight pressure of your fingers.

Fig. 5

5. Obtain the mass of the semimicropycnometer and record in your data table on the student answer page.

6. Obtain a sample of unknown liquid in a clean, dry crucible. Dip the capillary tube into the liquid in the crucible and tilt it slightly so that you are holding the larger portion of the pycnometer below the level of the liquid in the crucible. The pycnometer will fill. When it is completely filled, remove it from the crucible and hold it horizontally so that the liquid will not run out.

7. Obtain the mass of the filled pycnometer. Record this value on the student answer page.

8. Use the aspirator that your teacher has set up to remove the liquid from the pycnometer. Allow air to pull through it for approximately 30 seconds to dry it. Dispose of the remaining liquid as instructed by your teacher.

9. Dry the crucible with a paper towel. Half-fill it with distilled water. Fill the pycnometer as before with the distilled water. Obtain the mass of the pycnometer filled with water and record this value on the student answer page.

10. Calculate the specific gravity of the liquid. Determine the identity of the liquid.

Name _____

Period _____

Making a Semimicropycnometer
Determining the Specific Gravity of an Unknown Liquid

ANALYSIS

The pycnometer has capillary closures so that the volume of contained liquid is always the same. In that way, the mass of the unknown liquid may be divided by the mass of an equal volume of water. Since the volumes are the same, they cancel.

$$\frac{D_x}{D_{water}} = \frac{\dfrac{Mass_x}{Volume_{pycnometer}}}{\dfrac{Mass_{water}}{Volume_{pycnometer}}}$$

Possible choices:

Ethyl acetate	s.g. =0.90
Ethyl alcohol	s.g. =0.79
Hexane	s.g. =0.66
Dichloromethane	s.g. =1.33

DATA AND OBSERVATIONS

Mass of empty pycnometer	
Mass of pycnometer filled with unknown liquid	
Mass of unknown liquid	
Mass of pycnometer filled with water	
Mass of water	
Specific gravity of unknown liquid	
Identity of liquid	

CONCLUSION QUESTIONS

1. Why does the specific gravity have no units?

2. Your data indicate that the specific gravity of your liquid is 0.85. This is half-way between ethyl alcohol and ethyl acetate. Describe the steps you would take to determine the identity of the liquid.

3. When you fill your pycnometer with the unknown liquid, you get a bubble of air trapped in the pycnometer. Would the calculated value of the specific gravity be less than, more than, or the same as the actual value? Explain your answer.

Dimensional Analysis
Exercising Problem Solving Skills

OBJECTIVE
Students will become more proficient in solving mathematical science problems.

LEVEL
Chemistry

NATIONAL STANDARDS
UCP.1, UCP.3

TEKS
2(C), 3(A), 11(C)

CONNECTIONS TO AP
All four AP science courses use problem solving. Dimensional Analysis is an essential tool and students should be encouraged to practice the skill repeatedly. Some, but not all, of the AP connections are listed below.

AP Chemistry:
    III. Reactions B. Stoichiometry 3. Mass and volume relations with emphasis on the mole concept, including empirical formulas and limiting reactants

AP Physics:
    I. Newtonian mechanics II. Thermal physics III. Electricity and Magnetism IV. Waves and Optics

AP Environmental Science:
    II. Interdependence of Earth's Systems: Fundamental Principles and Concepts A. The Flow of Energy 1. forms and quality of energy

TIME FRAME
45 minutes

MATERIALS
calculator for each student (scientific, but not   dry erase marker for each student
    necessarily graphing)   paper towels
student white boards*

TEACHER NOTES

White boards are the modern version of the old-fashioned slate and chalk. They can be purchased from educational supply houses. However, they are expensive. A much cheaper alternative is to buy shower board from your local do-it-yourself store (Home Depot, Lowe's, etc.) and ask them to cut the board into 16" x 24" rectangles. Many stores will not charge for this service if they know that you are a teacher. The white boards can be used in classroom problem solving activities to engage and monitor the students. Students can solve a problem on the board and flip it up as soon as they get an answer. In that way you can assess each student and spot the students who are not participating or are having difficulty.

Dimensional analysis problem solving is also known as the factor label method. It relies on conversion factors that are thoroughly labeled with the proper units. Probably the most difficult part of the process is having the students carefully read the problem and write the factors out before beginning to solve the problem.

Here is a sample problem. You may want to use this problem or a similar one to model the technique for the students before asking them to try it on their own.

Example: Susan wants to drive her new car from San Antonio, Texas to her friend, Bill's house in Austin, Texas. It is 110. km from her house to Bill's. The speed limit on the highway is 38 m/s, and Susan would never speed. How many hours will it take Susan to reach Bill's house?

Step One: Read the problem carefully. Many students have difficulty with this first step and it is a skill that must be practiced.

Step Two: Determine the unit for the answer. In this case, it is hours.

Step Three: Estimate an answer. This keeps students from making silly errors or calculator errors. Certainly, if the answer came out to 0.0025 hours, the student should know that something was wrong. Certainly it would take more than one hour to go 110. km.

Step Four: Write down all factors and conversions needed under the problem. Make certain that students write the division line in a horizontal fashion and not as a diagonal slash. This seems like a picky thing, but it is the number one stumbling block for many students. For our problem, the factors would be: $110.\,\text{km}\ \dfrac{1000\,\text{m}}{\text{km}}\ \dfrac{38\,\text{m}}{\text{s}}\ \dfrac{60\,\text{s}}{\text{min}}\ \dfrac{60\,\text{min}}{\text{h}}$

Step Five: Write the unit for the answer at the far right hand side of the page. Precede it with an = sign and draw a straight line backwards across the page. It is recommended to work backwards because it gives students a place to start and better emphasizes the unit cancellation than if they start at the left-hand side of the page. $\underline{\hspace{5cm}} = \text{h}$

Step Six: Since the answer is to be in hours, we need to locate the factor that has this unit. The factor above is $\dfrac{60\,\text{min}}{\text{h}}$. However, the unit hour needs to be in the numerator. We must write the factor as $\dfrac{\text{h}}{60\,\text{min}}$. The reason that it is possible to invert the factors as needed often escapes students. You will need to point out that each factor is equal to 1, and that the reciprocal of 1 is still 1.

$$60\,\text{min} = 1\,\text{hour} \qquad \frac{60\,\cancel{\text{min}}}{\text{hour}} = \frac{1\,\cancel{\text{hour}}}{\cancel{\text{hour}}} = \frac{\text{hour}}{60\,\cancel{\text{min}}} = 1$$

Step Seven: Place the factors in position one at a time, until all that is left is the unit on the left.

$$\frac{110\,\cancel{\text{km}}}{}\times\frac{1000\,\cancel{\text{m}}}{\cancel{\text{km}}}\times\frac{\cancel{\text{s}}}{38\,\cancel{\text{m}}}\times\frac{\cancel{\text{min}}}{60\,\cancel{\text{s}}}\times\frac{\text{h}}{60\,\cancel{\text{min}}} = \text{h}$$

Step Eight: Use a calculator to do the arithmetic. If the number is on the top, the students should use the $\boxed{\times}$; if the number is on the bottom, the student should use the $\boxed{\div}$. It is best to work from factor to factor, rather than multiplying all the numbers in the numerator and then dividing by all the numbers in the denominator. If you work from factor to factor, the units will tell you what you have solved for at each step. Record the answer with the proper number of significant digits for the problem. Check for reasonableness.

ANSWERS TO EXERCISES

1. $\dfrac{349.5\,\cancel{\text{in}}}{}\times\dfrac{2.54\,\cancel{\text{cm}}}{\cancel{\text{in}}}\times\dfrac{\text{m}}{100\,\cancel{\text{cm}}}=8.877\,\text{m}$

2. $\dfrac{1.61\,\cancel{\text{km}}}{\cancel{\text{mile}}}\times\dfrac{1{,}000\,\text{m}}{\cancel{\text{km}}}\times\dfrac{55.0\,\cancel{\text{mile}}}{\cancel{\text{h}}}\times\dfrac{\cancel{\text{h}}}{60\,\cancel{\text{min}}}\times\dfrac{\cancel{\text{min}}}{60\,\text{s}}=\dfrac{24.6\,\text{m}}{\text{s}}$

3. $\dfrac{5.00\,\cancel{\text{grain}}}{}\times\dfrac{0.00229\,\cancel{\text{oz}}}{1.00\,\cancel{\text{grain}}}\times\dfrac{\cancel{\text{lb}}}{16\,\cancel{\text{oz}}}\times\dfrac{454\,\cancel{\text{g}}}{\cancel{\text{lb}}}\times\dfrac{1000\,\text{mg}}{\cancel{\text{g}}}=325\,\text{mg}$

4. $\dfrac{100.\,\cancel{\text{g}}}{}\times\dfrac{\text{mL}}{13.54\,\cancel{\text{g}}}=7.38\,\text{mL}$

5. $\dfrac{18.4\times10^{9}\,\cancel{\text{lb}}}{}\times\dfrac{1\,\cancel{\text{ton}}}{2000.\,\cancel{\text{lb}}}\times\dfrac{\$318}{\cancel{\text{ton}}}=\$2.93\times10^{9}$

In this "nonsense" problem, a dyne would be equivalent to a day. Let the students tell you whether the relationship would last or not.

6. $\dfrac{35.0\,\text{dig}}{} \times \dfrac{1000\,\text{m dig}}{\text{dig}} \times \dfrac{\text{zip}}{115\,\text{m dig}} \times \dfrac{\text{dyne}}{25\,\text{zip}} = 12.2\,\text{dyne}$

7. $\dfrac{15.0\,\text{mL}}{} \times \dfrac{1\,\text{cm}^3}{\text{mL}} \times \dfrac{19.2\,\text{g}}{\text{cm}^3} = 288\,\text{g}$

8. $\dfrac{3600\,\text{s}}{\text{h}} \times \dfrac{3.00 \times 10^{10}\,\text{cm}}{\text{s}} \times \dfrac{\text{m}}{100\,\text{cm}} \times \dfrac{\text{km}}{1000\,\text{m}} = 1.08 \times 10^{9}\,\dfrac{\text{km}}{\text{h}}$

9. $\dfrac{\text{h}}{3600\,\text{s}} \times \dfrac{112\,\text{km}}{\text{h}} \times \dfrac{1000\,\text{m}}{\text{km}} = \dfrac{31.1\,\text{m}}{\text{s}}$

Note that since multiplication is commutative, students may have their factors in different orders. This frequently happens when the unit for the answer has both a numerator and a denominator. In this case, the answer is in a form that has the first factor number in the denominator. Remind students that they must enter a 1 divided by 3600 to start on their calculator. The 1 (one) is understood.

10. $\dfrac{0.500\,\text{h}}{} \times \dfrac{3600\,\text{s}}{\text{h}} \times \dfrac{10.0\,\text{A}}{} \times \dfrac{\text{Coulomb}}{\text{s A}} \times \dfrac{\text{mol e}^-}{96{,}500\,\text{Coulomb}} \times \dfrac{\text{mol Au}}{3\,\text{mol e}^-} \times \dfrac{197\,\text{g Au}}{\text{mol Au}} = 12.2\,\text{g Au}$

Note that in this problem you have a complex factor that is developed from the fact that an ampere is equal to a coulomb/s. You may need to help the students develop this factor.

$1\,A = \dfrac{Coulomb}{s} \quad \therefore\; 1 = \dfrac{Coulomb}{s\,A}$ The difficulty can be avoided by substituting coulomb/s for ampere

before you start working the problem.

You might have noticed that the level of difficulty in the student problem set varies between being very easy to much more difficult. This gives students who are having difficulty a chance to be successful, while keeping the interest of students who already have proficiency in the skill. The last problem is about as difficult as chemistry problems get for a first year chemistry course.

You might want to use a song from Michael Offutt's Song Bag II, "Factor-Label It, Baby". It sets a lighter tone and makes a difficult lesson more palatable.

Dimensional Analysis
Exercising Problem Solving Skills

Many students have difficulty solving word problems. Your teacher will model a technique called dimensional analysis. You may have learned it in a previous class, so this will just refresh your skill. Even though you may be able to solve a problem easily in your head, you should practice this technique as it will come in useful when problems in chemistry and future science classes get more difficult.

PURPOSE

In this activity you will practice solving problems using the unit cancellation method called dimensional analysis.

MATERIALS

    calculator for each student (scientific, but   dry erase marker for each student
        not necessarily graphing)   paper towels
    student white boards

PROCEDURE

1. Your teacher will model the problem solving technique for you.

2. Steps for problem solving.
   a. Read the problem. Read the whole problem.
   b. Estimate an answer.
   c. Isolate the unknown with its units.
   d. Write down all the information that is given in correct dimensional analysis format. The division line must be horizontal and not diagonal.
   e. If additional conversion factors are needed, provide them or look them up.
   f. Set up the problem in dimensional analysis format. Work backwards from the unit of your answer.
   g. Use your calculator to do the arithmetic.
   h. Check your answer for units, reasonableness, and significant digits.

3. Work the problem out on the white board. When you have completed the problem, hold it up so that your teacher can see it.

Name _____

Period _____

Dimensional Analysis
Exercising Problem Solving Skills

1. The record long jump is 349.5 in. Convert this to meters. There are 2.54 cm in an inch.

2. A car is traveling 55.0 miles per hour. Convert this to meters per second. One mile is equal to 1.61 km.

3. How many mg are there in a 5.00 grain aspirin tablet? 1 grain = 0.00229 oz. There are 454 g/lb. There are 16.0 oz/lb.

4. Mercury has a mass density of 13.54 g/ml. How many milliliters would 100. grams fill?

5. In 1980 the US produced 18.4 billion (109) lb of phosphoric acid to be used in the manufacture of fertilizer. The average cost of the acid is $318/ton. (1 ton=2000 lb) What was the total value of the phosphoric acid produced?

6. On planet Zizzag, city Astric is 35.0 digs from city Betrek. The latest in teenage transportation is a Zeka which can travel a maximum of 115 millidigs/zip. On Zizzag the planet turns once on its axis each dyne. Their time system divides each dyne into 25.0 zips. How many dynes will it take Pezzi to get from Astric to Betrek to see his girlfriend? There is no telephone communication on zizzag. Do you think this relationship will last?

7. While prospecting in the North Woods, Joe found a gold nugget which had a mass density of 19.2 g/cm^3. When I dropped it into water in a graduated cylinder, the water level increased by 15.0 mL. How many grams of gold did I have?

8. Light travels at a speed of 3.00 x 10^{10} cm/s. What is the speed of light in km/h?

9. A cheetah has been clocked at 112 km/h over a 100-m distance. What is this speed in m/s?

10. An electric current of 10.0 amperes is passed through a solution of gold (III) chloride for a period of 0.500 hours. After this period of time, how much gold has plated out on the cathode? There are 96,500 coulombs/mol of electrons. A mol of gold has a mass of 197 g/mol. An ampere is equal to 1 coulomb/second. It is necessary to transfer 3.00 mol electron/mol gold.

Chromatography
Separating Metal Ions in Solution

OBJECTIVE

Students will use paper chromatography to physically separate and determine the metal ions in an unknown solution.

LEVEL

Chemistry

NATIONAL STANDARDS

UCP.1, UCP.2, UCP.3, A.1, B.2, B.3, G.2

TEKS

1(A), 2(A), 2(B), 2(C), 2(D), 2(E), 4(A), 4(C), 8(D)

CONNECTIONS TO AP

AP Chemistry:
I. Structure of Matter B. Chemical bonding 1. Binding forces b. Polarity of bonds

TIME FRAME

90 minutes

MATERIALS

(For a class of 28 working in pairs)

100 capillary tubes	chromatography paper and 12.5 cm #1 filter paper
14 scissors	14 metric rulers
14 unknown solutions	14 pencils
50 mL of each 0.5 M known solution: Cu^{2+}, Fe^{3+}, Ni^{2+}	freshly prepared solvent: 1 L of 90% acetone; 10% 6 M HCl
14 Petri dishes and 14 600 mL beakers	14 25 mL graduated cylinders
plastic wrap	21 paper clips
7 glass stirring rods	50 mL 15 M ammonia
50 mL 1% dimethylglyoxime in ethanol	

TEACHER NOTES

This lab can be performed using circular chromatography or strip chromatography. This lesson follows a discussion of classification of matter. Separation of ions is a physical separation of a homogeneous solution. Alternatively, this lab could be easily performed when studying bonding or solutions. It is suggested that you divide the class in half and experiment using both techniques and then compare the results and R_f factors. The student directions are written for both methods. If you choose to have students perform only the circular technique, conclusion questions 6 and 7 should be omitted. Students should be somewhat familiar with the process of chromatography from earlier courses. However, this may be their first encounter with chromatography that is not visible until the development reactions take place. The developing reactions that take place in this experiment are:

$$Fe^{3+} + OH^- \rightarrow Fe(OH)_3$$

A faint brown will be seen upon initial removal of the chromatogram, but a dark brown color will appear after the reaction with concentrated ammonia.

*Students may be confused by the production of the hydroxide ion from ammonia. Remind them that ammonia ionizes in water to form the hydroxide ion according to the following reaction:

$$NH_3 + HOH \leftrightarrow NH_4^+ + OH^-$$

$$Cu^{2+} + NH_3 \rightarrow [Cu(NH_3)_4]^{2+}$$

This copper complex ion forms a brilliant blue color upon developing in the presence of concentrated ammonia. The students must mark this color quickly because it fades rapidly.

After developing with concentrated ammonia and dimethylglyoxime, a bright strawberry red color will appear.

For greatest efficiency in performing this experiment:
- Both the known and the unknown chromatogram should be prepared simultaneously.
- The reaction chamber should be prepared, covered and allowed to sit while students prepare their chromatograms. This will allow the reaction chamber to become saturated with the solvent vapor which allows for an even movement of ions.
- Students must make spots small, yet concentrated, for maximum color upon development remembering to dry each spot between application.
- If the lab must be performed over a two day period, the first day should consist of preparing the chromatograms by spotting with the known and the unknown ions. All materials should be gathered and laid out for quick retrieval on the second day. Day two should consist of placing the solvent into the reaction chamber, covering for at least five minutes to saturate, and then developing the chromatograms. If students work efficiently, there should be plenty of time to perform the development procedure and use a pencil to mark the ion and solvent fronts. Once the fronts have been marked, any measurements and calculations can be performed later.

Preparation of solutions:

Chloride salts may be substituted for the nitrates salts listed, however, be sure to recalculate the required masses to ensure that a 0.5 M solution is prepared.

0.5 M Cu^{2+}:   Measure 6.04 grams of $Cu(NO_3)_2 \cdot 3\ H_2O$ and place into a 50.0 mL volumetric flask. Add enough distilled water to the line. Mix well.

0.5 M Fe^{3+}:   Measure 10.10 grams of $Fe(NO_3)_3 \cdot 9\ H_2O$ and place into a 50.0 mL volumetric flask. Add enough distilled water to the line. Mix well.

0.5 M Ni^{2+}:   Measure 7.27 grams of $NiNO_3 \cdot 6\ H_2O$ and place into a 50.0 mL volumetric flask. Add enough distilled water to the line. Mix well.

1% dimethylglyoxime: Measure 1.00 gram of dimethylglyoxime and place into a 100.0 mL volumetric flask and fill to the line with 95% ethanol. Mix well. Place into a spray container for easy use.

6 M HCl: Measure 50.0 mL of distilled water and place into a 100.0 mL volumetric flask. Measure 50.0 mL of concentrated HCl and slowly add it to the distilled water. Be sure to perform this step under the fume hood. The vapors of hydrochloric acid are very harmful to the lungs.

15 M ammonia: This is a very concentrated solution. Do not pour into the open Petri dish until the first students are ready for this step. Be sure to keep this under the fume hood with the fume hood turned on. Ammonia vapors are harmful to the respiratory tract.

Solvent: <u>Prepare this the day of the lab</u>. It must be fresh. Measure 900 mL of reagent grade acetone and 100 mL of 6M HCl. Mix these together.

Unknowns: The unknowns can be prepared by placing approximately 2.0 mL of each of the following ion solutions into small test tubes and then filling with distilled water to a total volume of about 5.0 mL. The following combination schemes can be set up and labeled so that students do not know that there are repeated unknowns within the same class period. For example, unknown #1 and #7 both contain only the Fe^{3+} ion.

<u>Ions:</u>	<u>Labeled:</u>		
Fe^{3+}	1	7	
Cu^{2+}	2	8	
Ni^{2+}	3	9	
Fe^{3+} / Cu^{2+}	4	10	13
Fe^{3+} / Ni^{2+}	5	11	14
Cu^{2+} / Ni^{2+}	6	12	

TEACHER PAGES

POSSIBLE ANSWERS TO THE CONCLUSION QUESTIONS AND SAMPLE DATA

DATA AND OBSERVATIONS

Color of ions in the known solution:

Fe^{3+} yellowish-orange

Cu^{2+} sky blue

Ni^{2+} lime green

Unknown Number: _____

Identity of Unknown: _____Answers will vary_____

Sketch each of your chromatograms in the space provided below:

    Known   Unknown

ANALYSIS

Note: Answers will vary. A sample set of data is given in the table below.

Chromatogram for Known Ions:		Distance Solvent traveled: 39 mm		
	Distance ion traveled	Distinguishing color	R_f	
Fe^{3+}	39 mm	light brown	1.0	
Cu^{2+}	32 mm	sky blue (after developing)	0.82	
Ni^{2+}	6.0 mm	hot pink (after developing)	0.15	

Chromatogram for Unknown Ions:		Distance Solvent traveled: 42mm Unknown number: 11	
	Distance ion traveled (or not present)	Distinguishing color	R_f
Fe^{3+}	42 mm	light brown	1.0
Cu^{2+}	None present	NA	NA
Ni^{2+}	5.0 mm	bright pink	0.12

$$R_f = \frac{\text{distance traveled by dye}}{\text{distance traveled by solvent}}$$

Show all work for R_f calculations below:

Known:

Fe^{3+}	Cu^{2+}	Ni^{2+}
$\dfrac{39\,mm}{39\,mm} = 1.0$	$\dfrac{32\,mm}{39\,mm} = 0.82$	$\dfrac{6.0\,mm}{39\,mm} = 0.15$

Unknown: Show a calculation for each ion found in your sample.

$$Fe^{3+} = \frac{42\,mm}{42\,mm} = 1.0 \qquad\qquad Ni^{2+} = \frac{5.0\,mm}{42\,mm} = 0.12$$

CONCLUSION QUESTIONS

1. Chromatography is a useful tool in the scientific world. Explain how it is used. Give at least two specific examples for its use today.
   - Chromatography allows a physical separation of components in a mixture in a fairly simple an inexpensive way.
   - Chromatography is often used today in screening for illegal drugs and in determining components of certain food products. It is also a useful technique to follow the products being produced in a chemical reaction.

2. Explain how you could experimentally determine if a solution contained Ni^{2+} ions.
   - A chromatogram could be prepared and run according to the procedure in this lab. After removing the chromatogram and marking the distance that the solvent traveled, the chromatogram could be developed by exposing it to concentrated ammonia and then spraying it with dimethylglyoxime. If a strawberry red color develops, then nickel (II) ions are present.
   - Alternatively, after developing, the ion front for nickel could be marked and the R_f value could be calculated and compared to a known R_f value.

3. Explain the purpose of preparing the reaction vessel at the beginning of the experiment and allowing it to sit covered until the chromatograms were prepared. Cite any errors that could be expected if this were not done.
   - Covering the reaction chamber with the solvent allows for the vapor phase inside the container to become saturated with solvent. This will help to ensure an even movement of ions along the chromatogram.

4. The instructions in the lab told you to mark the solvent front immediately upon removal from the reaction chamber. Explain why this is important.
   - The solvent was largely composed of acetone which evaporates readily. If the solvent front were not marked immediately, it might be difficult to detect. The solvent might also move a bit more up the paper after removal from the chamber.

5. What would be expected in terms of R_f values if the solvent was replaced with another substance?
   - R_f values do depend largely on the type of solvent used in an experiment. The values are relative within a given experiment. So, though the R_f values might not be the exact same number, you could still expect that the known and the unknown would have the same values if both experiments were run with the same solvent and same type of paper at the same temperature.

6. Compare the results of the circular technique with the strip technique by comparing with other groups. Comment on similarities and differences along with possible explanations for each.
   - Answers will vary here, but students should include information such as: the relative movement of the ions on the chromatogram, the comparison of the R_f values for knowns, the time required for each technique, and the ion fronts produced by each.

7. A student performed a paper chromatography separation using the strip technique. After completing the experiment and cleaning the lab area, the student realized that although he measured the solvent front immediately upon removal, measurements were not taken for the ion fronts of the metals. Immediately these were measured and recorded. How will this time lapse affect the calculated R_f values?

- The calculated R_f values will be larger than they should be. By not measuring immediately, the ions had the opportunity to move further along the paper, therefore, the distance traveled by the ion is greater than it should be while the distance traveled by the solvent is correct. When the numerator is larger, the final answer appears larger.

Chromatography of Metal Ion
Possible Student Data for Chromatograms
of Known Solutions After Developing Reactions

Circular Technique

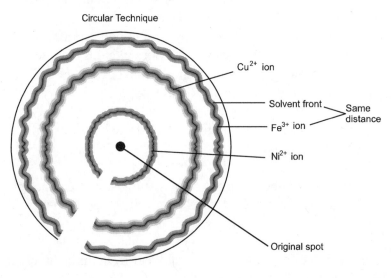

Strip Technique

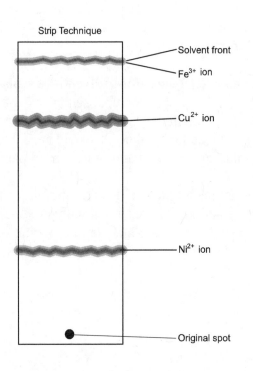

Laying the Foundation in Chemistry

Chromatography
Separating Metal Ions in Solution

Chromatography is a laboratory technique often used to separate mixtures into their component parts. The first recorded history of this technique dates back to 1906 when the Russian botanist, Twsett, separated colored pigments in leaves. Because different bands of color were produced, he named the separation chromatography. Chromatography literally comes from the Greek *chroma* meaning color and *graphein*, to write. However, today we realize that color is not a necessary property needed to achieve a separation of components in a mixture. Separation is always based on a physical characteristic of the compounds, such as size or polarity. Because of its simplistic nature, chromatography is widely used today in the medical and legal fields to screen for illegal drug use. It is also used to isolate the compounds that give food its characteristic flavors so that the food industry can determine how to synthesize flavor compounds artificially.

While there are many different types of chromatography, they all have a stationary phase (that which does not move) and a mobile phase (that which moves). If each mixture component has a different affinity for the mobile and stationary phases, they can be separated. The art of chromatography is in the selection of the correct stationary and mobile phases to use. In paper chromatography, the stationary phase is paper and the mobile phase is a liquid solvent. The mixture to be separated is called the analyte. The analyte is placed as a small spot on the paper and then placed into a chamber containing the mobile phase at the bottom. Capillary action draws the mobile phase up the paper. If a component has a strong attraction for the mobile phase, it tends to move with it. If a component has a strong attraction for the paper, it tends to stay behind. These differences in attraction result in different rates of movement and thus, complex mixtures can be separated. The act of placing the paper into the solvent and allowing the solvent to travel up the paper by capillary action results in a chromatogram. The identity of unknowns in a solution can be determined by comparing the chromatogram of the unknown solution to a known chromatogram or by the distance the components travel. The ratio of the distance traveled by the component to the distance the solvent traveled is called its R_f value. R_f values depend on many variables such as the solvent used, the type of paper and the temperature. Controlling these factors within a given experiment should lead to fairly accurate results.

The ions to be separated in this lab do not all impart a color upon initial separation. In order to visualize the distance that some of the ions travel, the chromatogram will need to undergo a few chemical reactions. Characterizations of the ions in this lab include the following:

Fe^{3+} will impart a rusty brown color upon initial developing of the chromatogram and will form a dark rust color when reacted with concentrated ammonia as iron (III) hydroxide forms as a precipitate.

Cu^{2+} may impart a faint blue color but will readily be detected after reacting with concentrated ammonia. The copper (II) ion reacts with concentrated ammonia to form a complex ion, $[Cu(NH_3)_4]^{2+}$, tetraamminecopper (II) ion, which yields a brilliant blue color that is easily distinguishable.

Ni^{2+} reacts with the organic reagent, dimethylglyoxime, to form a bright strawberry red color.

PURPOSE

In this experiment you will physically separate a mixture of known metal ions using paper chromatography and determine the R_f values for each known ion. Developing reactions will be performed to identify each of the ions in solution since colors will not be visible upon separation. You will then perform the same procedure given an unknown and determine the identities of each component in the solution.

MATERIALS

capillary tubes	chromatography paper or filter paper
scissors	metric ruler
unknown solution	pencil
known solutions of: Cu^{2+}, Fe^{3+}, Ni^{2+}	solvent
2 Petri dishes or 2 600 mL beakers	25 mL graduated cylinder
plastic wrap	paper clips
2 stirring rods	

Safety Alert

1. The vapors of the solvent and the concentrated ammonia in this lab can be harmful if inhaled.
2. The solvent contains hydrochloric acid which is an irritant. Flush with water if any solvent comes in contact with skin.
3. Acetone is flammable and must be kept away from open flames.
4. The iron solution can stain clothing.
5. Always wear goggles and aprons.
6. Wash hands thoroughly at the end of the experiment.

PROCEDURE: CIRCULAR TECHNIQUE

1. Prepare two reaction chambers by obtaining two Petri dishes or two evaporating dishes as directed by your teacher. Fill each dish with solvent to a depth between 5 and 7 mm. Cover each dish with plastic wrap.

2. Obtain two pieces of filter paper and cut a wick approximately 1 cm in width on each paper. (see Figure 1)

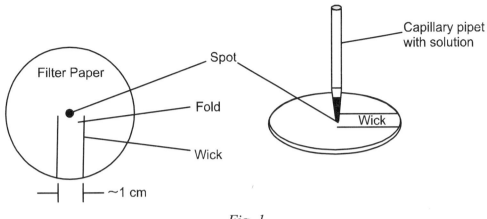

Fig. 1

3. Using a pencil, place a dot in the center of each paper circle. Label one paper "known" and the other "unknown".

4. Obtain a small sample of each of the known solutions, Fe^{3+}, Ni^{2+} and Cu^{2+}. Record the color of each solution on the student answer page.
   a. Using a capillary tube, place a drop of the iron solution directly on the pencil dot on the paper labeled "known". Allow this spot to dry thoroughly and repeat this same process by spotting the nickel and the copper solutions directly on top of the same pencil dot. Be sure to allow the spot to dry between each application.
   b. Repeat this same procedure with your unknown solution and the paper labeled "unknown". Be sure to record your unknown number on the student answer page.

5. Bend the wicks of each filter paper in a downward slope and carefully place the spotted filter paper on the rim of the reaction chamber with the wick just touching the solvent. Quickly and carefully replace the plastic wrap around the dish with solvent and chromatography paper. Seal from the external environment.
   a. A sealed reaction vessel will allow for an even movement of ions. When the vapor is saturated with the solvent it will travel across the paper in an even fashion permitting a more even separation of ions.
   b. Do not disturb the reaction chambers as the solvent travels.

6. When the solvent front almost reaches the edge of the paper, remove the chromatogram and quickly mark the solvent front with a pencil.

7. If there are any visible rings from any of the ions mark their position with a pencil. (Hint: There should be one ion detected on the known chromatogram. Can you tell which one this is?)

8. Carry your chromatograms to the fume hood and lay them one at a time over the open container of concentrated ammonia. Be sure not to get the chromatogram wet in the ammonia. The ammonia vapors are the reactant involved. As the ring fronts appear, quickly mark them with pencil.

9. Obtain the spray bottle of dimethylglyoxime and spray each of your chromatograms. Use the pencil to mark the ring front that appears.

10. Measure the distances (in millimeters) for the solvent front and ring fronts in each chromatogram. Record your values in the data table on the student answer page.

11. Make a sketch of each of your chromatograms in the space provided on the student answer page.

12. Calculate the R_f values for each of the ions in the known and the unknown. Be sure to show all work in the space provided on the student answer page. Identify the ion/ions in your unknown and record your answer in the space provided.

PROCEDURE: STRIP TECHNIQUE

1. Prepare two reaction chambers by obtaining two 600 mL beakers. Fill each beaker with solvent to a depth between 10 and 15 mm. Cover each beaker with plastic wrap.

2. Obtain two strips of chromatography paper approximately 15 cm in length.

3. Using a pencil and a metric ruler, draw a pencil line 1 cm from the bottom of each strip. Place a pencil dot in the center of each pencil line on each strip. Label one paper "known" and the other "unknown".

4. Obtain a small sample of each of the known solutions, Fe^{3+}, Ni^{2+} and Cu^{2+}. Record the color of each solution on the student answer page.
   a. Using a capillary tube, place a drop of the iron solution directly on the pencil dot on the paper labeled "known". Allow this spot to dry thoroughly and repeat this same process by spotting the nickel and the copper solutions directly on top of the same pencil dot. Be sure to allow the spot to dry between each application.
   b. Repeat this same procedure with your unknown solution and the paper labeled "unknown". Be sure to record your unknown number on the student answer page.

5. Carefully place the spotted filter paper onto a glass stirring rod and paper clip in place. Place the chromatography strip into the reaction chamber with the end of the paper just touching the solvent. Quickly and carefully place the plastic wrap around the reaction chamber and seal from the external environment. See Figure 2.

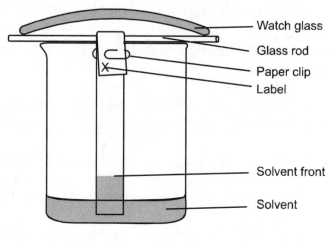

Fig. 2

a. A sealed reaction vessel will allow for an even movement of ions. When the vapor is saturated with the solvent it will travel across the paper in an even fashion permitting a more even separation of ions.

b. Do not disturb the reaction chambers as the solvent travels.

6. When the solvent front almost reaches the edge of the paper, remove the chromatogram and quickly mark the solvent front with a pencil.

7. If there are any visible rings from any of the ions mark their position with a pencil. (Hint: There should be one ion detected on the known chromatogram. Can you tell which one this is?)

8. Carry your chromatograms to the fume hood and lay them one at a time over the open container of concentrated ammonia. Be sure not to get the chromatogram wet in the ammonia. The ammonia vapors are the reactant involved. As the ion fronts appear, quickly mark them with pencil.

9. Obtain the spray bottle of dimethylglyoxime and spray each of your chromatograms. Use the pencil to mark the ion front that appears.

10. Measure the distances (in millimeters) for the solvent front and ion fronts in each chromatogram. Record your values in the data table on the student answer page.

11. Make a sketch of each of your chromatograms in the space provided on the student answer page.

12. Calculate the R_f values for each of the ions in the known and the unknown. Be sure to show all work in the space provided on the student answer page. Identify the ion/ions in your unknown and record your answer in the space provided.

Name _____

Period _____

Chromatography
Separating Metal Ions in Solution

DATA AND OBSERVATIONS

Color of ions in the known solution:

Fe^{3+} _____

Cu^{2+} _____

Ni^{2+} _____

Unknown Number: _____

Identity of Unknown: _____

Sketch each of your chromatograms in the space provided below:

     Known   Unknown

DATA AND OBSERVATIONS

Chromatogram for Known Ions:		Distance Solvent traveled: _____mm	
	Distance ion traveled	Distinguishing color	R_f
Fe^{3+}			
Cu^{2+}			
Ni^{2+}			

Chromatogram for Unknown Ions:		Distance Solvent traveled: _____mm Unknown number: _____	
	Distance ion traveled (or not present)	Distinguishing color	R_f
Fe^{3+}			
Cu^{2+}			
Ni^{2+}			

$$R_f = \frac{\text{distance traveled by dye}}{\text{distance traveled by solvent}}$$

Show all work for R_f calculations below:

Known:

_____Fe^{3+}_____Cu^{2+}_____Ni^{2+}_____

Unknown: Show a calculation for each ion found in your sample.

3 *Chromatography*

CONCLUSION QUESTIONS

1. Chromatography is a useful tool in the scientific world. Explain how it is used. Give at least two specific examples for its use today.

2. Explain how you could experimentally determine if a solution contained Ni^{2+} ions.

3. Explain the purpose of preparing the reaction vessel at the beginning of the experiment and allowing it to sit covered until the chromatograms were prepared. Cite any errors that could be expected if this were not done.

4. The instructions in the lab told you to mark the solvent front immediately upon removal from the reaction chamber. Explain why this is important.

5. What would be expected in terms of R_f values if the solvent was replaced with another substance?

6. Compare the results of the circular technique with the strip technique by comparing with other groups. Comment on similarities and differences along with possible explanations for each.

Laying the Foundation in Chemistry

7. A student performed a paper chromatography separation using the strip technique. After completing the experiment and cleaning the lab area, the student realized that although he measured the solvent front immediately upon removal, measurements were not taken for the ion fronts of the metals. Immediately these were measured and recorded. How will this time lapse affect the R_f calculated values?

Matter Waves
An Exercise in Literal Equations

OBJECTIVE
This is one of several small exercises to enrich a unit on atomic structure and improve student involvement. Students will use the traditional Einsteinian equations to derive the de Broglie equation for matter waves.

LEVEL
Chemistry

NATIONAL STANDARDS
UCP.1, UCP.2, B.1, B.2, B.6, G.2

TEKS
6(A)

CONNECTIONS TO AP
AP Chemistry:
 Atomic Structure
AP Physics:
 Literal Equations

TIME FRAME
10 minutes

MATERIALS
 Student activity sheets   calculator

TEACHER NOTES
This activity takes little time, but allows students to work with literal equations while enriching the unit on atomic structure. The concept of electron waves is one that is difficult for students to grasp. Introducing this concept in Pre-AP Chemistry makes it easier for students to understand concepts in AP Chemistry.

Einstein suggested that electromagnetic radiation interacted with matter in small, discrete particles called *photons*. These photons have energy related to the frequency of the electromagnetic radiation. The important equation is $E = h\upsilon$, where E is energy in joules, h is Planck's constant (6.63×10^{-34} J•s), and υ is the frequency of the radiation in cycles per second or Hertz. Einstein also developed his famous equation $E = mc^2$ where m is the mass in kg, and c is the speed of light in a vacuum (3.0×10^8 m/s). The universal wave equation, $c = \lambda\upsilon$, had been known for some time. In this equation, λ is the wavelength of the radiation measured in meters.

Louis de Broglie, examining Einstein's work, realized that if light had particle characteristics, then matter most likely had wave properties. Using the equations above, and realizing that matter could not be traveling at the speed of light, but did indeed have a velocity, v, which could be substituted for the speed of light, de Broglie derived the equation for matter waves: $\lambda = h/mv$.

Note: The symbol "υ" (nu) is reserved for the frequency of electromagnetic radiation.

POSSIBLE ANSWERS TO THE QUESTIONS AND SAMPLE DATA

1. Remember that matter cannot travel at the speed of light, so the symbol for velocity, v, should be substituted for c. Use the available equations to solve for the de Broglie equation.

   - $E = h\nu$ $E = mc^2$ $c = \lambda\nu$

   - According to the transitive property $h\nu = mc^2$

   - Using $c = \lambda\nu$ solve for ν: $c/\lambda = \nu$

   - Substitute the value for ν in to $h\nu = mc^2$. This will yield $hc/\lambda = mc^2$.

   - Of course, matter cannot travel at the speed of light, so the c in the equation needs to be changed to v for velocity. $h\nu/\lambda = mv^2$.

   - Rearrange this equation by solving for λ. This will yield the de Broglie matter wave equation of $\lambda = h/mv$

   - After students have a chance to work with these equations to produce the de Broglie equation, you may have to walk them through the process. Success in independently solving this problem is dependent upon the math level and skill of the students.

2. Use the de Broglie equation to solve for the wavelength of a track star with a mass of 65 kg who is jogging around the track at a speed of 6.0 m/s. The value of Planck's constant is 6.63×10^{-34} J•s.

   - $\lambda = \dfrac{6.63 \times 10^{-34} \text{J} \cdot \text{s}}{(65\,\text{kg})(6.0\,\text{m/s})} = 1.7 \times 10^{-36}\,\text{m}$

Matter Waves
An Exercise in Literal Equations

Einstein suggested that electromagnetic radiation interacted with matter in small, discrete particles called *photons*. These photons have energy related to the frequency of the electromagnetic radiation. The important equation is $E = h\upsilon$, where E is energy in joules, h is Planck's constant (6.63×10^{-34} J•s), and υ is the frequency of the radiation in cycles per second or Hertz. Einstein also developed his famous equation $E=mc^2$ where m is the mass in kg, and c is the speed of light in a vacuum (3.0×10^8 m/s). The universal wave equation, $c=\lambda\upsilon$, had been known for some time. In this equation, λ is the wavelength of the radiation measured in meters.

Louis de Broglie, examining Einstein's work, realized that if light had particle characteristics, then matter most likely had wave properties. Using the equations above, and realizing that matter could not be traveling at the speed of light, but did indeed have a velocity, v, which could be substituted for the speed of light, de Broglie derived the equation for matter waves: $\lambda = h/mv$.

PURPOSE
You will use the equations in the first paragraph to derive the de Broglie equation.

MATERIALS
Student answer page   calculator

PROCEDURE
1. Work with a partner to derive the de Broglie equation. Show all work on the student answer page.

2. Use the de Broglie equation to solve the problems.

Name _____

Period _____

Matter Waves
An Exercise in Literal Equations

AVAILABLE EQUATIONS

$E = h\nu$ $\qquad\qquad$ $E = mc^2$ $\qquad\qquad$ $c = \lambda\nu$

DE BROGLIE EQUATION

$\lambda = h/mv$

QUESTIONS

1. Remember that matter cannot travel at the speed of light, so the symbol for velocity, v, should be substituted for c. Use the available equations to solve for the de Broglie equation.

2. Use the de Broglie equation to solve for the wavelength of a track star with a mass of 65 kg who is jogging around the track at a speed of 6.0 m/s. The value of Planck's constant is 6.63×10^{-34} J•s.

Electron Configurations, Orbital Notation and Quantum Numbers
Understanding Electron Arrangement and Oxidation States

OBJECTIVE

Students will learn to write correct electron configurations, orbital notations and quantum numbers for the valence electron in certain elements. Students will learn to justify oxidation states based on electron configurations which will enhance their understanding of chemical formula writing.

LEVEL

Chemistry

NATIONAL STANDARDS

UCP.1, UCP.2, B.1, B.2

TEKS

4(D), 6(C)

CONNECTIONS TO AP

AP Chemistry:

I. Structure of Matter A. Atomic theory and atomic structure 4. Electron energy levels: atomic spectra, quantum numbers, atomic orbitals

TIME FRAME

90 minutes

MATERIALS

Transparency of Periodic Table [unless one hangs on your classroom wall]
Transparency of the Diagonal Rule

TEACHER NOTES

This lesson should be taught after students have mastered basic atomic structure and layout of the periodic table. This lesson will introduce skills necessary for formula writing and nomenclature units. Writing electron configurations, orbital notations and quantum numbers are fundamental to a first-year foundational course. All serve to help students understand the logic of writing chemical formulas and predicting oxidation states.

Suggested Teaching Procedure:

1. Before presenting to students, complete the student activity yourself. You will need a copy of the periodic table included with this activity.

2. Once you have mastered the skill, introduce the lesson by demonstrating your predicting prowess as to engage the students' curiosity.

3. Keep the following points in mind during your introduction:
   - Use the periodic table below only for you to make this task easier. Do not share this table with the students, keep a copy handy as you impress them with your knowledge of electron configurations and engage their curiosity. Hopefully, by the end of this activity, they will deduce these patterns for themselves.
   - The periodic table is arranged according to electron configuration. With practice, you'll be able to give the ending term of the electron configuration within a few seconds
   - Study the periodic table below that shows the patterns of the electron configurations.
   - Period numbers represent energy levels.
   - Notice that there are only two elements in the first period, which represents the first energy level.
   - The second period is separated so that two elements are together on the left and six are together on the right. The second energy level has two sublevels s and p; s-sublevels contain two electrons while full while p-sublevels contain six electrons. You will notice there are two boxes grouped together on the left, called the s-block, and 6 boxes on the right for the p-block.

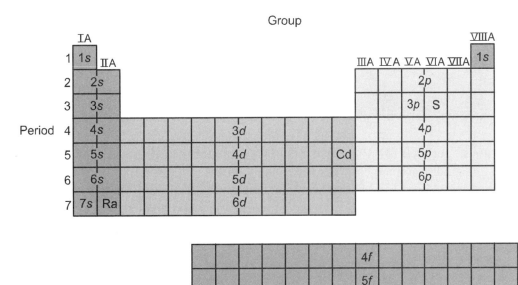

   - Notice the third period looks like the second.
   - The third energy level is the first energy level able to contain d sublevels. You would expect to see the insertion of the d sublevel here, but it is delayed until the 4th period.

- The delay is because of the interference all of the electrons have with each other. The fact is that it takes less energy for an electron to be placed in the 4s sublevel rather than in the 3d; the 4s fills before the 3d. Therefore, on the table the 10 electrons representing 3d are placed immediately after the 4s and begin the section called the transition metals. Students often mis-number the d-block elements. You must remind them that the first time you get a d-sublevel at all, it is in the <u>third</u> energy level so that top row of d-elements must be <u>3d</u> while the one below must be 4d, etc.

- Finally, look at the rare earth elements at the bottom of the table. There are 14 of them which correspond to the filling of the f sublevels. There are only 2 rows in the f-block; the top row must be 4f since that's the first energy level with an f-sublevel and the bottom row must be 5f.

A practice exercise:
Using the periodic table, determine the electron configuration for the valence electron in sulfur.
- First locate sulfur on the periodic table; it is in the 3rd period in the p-block of elements.
- Count from left to right within the p-block and you determine the valence electron has an ending configuration of 3p^4. The Aufbau principle states that all lower energy sublevels must be full, so its entire electron configuration is 1s^{2}2s^{2}2p^{6}3s^{2}3p^4.
- The noble gas short hand allows us to substitute the noble gas symbol for the core electrons. In the case of sulfur it would be [Ne] 3s^{2}3p^4. It is up to you whether or not you want students to short-hand the electron configurations on their answer page.

4. Start class by introducing the student handout. An example for determining the electron configuration for sulfur is given in the student directions. The electron configuration and orbital notation for oxygen is given as an example in the "time to get the lingo straight section".

5. This is where you show off your skills—remember to keep your periodic table with the sublevels labeled handy for moments of panic. Ask students to practice by writing the electron configurations for the following elements in class before they begin their answer sheet exercise. As you announce each element, also announce your prediction for the ending electron configuration—do this by using the element's position on the periodic table. As you gain confidence, let the students pick their favorite elements and try to stump you. Do not divulge your secrets. Perhaps offer extra credit to the first student that can explain the source of your wisdom to the class. If challenged, students will deduce this pattern on their own.

Element	ENDING Valence electron configuration
Na	3s^1
Ni	3d^8
C	2p^2
I	6p^5
Y	4d^1
Es	5f^{10}

ANSWERS TO THE CONCLUSION QUESTIONS

Element	Electron configuration	Valence Orbital notation [only the outermost orbitals are drawn]	Set of Quantum Numbers for the LAST Valence electron to fill
K	[Ar] $4s^1$	↑ 4s	4, 0, 0, +½
Fe	[Ar] $4s^2 3d^6$	↑↓ ↑↓ ↑ ↑ ↑ ↑ 4s 3d	3, 2, -2, -½
N	$1s^2 2s^2 2p^3$	↑↓ ↑↓ ↑ ↑ ↑ 1s 2s 2p	2, 1, 1, +½
Sn	[Kr] $5s^2 4d^{10} 5p^2$	↑↓ ↑↓ ↑↓ ↑↓ ↑↓ ↑↓ ↑ ↑ __ 5s 4d 5p	5, 1, 0, +½
Br	[Ar] $4s^2 3d^{10} 4p^5$	↑↓ ↑↓ ↑↓ ↑↓ ↑↓ ↑↓ ↑↓ ↑↓ ↑ 4s 3d 4p	4, 1, 0, -½
Ba	[Xe] $6s^2$	↑↓ 6s	6, 0, 0, -½
Ni	[Ar] $4s^2 3d^8$	↑↓ ↑↓ ↑↓ ↑↓ ↑ ↑ 4s 3d	3, 2, 0, -½
P	[Ne] $3s^2 3p^3$	↑↓ ↑ ↑ ↑ 3s 3p	3, 1, 1, +½
Zr	[Kr] $5s^2 4d^2$	↑↓ ↑ ↑ __ __ __ 5s 4d	4, 2, -1, +½
U*	[Rn] $7s^2 5f^4$	↑↓ ↑ ↑ ↑ ↑ __ __ __ 7s 5f	5, 3, 0, +½
Ag**	[Kr] $5s^2 4d^9$	↑↓ ↑↓ ↑↓ ↑↓ ↑↓ ↑ 5s 4d	4, 2, 1, -½*
Mg	$1s^2 2s^2 2p^6 3s^2$	↑↓ 3s	3, 0, 0, -½
Kr	[Ar] $5s^2 5p^6$	↑↓ ↑↓ ↑↓ ↑↓ 5s 5p	5, 1, 1, -½
As	[Ar] $4s^2 3d^{10} 4p^3$	↑↓ ↑↓ ↑↓ ↑↓ ↑↓ ↑↓ ↑ ↑ ↑ 4s 3d 4p	4, 1, 1, +½
W	[Xe] $6s^2 4f^{14} 5d^4$	↑↓ ↑↓ ↑↓ ↑↓ ↑↓ ↑↓ ↑↓ ↑↓ 6s 4f ↑ ↑ ↑ ↑ __ 5d	5, 2, 1, +½
Fr	[Rn] $7s^2$	↑↓ 7s	7, 0, 0, -½
*Pu	[Rn] $7s^2 5f^6$	↑↓ ↑ ↑ ↑ ↑ ↑ ↑ __ 7s 5f	5, 3, 2, +½
B	$1s^2 2s^2 2p^1$	↑↓ ↑ __ __ 2s 2p	2, 1, -1, +½
Mn	[Ar] $4s^2 3d^5$	↑↓ ↑ ↑ ↑ ↑ ↑ 4s 3d	3, 2, 2, +½
I	[Kr] $5s^2 4d^{10} 5p^5$	↑↓ ↑↓ ↑↓ ↑↓ ↑↓ ↑↓ ↑↓ ↑↓ ↑ 5s 4d 5p	5, 1, 1, -½

**Students may also write [Kr] $5s^1 4d^{10}$ which is how it really exists in which case its quantum number set would be 4, 2, 2, -½.

* Some books will list the rare earth metals as having an electron in the d-sublevel before beginning to fill the f-sublevel. This will reduce the number of electrons shown in the table above by one, but the one electron placed in the previous d-sublevel should be shown in order for credit to be given. You will find that different periodic tables break the f-block differently. Some place the break before La, others after La. Some have 14 boxes for the f-block at the bottom of the table, some have 15.

6. Iron has two common oxidation states, +2 and +3. Justify each of these oxidation states. Draw the orbital notation of the neutral atom and each oxidation state as part of your justification.

   Neutral atom: ⇅   ⇅ ↑ ↑ ↑ ↑
                  4s   3d

   +2 oxidation state — 2 electrons are lost and they come from the outermost 4s leaving:

   ___   ⇅ ↑ ↑ ↑ ↑
   4s   3d

   +3 oxidation state — 3 electrons are lost. Losing one of the d-electrons minimized electron-electron repulsions:

   ___   ↑ ↑ ↑ ↑ ↑
   4s   3d

7. Nitrogen has a common oxidation state of –3. Justify this oxidation state. Draw the orbital notation for the neutral atom and oxidation state as part of your justification.

   Neutral atom: ⇅   ⇅   ↑ ↑ ↑
                 1s   2s   2p

   -3 oxidation state, three electrons are gained forming a stable octet: ⇅   ⇅   ⇅ ⇅ ⇅
                                                                          1s   2s   2p

8. Silver has only one oxidation state, +1. Justify this oxidation state even though most transition metals have an oxidation state of +2 among others. Draw the orbital notation for the neutral atom and oxidation state as part of your justification.

   Neutral atom: ⇅   ⇅ ⇅ ⇅ ⇅ ↑
                 5s   4d

   +1 oxidation state — one electron is lost. This electron will come from the outermost sublevel which is 5s. The movement of the remaining lone electron further stabilizes the ion:

   ___ ⇅ ⇅ ⇅ ⇅ ⇅
   5s   4d

   There is only a slight energy difference between s and d electrons. The further from the nucleus, the less this difference becomes. It is not uncommon to see electrons shift to fill the d-orbitals.

9. Manganese has a common oxidation state of +7. Justify this oxidation state. Draw the orbital notation for the neutral atom and oxidation state as part of your justification.

Neutral atom: ↑↓   ↑ ↑ ↑ ↑ ↑
            4s   3d

+7 oxidation state —7 electrons are lost, which is the entire contents of the 4s and 3d orbitals to create the same electron configuration as argon: ___   __ __ __ __ __
                                                          4s   3d

TEACHER PAGES

Electron Configurations, Orbital Notation and Quantum Numbers
Understanding Electron Arrangement and Oxidation States

Chemical properties depend on the number and arrangement of electrons in an atom. Usually, only the valence or outermost electrons are involved in chemical reactions. The electron cloud is compartmentalized. We model this compartmentalization through the use of electron configurations and orbital notations. The compartmentalization is as follows, energy levels have sublevels which have orbitals within them. We can use an apartment building as an analogy. The atom is the building, the floors of the apartment building are the energy levels, the apartments on a given floor are the orbitals and electrons reside inside the orbitals. There are two governing rules to consider when assigning electron configurations and orbital notations. Along with these rules, you must remember electrons are lazy and they hate each other, they will fill the lowest energy states first AND electrons repel each other since like charges repel.

Rule 1: The Pauli Exclusion Principle
In 1925, Wolfgang Pauli stated: N*o two electrons in an atom can have the same set of four quantum numbers*. This means no atomic orbital can contain more than TWO electrons and the electrons must be of opposite spin if they are to form a pair within an orbital.

Rule 2: Hunds Rule
The *most stable* arrangement of electrons is one with the maximum number of unpaired electrons. It *minimizes electron-electron repulsions* and stabilizes the atom. Here is an analogy. In large families with several children, it is a luxury for each child to have their own room. There is far less fussing and fighting if siblings are not forced to share living quarters. The entire household experiences a lower, less frazzled energy state. Electrons find each other very repulsive, so they too, are in a lower energy state if each "gets their own room" or in this case orbital. Electrons will fill an orbital singly, before pairing up in order to minimize electron-electron repulsions. All of the electrons that are single occupants of orbitals have parallel (same direction) spins and are assigned an up arrow. The second electron to enter the orbital, thus forming an electron pair, is assigned a down arrow to represent opposite spin.

PURPOSE
In this activity you will acquire an ability to write electron configurations, orbital notations and a set of quantum numbers for electrons within elements on the periodic table. You will also be able to justify oxidation or valence states using electron configurations and orbital notations.

MATERIALS
Periodic Table found at the end of this activity

To write electron configurations and orbital notations successfully, you must formulate a plan of attack—learn the following relationships:

ELECTRON CONFIGURATIONS

1. Each main energy level has *n* sublevels, where *n* equals the number of the energy level. That means the first energy level has one sublevel, the second has two, the third has three….

2. The sublevels are named s, p, d, f, g . . . and continue alphabetically. The modern periodic table does not have enough elements to necessitate use of sublevels beyond f. Why s, p, d, f? Early on in the development of this model, the names of the sublevels came from sharp, principle, diffuse and fundamental, words used in describing spectral lines of hydrogen.

3. It may be easier for you to understand this by studying the table presented below:

Energy level	Number of sublevels	Names of sublevels
1	1	s
2	2	s, p
3	3	s, p, d
4	4	s, p d, f
5	5	s, p, d, f, g

Sublevel Name	s	p	d	f
Number of Orbitals	1	3	5	7
Maximum number of electrons	2	6	10	14

4. Each sublevel has increasing odd numbers of orbitals available. s = 1, p = 3, d = 5, f = 7. Each orbital can hold *only two electrons* and they *must be of opposite spin*. An s-sublevel holds 2 electrons, a p-sublevel holds 6 electrons, a d-sublevel holds 10 electrons, and an f-sublevel holds 14 electrons.

5. The filling of the orbitals is related to energy. Remember, electrons are lazy, much like us! Just as you would place objects on a bottom shelf in an empty store room rather than climb a ladder to place them on a top shelf, expending more energy—electrons fill the lowest sublevel available to them. Use the diagonal rule as your map as you determine the outermost or valence electron configurations for any of the elements.

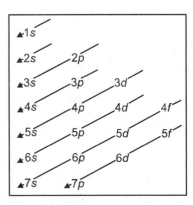

Using the diagonal rule you can quickly determine the electron configuration for the outermost valence electron in sulfur. First locate sulfur on the periodic table and notice that the atomic number of sulfur is 16. That means it has 16 protons and 16 electrons in a neutral atom. The first two electrons go into the 1s sublevel and fill it, the next two go into the 2s sublevel and fill it. That leaves 12 more electrons to place. The next six go into the 2p sublevel, filling it and leaving six more. Two of them go into the 3s sublevel, filling it and the remaining four go into the 3p sublevel. The completed electron configuration looks like this: $1s^2 2s^2 2p^6 3s^2 3p^4$.

6. Complete the electron configuration portion of the table on your student answer sheet.

ORBITAL NOTATION

Orbital notation is a drawing of the electron configuration. It is very useful in determining electron pairing and thus predicting oxidation numbers. The orbital notation for sulfur would be represented as follows:

1 2	3 4	5 8	6 9	7 10		11 12		13 16	14	15
↑↓	↑↓	↑↓	↑↓	↑↓		↑↓		↑↓	↑	↑
1s	2s		2p			3s			3p	

The electrons are numbered as to the filling order. Notice electrons 5, 6, and 7 went into their own orbitals before electrons 8, 9, and 10 forced a pairing to fill the 2p sublevel. This is an application of Hund's rule which minimizes electron-electron repulsions. The same filling order is repeated in the 3p sublevel.

It's time to get the lingo straight!

Electron configurations

Group the 1's, 2's, etc. TOGETHER and it looks like this:

$$1s^2 2s^2 2p^4$$

Which element has this electron configuration?

Orbital notations

Use blanks to represent orbitals and arrows to represent electrons and looks like this:

```
 1 2   3 4    5 8  6   7
 ↑↓    ↑↓     ↑↓  ↑   ↑     The electrons are numbered as to the filling order.
 1s    2s        2p         Notice electrons 5,6,7 went into their own orbitals before
                            electron 8 forced a pairing.  This minimizes repulsion.
```

Which element has this orbital notation?

7. Complete the orbital notation column on your student answer page.

JUSTIFYING OXIDATION STATES

Elements in compounds have oxidation states. These oxidation states determine their behavior in the company of other elements. Your understanding of oxidation states will become very important as you learn to write correct chemical formulas for compounds. Some elements have only one oxidation state, while others have several. In general, the representative elements, those groups or families numbered as 1 - 8A have the oxidation states listed on the periodic table below.

The transition metals generally have several oxidation states possible.

IA	IIA											IIIA	IVA	VA	VIA	VIIA	VIIIA
Li^+														N^{3-}	O^{2-}	F^-	
Na^+	Mg^{2+}											Al^{3+}			S^{2-}	Cl^-	
K^+	Ca^{2+}			Cr^{2+} Cr^{3+}	Mn^{2+} Mn^{3+}	Fe^{2+} Fe^{3+}	Co^{2+} Co^{3+}		Cu^+ Cu^{2+}	Zn^{2+}						Br^-	
Rb^+	Sr^{2+}								Ag^+	Cd^{2+}			Sn^{2+} Sn^{4+}			I^-	
Cs^+	Ba^{2+}										Hg_2^{2+} Hg^{2+}		Pb^{2+} Pb^{4+}				

Learn the following; it will help you make your predictions:
*   Metals (found to the left of the stair-step line) lose electrons to either minimize electron-electron repulsions or eliminate their valence electrons entirely.
*   Nonmetals tend to gain electrons to acquire an octet of electrons. An octet means the atom has eight valence electrons arranged as $ns^2 np^6$ where n corresponds to the main energy level.

- Transition metals generally have an oxidation state of +2 since they lose the s^2 that was filled just before the d-sublevel began filling.
- Electrons in the d-sublevels are very similar in energy to those in the s-sublevel preceding them. This means that 3d electrons are similar in energy to 4s electrons and 4d are similar to 5s, etc.
- Noble gases have an octet naturally, so they generally do not react.

Let's practice.
- Sulfur has many oxidation states. Use an oribital notation to justify its most common -2 oxidation state:

Sulfur has a valence electron configuration of $3s^2 3p^4$. Start by drawing its orbital notation for the outermost, valence electrons.

[Ne] ↑↓ ↑↓ ↑ ↑
    3s   3p

Sulfur is a nonmetal and tends to gain electrons, creating the -2 charge. Gaining two electrons gives it an octet of $3s^2 3p^6$.

- Copper has two common oxidation states, +1 and +2. Justify both oxidation states:

Copper has an ending electron configuration of $4s^2 3d^9$. Start by drawing its orbital notation for the outermost, valence electrons.

[Ar] ↑↓ ↑↓ ↑↓ ↑↓ ↑↓ ↑
    4s   3d

Since copper is a transition metal, the +2 oxidation state comes from losing the 4s electrons leaving $4s^0 3d^9$. Almost all of the transition metals lose the s sublevel and have an oxidation state of +2. [Silver is an exception and only makes an oxidation state of +1.] The +1 oxidation state for copper comes from transferring one of the s electrons to the d orbitals to fill that sublevel and then losing the remaining s electron to form $4s^0 3d^{10}$.

8. Complete the conclusion questions that justify oxidation states on your student answer page.

QUANTUM NUMBERS AND ATOMIC ORBITALS

Principal quantum number (n) **1, 2, 3, 4, 5, etc.**	Determines the total energy of the electron. Describes the energy level of the electron and refers to the average distance of the electron from the nucleus. $2n^2$ electrons may be assigned to an energy level. For n = 1, 2 electrons. For n = 2, 8 electrons, etc.
Angular momentum or azimuthal quantum number (ℓ) **0, 1, 2, 3…**	Refers to the sublevels that occur within each principal level and determines the shape of the orbital. Corresponds to the s, p, d, f [in order of increasing energy]. Each ℓ is a different orbital shape or orbital type. This quantum number has integral values from 0 up to n-1.
Magnetic quantum number (m_ℓ) **…-2, -1, 0, 1, 2, …**	Specifies which orbital within a sublevel you are likely to find the electron. It determines the orientation of the orbital in space relative to the other orbitals in the atom. This quantum number has values from -ℓ through zero to +ℓ.
Spin quantum number (m_s) **+ ½ or -½**	Specifies the value for the spin. Only two possibilities: +½ and -½. No more than two electrons can occupy an orbital. In order for two electrons to occupy an orbital, they must have opposite spins.

Determining Quantum Numbers

Now that we know the electron configuration of the valence electron in sulfur is $3p^4$ based on its position in the periodic table, and we have a picture of how those p electrons are filling the p sublevel, the set of quantum numbers for this valence electron are extremely easy to obtain. First, n = 3 since it is a **3**p electron. Next it is a **p** electron and p sublevels have an ℓ value of 1. So far we know 3,1. To get the m_ℓ quantum number we go back to the orbital notation for the valence electron and focus on the 3p sublevel alone. It looks like this:

$$\underset{-1}{\uparrow\downarrow} \quad \underset{0}{\uparrow} \quad \underset{+1}{\uparrow}$$

Simply number the blanks with a zero assigned to the center blank and increasing negative numbers to the left and increasing positive to the right of the zero. The last electron was number 16 and "landed" in the first blank as a down arrow. This picture gives us the last two quantum numbers of m_ℓ = -1 and m_s = -½ since it is the second electron to be placed in the orbital.

In summary:

Energy Level	1	2	3	4	5	6	7....
# of sublevels	1	2	3	4	5	6	7....
Names of sublevels	s	s, p	s,p,d	s,p,d,f	s,p,d,f,g	s,p,d,f,g,h	s,p,d,f,g,h,i
n, principal quantum number	1	2	3	4	5	6	7....
Name of sublevel	s	p	d	f	g	h	i...
ℓ, angular momentum quantum number [= n-1]	0	1	2	3	4	5	6
# of orbitals [= -ℓ to +ℓ]	1	3	5	7	9	11	13
m_ℓ for each orbital within a sublevel	$\overline{\ 0\ }$	$\overline{-1}\ \overline{\ 0\ }\ \overline{+1}$	$\overline{-2}\ \overline{-1}\ \overline{\ 0\ }\ \overline{+1}\ \overline{+2}$	And so on, just pretend you're in elementary school and make a number line with ZERO in the middle and obviously, negative numbers to the left and positive to the right. Make as many blanks as there are orbitals for a given sublevel.			

For assigning m_s, the first electron placed in an orbital [the up arrow] gets the +½ and the second one [the down arrow] gets the -½.

Try working backwards. Which element has this set of quantum numbers 5, 1, -1, -½? First think about the electron configuration. $n = 5$ and $\ell = 1$, so it must be a 5 p electron. The m_s quantum number corresponds to this orbital notation picture $\underset{-1}{\downarrow}\ \underset{0}{\_}\ \underset{1}{\_}$. Be sure and number the blanks and realize that the -½ means it is a pairing electron, so the orbital had to be half-filled before pairing could occur, thus for the electron to occupy the -1 position, it must be a p^4 electron. The element has a configuration of $5p^4$ so it must be Tellurium.

Quantum numbers are a set of the 4 numbers that describe an electron's position within an atom. They are quite easy to determine if you start with the electron configuration. The set of quantum numbers for the $2p^4$ electron would be 2, 1, -1, +½. Each electron in the atom has a set of quantum numbers, but you will most often be asked for the set describing the valence electron.

Name_____

Period _____

Electron Configurations, Orbital Notation and Quantum Numbers
Understanding Electron Arrangement and Oxidation States

ANALYSIS

Complete this table:

Element	Electron configuration	Valence Orbital notation [only the outermost orbitals are drawn]	Set of Quantum Numbers for the LAST Valence electron to fill
		↑ 4s	
Fe			
	$1s^2 2s^2 2p^3$		
		↑↓ ↑↓ ↑↓ ↑↓ ↑↓ ↑↓ ↑ ↑ __ 5s 4d 5p	
Br			
		↑↓ 6s	
	$[Ar]\ 4s^2 3d^8$		
P			
		↑↓ ↑ ↑ __ __ __ 5s 4d	
U			
	$[Kr]\ 5s^2\ 4d^{10}$		
		↑↓ 3s	
	$[Ar]\ 5s^2\ 5p^6$		
		↑↓ ↑↓ ↑↓ ↑↓ ↑↓ ↑↓ ↑ ↑ ↑ 4s 3d 4p	
W			
	$[Rn]\ 7s^2$		
Pu			
	$1s^2 2s^2 2p^1$		
		↑↓ ↑ ↑ ↑ ↑ ↑ 4s 3d	
I			

CONCLUSION QUESTIONS

1. Iron has two common oxidation states, +2 and +3. Justify each of these oxidation states. Draw the orbital notation of the neutral atom and each oxidation state as part of your justification.

2. Nitrogen has a common oxidation state of -3. Justify this oxidation state. Draw the orbital notation for the neutral atom and oxidation state as part of your justification.

3. Silver has only one oxidation state, +1. Justify this oxidation state even though most transition metals have an oxidation state of +2 among others. Draw the orbital notation for the neutral atom and oxidation state as part of your justification.

4. Manganese has a common oxidation state of +7. Justify this oxidation state. Draw the orbital notation for the neutral atom and oxidation state as part of your justification.

The Diagonal Rule or Aufbau Series

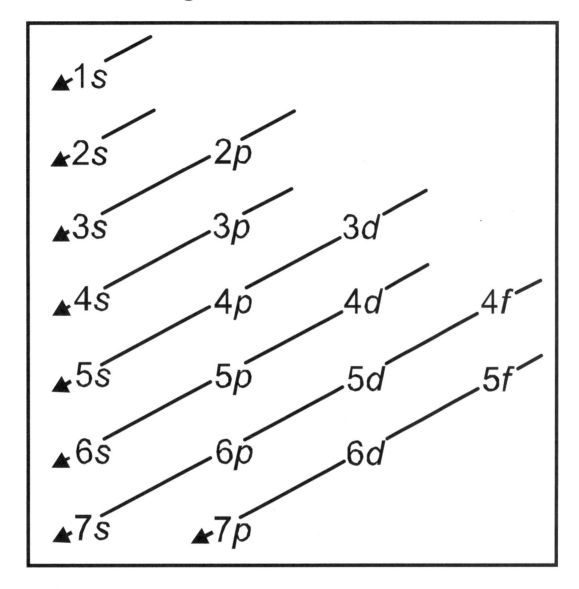

Alkali metals → 1 IA	Alkaline earth metals → 2 IIA	3	4	5	6	7	8	9	10	11	12	13 IIIA	14 IVA	15 VA	16 VIA	Halogens → 17 VIIA	Noble gases → 18 VIIIA
1 H						Transition metals											2 He
3 Li	4 Be											5 B	6 C	7 N	8 O	9 F	10 Ne
11 Na	12 Mg											13 Al	14 Si	15 P	16 S	17 Cl	18 Ar
19 K	20 Ca	21 Sc	22 Ti	23 V	24 Cr	25 Mn	26 Fe	27 Co	28 Ni	29 Cu	30 Zn	31 Ga	32 Ge	33 As	34 Se	35 Br	36 Kr
37 Rb	38 Sr	39 Y	40 Zr	41 Nb	42 Mo	43 Tc	44 Ru	45 Rh	46 Pd	47 Ag	48 Cd	49 In	50 Sn	51 Sb	52 Te	53 I	54 Xe
55 Cs	56 Ba	71 Lu	72 Hf	73 Ta	74 W	75 Re	76 Os	77 Ir	78 Pt	79 Au	80 Hg	81 Tl	82 Pb	83 Bi	84 Po	85 At	86 Rn
87 Fr	88 Ra	103 Lr	104 Rf	105 Db	106 Sg	107 Bh	108 Hs	109 Mt	110 Uun	111 Uuu	112 Uub						

Rare earth metals

*Lanthanides	57 La*	58 Ce	59 Pr	60 Nd	61 Pm	62 Sm	63 Eu	64 Gd	65 Tb	66 Dy	67 Ho	68 Er	69 Tm	70 Yb
†Actinides	89 Ac†	90 Th	91 Pa	92 U	93 Np	94 Pu	95 Am	96 Cm	97 Bk	98 Cf	99 Es	100 Fm	101 Md	102 No

Laser Light
Determining the Wavelength of Light

OBJECTIVE
Students will use a diffraction grating to determine the wavelength of laser light.

LEVEL
Chemistry

NATIONAL STANDARDS
UCP.1, UCP.2, UCP.3, B.2, E.1, E.2

TEKS
2(B), 2(C), 2(D), 2(E)

CONNECTIONS TO AP
AP Chemistry:
    II. States of Matter B. Liquids and solids 1. Liquids and solids from the molecular viewpoint
    4. Structure of solids

TIME FRAME
45 minutes

MATERIALS
(For a class of 28 working in groups of 3 or 4)

The limiting factor in this lab is the laser. Laser pointers will not work. A more expensive laser is necessary. Physics departments often have lasers you may borrow. The data only takes a few minutes to collect, so groups may rotate through the apparatus collecting their data. The material below is for one laser set up.

2 meter sticks	screen support (Frey Scientific 15599103
laser	$1.35 each)
1 3" x 12" white card stock (poster board or	diffraction gratings
Bristol board)	masking tape
	ruler

TEACHER NOTES
This is one of a series of small activities that can be used to enrich a unit on atomic structure. Depending upon the number of lasers available, the lab data can be collected in a matter of minutes. It could also be done as a demonstration, with various students measuring the distances and reading them out to the class.

The diffraction grating may be purchased from a scientific supply company such as Edmond Scientific (Item #D30013-07 was $7.95 for a pack of 15 at the time of this printing). The diffraction gratings may be used in other labs, such as examining emission spectra.

Below is a diagram of what occurs when light passes through a diffraction grating. You may use it to help students understand what is happening as light passes through the diffraction grating. From this diagram you can see that **nλ= d sin θ.** This is called the **Bragg equation**. It has many uses and it is used in Advanced Placement Chemistry in the study of x-ray crystallography and the solid state. In this equation d is the distance between the grooves on the grating and n is the order of the image. Students will be measuring the first order image, so n = 1.

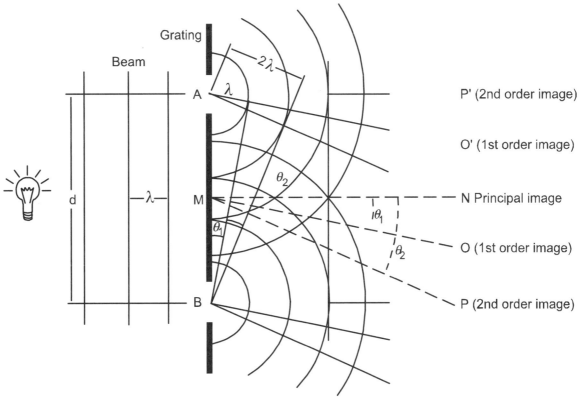

(Reference, *Modern Physics*, Williams, Trinklein, and Metcalfe, Holt, New York, 1980)

POSSIBLE ANSWERS TO THE CONCLUSION QUESTIONS AND SAMPLE DATA
(Using a He/Ne laser with a wavelength of 633 nm.)

Trial	X value (cm) Measured	Y value (cm) Measured	Z value (cm) Calculated	Sin θ x/z	Wavelength (nm) Calculated
1	6.30	17.4	18.5	0.34	641
2	5.80	17.8	18.7	0.31	583
3	3.30	8.83	9.42	0.35	659
4	4.30	11.9	12.6	0.34	640
5	4.95	12.8	13.8	0.36	677

CALCULATIONS AND ANALYSIS

1. Calculate the value for z to the correct number of significant figures by using the Pythagorean formula.
   - $a^2 + b^2 = c^2$

2. Calculate d, the distance between two grooves (in nm) on the diffraction grating. Show your calculations.
   - $\dfrac{2.54\,\text{cm}}{\text{in}} \times \dfrac{\text{in}}{13{,}500\ \text{lines}} \times \dfrac{1 \times 10^7\ \text{nm}}{\text{cm}} = 1.88 \times 10^3\ \dfrac{\text{nm}}{\text{line}}$

3. **Sin θ** is **x** (cm)/**z** (cm).
   - See table for sample data.

4. Calculate the wavelength (λ) for each of the 5 trials. Remember, $\lambda = \mathbf{d\ \sin\theta}$.
   - See table for sample data and answers.

5. Once you have calculated the wavelengths, omit the largest and smallest and average the remaining

   - Answers will vary. The average wavelength for the sample data is 647 nm

6. Your teacher will tell you the actual wavelength of your laser. What is your % error (relative error)?
   - Most He/Ne lasers have wavelengths in the neighborhood of 633 nm. You will need to provide the wavelength of whatever laser you are using.

   - $\dfrac{|647 - 633|}{633} \times 100\% = 2.21\%$

7. How many significant digits should be reported in the answer for the wavelength?
   - Depending upon the measurements, either 3 or 4. If any of the distances fall below 10.00 cm, the number of significant digits will be reduced to 3.

8. What are the possible sources of error(s) which could account for any differences between the experimental and accepted values for the wavelength?
   - Answers will vary.
   - Not having the screen perpendicular to the laser beam is the most serious.
   - Not measuring properly.
   - Not being able to determine the center of the dot.

Laser Light
Determining the Wavelength of Light

Laser stands for Light Amplification by Stimulated Emission of Radiation. It is a process by which all of the photons that are emitted from the excited atoms in the laser are stimulated to emit in phase, with the same frequency, and in a directional column. There are four important characteristics of laser light:

1. **Small divergence**: The beam from a laser does not spread out (diverge) much from a laser source. Thus, instead of being dissipated rapidly, the energy is concentrated in a narrow beam.

2. **Monochromatic**: The laser light is said to be monochromatic because it is mostly of one color, or one wavelength. This characteristic is important for this lab, as we will be determining the wavelength of the light emitted from the laser. Your teacher will tell you the actual wavelength of the laser light so that you may calculate your % error.

3. **Coherent**: Ordinary light is incoherent with crests and troughs being emitted at random from different parts of the light source. Laser light, however, is coherent with almost all crests and troughs in phase regardless of where they were generated in the laser tube. That is a function of a stimulated emission.

4. **High Intensity**: Laser light is very intense because all of its energy is concentrated. Although light from a powerful ruby laser or carbon dioxide laser can be made to burn through concrete or steel, the light from the typical classroom HeNe laser is relatively safe and even if focused on the hands, cannot be felt. However, it is powerful enough to harm the retina of the eye if you look directly into the laser or if you look at the reflection of the beam from a shiny object. This should never be done! (Information taken from Metrologic Laser Workbook & Catalog.)

The spreading of light into a region behind an obstruction is called **diffraction**. A slit opening, a fine wire, a sharp edged object, or a pinhole can serve as a suitable obstruction in the path of a beam of light from a point source. You can observe this phenomenon anywhere. Bring two fingers together close to one eye. As the fingers get closer together, you will observe dark fringes in the light between the two fingers. This is most noticeable right before the fingers block out all of the light.

Shining a light through a plastic surface with many closely spaced parallel lines can produce very useful diffraction patterns. This is known as a **transmission diffraction grating**. A transmission grating placed in the path of plane waves disturbs the wave front because the ruled lines are opaque to light and the narrow spacing between the lines are transparent. These spaces provide a large number of fine, closely spaced transmission slits. The grating that you are using has **13,500 lines/in**. There are 2.54 cm/in. (Reference, *Modern Physics*, Williams, Trinklein, and Metcalfe, Holt, New York, 1980)

The equation you will be using is the Bragg equation. The equation is $n\lambda = d \sin \theta$. It has many uses and it is used in Advanced Placement Chemistry in the study of x-ray crystallography and the solid state. In this equation d is the distance between the grooves on the grating and n is the order of the image. You will be measuring the first order image, so $n = 1$.

PURPOSE

In this activity you will measure the first order diffraction image and use the Bragg equation to determine the wavelength of the laser light.

MATERIALS

2 meter sticks
laser
1 3" x 12" white card stock (poster board or
   Bristol board)

diffraction grating (Edmond Scientific)
masking tape
ruler

Safety Alert
1. Laser light is very intense and will damage eyes. You must NOT look into the laser beam or allow the laser light to reflect off shiny surfaces.

PROCEDURE

Assemble the apparatus as shown in the diagram below. Use tape to secure the white cardboard to the meter stick for support. Be sure the laser is lined up parallel with the meter stick holding the diffraction grating and perpendicular to the meter stick holding the screen.

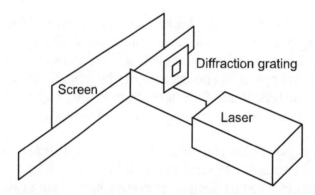

1. Adjust the diffraction grating and the screen so that you can see both the principal image and the first order diffraction spot. The principle image is the bright spot in the center. The first order image is the dimmer spot next to it. Measure the distance between the principal image and the first order image. Record your data in cm. This distance is x. Measure the distance from the screen to the diffraction grating. This value is y.

2. Move the diffraction grating to another position on the meter stick and take the same two measurements. Repeat until you have at least 5 sets of data.

3. Turn off the laser.

Name _____

Period _____

Laser Light
Determining the Wavelength of Light

DATA AND OBSERVATIONS

Trial	X value (cm) Measured	Y value (cm) Measured	Z value (cm) Calculated	Wavelength (nm) Calculated
1				
2				
3				
4				
5				

CALCULATIONS AND ANALYSIS

1. Calculate the value for z to the correct number of significant figures by using the Pythagorean formula.

2. Calculate d, the distance between two grooves (in nm) on the diffraction grating. Show your calculations.

3. $Sin\ \theta$ is x (cm)/z (cm).

4. Calculate the wavelength (λ) for each of the 5 trials. Remember, $\lambda = d \sin \theta$

5. Once you have calculated the wavelengths, omit the largest and smallest and average the remaining

6. Your teacher will tell you the actual wavelength of your laser. What is your % error (relative error)?

7. How many significant digits should be reported in the answer for the wavelength?

8. What are the possible sources of error(s) which could account for any differences between the experimental and accepted values for the wavelength?

Write Your Notes and Ideas Here!

Why Do They Call It a Periodic Table?
Investigating and Graphing Periodic Trends

OBJECTIVE
Students will graph various periodic properties to examine the trends of the elements as they are arranged on the periodic table.

LEVEL
Chemistry

NATIONAL STANDARDS
UCP.1, UCP.2, A.2, B.2, (possibly E.1, E.2 if using technology instead of paper), G.2

TEKS
4(D), 6(C)

CONNECTIONS TO AP
AP Chemistry:

    I. Structure of Matter 5. Periodic relationships including for example, atomic radii, ionization energies, electron affinities, oxidation states

    IV. Descriptive Chemistry 2. Relationships in the periodic table: horizontal, vertical, and diagonal with examples from alkali metals, alkaline earth metals, halogens, and the first series of transition elements

TIME FRAME
45 minutes

MATERIALS
(For a class of 28 students working individually (at home) and in pairs for the graphing activities.)

28 student sets of periodic trend cards	14 scissors
tape	extra paper
graph paper*	classroom computers if the graphing is to be done on the computer

TEACHER NOTES
The day before the activity, distribute the page containing the elements with the atomic masses and the physical/chemical changes. Instruct the students to cut the cards apart and, using the properties, arrange the cards in some order. They should tape the cards to a piece of paper to secure the order for the in-class discussion.

For the graphing activities, students should work in pairs and turn in one lab report for each student group.

*An alternate method for doing this activity is to use computers and computer graphing software such as Graphical Analysis™, LoggerPro™, or Excel. This reduces the time needed for graphing and allows more time for discussion. Computer availability and software may dictate which method you will use.

Students need a thorough understanding of atomic structure and periodic trends to be successful on the AP Chemistry examination and indeed, to be successful in the field of chemistry in general. This exercise allows you to explain the reasons behind the trends.

When explaining the reasons behind the trends, it is helpful to start with atomic radii or size of the atoms. This is the easiest for students to understand and, once students grasp this concept, other concepts such as ionization energy, electron affinity, and electronegativity are easily understood.

Size: As one moves down a family, additional energy levels are added and the atoms get larger. As more levels are added, there are more electrons between the nucleus and the outer energy level. These electrons "shield" the nucleus from holding tightly to the outer electrons. As one moves across a period from left to right, the electrons are in the same energy level, but each successive atom has one more proton in the nucleus. The additional proton increases the nuclear charge or it increases the Z_{eff} (Z effective) without increasing any inner shielding. This causes the electrons to be drawn closer to the nucleus.

Ionization Energy: Defined as the amount of energy needed to remove an electron from an atom forming a +1 ion. Ionization energy can be expressed in equation form: $M \rightarrow M^+ + e^-$. The reasons for the general trends in IE are the same reasons as the trend for size. As you move down a family, electrons are further from the nucleus and have more inner electrons shielding them from the nuclear pull, therefore the amount of energy required to remove an electron decreases. As you move across a period from left to right, the Z_{eff} for that period gets greater, the electrons are held more tightly and therefore electrons are more difficult to remove. IE gets larger.

As you look at the graph of ionization energies across a period, there are clearly some anomalies. Looking at the second period, the first observed irregularity is the lower ionization energy from Be to B. The simplest explanation is that the first electron to be removed from boron is coming from a 2p orbital which is higher energy than the electron from beryllium which is in a 2s and is of lower energy. There are certainly more sophisticated arguments which may be dealt with in the AP course. **Under no circumstance should a student be given the impression that the reason the 2s electron is harder to remove is that it is being taken from a full subshell that has an intrinsic stability!** The second anomaly occurs when one moves from nitrogen to oxygen. The increasing nuclear charge should dictate that the first IE of oxygen should be higher than that of nitrogen. However, oxygen has a lower first ionization energy than nitrogen due to the spin-spin repulsion of two electrons in the same orbital.

Again, there is no increased stability due to a half-filled subshell. **Even though this is written in many textbooks, it is not true and students should not be taught that there is some magic stability to half and totally filled subshells!**

Electron affinity: Defined as the energy change involved in forcing a gaseous atom to accept an electron to form a -1 ion. Electron affinity is represented by $X + e^- \rightarrow X^-$. Again, the trends are dictated by the same factors—distance from the nucleus, inner electron screening, and increasing nuclear charge (Z_{eff} in the same period). Therefore, electron affinity decreases as you move down a family and increases as you move from left to right across a period.

Electronegativity (Pauling's): Electronegativity actually combines several factors and gives a relative number expressing attraction for electrons within a chemical bond. It is relative and, while it is very useful, there are many factors which affect this attraction for electrons within compounds. You may choose to introduce it at this point, because the factors which influence the trends are the same as the others that we have been discussing. Or you may choose to introduce it when you talk about bonding.

Metallic trends: Metals are really defined by loosely held, mobile electrons. The elements with the most loosely held electrons, of course, are in the lower left portion of the periodic table (largest and furthest from the nucleus with the lowest Z_{eff}) and the trend gets less as you move diagonally from lower left to upper right. After students understand the other periodic trends, and what really defines a metal, it will be easier for them to see why the zigzag line divides the metals from the non-metals.

In teaching students to discuss various periodic trends, it is very important that they not answer a "why" question with a "what" answer. For example, if they are asked why the first ionization energy for magnesium is greater than the first ionization energy for sodium, they must NOT say that the trend is for ionization energies to increase from left to right across the periodic table. Instead they must address the fact that the electrons are in the same energy level, but the number of protons in magnesium is greater than the number of protons in sodium and therefore the Z_{eff} is greater. They must also clearly address both chemical species (i.e. magnesium and sodium) in their answer.

POSSIBLE ANSWERS TO THE CONCLUSION QUESTIONS AND SAMPLE DATA

After the students have made the arrangement of their cards and discussed their ideas with their partner, you will need to tell them the actual name and atomic number represented by each of the cards. The chart below will help you do so without having to actually match each element card with the periodic table.

Atomic Mass 1 Hydrogen Atomic number: 1	Atomic Mass 19 Fluorine Atomic number: 9	Atomic Mass 36 Chlorine Atomic number: 17	Atomic Mass 9 Beryllium Atomic number: 4	Atomic Mass 20 Neon Atomic number 10
Atomic Mass 27 Aluminum Atomic number: 13	Atomic Mass 16 Oxygen Atomic number: 8	Atomic Mass 40 Argon Atomic number: 20	Atomic Mass 7 Lithium Atomic number: 3	Atomic Mass 32 Sulfur Atomic number 16
Atomic Mass 23 Sodium Atomic number: 11	Atomic Mass 39 Potassium Atomic number: 19	Atomic Mass 12 Carbon Atomic number: 6	Atomic Mass 4 Helium Atomic number: 2	Atomic Mass 40 Calcium Atomic number: 20
Atomic Mass 31 Phosphorus Atomic number: 15	Atomic Mass 14 Nitrogen Atomic number: 7	Atomic Mass 11 Boron Atomic number: 5	Atomic Mass 28 Silicon Atomic number: 14	Atomic Mass 24 Magnesium Atomic number: 12

T E A C H E R   P A G E S

7

1. Compare the following graphs. Do any of the graphs show a repeating, or cyclic, pattern? Focus on elements with very large or very small values.
   a. Oxygen atoms vs. atomic number and Chlorine atoms vs. atomic number
      • These graphs show a rise and fall within the period, but definitely a repeating pattern.

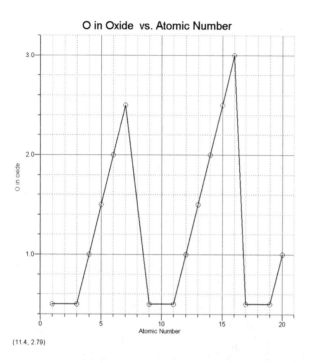

(11.4, 2.79)

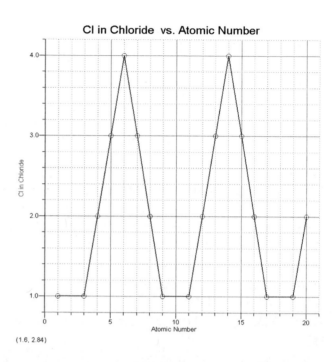

(1.6, 2.84)

   b. Melting point vs. atomic number and Boiling point vs. atomic number
      • Again, these graph show a rise and fall within the period with the highest melting points and boiling points being toward the middle of the period.

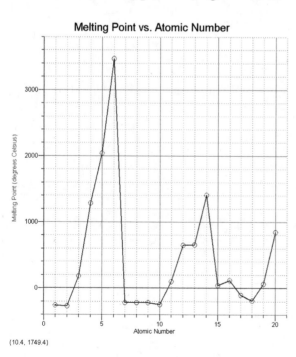

(10.4, 1749.4)

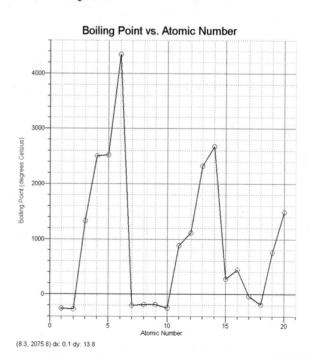

(8.3, 2075.8) dx: 0.1 dy: 13.8

c. Electronegativity vs. atomic number and Ionization energy vs. atomic number
  • These graphs show an increase across the period and falling as a new period starts.

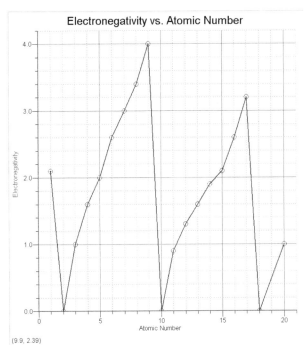

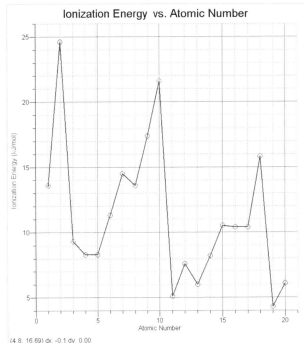

2. Are these graphs consistent with patterns found in your earlier grouping of elements? Explain.
  • Student answers will vary based on their original arrangements. Focus on the logic of the answers and discourage any assignment of "right" or "wrong" patterns at this time. After all, there are still many versions of periodic tables that are being used. You might even encourage students to find other models of the periodic table to share with the class.

3. Based on your graphs, why is the chemist's organization of elements called a periodic table?
  • Because the trends repeat themselves in an organized repeating pattern, or "periodically".

4. Where are the elements with the highest oxide numbers located on the periodic table? How about the elements with the highest chloride numbers?
  • Highest oxide and chloride numbers are located toward the middle of the periods.

5. Predict which element should have the lowest boiling point: selenium (Se), bromine (Br), or krypton (Kr)? Explain your prediction.
  • Krypton will have the lowest boiling point. It is a noble gas and is located at the end of the period. The boiling point trend shows a decrease in boiling point toward the end of a period.

6. Which should have the lowest ionization energy: rubidium (Rb) or cesium (Cs)? Explain your prediction.

   - Cesium will have a lower ionization energy because its outer electrons are further from the nucleus. Rb has only 6 energy levels where Cs has 7 energy levels. The extra layer of electrons will cause the nucleus of Cs to be more shielded than in Rb, therefore, the outer electrons will not be held as tightly, resulting in a lower ionization energy.

7. Which would have the greatest ionization energy: antimony (Sb) or tellurium (Te)? Explain your prediction.

   - This is tricky. Both elements are on the 5th period. Tellurium has a higher Zeff but has its last electron going into a p orbital that already has an electron in it. Spin-spin repulsion will cause that electron to be more easily removed and therefore antimony will have the higher ionization energy.

Why Do They Call It a Periodic Table?
Investigating and Graphing Periodic Trends

Have you ever wondered why chemists call that big chart in the front of the room that contains all of the elements, the *periodic* table? For years scientists attempted to find some order to the elements that they knew existed. They looked at various properties and tried out many arrangements. You are going to attempt to do a similar task. You will be given a set of cards that contain certain properties. Without referring to the actual periodic table you will attempt to put the cards in some order that makes sense to you. You will be asked to explain your reasoning to your classmates.

PURPOSE
In this activity your will graph various periodic properties to ascertain the order. In so doing, you will see the trends of the periodic table of the elements.

MATERIALS
    sets of periodic trend cards   scissors
    tape   extra paper
    graph paper or computer as teacher instructs

AT HOME PREPARATION
Cut along the lines of the periodic trends cards. Try to arrange them in an order. Try several different patterns to see which makes the most sense to you. When you have an arrangement that you like, tape them in order to a piece of paper. Justify your arrangement on the student answer page.

Periodic Trend Cards

Atomic Mass 1	Atomic Mass 19	Atomic Mass 36	Atomic Mass 9	Atomic Mass 20
Melting Point (°C) -259	Melting Point (°C) -218	Melting Point (°C) -101	Melting Point (°C) 1287	Melting Point (°C) -248
Boiling point (°C) -252	Boiling point (°C) -188	Boiling point (°C) -34	Boiling point (°C) 2507	Boiling point (°C) -246
O in oxide 0.5	O in oxide 0.5	O in oxide 0.5	O in oxide 1	O in oxide --
Cl in chloride 1	Cl in chloride 1	Cl in chloride 1	Cl in chloride 2	Cl in chloride --
Electronegativity 2.1	Electronegativity 4.0	Electronegativity 3.2	Electronegativity 1.6	Electronegativity --
Ionization Energy 13.6	Ionization Energy 17.4	Ionization Energy 10.4	Ionization Energy 8.3	Ionization Energy 21.6

Atomic Mass 27	Atomic Mass 16	Atomic Mass 40	Atomic Mass 7	Atomic Mass 32
Melting Point (°C) 659	Melting Point (°C) -219	Melting Point (°C) -189	Melting Point (°C) 179	Melting Point (°C) 119
Boiling point (°C) 2327	Boiling point (°C) -183	Melting Point (°C)) 186	Boiling point (°C) 1327	Boiling point (°C) 445
O in oxide 1.5	O in oxide --	O in oxide --	O in oxide 0.5	O in oxide 3
Cl in chloride 3	Cl in chloride 2	Cl in chloride --	Cl in chloride 1	Cl in chloride 2
Electronegativity 1.6	Electronegativity 3.4	Electronegativity --	Electronegativity 1.0	Electronegativity 2.6
Ionization Energy 6.0	Ionization Energy 13.6	Ionization Energy 15.8	Ionization Energy 9.3	Ionization Energy 10.4

Atomic Mass 23	Atomic Mass 39	Atomic Mass 12	Atomic Mass 4	Atomic Mass 40
Melting Point (°C) 97	Melting Point (°C) 64	Melting Point (°C) 3470	Melting Point (°C) -272	Melting Point (°C) 851
Boiling point (°C) 889	Boiling point (°C) 757	Boiling point (°C) 4347	Boiling point (°C) -269	Boiling point (°C) 1487
O in oxide 0.5	O in oxide 0.5	O in oxide 2	O in oxide --	O in oxide 1
Cl in chloride 1	Cl in chloride 1	Cl in chloride 4	Cl in chloride --	Cl in chloride 2
Electronegativity 0.9	Electronegativity 0.8	Electronegativity 2.6	Electronegativity --	Electronegativity 1.0
Ionization Energy 5.1	Ionization Energy 4.3	Ionization Energy 11.3	Ionization Energy 24.6	Ionization Energy 6.1

Atomic Mass 31	Atomic Mass 14	Atomic Mass 11	Atomic Mass 28	Atomic Mass 24
Melting Point (°C) 44	Melting Point (°C) -210	Melting Point (°C) 2037	Melting Point (°C) 1407	Melting Point (°C) 650
Boiling point (°C) 280	Boiling point (°C) -196	Boiling point (°C) 2527	Boiling point (°C) 2677	Boiling point (°C) 1117
O in oxide 2.5	O in oxide 2.5	O in oxide 1.5	O in oxide 2	O in oxide 1
Cl in chloride 3	Cl in chloride 3	Cl in chloride 3	Cl in chloride 4	Cl in chloride 2
Electronegativity 2.1	Electronegativity 3.0	Electronegativity 2.0	Electronegativity 1.9	Electronegativity 1.3
Ionization Energy 10.5	Ionization Energy 14.5	Ionization Energy 8.3	Ionization Energy 8.2	Ionization Energy 7.6

PROCEDURE

1. Compare your arrangement with your classmates' results. Are there differences? Are there similarities? If there are differences, try to resolve them.

2. Now your teacher will give you the atomic number and the symbol for each of your cards. Work with your partner to arrange the cards in order of atomic number.

3. Use two sheets of graph paper. Label each "Trends in Chemical Properties". Label the x-axis "Atomic Numbers" and number it from 1 to 20. One of you should label your y-axis "Oxygen atoms per atom of element". The other should have the y-axis labeled "Chlorine atoms per atom of element". Determine a proper scale for each. Construct a bar graph for each.

4. Get two more sheets of graph paper. Label one "Boiling Points vs. Atomic Number". Label the other "Melting Points vs. Atomic Number". Number the x-axes with atomic numbers from 1 to 20 as before. Determine an appropriate scale for the temperatures along the y-axes. Do not graph element #6. Construct bar graphs for each.

5. Construct two more graphs. Label one "Ionization Energy vs. Atomic Number". The units for ionization energy are kJ/mol. Be sure to include the units on your graph. The other should be labeled "Electronegativity vs. Atomic Number". Electronegativity is a relative number and has no units. As you did before, label the x-axis "Atomic Number" and scale from 1 to 20. Use a line graph to plot these properties. Look at all six graphs, and answer the conclusion questions.

Name _____

Period _____

Why Do They Call It a Periodic Table?
Investigating and Graphing Periodic Trends

JUSTIFICATION OF ARRANGEMENT

Explain why you arranged the cards in the particular order that you chose.

DATA AND OBSERVATIONS

Include your graphs with your report.

ANALYSIS AND CONCLUSION QUESTIONS

1. Compare the following graphs. Do any of the graphs show a repeating, or cyclic, pattern? Focus on elements with very large or very small values. If the value is not given, skip that number or enter "–" on the computer.

2. Oxygen atoms vs. atomic number and Chlorine atoms vs. atomic number

3. Melting point vs. atomic number and Boiling point vs. atomic number

4. Electronegativity vs. atomic number and Ionization energy vs. atomic number

5. Are these graphs consistent with patterns found in your earlier grouping of elements? Explain.

6. Based on your graphs, why is the chemist's organization of elements called a Periodic Table?

7. Where are the elements with the highest oxide numbers located on the Periodic Table? How about the elements with the highest chloride numbers?

8. Predict which element should have the lowest boiling point: selenium (Se), bromine (Br), or krypton (Kr)? Explain your prediction.

9. Which should have the lowest ionization energy: rubidium (Rb) or cesium (Cs)? Explain your prediction.

10. Which would have the greatest ionization energy: antimony (Sb) or tellurium (Te)? Explain your prediction.

Isotopic Pennies
Finding the Percent Abundance of Isotopes

OBJECTIVE
Students will determine the percent abundance of a sample of pennies and then practice calculations involving abundance, isotopes, and atomic mass.

LEVEL
Chemistry

NATIONAL STANDARDS
UCP.2, UCP.3, B.1, B.2, G.2

TEKS
1(B), 6(A)

CONNECTIONS TO AP
AP Chemistry:
> I. Structure of Matter A. Atomic Theory and Atomic Structure 2. Atomic masses; determined by chemical and physical means 3. Atomic number and mass number; isotopes

TIME FRAME
20 minutes

MATERIALS
(For a class of 28 students working in pairs)

160 pennies approximately ½ pre-1982	balances
14 empty film canisters	28 paper labels

TEACHER NOTES
This is one of a series of exercises that can be used to reinforce concepts of atomic structure. These activities take between 10 to 20 minutes and are meant to be a portion of a lesson on atomic structure rather than a stand-alone lesson.

Set up: Place two labels on each film canister. Mass the canister and record the mass on one label. Put a sample number on the other label. Place 10 pennies in each canister, recording the number of pre-1982 pennies and the number of post-1982 pennies for your answer key. Pennies produced before 1982 were made of solid copper and have a mass of approximately 3.1 g/penny. Pennies produced after 1982 are a copper clad zinc and have a mass of approximately 2.5 g/penny.

Here is an example of the set up.

Identification #	Pre-1892	Post-1982
1	3	7
2	4	6
3	6	4
4	7	3
5	4	6
6	2	8
7	3	7
8	8	2
9	5	5
10	5	5
11	8	2
12	4	6
13	5	5
14	2	8

POSSIBLE ANSWERS TO THE CONCLUSION QUESTIONS AND SAMPLE DATA

Sample number	#6
Mass of container	6.84 g
Mass of container and pennies	33.04 g
Mass of pennies in container	26.20 g

ANALYSIS

Let x = Pre-1982 pennies; let y = Post-1982 pennies

$x + y = 10$

$26.20 = 3.1x + 2.5y$

$26.20 = 3.1x + 2.5(10 - x)$

$x = 2; y = 8$

CONCLUSION QUESTIONS

1. Number of Pre-1982 pennies _____ Number of Post-1982 pennies _____
   - Answers will vary. Compare to your answer key.

2. There are three naturally occurring isotopes of magnesium: Mg-24 (23.98 amu) which is found in 78.70% abundance; Mg-25 (24.99 amu) which is 10.13% abundant; and Mg-26 (25.98 amu) which is 11.17% abundant. Calculate the atomic mass of magnesium.
   - $(23.98)(0.7870) + (24.99)(0.1013) + (25.98)(0.1117) = 24.31 \text{ g/mol}$

3. The atomic mass of boron is 10.81 g/mol. The two naturally occurring isotopes are B-10 with a mass of 10.0129 amu and B-11 with a mass of 11.0093 amu. Determine the percent of naturally occurring B-10.
   $$10.81 = (10.0129)(x) + (11.0093)(1-x)$$
   - $x = 0.2000$ or 20.00% $^{10}_{5}B$

     $1-x = 0.8000$ or 80.00% $^{11}_{5}B$

4. In the ANALYSIS section, it says that "pennies are quantized." What is meant by that statement?
   - The term, "quantized" refers to things that are discrete packets. A penny is a discrete packet and cannot be divided.

Isotopic Pennies
Finding the Percent Abundance of Isotopes

You will be given a film canister containing a mixture of 10 pre-1982 and post-1982 pennies. These pennies represent the two isotopes of the element "Pennium". The canister is labeled with a number and has the mass of the container itself written on the label. Your canister might hold any combination of the two different isotopes. The average mass of the pre-1982 isotope is 3.1 g; the average mass of the post-1982 isotope is 2.5 g.

Basic Vocabulary:

Atomic number: the number of protons in the nucleus of an atom.

Mass number: the total number of protons and neutrons in the nucleus of an atom.

Atomic mass: sometimes referred to as atomic weight. The weighted average of naturally occurring isotopes of an element.

Isotope: atoms of the same elements that have the same atomic number but different atomic masses due to a different number of neutrons.

PURPOSE
In this activity you will determine the number of each type of penny in the canister. Opening the canister will result in an automatic 0 (zero) for this assignment.

MATERIALS
film canister containing 10 pennies   balance
                                             calculator

PROCEDURE
The mass of the film canister is recorded on the label. Determine the mass of the ten (10) pennies contained in the canister. Record your results.

Isotopic Pennies
Finding the Percent Abundance of Isotopes

DATA AND OBSERVATIONS

Sample number	
Mass of container	
Mass of container and pennies	
Mass of pennies in container	

ANALYSIS

Calculate the number of each type of "isotope" in your sample. Round your results to the nearest whole number. (Pennies are quantized.) Show all calculations.

CONCLUSION QUESTIONS

1. Number of Pre-1982 pennies _____ Number of Post-1982 pennies _____

2. There are three naturally occurring isotopes of magnesium: Mg-24 (23.98 amu) which is found in 78.70% abundance; Mg-25 (24.99 amu) which is 10.13% abundant; and Mg-26 (25.98 amu) which is 11.17% abundant. Calculate the atomic mass of magnesium.

3. The atomic mass of boron is 10.81 g/mol. The two naturally occurring isotopes are B-10 with a mass of 10.0129 amu and B-11 with a mass of 11.0093 amu. Determine the percent of naturally occurring B-10.

4. In the ANALYSIS section, it says that "pennies are quantized." What is meant by that statement?

Red Hot Half-Life
Modeling Nuclear Decay

OBJECTIVE
Students will model a system of nuclear decay using red hots. They will analyze the data using a graphing calculator.

LEVEL
Chemistry

NATIONAL STANDARDS
UCP.2, B.1, B.2, E.1, E.2

TEKS
Chemistry 6(B)
Physics 2(E)

CONNECTIONS TO AP
AP Chemistry:
    I. Structure of Matter A. Atomic Theory and Atomic Structure 4. Electron energy levels; atomic spectra, quantum numbers, atomic orbitals
AP Physics:
    graphing and transformation of equations

TIME FRAME
45 minutes

MATERIALS
(For a class of 28 working in pairs)

    6 packages of red hots (1 lb each)   14 paper plates
    14 paper cups   graphing calculators

TEACHER NOTES
Before class, use a marker to divide each paper plate into equal sections. Some plates should be divided in thirds, some in fourths, some in sixths, some in eighths. Put a star in one of the sections. Each group should have a sample of red hots containing approximately 150-180 red hots. Put the candy into the paper cups for distribution. Depending upon your own lab procedure, you may allow students to eat the candy after the lab is complete.

The students count the number of red hots in the original sample and record. They then pour the red hots onto the paper plate and remove the candy that falls into the starred section. The students count the remaining red hots and record. Each trial is considered to be 10 seconds. The procedure is continued for ten trials or until the number of red hots is reduced to less than 5, whichever comes first.

Students will analyze the data collected with a graphing calculator. You may refer to *Foundation Lesson VI: Use of the Graphing Calculator* for help. While this lesson has been designed as a calculator lesson, it may also be done as a paper/pencil graphing exercise with simple modifications.

POSSIBLE ANSWERS TO THE CONCLUSION QUESTIONS AND SAMPLE DATA

Starting with 180 red hots and a plate divided into 6 sections.

Time (s)	Number of red hots that decayed	Number of red hots remaining
0.0	0	180
10	30	150
20	25	125
30	22	103
40	17	86
50	15	71
60	12	59
70	10	49
80	8	41
90	7	34
100	5	29

1. Enter the "time" in L_1 and the red hots remaining in L_2. Be sure that your Stat Plot is set up correctly to display a scatterplot of L_1 and L_2 data. Describe the shape of the graph.
   - The graph is curved downward.

2. Look at graphs from at least two other groups. What is the same about the graphs? What is different about the graphs?
   - The graphs are all curved downward, but some of them are more sharply curved than others.

3. Press [STAT] [▶]to Calc. Arrow down to **0: ExpReg**. [ENTER] [VARS] [▶] [ENTER] [ENTER] [ENTER]. Look at the graph by going to [ZOOM] [9]. Write the equation for the line below. Since this is not a linear equation, it will not be in the form of y= mx + b.
   - Equations will vary. This an exponential decay function, rather than a linear equation, so you may have to provide a considerable amount of assistance to the students at this point. Point students to the form of the equation that is at the top of screen. It will look like $y = a*b^x$. You might want to write the example on the board. The equation for the sample data is $y=(179)(0.98)^x$

4. Half-life is defined as the amount of time it takes for ½ of a sample to decay. Open [Y=] and arrow down to Y_2. Enter the original number of red hots [÷] [2]. This corresponds to the time it takes for half of the original sample to decay. Go back to [ZOOM] [9]. Go to [2nd] [TRACE] [▼] [▼] [▼] [▼] [ENTER] [ENTER] [ENTER] [ENTER] to get the intersection of these two lines. It will be shown at the bottom of your calculator screen. What is the half-life of this system?
   - Answers will vary. Answers will vary depending upon the number of sections on the plate. The half-life for the sample data (plate divided into 6 sections) is 38 seconds. Since half-life is

defined as the time it takes for ½ of a sample to decay, when the number falls to ½ the original number, the x-coordinate will be the half-life.
- The half-life for the sample data is 38 seconds.

5. Take the number of red hots that you had at the half-life point. Divide it by 2. Open $\boxed{Y=}$ and arrow down to Y_2. Clear the number and enter the new number. Repeat the procedure that you did in question 4. Is the new time close to two times the half-life (2 half-lives)?
- When ¼ of the red hots are left, it should be approximately two half-lives. Remember this is real data, so it may not be exact.

6. Refer back to question 3. The equation is not linear but instead is in the form of $y=(a)(b)^x$. This is an exponential function. In science we like to have equations in the form of $y=mx + b$. Manipulating the equation, we can do this. It is called a linear transform. In this case, we would take the natural log (ln) of both sides of the equation. Do this with the equation that you have found in question 3. Your teacher will assist you if necessary. Write the equation below.
- To do a linear transform on $y=(a)(b)^x$, you will need to take the ln of both sides of the equation. You will end up with [ln y = x + ln(ab)].

7. What variables will be used from the equation above to obtain a linear graph?
- To get a linear graph, graph ln (y) vs. x.

8. Go back to the statistical function, $\boxed{STAT}$, Edit. You want to transform the data in L_2. Arrow up so the cursor is sitting on the L_2 icon. Enter $\boxed{LN}$ $\boxed{2nd}$ $\boxed{2}$ $\boxed{ENTER}$. Go to $\boxed{ZOOM}$ $\boxed{9}$ to look at the graph. Is it linear?
- The graph should be linear.

9. Go to $\boxed{Y=}$ and clear it. Go to $\boxed{STAT}$ $\boxed{\blacktriangleright}$ to Calc. Arrow down to **4: LinReg**. $\boxed{ENTER}$ $\boxed{VARS}$ $\boxed{\blacktriangleright}$ $\boxed{ENTER}$ $\boxed{ENTER}$ $\boxed{ENTER}$. Look at the graph by going to $\boxed{ZOOM}$ $\boxed{9}$. Write the equation for the line below.
- Answers will vary. The sample data give y=-0.018x + 5.2.

10. For this type of equation, the absolute value of the slope of the line is equal to a constant that we will refer to as k. The half-life is found by the following equation: $t_{1/2} = \dfrac{\ln(2)}{k}$. Divide ln(2) by the absolute value of your slope. Is it close to the half-life that you found in question 4? Hint: Do not forget to put ( ) around the 2.
- $t_{1/2} = \dfrac{\ln(2)}{0.018} = 38.5 \text{ minutes}$

Red Hot Half-Life
An Exercise in Nuclear Decay

Some atoms have unstable nuclei. They will undergo radioactive decay to become more stable. The amount of time it takes for a sample to decay is specific to the type of atom that is decaying. The amount of time it takes for one half of a radioactive sample to decay is called its half-life.

PURPOSE
In this activity you will model radioactive decay with red hots. The analysis of data and the determination of half-life will be done by graphical means.

MATERIALS
    sample of red hots contained in a paper cup.   graphing calculators
    1 paper plate marked in sections

PROCEDURE
1. Count the red hots in your cup. Record the number in the data table. This is the amount for 0.0 seconds.

2. Put the red hots in the cup and then pour them out on the paper plate. Remove the red hots that landed in the starred section. These red hots will be considered decayed. Count the remaining red hots and record. Each trial is to be counted as 10 seconds.

3. Continue this procedure until you have ten trials or until you have fewer than 5 red hots left, whichever comes first. Record each trial in the data table.

4. Answer the questions in the graphing analysis section.

Name _____

Period _____

Red Hot Half-Life
An Exercise in Nuclear Decay

DATA AND OBSERVATIONS

Time (s)	Number of red hots that decayed	Number of red hots remaining
0.0		
10		
20		
30		
40		
50		
60		
70		
80		
90		
100		

GRAPHING ANALYSIS

1. Enter the "time" in L_1 and the red hots remaining in L_2. Be sure that your Stat Plot is set up correctly to display a scatterplot of L_1 and L_2 data. Describe the shape of the graph.

2. Look at graphs from at least two other groups. What is the same about the graphs? What is different about the graphs?

3. Press ⌑STAT⌑ ⌑▶⌑to Calc. Arrow down to **0: ExpReg**. ⌑ENTER⌑ ⌑VARS⌑ ⌑▶⌑ ⌑ENTER⌑ ⌑ENTER⌑ ⌑ENTER⌑. Look at the graph by going to ⌑ZOOM⌑ ⌑9⌑. Write the equation for the line below. Since this is not a linear equation, it will not be in the form of y = mx + b.

4. Half-life is defined as the amount of time it takes for ½ of a sample to decay. Open ⌑Y=⌑ and arrow down to Y_2. Enter the original number of red hots ⌑÷⌑ ⌑2⌑. Go back to ⌑ZOOM⌑ ⌑9⌑. Go to ⌑2nd⌑ ⌑TRACE⌑ ⌑▼⌑ ⌑▼⌑ ⌑▼⌑ ⌑▼⌑ ⌑ENTER⌑ ⌑ENTER⌑ ⌑ENTER⌑ ⌑ENTER⌑ to get the intersection of these two lines. It will be shown at the bottom of your calculator screen. Since half-life is defined as the time it takes for ½ of a sample to decay, when the number falls to ½ the original number, the x-coordinate will be the half-life. What is the half-life of this system?

5. Take the number of red hots that you had at the half-life point. Divide it by 2. Open ⌑Y=⌑ and arrow down to Y_2. Clear the number and enter the new number. Repeat the procedure that you did in question 4. Is the new time close to two times the half-life (2 half-lives)?

6. Refer back to question 3. The equation is not linear but instead is in the form of $y=(a)(b)^x$. This is an exponential function. In science we like to have equations in the form of y=mx + b. Manipulating the equation, we can do this. It is called a linear transform. In this case, we would take the natural log (ln) of both sides of the equation. Do this with the equation that you have found in question 3. Your teacher will assist you if necessary. Write the equation below.

7. What variables will be used from the equation above to obtain a linear graph?

8. Go back to the statistical function, ⌑STAT⌑, Edit. You want to transform the data in L_2. Arrow up so the cursor is sitting on the L_2 icon. Enter ⌑LN⌑ ⌑2nd⌑ ⌑2⌑ ⌑ENTER⌑ Go to ⌑ZOOM⌑ ⌑9⌑ to look at the graph. Is it linear?

9. Go to ⌑Y=⌑ and clear it. Go to ⌑STAT⌑ ⌑▶⌑ to Calc. Arrow down to **4: LinReg**. ⌑ENTER⌑ ⌑VARS⌑ ⌑▶⌑ ⌑ENTER⌑ ⌑ENTER⌑ ⌑ENTER⌑. Look at the graph by going to ⌑ZOOM⌑ ⌑9⌑. Write the equation for the line below.

10. For this type of equation, the absolute value of the slope of the line is equal to a constant that we will refer to as k. The half-life is found by the following equation: $t_{1/2} = \dfrac{\ln(2)}{k}$. Divide ln(2) by the absolute value of your slope. Is it close to the half-life that you found in question 4? Hint: Do not forget to put ( ) around the 2.

Chemical Bonding
Identifying Characteristics and Drawing Structures

OBJECTIVE
Students will identify characteristics for the three most common types of chemical bonds: ionic, covalent and metallic. Structures will be drawn, shapes and hybridizations will be determined and properties will be discussed.

LEVEL
Chemistry

NATIONAL STANDARDS
UCP.2, UCP.5, B.2

TEKS
8(A), 8(C)

CONNECTIONS TO AP
AP Chemistry:
I. Structure of Matter B. Chemical bonding 1. Binding forces a. Types: ionic, covalent, metallic, hydrogen bonding, van der Waals (including London dispersion forces) b. Relationships to states, structure, and properties of matter c. Polarity of bonds 2. Molecular models a. Lewis structures b. Valence bond: hybridization of orbitals, resonance, sigma and pi bonds c. VSEPR 3. Geometry of molecules and ions

AP Biology:
I. Molecules and Cells A. Chemistry of Life

TIME FRAME
90 minutes class lecture; 45 minutes homework

MATERIALS
periodic table   molecular models

TEACHER NOTES
This lesson should precede the *Molecular Geometry* activity found in this guide. This lesson equips students with the necessary skills to predict molecular geometries and hybridization. The students should actively participate as you introduce the different types of bonds and examples of each bond type. For each of your example compounds, be sure to point out each element's placement on the periodic table. Students should be able to deduce a good deal about the type of chemical bond formed within a compound by examining the positions of the elements on the periodic table. If the elements are from opposite sides of the periodic table, the compound tends to be ionic; elements found closer together (most often both nonmetals) tend to form covalent compounds.

Students should also have an awareness regarding the strength of bonds. Specifically, that ionic attractions are among the strongest. Students may have questions regarding the strength or hardness of diamond. Diamonds are covalently bonded as a network, which makes them an exception to the general trends in bonding discussed here. Diamonds have very strong directional covalent bonds and are actually grouped into a class known as atomic molecular solids. Some metals also have very strong attractions and are therefore very strong, like iron, while others metals have weaker attractions, like mercury, which is a liquid at room temperature.

Examples for types of compounds:
Ionic: table salt, sodium chloride, $NaCl$; baking soda, sodium bicarbonate, $NaHCO_3$; copper (II) sulfate, $CuSO_4$ (available at hardware stores as a root killer); lye, sodium hydroxide, $NaOH$ (available at hardware stores).

Covalent: table sugar, sucrose, $C_{12}H_{22}O_{11}$; ammonia, NH_3; water, H_2O; any plastic material (chains of carbon atoms with hydrogen atoms attached to each carbon).

Metallic: aluminum foil, Al; gold, Au; silver, Ag; lead, Pb (point out that pencil "lead" is not Pb but is really graphite, carbon, C).

Lewis Structures and Molecular Geometries
When discussing Lewis structures, be sure to continually refer to the periodic table and point out that the Roman numeral at the head of each column for the main group (or representative) elements is in fact the number of valence electrons. Demonstrate the regions of electron density on the chalkboard, overhead or with models so that students have a concrete understanding of this concept. Remember that a region of electron density may consist of a bond or lone pair and that double and triple bonds count as only one region of electron density.

Molecular models representing each type of molecular geometry mentioned in the notes should be prepared ahead of time. Students benefit from seeing the models before being asked to make their own. Molecular model kits may be purchased from any science supply company or may be constructed using gumdrops and toothpicks. The advantage to the purchased model sets is that angles are in the appropriate places.

Intermolecular attractive forces: IMF's
Hydrogen bonds—strongest IMF—exist between H and unshared electron pair on F, O, and N. Gives rise to many unique properties. As a result of these attractions, water has high boiling point, high specific heat, etc.
- **Dipole-dipole**—forces of attraction between polar molecules. Polar molecules are those which have an uneven charge distribution. Hydrochloric acid molecules are held to each other by this type of force. HCl—the chlorine pulls the electrons in the bond with greater force than hydrogen so the molecule is polar in terms of electron distribution. Two neighboring HCl molecules will align their oppositely charged ends and attract one another.
- **Dipole-Induced dipole**—the force of attraction that exists between a polar molecule and a nonpolar molecule (oil and water do not mix well with each other).

London Dispersion forces—LDF or **Induced dipole-Induced dipole forces**—the force of attraction that exists between two nonpolar molecules. Liquid nitrogen, N_2, used to "burn" off warts is held by this type of force. Although easily broken, this is the predominant attractive force and exists between all types of molecules. The strength of this force increases as the number of electrons increases. We say that the molecule increases in polarizability; the chance for electron distribution to become unbalanced.

Students should be able to look at the structures that are drawn and determine whether they are polar (unequal charge distribution around the central atom) or nonpolar (equal charge distribution around the central atom). Therefore, the type of IMF can be predicted.

POSSIBLE ANSWERS TO THE SELF-CHECK EXERCISES

Self-Check #1:
Predict the type of bonding found in the following compounds:
1. NaCl — ionic (a metal with a nonmetal)
2. H_2O — covalent (a nonmetal with a nonmetal)
3. Ca — metallic (only one type of atom and it is a metal)

Self-Check #2:
Predict the number of valence electrons for the following elements: Remind students that the Roman numeral Group number of the representative elements will be the number of valence electrons. You may have to remind them to use the electron configurations as a sure alternative.

Element	Valence Electrons	Electron Configuration
Li	1	$1s^2 2s^1$
Ba	2	$[Xe]6s^2$
B	3	$1s^2 2s^2 2p^1$
Si	4	$[Ne]3s^2 3p^2$
N	5	$1s^2 2s^2 2p^3$
S	6	$[Ne]3s^2 3p^4$
Br	7	$[Ar]4s^2 3d^{10} 4p^5$
Ne	8	$1s^2 2s^2 2p^6$

Self-Check #3:
Draw Lewis structures for the following:

1. CCl_4 Tally the valence electrons

C = 1 x 4 = 4
Cl = 4 x 7 = 28
Total: 32 valence electrons or 16 pair

C is in center with 4 Cl's around-this uses 8 electrons or 4 pair. We still have 12 pair—place 3 pair on each Cl atom. Count and make sure that all atoms have 8 electrons.
The shape is tetrahedral. The molecule is nonpolar with sp^3 hybridization. LDF

2. BF$_3$ Tally the valence electrons.   B = 1 x 3 = 3
                                             F = 3 x 7 = 21
                                             Total: 24 valence electrons or 12 pair

B is in the center with the 3 F's around at angles of 120°. This uses 6 electrons or 3 pairs—use the other 9 pairs and place them around each fluorine. Boron has only 6 electrons. It is often an octet exception. Each F is filled with eight.
The shape is trigonal planar. The molecule is nonpolar with sp^2 hybridization. LDF

3. OH$^-$ Tally the valence electrons.   O = 1 x 6 = 6
                                             H = 1 x 1 = 1
                                             Negative one charge = 1
                                             Total: 8 valence electrons or 4 pair

O bonds to H with a single bond. This uses one pair. The other 3 pairs are found around the O atom. Place the entire ion in brackets and place the negative sign as a superscript on the right hand bracket. The shape may be recorded as linear. The ion is polar.

DRAWING LEWIS STRUCTURES AND IDENTIFYING SHAPES OF MOLECULES

For each of the following molecules, draw a Lewis diagram and predict the shape, overall polarity and hybridization of the molecule or ion. Draw your final diagram in the best 3-dimensional structure possible. The following shapes may be used: linear, trigonal planar, tetrahedral, trigonal bipyramidal, octahedral and the variations of tetrahedral—trigonal pyramidal and bent or v-shape.

1. BeH$_2$ Valence electrons: Shape: linear; nonpolar; sp; LDF

    H——Be——H

                   Be = 1 x 2 = 2

                   H = 2 x 1 = 2

                   Total: 4 electrons or 2 pair

2. SiBr$_4$ Valence electrons: Shape: tetrahedral; nonpolar; sp^3;
                                                     LDF

                   Si = 1 x 4 = 4

                   Br = 4 x 7 = 28

                   Total: 32 electrons or 16pair

3. BF_3

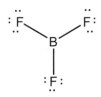

Valence electrons:

B = 1 x 3 = 3

F = 3 x 7 = 21

Total: 24 electrons or 12 pair

Shape: trigonal planar; nonpolar; sp^2; LDF

4. CO_2

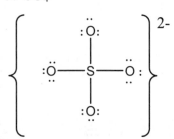

Valence electrons:

C = 1 x 4 = 4

O = 2 x 6 = 12

Total: 16 electrons or 8 pair

Shape: linear; nonpolar; sp; LDF

5. SO_4^{2-}

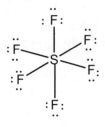

Valence electrons:

S = 1 x 6 = 6

O = 4 x 6 = 24

2- charge= 2

Total: 32 electrons or 16 pair

Shape: tetrahedral; nonpolar; sp^3

6. SF_6

Valence electrons:

S = 1 x 6 = 6

F = 6 x 7 = 42

Total: 48 electrons or 24 pair

Shape: octahedral; nonpolar; sp^3d^2; LDF

7. NH_3

Valence electrons:

N = 1 x 5 = 5

H = 3 x 1 = 3

Total: 8 electrons or 4 pair

Shape: trigonal pyramidal; polar; sp^3; dipole-dipole

8. H_2S

Valence electrons:

H = 2 x 1 = 2

S = 1 x 6 = 6

Total: 8 electrons or 4 pair

Shape: bent or v-shape; polar; sp^3; dipole-dipole

9. PCl$_5$

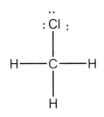

Valence electrons:

P = 1 x 5 = 5

Cl = 5 x 7 = 35

Total: 40 electrons or 20 pair

Shape: trigonal bipyramidal; nonpolar; sp^{3}d; LDF

10. CH$_3$Cl

Valence electron:

C = 1 x 4 = 4

H = 3 x 1 = 3

Cl = 1 x 7 = 7

Total: 14 electrons or 7 pair

Shape: tetrahedral; polar; sp^3; dipole-dipole

Chemical Bonding
Identifying Characteristics and Drawing Structures

Have you ever wondered why some substances exist as solids, others as liquids and still others as gases at room temperature? The *intramolecular* and *intermolecular* forces of attraction can explain almost every observable property of a substance. The prefix "intra" means "within", thus *intra*molecular forces of attraction are otherwise known as chemical bonds. Intramolecular forces are the attractive forces between atoms within a compound. The prefix "inter" means "between", thus *inter*molecular forces of attraction are those occurring between molecules and are ultimately responsible for properties such as boiling point, freezing point, and physical state.

PURPOSE
In this lesson you will study the main types of intramolecular forces that exist within a molecule. These forces bind atoms together. You will also be introduced to the types of intermolecular forces of attraction that occur between molecules. You will learn to draw Lewis structures, predict the shape, hybridization and polarity of molecules drawn.

MATERIALS
periodic table   molecular models

CLASS NOTES
What is a chemical bond?
Bonds are forces of attraction that hold atoms or groups of atoms together and allow them to function as one unit. The type of bond formed between atoms within a molecule relates to physical properties such as melting point, hardness, electrical and thermal conductivity, as well as solubility characteristics. If you think about it, most of the chemical substances you can name or identify are *not* elements. They are compounds. Matter tends to favor systems that have positions of lowest energy. That means being bound requires less energy than existing in the elemental form. It also means that energy *was released* from the system when the atoms joined together. This is a *huge* misconception that most students have—it takes energy to break a bond, not make a bond! Energy is *released* when a bond is formed (exothermic); therefore, it *requires* energy (endothermic) to break a bond. Just remember "Breaking Up is Hard to Do" and this will help you remember.

Types of Chemical Bonds and Identifying Characteristics

Ionic
Characteristics of ionic substances usually include:
- electrons that are transferred between atoms having high differences in electronegativity (greater than 1.67)
- compounds containing a metal and a nonmetal (Remember that metals are on the left side of the periodic table)
- strong electrostatic attractions between positive and negative ions
- formulas given in the simplest ratio of elements (empirical formula; NaCl)
- crystalline structures that are solids at room temperature

- ions that form a crystal lattice structure as pictured in Figure 1`
- compounds that melt at high temperatures
- substances that are good conductors of electricity in the molten or dissolved state

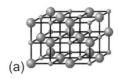

(a)

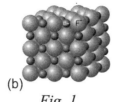

(b)

Fig. 1

Covalent

Characteristics of covalent substances usually include:

- the sharing of electrons between atoms having small differences in electronegativities (less than 1.67)
- nonmetals attracted to other nonmetals
- formulas that are given in the true ratios of atoms (molecular formulas; $C_6H_{12}O_6$)
- substances that may exist in any state of matter at room temperature (solid, liquid, or gas)
- compounds that melt at low temperatures
- substances that are nonconductors of electricity

Metallic

Characteristics of metallic substances usually include:

- substances that are metals
- a "sea" of electrons surrounding a positively charged metal center
- an attraction between metal ions and surrounding electrons
- formulas written as an atom (Mg, Pb)
- solids with a crystalline structure at room temperature
- a range of melting points—usually depending on the number of valence electrons
- substances that are excellent conductors of electricity since the electrons in the "sea" are free to move

Most chemical bonds are in fact somewhere between purely ionic and purely covalent.

Complete Self-Check #1 on the student answer page.

What part of the atom is involved in bonding?
The valence electrons, the outermost electrons, are the only subatomic particle involved in bonding. The electron configurations allow us to predict the number of valence electrons that each element has available for bonding. The periodic table is arranged according to the electron configuration, so it also becomes a most useful tool in drawing proper Lewis structures.

Complete Self-Check #2 on the student answer page:

Drawing Lewis Structures

Lewis structures are only drawn for covalently bonded molecules. Ionically bonded substances do not share electrons. They are the result of strong electrostatic attractions between + and – ions and usually form very large 3-dimensional structures. Metals are also not represented by Lewis structures. There is no true sharing of electrons because the sea of electrons is available for any positive metal center.

Types of Covalent Bonds:

- **Single bond**—one pair of electrons shared; represented by a single line drawn connecting the two atoms. Example: C——C
- **Double bond**—two pairs of electrons shared; represented by two lines drawn connecting the two atoms. Example: C══C
- **Triple bond**—three pairs of electrons shared; represented by three lines drawn connecting the two atoms. Example: C≡≡C
- **Multiple bonds** are commonly formed by carbon, nitrogen, oxygen, phosphorous and sulfur. Just remember, "C-NOPS" for the symbols of those elements that can multiple bond.
- **Triple bonds** are stronger than double bonds and double bonds are stronger than single bonds. Consequently, it takes more energy to break a double bond than a single bond, a triple bond takes more energy to break than a double or a single bond.
- **Multiple bonds** increase the electron density between two nuclei. As the electron density increases, the repulsion between the two nuclei decreases. An increase in electron density also increases the attraction each nucleus has for the additional bonding electron pairs. Either way, the nuclei move closer together and the bond length is shorter for a double bond than a single bond. Of course that means the triple bond is the shortest of all!

Predicting the Arrangement of Atoms within a Molecule:

Rules:
1. Atoms are trying to achieve a stable configuration of eight electrons by sharing. This is known as the octet rule. (There are a few exceptions discussed below.)
2. Hydrogen is always a terminal atom. It may only connect to one other atom.
3. The element with the lowest electronegativity is the central atom in the molecule.
4. Nature tends toward symmetry. Make molecules as symmetrical as possible.
5. Find the total number of valence electrons in the molecule or ion. For negative ions **add** the number of electrons equal to the charge. For positive ions **subtract** electrons equal to the positive charge.
6. Divide the total number of electrons by 2 to get the number of electron pairs available.
7. Place one pair of electrons, a single bond, between each pair of bonded atoms.
8. Subtract the number of pairs used from the total the number of bonds you had available.
9. Place lone pairs around each terminal atom (except for H) to satisfy the octet rule. Left over pairs are assigned to the central atom. (If the central atom is from periods 3-7, it can accommodate more than four electron pairs due to the presence of d orbitals.)
10. If the central atom is not yet surrounded by four electron pairs, convert one or more terminal atom lone pairs into double or triple bonds. Remember, not all elements form multiple bonds; Only C, N, O, P, and S.
11. Count regions of electron density around the central atom. (A multiple bond counts as one area of electron density.) Count the number of unshared pairs (lone pairs) of electrons on the central atom.

12. Identify the shape and hybridization of the molecule.
13. Identify the polarity of the molecule. A molecule is considered polar if there are uneven electron forces acting on the central atom. Uneven forces are generally caused by the presence of a lone pair on the central atom or different kinds of atoms bonded to the central atom.

Exceptions to the Octet Rule

A few elements will have fewer than eight electrons. Hydrogen, as mentioned above, has only two electrons. Beryllium has a maximum of four valence electrons. Boron compounds are stable with only six valence electrons.

Also mentioned above, elements in periods 3 through 7 are often able to expand their octet and may have ten or twelve electrons around the central atom. This is due to the availability of d orbitals that begins on the third energy level.

Predicting the Molecular Shape and Hybridization

The atoms involved in bonding will arrange themselves to minimize electron pair repulsions. This is known as the **VSEPR Theory**—valence shell electron pair repulsion theory. The molecular shape changes with the areas of electron density around the central atom. The molecular geometry changes when one or more of the areas of electron density are occupied by an unshared electron pair (also called a lone pair) on the central atom. Unshared electron pairs have greater repulsive force than shared electron pairs. Unshared electron pairs are bound to only one nuclei and thus, "push" the shared pairs of electrons closer together.

The basic shapes with no unshared electron pairs:

Linear

Trigonal Planar

Tetrahedral

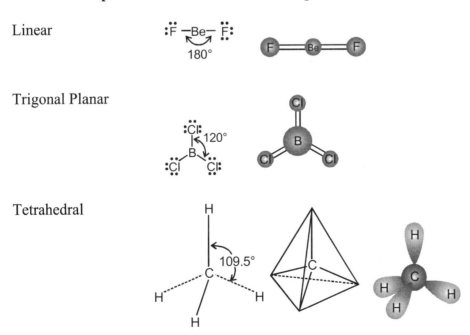

Trigonal bipyramidal

Octahedral

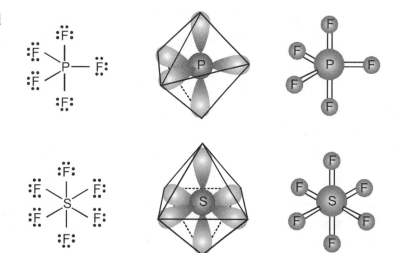

Variations to Shape as a Result of Lone Pair Electrons:

Ammonia, NH_3, is predicted to have a tetrahedral shape because there are 4 electron regions around the central atom, however, in reality it exhibits a structure known as trigonal pyramidal. The lone pair on nitrogen causes the bond angle, H-N-H to decrease from 109.5° to 107°.

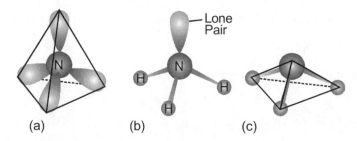

Water, H_2O, is also predicted to have a tetrahedral shape but exhibits a variation known as V-shape or bent. The two lone pairs "gang up" on the shared pairs and decrease the H-O-H bond angle from 109.5° to 104.5°.

In summary, lone pairs have greater repulsive force than shared pairs and cause distortion to the remaining bond angles.

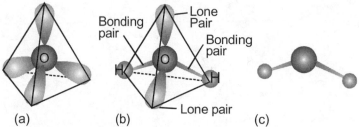

The same type of phenomenon occurs with unshared pairs occupy positions in the trigonal bipyramidal and octahedral geometries. The names of the molecular shapes are shown in Table 10.1.

Valence Bond Theory:
Another bonding theory which helps to describe *how* the bonding occurs is known as the **valence bond theory**. The **VB** theory explains which atomic orbitals must have overlapped in order to obtain a particular geometry where all the bonds are created equal. The two theories compliment each other to better describe molecular structure.

Consider a molecule of $BeCl_2$. In writing the electron configuration for Be $-1s^2 2s^2$ we see that the valence electrons are the 2s electrons. However, the orbital is already full—how can there be any sharing? The valence bond theory explains just this—the second energy level has two types of orbitals available for bonding, s and p. By taking one s orbital and one p orbital we can create two spaces for each electron to occupy. The new name for each of the 2 orbitals of equal energy is sp.

The idea is similar to mixing pigments—red and white yields pink—a hybrid. The sp orbitals are not **s** and not **p** they are a hybrid known as **sp**.

This same idea is found with all of the electronic geometries. Consider BCl_3 —the electron configuration for B is $1s^2 2s^2 2p^1$. This time the valence electrons are not even in the same sublevel! How are three equal bonds explained here? Again, by mixing orbitals we can create a hybrid where all bonding electrons occupy areas of equal energy resulting in minimum overall energy usage.

What do you suppose happens with a molecule like CCl_4? Carbon's electron configuration is $1s^2 2s^2 2p^2$. This would lead to unequal bonds since the s and p orbitals are not of the same energy. Predict what happens to create 4 orbitals of equal energy.

If you guessed that one s combines with 3 p's to yield 4 sp^3 hybrid orbitals, you are correct! The same pattern occurs when expanding the octet for trigonal bipyramidal and octahedral geometries. Once the s is used and the 3 p's are used, the d orbitals are brought into the mix to yield 5 sp^3d orbitals and 6 sp^3d^2 orbitals, respectively. An easy way to remember this is to count the regions of electron density around the central atom and then use that number of orbitals to build your hybrid. Use the Table 10.1 as a guide to predict the geometry and hybridization of various molecules.

Table 10.1

Areas of Electron Density on Central Atom	Number of Lone Pairs on Central Atom	Predicted Geometry of the Molecule	Hybridization
2	0	Linear	sp
3	0	Trigonal planar	sp^2
4	0	Tetrahedral (or pyramidal)	sp^3
4	1	Trigonal pyramidal	sp^3
4	2	V-shape (or bent or angular)	sp^3
5	0	Trigonal bipyramidal	sp^3d
5	1	See-saw	sp^3d
5	2	T-shape	sp^3d
5	3	Linear	sp^3d
6	0	Octahedral	sp^3d^2
6	1	Square pyramidal	sp^3d^2
6	2	Square planar	sp^3d^2

Complete Self-Check #3 on the student answer page

Intermolecular Attractive Forces (IMF's)
Intermolecular forces are those forces that exist *between* molecules. The prefix "inter-" means between —think of the term "interstate"—highways that connect one state to another. Intermolecular attractive forces are *not* bonds. They are simply forces of attraction between molecules.

Properties such as melting points and boiling points can be attributed to the strength of the intermolecular attractions. It works like this: the lower the boiling point, the weaker the intermolecular attractions; the higher the boiling point, the stronger the intermolecular attractions. For example, gasoline evaporates much more quickly than water. Therefore, the intermolecular attractive forces that hold one gasoline molecule to another are much weaker than the force of attraction that holds one water molecule to another water molecule. In fact, water molecules are held together by the strongest of the intermolecular attractive forces, hydrogen bonds. Hydrogen bonds are not true bonds—they are just forces of attraction that exist between hydrogen on one molecule and the unshared electron pair on fluorine, oxygen or nitrogen of another molecule. The strands of DNA that make up our genetic code are held together by this type of intermolecular attraction. Other attractive forces that exist to hold molecules to each other are:

- London dispersion forces—predominant in causing two nonpolar molecules to attract each other
- Dipole-induced dipole—the force of attraction that would enable a polar molecule and a nonpolar molecule to be attracted to one another

- Dipole-dipole—the force of attraction that enables two polar molecules to attract to one another
- Hydrogen bonding—the force of attraction that holds hydrogen from one molecule to an unshared electron pair on F, O, and N of a neighboring molecule (a special case of dipole-dipole).

Name _____

Period _____

Chemical Bonding
Identifying Characteristics and Drawing Structures

SELF-CHECK EXERCISES

Self-Check #1:
Predict the bond type found in the following compound:

1. NaCl _____

2. H_2O _____

3. Ca _____

Self-Check #2:
1. Predict the number of valence electrons for an atom of the following elements:

Li _____

Ba _____

B _____

Si _____

N _____

S _____

Br _____

Ne _____

Self-Check #3:
Draw Lewis structures and identify the shapes for the following:
1. CCl_4

2. BF_3

3. OH^-

Laying the Foundation in Chemistry

DRAWING LEWIS STRUCTURES AND IDENTIFYING SHAPES OF MOLECULES

For each of the following molecules, draw a Lewis diagram and predict the shape, overall polarity and hybridization of the molecule. Draw your final diagram as a 3-dimensional structure. The following shapes may be used: linear, trigonal planar, tetrahedral, trigonal bipyramidal, octahedral and the variations of tetrahedral—trigonal pyramidal and bent or v-shape.

Initial Lewis Structure   **Shape & Hybridization**   **Final 3-D Structure**

1. BeH_2

    Shape: _____

    Polarity: _____

    Hybridization: _____

2. $SiBr_4$

    Shape: _____

    Polarity: _____

    Hybridization: _____

3. BF_3

    Shape: _____

    Polarity: _____

    Hybridization: _____

4. CO_2

    Shape: _____

    Polarity: _____

    Hybridization: _____

5. SO_4^{2-}

    Shape: _____

    Polarity: _____

    Hybridization: _____

6. SF_6

Shape: _____

Polarity: _____

Hybridization: _____

7. NH_3

Shape: _____

Polarity: _____

Hybridization: _____

8. H_2S

Shape: _____

Polarity: _____

Hybridization: _____

9. PCl_5

Shape: _____

Polarity: _____

Hybridization: _____

10. CH_3Cl

Shape: _____

Polarity: _____

Hybridization: _____

Write Your Notes and Ideas Here!

Don't Flip Your Lid
Comparing Intermolecular Forces

OBJECTIVE
Students will use their knowledge of intermolecular forces (IMF) to explain relative differences in melting points.

LEVEL
Chemistry

NATIONAL STANDARDS
UCP.2, UCP.3, A.1, A.2, B.2, B.4

TEKS
Chemistry: 2(A), 2(D), 3(A), 8(A), 8(B), 8(D)

CONNECTIONS TO AP
AP Chemistry:
   I. Structure of Matter B. Chemical bonding 1. Binding forces a. Types: ionic, covalent, metallic, hydrogen bonding, van der Waals (including London dispersion forces) b. Relationships to states, structure, and properties of matter c. Polarity of bonds, electronegativities

TIME FRAME
20 minutes for lab; 25 minutes for discussion and write-up

MATERIALS
(For a class of 28 working in pairs)

14 ring stands	14 rings
wooden splints	14 Bunsen burners
samples of the following: NaCl, $C_6H_{12}O_6$, paraffin, I_2	14 can lid #2 ½. These are the lids that are on 28 oz vegetable cans
condiment cups with lids	28 student answer pages
14 grease pencils or permanent markers	

TEACHER NOTES
This is a simple activity that takes little time. However, a good explanation requires students to have a relatively thorough understanding of intermolecular forces and the role they play in physical properties such as melting point.

The four compounds tested are NaCl (an ionic compound), paraffin (a medium sized non-polar compound), glucose (a medium sized polar compound), and iodine (a larger non-polar molecule). They are placed on the outer rim of the can lid and the Bunsen burner is moved under the center of the lid. The compounds melt in this order: paraffin, iodine, glucose. The sodium chloride does not melt.

POSSIBLE ANSWERS TO THE CONCLUSION QUESTIONS AND SAMPLE DATA

ANALYSIS

From what you know about intermolecular forces, explain the relative order of melting points.

Student answers will vary, however a good answer will include the following points.

- Ionic bonds are very strong. They are electrostatic and go in all directions. Therefore the sodium chloride is the last to melt.
- The other issues involved have to do with whether the molecule is polar or nonpolar and the number of electrons the molecule has. All the molecules have electrons and therefore have London forces. However, glucose is also polar and can form hydrogen bonds. Therefore it has stronger IMFs and a higher melting point. Between the remaining two molecules, paraffin and iodine, both have only London forces. However, iodine has more electrons and a larger, more polarizable electron cloud. It, therefore, has stronger intermolecular forces and a higher melting point.
- Note: Be sure that students do not think that the more massive molecule has stronger IMFs. The problem with this argument is that it confuses gravitational attraction with London forces. Gravitational attraction is a function of the mass of the particle, but it is very weak and does not really play a part in the attraction between molecules. London forces are the electrostatic forces that arise from the distortion of the electron cloud. The more electrons the molecule has, and the more loosely the electrons are held, the stronger the IMFs. You might think of this as the "squishiness" of the electron cloud. This is referred to as polarizability.

CONCLUSION QUESTIONS

1. Glucose, $C_6H_{12}O_6$, and sucrose, $C_{12}H_{22}O_{11}$, are both sugars with similar polarities. Which do you think would have the higher melting point? Justify your answer.
    - Sucrose has a higher melting point. It has more electrons and therefore its electron cloud is more polarizable and will have greater IMFs.

2. Hydrogen sulfide is a gas at room temperature, while water is a liquid, yet hydrogen sulfide has more electrons than water. Explain this anomaly.
    - H_2S and H_2O are both bent molecules and are polar. However, H_2O has the hydrogen bonded directly to a small, electronegative atom—oxygen. Therefore hydrogen bonds are formed between water molecules.

Hydrogen bond

3. Consider the halogens at room temperature and 1 atmosphere of pressure. Why are fluorine and chlorine gases at room temperature, while bromine is a liquid and iodine a solid?

Don't Flip Your Lid
Comparing Intermolecular Forces

The forces that hold one molecule to another molecule are referred to as *intermolecular forces*. These forces arise from unequal distribution of the electrons in the molecule and the electrostatic attraction between oppositely charged portions of molecules.

Van der Waals forces are a function of the number of electrons in a given molecule and how tightly those electrons are held. Let us assume that the molecule involved is nonpolar. A good example would be O_2. Pretend that the molecule is all alone in the universe. If that were the case, the electrons in the molecule would be perfectly symmetrical. However, the molecule is not really alone. It is surrounded by other molecules that are constantly colliding with it. When these collisions occur, the electron cloud around the molecule is distorted. This produces a momentary induced dipole within the molecule. The amount of distortion of the electron cloud is referred to as polarizability. Since the molecule now has a positive side and a negative side, it can be attracted to other molecules. This attractive force is called a London force or a dispersion force. Since all molecules have electrons, all molecules have London forces. These forces range from 5-40 kJ/mol.

Some molecules also are naturally polar. Therefore they can also have a permanent dipole which attracts other molecules that also have dipoles (either induced or permanent).

A third type of intermolecular force is *hydrogen bonding*. When hydrogen is covalently bonded to a small electronegative atom like nitrogen, oxygen, or fluorine, the electron cloud on the hydrogen is very distorted and pulled toward the electronegative atom. Since hydrogen has no inner core electrons, the nuclear charge is somewhat exposed. This sets up the potential for a reasonably strong attraction between this hydrogen and an electronegative atom in another molecule. Hydrogen bonds are significant in determining such factors as the high boiling point of water, solubility of acetone in water, and the shape and structure of proteins and DNA.

Ionic bonds are formed between charged particles. Ionic compounds do not form molecules, but rather are held together in a crystal by electrostatic attraction. These electrostatic attractions are strong and are omnidirectional.

PURPOSE
In this laboratory activity you will determine the relative melting points of four substances, paraffin, NaCl (table salt), $C_6H_{12}O_6$, (glucose), and iodine, I_2. Paraffin is a medium size nonpolar molecule, NaCl is an ionic compound. Glucose is a medium size polar molecule. Iodine is large nonpolar molecule.

MATERIALS
ring stand	ring
wooden splints	Bunsen burners
samples of the following: NaCl, $C_6H_{12}O_6$, paraffin, I_2	can lid #2 ½. These are the lids that are on 28 oz vegetable cans
student answer page	grease pencil or permanent marker

Safety Alert
1. Wear your goggles.
2. Do not inhale the iodine vapors.
3. Allow the can lid to cool before disposal.

PROCEDURE

1. In the space marked HYPOTHESIS on your student answer page, write a statement predicting in which order the salt, paraffin, iodine, and glucose will melt.

2. Turn the can lid over so that the grooves around the edge form a trough. Use the grease pencil or marker to divide the lid into four quadrants as shown in the picture.

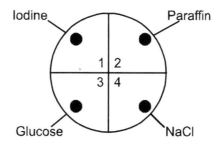

3. Use wooden splints to transfer small samples of each of the four compounds to the outermost groove of the can lid. You only need a very small sample—just enough to be able to see the compound. For the iodine, use only one crystal.

4. Carefully lift the can lid to the ring on the ring stand as shown in the picture. Light the Bunsen burner and place it so that the flame is directly in the middle of the can lid.

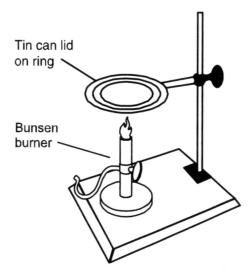

5. Watch carefully as the compounds melt. You will need to record the relative order of melting. As soon as three of the compounds melt, turn off the burner.

1. Allow the can lid to cool and dispose of it in a trash can.

2. Answer the conclusion questions on your student answer page.

Name _____

Period _____

Don't Flip Your Lid
Comparing Intermolecular Forces

HYPOTHESIS

DATA AND OBSERVATIONS

Record the relative order of melting for the four compounds.

ANALYSIS

From what you know about intermolecular forces, explain the relative order of melting points.

CONCLUSION QUESTIONS

1. Glucose, $C_6H_{12}O_6$, and sucrose, $C_{12}H_{22}O_{11}$, are both sugars with similar polarities. Which do you think would have the higher melting point? Justify your answer.

2. Hydrogen sulfide is a gas at room temperature, while water is a liquid, yet hydrogen sulfide has more electrons than water. Explain this anomaly.

3. Consider the halogens at room temperature and 1 atmosphere of pressure. Why are fluorine and chlorine gases at room temperature, while bromine is a liquid and iodine a solid?

Write Your Notes and Ideas Here!

Molecular Geometry
Investigating Molecular Shapes with VSEPR

OBJECTIVE

Students will draw Lewis (electron) dot diagrams and then build the models with toothpicks and gumdrops. They will use their models to visualize the shape of the molecule and determine the hybridization.

LEVEL

Chemistry

NATIONAL STANDARDS

UCP.2, UCP.5, B.2

TEKS

8(A), 8(B), 8(C)

CONNECTIONS TO AP

AP Chemistry:

    I. Structure of Matter B. Chemical Bonding 2. Molecular models a. Lewis structures b. Valence bond; hybridization of orbitals, resonance, sigma and pi bonds c. VSEPR

TIME FRAME

45 minutes

MATERIALS

(For a class of 28 working in pairs)

14 resealable plastic sandwich or snack bags	12 small gumdrops or spice drops per bag
several boxes of toothpicks	roll of paper towels

TEACHER NOTES

Each student pair will need a kit containing 12 gumdrops—two must be the same color and 10 of a different color. If you have model kits with 4, 5, and 6 holed central atoms, they may also be used. If you are using model kits it is a good idea to explain the relationship between the number of holes on the central atom and the sites of electron density in a Lewis structure. Constructing double bonds should also be discussed.

This lesson is designed to follow an introduction to Lewis structures for covalent compounds. Students should also have been introduced to the concept of hybridization. During a pre-lab discussion you should demonstrate the Lewis structures and corresponding geometries for several of the example compounds in the reference table on the student pages.

VSEPR (Valence Shell Electron Pair Repulsion) is a simple model that employs the concept that electrons, being negatively charged, are repulsive. Therefore, regions of electron densities will attempt to position themselves as far away from one another as possible. Regions of electron density are as follows:

- Single bond
- Double bond
- Triple bond
- Lone pair

These regions get increasingly more repulsive moving down the list. You will find a table of basic VSEPR shapes, along with examples of molecular species that exhibit that shape, on the student instruction page. Note that lone pairs are more repulsive than any of the bonds. This is because they are only influenced by one nucleus rather than two nuclei. For this reason, lone pairs take up more space and will cause the other bond angles to be smaller. In general, each lone pair will collapse the bond angle by approximately 2°/lone pair.

POSSIBLE ANSWERS TO THE STUDENT ANSWER PAGE

1. CO_2

O==C==O

Shape: linear

Hybridization: sp

2. BF_4

Shape: tetrahedral

Hybridization: sp^3

3. H_2CO

Shape: trigonal planar

Hybridization: sp^2

4. PF_5

Shape: trigonal bipyramid

Hybridization: sp^3d

5. SiH_4

Shape: tetrahedral

Hybridization: sp^3

6. SeF_6

Shape: octahedral

Hybridization: sp^3d^2

7. IF_4^-

Shape: square planar

Hybridization: sp^3d^2

8. F_2CO

Shape: trigonal planar

Hybridization: sp^2

9. XeF_4

Shape: square planar

Hybridization: sp^3d^2

10. NO_2^-

Shape: bent (2 resonance structures)

Hybridization: sp^2

11. O_3

Shape: bent (2 resonance structures)

Hybridization: sp^2

12. ClO_3^-

Shape: trigonal pyramid

Hybridization: sp^3

13. I_3^-

Shape: linear

Hybridization: sp^3d

14. IOF_5 (I is the central atom)

Shape: octahedral

Hybridization: sp^3d^2

15. NH_2^-

$$\left[\begin{array}{c} \overset{\displaystyle \cdot\cdot}{\underset{\displaystyle \cdot\cdot}{N}} \\ H \qquad H \end{array} \right]^{-}$$

Shape: bent

Hybridization: sp^3

Molecular Geometry
Investigating Molecular Shapes with VSEPR

The shape of a molecule will dictate many physical and chemical properties of a substance. In biological systems many reactions are controlled by how a molecule fits into the substrate that it is reacting with. Physical properties of substances, such as solubility and boiling point are also influenced by molecular geometry.

PURPOSE
In this activity you will draw Lewis structures for a set of molecules and ions. You will then build the molecules from gumdrops and toothpicks to model the molecular geometry and determine the hybridization of the molecules.

MATERIALS
kit containing 12 gumdrops and toothpicks   paper towel

Safety Alert
1. Keep the gumdrops on the paper towel at all times. When you are finished, you may eat the candy if your teacher allows.

PROCEDURE
1. All of the following are covalent molecules or ions.

2. Draw Lewis dot structures in the space provided on the student answer page. Use the VSEPR theory to predict the shape of each molecule or ion listed on your student answer page.

3. Use the gumdrops and toothpicks provided to build each chemical species. Be sure that you use one toothpick for each pair of electrons on the central atom.

4. Write the hybridization of the orbitals in the space provided below the shape for each substance.

Name _____

Period _____

Molecular Geometry
Investigating Molecular Shapes with VSEPR

VSEPR (Valence Shell Electron Pair Repulsion) is a simple model that employs the concept that electrons, being negatively charged, are repulsive. Therefore, regions of electron densities will attempt to position themselves as far away from one another as possible. Regions of electron density are as follows:

- Single bond
- Double bond
- Triple bond
- Lone pair

These regions get increasingly more repulsive moving down the list. The following table is provided as a reference for basic VSEPR molecular shapes.

Regions of Electron Density	Representative formula	Example	Shape	Hybridization
2	MX_2	CO_2	Linear (180°)	sp
3	MX_3	BF_3	Trigonal planar (120°)	sp^2
3	MX_2E	SO_2	Bent (118°)	sp^2
4	MX_4	CH_4	Tetrahedral (109.5°)	sp^3
4	MX_3E	NH_3	Trigonal pyramidal (107°)	sp^3
4	MX_2E_2	H_2O	Bent (105°)	sp^3
5	MX_5	PF_5	Trigonal bipyramidal (TBP)	sp^3d
5	MX_4E	SF_4	See Saw	sp^3d
5	MX_3E_2	ICl_3	T-shaped	sp^3d
5	MX_2E_3	I_3^-	Linear	sp^3d
6	MX_6	SCl_6	Octahedral	sp^3d^2
6	MX_5E	XeF_5^+	Square pyramidal	sp^3d^2
6	MX_4E_2	ICl_4^-	Square planar	sp^3d^2

1. CO_2

Shape_____

Hybridization_____

2. BF_4^-

Shape_____

Hybridization_____

3. H_2CO

Shape_____

Hybridization_____

4. PF_5

Shape_____

Hybridization_____

5. SiH_4

Shape_____

Hybridization_____

6. SeF_6

Shape_____

Hybridization_____

7. IF_4^-

Shape_____

Hybridization_____

8. F_2CO

Shape_____

Hybridization_____

9. XeF_4

Shape_____

Hybridization_____

10. NO_2^-

Shape_____

Hybridization_____

11. O_3

Shape_____

Hybridization_____

12. ClO_3^-

Shape_____

Hybridization_____

13. I_3^-

Shape_____

Hybridization_____

14. IOF_5 (I is the central atom)

Shape_____

Hybridization_____

15. NH_2^-

Shape_____

Hybridization_____

Chemical Nomenclature
Naming and Writing Chemical Formulas

OBJECTIVE
Students will learn to apply the rules of chemical nomenclature.

LEVEL
Chemistry

NATIONAL STANDARDS
UCP.1, UCP.2, B.2

TEKS
11(A)

CONNECTIONS TO AP
AP Chemistry:
Chemical nomenclature is fundamental to the AP Chemistry, AP Biology and AP Environmental Science courses

TIME FRAME
180 minutes

MATERIALS

copy paper or card stock for making a
   demonstration set of the manipulatives
different colored markers
roll of sticky magnetic tape or masking tape

Optional: white flash boards for each student
scissors
resealable plastic sandwich bags

TEACHER NOTES
This activity is designed to enhance a student's ability to write and name chemical formulas. The students must be assigned the list of polyatomic ions for memorization ideally as a summer assignment or at least a few weeks before this activity. Quiz students over the list of ions each class period to encourage their participation. Once they have mastered the memory work, continue quizzing them weekly to ensure retention. Use the student handout provided to teach the lesson.

The students will use the templates provided in the student section of this activity to create a set of manipulatives. These manipulatives will help students determine correct chemical formulas. It is recommended that you enlarge the template on a copy machine and make a demonstration set of the manipulatives for use on the chalk or white board in your classroom. If possible, make your set of manipulatives out of colored card stock and laminate them for durability. If the chalkboard or white board surface is magnetic, a small piece of adhesive magnetic strip may be added to the back of the manipulatives. Magnetic strip is readily available in hobby stores and relatively inexpensive. If the board surface is not magnetic, masking tape may be used but must be replaced regularly.

This activity instructs the students to carefully cut out the pieces of Template 1 and arrange them by similar valence or oxidation state. They are to then trace over the element's symbol and oxidation state with colored markers using the chart given. You may want to use colored card stock for your enlarged set of demonstration manipulatives. Alternatively, you may use the colored markers in the same way that the students do. Template 2 is used during the organic section and does not need to be colored.

Give each student a copy of the activity. The background information is essential to their success.

POSSIBLE ANSWERS TO THE CONCLUSION QUESTIONS AND SAMPLE DATA

Table A	
Acid Formula:	Acid Name:
HCl	hydrochloric acid
HClO	hypochlorous acid
$HClO_2$	chlorous acid
$HClO_3$	chloric acid
$HClO_4$	perchloric acid [or 'hyperchloric acid]
HNO_3	nitric acid
HBr	hydrobromic acid
H_3PO_4	phosphoric acid
H_3PO_3	phosphorous acid
HCN	hydrocyanic acid
$HC_2H_3O_2$	acetic acid
H_2CO_3	carbonic acid
HI	hydroiodic acid
HF	hydrofluoric acid

Table B					
# of carbon atoms = n	prefix or stem	-ane C_nH_{2n+2}	-ene C_nH_{2n}	-yne C_nH_{2n-2}	-anol $C_nH_{2n+1}OH$
1	meth-	CH_4 methane	cannot form		CH_3OH methanol
2	eth-	C_2H_6 Ethane	C_2H_4 ethene	C_2H_2 ethyne	C_2H_5OH Ethanol
3	prop-	C_3H_8 propane	C_3H_6 propene	C_3H_4 propyne	C_3H_7OH propanol
4	but-	C_4H_{10} butane	C_4H_8 butene	C_4H_6 butyne	C_4H_9OH butanol
5	pent-	C_5H_{12} pentane	C_5H_{10} pentene	C_5H_8 pentyne	$C_5H_{11}OH$ pentanol
6	hex-	C_6H_{14} hexane	C_6H_{12} hexene	C_6H_{10} hexyne	$C_6H_{13}OH$ hexanol
7	hept-	C_7H_{16} heptane	C_7H_{14} heptene	C_7H_{12} heptyne	$C_7H_{15}OH$ heptanol
8	oct-	C_8H_{18} octane	C_8H_{16} octene	C_8H_{14} octyne	$C_8H_{17}OH$ octanol

Table B					
9	non-	C_9H_{20} nonane	C_9H_{28} nonene	C_9H_{16} nonyne	$C_9H_{19}OH$ nonanol
10	dec-	$C_{10}H_{22}$ decane	$C_{10}H_{20}$ decene	$C_{10}H_{18}$ decyne	$C_{10}H_{21}OH$ decanol

Table C							
	Ag	Pb^{2+}	Cu^+	Ba	NH_4	Al	Mn^{2+}
N	Ag_3N silver nitride	Pb_3N_2 lead (II) nitride	Cu_3N copper (I) nitride	Ba_3N_2 barium nitride	$(NH_4)_3N$ ammonium nitride	AlN aluminum nitride	Mn_3N_2 manganese (II) nitride
O	Ag_2O silver oxide	PbO lead (II) oxide	Cu_2O copper (I) oxide	BaO barium oxide	$(NH_4)_2O$ ammonium oxide	Al_2O_3 Aluminum oxide	MnO manganese (II) oxide
Br	$AgBr$ silver bromide	$PbBr_2$ lead (II) bromide	$CuBr$ copper (I) bromide	$BaBr_2$ barium bromide	NH_4Br ammonium bromide	$AlBr_3$ aluminum bromide	$MnBr_2$ manganese (II) bromide
S	Ag_2S silver sulfide	PbS lead (II) sulfide	Cu_2S copper (I) sulfide	BaS barium sulfide	$(NH_4)_2S$ ammonium sulfide	Al_2S_3 aluminum sulfide	MnS manganese (II) sulfide
SO_4	Ag_2SO_4 silver sulfate	$PbSO_4$ lead (II) sulfate	Cu_2SO_4 copper (I) sulfate	$BaSO_4$ barium sulfate	$(NH_4)_2SO_4$ ammonium sulfate	$Al_2(SO_4)_3$ aluminum sulfate	$Mn(SO_4)$ manganese (II) sulfate
ClO_2	$AgClO_2$ silver chlorite	$Pb(ClO_2)_2$ lead (II) chlorite	$CuClO_2$ copper (I) chlorite	$Ba(ClO_2)_2$ barium chlorite	NH_4ClO_2 ammonium chlorite	$Al(ClO_2)_3$ aluminum chlorite	$Mn(ClO_2)_2$ manganese(II) chlorite
PO_3	$Ag(PO_3)_3$ silver phosphite	$Pb_3(PO_3)_2$ lead (II) phosphite	Cu_3PO_3 copper (I) phosphite	$Ba_3(PO_3)_2$ barium phosphite	$(NH_4)_3PO_3$ ammonium phosphite	$AlPO_3$ aluminum phosphite	$Mn_3(PO_3)_2$ manganese (II) phosphite

Table D	
Formula	Name
$[Al(OH)_4]^-$	tetrahydroxoaluminate (III) ion
$[Ag(NH_3)_2]^+$	diamminesilver(I) ion
$[Zn(OH)_4]^{2-}$	tetrahydroxozincate (II) ion
$[Zn(NH_3)_4]^{2+}$	tetramminezinc (II) ion
$[Cu(NH_3)_4]^{2+}$	tetramminecopper(II) ion
$[FeSCN]^{2+}$	thiocyanatoiron(III) ion*
$[FeSCN]Cl_2$	thiocyanoiron(III) chloride
$[Cd(NH_3)_4]^{2+}$	tetramminecadmium(II) ion
$[Ag(CN)_2]^-$	dicyanoargentate(I) ion
$Mg\{[Ag(CN)_2]^-\}_2$	magnesium dicyanoargentate(I)
$[Cu(Cl_2Br_2I_2)]^{4-}$	dibromodichlorodiiodocuprate(II) ion
$[Co(NH_3)_5Cl]Cl_2$	pentaamminechlrocobalt(III) chloride
$K_3[Fe(CN)_6]$	potassium hexacyanoferrate(III)
$[Pt(NH_3)_3Br]Cl$	triamminebromoplatinum(II) chloride
$[Cu(Cl_2Br_2INH_3)]^{4-}$	amminedibromodichloroiodocuprate(I) ion
$K_3[Co(F)_6]$	potassium hexafluorocobaltate(III)
$[Co(NH_3)_6]Cl_2$	hexaamminecobalt(II) chloride
$[Fe(CN)_6]^{4-}$	hexacyanoferrate(II)

* actually it is the nitrogen that is the donor atom so you may see SCN⁻ written as NSC⁻ which makes this complex $[FeNSC]^{2+}$ and named isothiocyanatoiron(III) ion.

Chemical Nomenclature
Naming and Writing Chemical Formulas

Writing chemical formulas will open your eyes to the chemical world! Once you are able to write correct chemical formulas, there are four naming systems you will need to master. The trick lies in recognizing *which* naming system to use! Use the following guidelines when making your decision about how to name compounds.

- If the compound starts with H, it is an acid. Use the Naming Acids rules.
- If the compound starts with C and contains quite a few H's and perhaps some O, it is organic. Use the Naming Organic Compound rules.
- If the compound starts with a metal it is most likely ionic. Use the Naming Binary Ionic Compounds rules.
- If the compound starts with a nonmetal other than H or C, use the Naming Binary Molecular Compounds rules.

It is *essential* that you memorize some common polyatomic ions. Polyatomic ions are groups of atoms that behave as a unit and possess an overall charge. *If more than one group is needed to balance a formula, the group must be enclosed in parentheses before adding the subscripts.* You need to know their names, formulas and charges. If you learn the nine that follow, you can get many others from applying two simple patterns! These nine are grouped by charge to make the memory work easier.

Name of Polyatomic Ion:	Formula & Charge:
Ammonium ion	NH_4^+
Acetate ion	$C_2H_3O_2^-$
Cyanide ion	CN^-
Hydroxide ion	OH^-
Nitrate ion	NO_3^-
Chlorate ion	ClO_3^-
Sulfate ion	SO_4^{2-}
Carbonate ion	CO_3^{2-}
Phosphate ion	PO_4^{3-}

Pattern 1: The -ates "*ate*" one more oxygen than the –ites **however,** their charge doesn't change as a result. For instance, if you know nitrate is NO_3^-, then nitrite is NO_2^-. If you know phosphate is PO_4^{3-}, then phosphite is PO_3^{3-}. You can also use the prefixes *hypo-* and *per-* with the chlorate series. Perchlorate, ClO_4^-, was really "*hyper and -ate yet another oxygen*" when compared to chlorate, ClO_3^-. Hyopchlorite is a double whammy. It is –ite and therefore "*ate*" one less oxygen than chlorate **and** it is hypo- which means "below" so it "*ate*" even one less oxygen than plain chlorite so its formula is ClO^-. You can substitute the other halogens for chlorine and make similar sets of this series.

Pattern 2: The -ates with charges less than negative one, meaning ions with charges of -2, -3, etc., can have an H added to them to form new polyatomic ions. For each H added the charge is increased by a +1. For instance, CO_3^{2-} can have an H added and become HCO_3^-. HCO_3^- is called either the bicarbonate ion or the hydrogen carbonate ion. Since phosphate is negative three, you can add one or two hydrogens to make two new polyatomic ions, HPO_4^{2-} and $H_2PO_4^-$. The names are hydrogen phosphate and dihydrogen phosphate, respectively. If you continue adding hydrogen ions until you reach neutral, you've made an acid! That means you need to see the **Naming Acids** rules.

Pattern 3: Use of the following periodic table will also come in handy. Notice the simple patterns for determining the most common oxidation states of the elements based on their family's position on the periodic table. Notice the IA family is +1 while the IIA family is +2. Skip across to the IIIA family, and notice that aluminum is +3. Working backwards from the halogens or VIIA family, they are most commonly -1 while the VI A family is -2 and the VA family is -3. The IV A family is "wishy-washy", and can be several oxidation states, the most common being ± 4.

Common type I cations   Common type III cations

Common type II cations

NAMING ACIDS

How do I know it's an acid? The compound's formula begins with a hydrogen, H, and water doesn't count. Naming acids is extremely easy, if you know your polyatomic ions. There are three rules to follow:

- H + element: If the acid has only one element following the H, then use the prefix hydro- followed by the element's root name and an -ic ending. HCl is hydrochloric acid. H_2S is hydrosulfuric acid. When you see an acid name beginning with "hydro", think "Caution, element approaching!" (HCN is an exception since it is a polyatomic ion without oxygen, and it is named hydrocyanic acid.)
- H + -ate polyatomic ion: If the acid has an "-ate" polyatomic ion after the H, then it makes an "-ic" acid. H_2SO_4 is sulfuric acid.
- H + -ite polyatomic ion: If the acid has an -"ite" polyatomic ion after the H, then it makes an "-ous" acid. H_2SO_3 is sulfurous acid.

When writing formulas for acids you must have enough H^+ added to the anion to make the compound neutral.

NAMING ORGANIC COMPOUNDS

How do I know it's organic? The formula will start with a C followed by hydrogens and may even contain some oxygen. Most of the organic carbons you will encounter will be either hydrocarbons or alcohols. These are the simplest of all to name! Memorize the list of prefixes in Table B found in the conclusion questions. The prefixes correspond to the number of carbons present in the compound and will be the stem for each organic compound. Notice that the prefixes are standard geometric prefixes once you pass the first four carbons. This silly statement will help you remember the order of the first four prefixes since they are not ones you are familiar with: "Me Eat Peanut Butter." This corresponds to meth-, eth-, prop-, and but- which correspond to 1, 2, 3, and 4 carbons, respectively. Now that we have a stem, we need an ending. There are four common hydrocarbon endings that you will need to know. The ending changes depending on the structure of the molecule.

- -ane - alkane (all single bonds & saturated) C_nH2_{n+2}; The term saturated means that the compound contains the maximum number of hydrogen atoms.
- -ene = alkene (contains one double bond & unsaturated) C_nH_{2n}; The term unsaturated means that in the compound hydrogens have been removed in order to create a multiple bond.
- -yne ≡ alkyne (contains one triple bond & unsaturated) C_nH_{2n-2}; The term polyunsaturated means that the compound contains more than one double or triple bond.
- -ol – alcohol (one H is replaced with a hydroxyl group, -OH group, to form an alcohol) C_nH_{2n+1}; Do not be fooled—this looks like a hydroxide group, but is not! It does not make this hydrocarbon an alkaline or basic compound. Do not name these as a hydroxide! C_2H_6 is ethane while C_2H_5OH is ethanol.

NAMING BINARY IONIC COMPOUNDS

How do I know it is ionic? The formula will begin with a metal cation or the ammonium cation. The ending is often a polyatomic anion. If only two elements are present, they are usually from opposite sides of the periodic table, like KCl. If the metal is a transition metal, be prepared to use a Roman numeral indicating which oxidation state the metal is exhibiting. Silver (Ag), cadmium (Cd) and zinc (Zn) are exceptions to the Roman numeral rule because their charges are constant. In order to remember their charges, place their symbols in alphabetical order and then they are in order; +1, +2, +2.

In order to name these compounds, first let's name the ions.

Naming positive ions: Metals commonly form cations.

- Monatomic positive ions in Group A are named by simply writing the name of the metal from which it is derived. Al^{3+} is the aluminum ion.
- Transition metals often form more than one positive ion so Roman numerals (in parentheses) follow the ion=s name. Cu^{2+} is copper (II) ion. Remember the exceptions listed above.
- NH^{4+} is the ammonium ion. It is the only positive polyatomic ion that you will encounter.

Naming negative ions: Nonmetals commonly form anions. Most of the polyatomic ions are also negatively-charged.

- Monatomic negative ions are named by adding the suffix -ide to the stem of the nonmetal's name. Halogens are called the halides. Cl^- is the chloride ion.
- Polyatomic anions are given the names of the polyatomic ion. You must memorize these. NO_2^- is the nitrite ion.

Naming the Compound: The + ion (cation) name is given *first* remembering that transition metals will contain the Roman numeral indicating its charge followed by the name of the negative ion (anion). No prefixes are used.

NAMING BINARY MOLECULAR COMPOUNDS

How will I know it is a molecular compound? It will be a combination of nonmetals, both lying near each other on the periodic table. No polyatomic ions will be present! Use the following set of prefixes when naming molecular compounds.

Subscript	Prefix
1	Mono- [usually used only on the second element; such as carbon monoxide or nitrogen monoxide]
2	di-
3	tri-
4	tetra-
5	penta-
6	hexa-
7	hepta-
8	octa-
9	nona-
10	deca-

Naming the Compound: The + ion name is given *first* followed by the name of the negative ion. Use prefixes to indicate the number of atoms of each element. Don't forget the –ide ending! If the second element's name begins with a vowel, then the "a" at the end of the prefix is usually dropped. N_2O_5 is dinitrogen pentoxide not dinitrogen pentaoxide. PCl_5 is phosphorous pentachloride not phosphorous pentchloride.

FORMULA WRITING

Naming is the trickiest part! Once you've been given the name, the formula writing is easy *as long as you have memorized the formulas and charges of the polyatomic ions*. The prefixes of a molecular compound make it really easy to write the formula since the prefix tells you how many atoms are present for each element. Roman numerals are your friend; they tell you the charge on the transition metal. Remember that Ag, Cd, & Zn are usually not written with a Roman numeral; you must know their charges. The most important thing to remember is that, the sum of the charges must add up to zero in order to form a neutral compound. The *crisscross method* is very useful—the charge on one ion becomes the subscript on the other. If you use this method, you must always check to see that the subscripts are in their lowest ratio! Here are some examples:

potassium oxide → K^{1+} O^{2-} → $K_1 \times O_2$ → K_2O

iron (III) chlorate → Fe^{3+} ClO_3^{1-} → $Fe_3 \times ClO_3{}_1$ → $Fe(ClO_3)_3$

tin (IV) sulfite → Sn^{4+} SO_3^{2-} → $Sn_4 \times SO_3{}_2$ → $Sn_2(SO_3)_4$ → $Sn(SO_3)_2$

zinc acetate → Zn^{2+} $C_2H_3O_2^{1-}$ → $Zn_2 \times C_2H_3O_2{}_1$ → $Zn(C_2H_3O_2)_2$

COORDINATION CHEMISTRY NOMENCLATURE

These are actually quite fun! The rules are simple and you will really feel like you are speaking chemistry. Square brackets are used to enclose a complex ion or neutral coordination species. A complex ion is composed of a single central atom or ion with other atoms or molecules attached. The atoms or molecules attached are known as **ligands**. The number of ligands attached is called the coordination number of the complex ion. The naming of complex cations and complex anions is similar, except that anions are always made to end in **-ate**. Coordination compounds, like other ionic compounds, are named with the cation preceding the anion regardless of which (if either) one of them is a complex ion.

The rules for naming complex ions or compounds are as follows:
- As with any ionic compound, the cation is named before the anion.
- In naming a complex ion, the ligands are named before the central metal ion.
- In naming ligands, an "-o" is added to the root name of any anion. For example, the halides as ligands are called fluoro, chloro, bromo, and iodo; hydroxide is hydroxo; cyanide is cyano; nitrite is nitrito, etc. If the ligands are neutral, omit the "-o" ending. Neutral ligands take the name they normally use as neutral molecules. There are four exceptions which must be memorized: H_2O as a ligand is known as aqua, NH_3 is named ammine [note the "mm" in the spelling so it is not confused with the functional group $-NH_2$, an amine group], CO is named carbonyl, and NO is nitrosyl.
- The number of each kind of ligand is specified by the usual Greek prefix: mono-, di-, tri-, tetra-, penta-, and hexa-.
- The oxidation state of the central metal atom is designated by a Roman numeral in parentheses.

- When more than one type of ligand is present, they are named in alphabetic order with no regard for the Greek (numerical) prefix.
- If the complex has a negative charge, the suffix **-ate** is added to the name of the metal. When a Latin symbol is used for the element, the element takes the Latin name in complex **anions** but **not in** complex **cations**. For example, $[Cu(NH_3)_4]^{2+}$ is called the tetraamminecopper(II) ion but $[Cu(CN)_6]^{4-}$ is called the hexacyanocuprate(II) ion. Likewise $[Al(NH_3)_6]^{3+}$ is called the hexaamminealuminum(III) ion but $[Al(OH)_4]^-$ is called the tetrahydroxoaluminate(III) ion.

Common Neutral Ligands	
Formula	Name
H_2O	aqua
NH_3	ammine
CO	carbonyl
NO	nitrosyl
Common Anion Ligands	
Formula	Name
F^-	fluoro
Cl^-	chloro
Br-	bromo
I^-	iodo
OH^-	hydroxo
CN^-	cyano
SCN^-	thiocyano
$S_2O_3^{2-}$	thiosulfate
$C_2O_4^{2-}$	oxalato
Latin Names Used for Some Metal Ions in Anionic Complex Ions	
Iron	ferrate
Copper	cuprate
Lead	plumbate
Silver	argentate
Gold	aurate
Tin	stannate

PURPOSE
To master the skill of writing and naming chemical formulas.

MATERIALS
scissors
colored markers: blue, red, yellow, green, purple, pink

Safety Alert
1. Use care when handling scissors.

PROCEDURE

1. Carefully cut out the shapes on Template 1. Group them by similar charge or oxidation state.

2. Trace over the symbol and oxidation state of each element using colored markers and apply the color scheme below:

Color	Oxidation State
Blue	+1
Red	-1
Yellow	+2
Green	-2
Purple	+3
Pink	-3

3. Notice how the models fit together. If an element has a 3^+ oxidation state, it requires three elements with a 1^- oxidation state to create a complete compound.

4. Carefully cut out the shapes on Template 2. Each carbon model has 4 inward notches. The model "bonds" found on Template 2 are for connecting the carbons. These shapes will be used to help you with organic compounds. There is no need to color them.

5. Review the rules for naming acids and complete Table A on your student answer sheet. Use the models you created from Template 1 as needed. Supply either the acid's name or its formula to complete Table A.

6. Review the rules for naming organic hydrocarbons and alcohols. Use your models from Template 2 as needed. Fill in the missing formulas **and** names for each compound in Table B.

7. Review the rules for naming binary ionic **and** molecular compounds and complete the table below. Use the models you created from the template as needed. Supply the compound's formula **and** **name** to complete Table C. If the charge or oxidation state is missing from the table, it is because you should already know them or be able to determine them due to their position in the periodic table.

Name _____

Period _____

Chemical Nomenclature
Naming and Writing Chemical Formulas

All You Really Need to Know About Chemical Names and Formulas SUMMARIZED

In this flowchart, D and J in the general formula D_xJ_y can represent atoms, monatomic ions, or polyatomic ions.

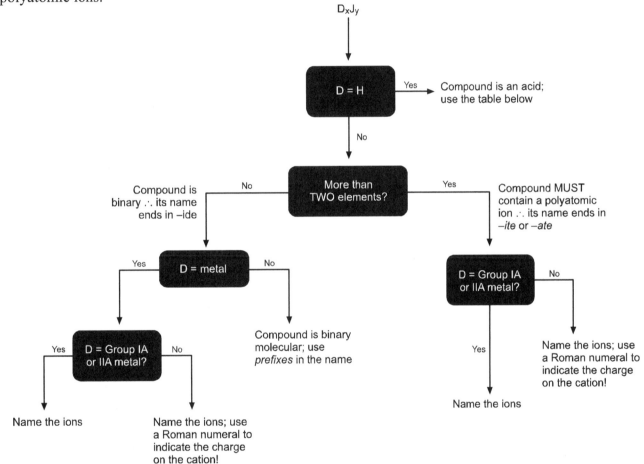

Naming Acids			
Anion ending	Example	Acid name	Example
–ide	S^{2-} sulfide	Hydro-(stem)-ic acid	hydrosulfuric acid
–ite	SO_3^{2-} sulfite	(stem)-*ous* acid	sulfurous acid
–ate	SO_4^{2-} sulfate	(stem)-*ic* acid	sulfuric acid

CONCLUSION QUESTIONS

Table A	
Acid Formula	Acid Name:
HCl	
	hypochlorous acid
	chlorous acid
	chloric acid
	perchloric acid ('hyperchloric acid)
HNO_3	
	hydrobromic acid
H_3PO_4	
H_3PO_3	
	hydrocyanic acid
$HC_2H_3O_2$	
	carbonic acid
	hydroiodic acid
HF	

Table B					
# of carbon atoms = n	prefix or stem	-ane C_nH_{2n+2}	-ene C_nH_{2n}	-yne C_nH_{2n-2}	-anol $C_nH_{2n+1}OH$
1	meth-		None here because you must have at least 2 carbons !		CH_3OH
2	eth-				
3	prop-		C_3H_6		
4	but-				
5	pent-	C_5H_{12}			
6	hex-				
7	hept-				$C_7H_{15}OH$
8	oct-			C_8H_{14}	
9	non-				
10	dec-				

Table C

	Ag^+	Pb^{2+}	Cu^+	Ba	NH_4	Al^{3+}	Mn^{2+}
N							
O							
Br							
S							
SO_4							
ClO_2							
PO_3							

Table D	
Formula	Name
	tetrahydroxoaluminate (III) ion
$[Ag(NH_3)_2]^+$	
	tetrahydroxozincate (II) ion
$[Zn(NH_3)_4]^{2+}$	
	tetramminecopper(II) ion
$[FeSCN]^{2+}$	
$[FeSCN]Cl_2$	
	tetramminecadmium(II) ion
$[Ag(CN)_2]^-$	
$Mg[Ag(CN)_2]_2$	
	dibromodichlorodiiodocuprate(II) ion
$[Co(NH_3)_5Cl]Cl_2$	
	potassium hexacyanoferrate(III)
	triamminebromoplatinum(II) chloride
$[Cu(Cl_2Br_2INH_3)]^{4-}$	
$K_3[Co(F)_6]$	
$[Co(NH_3)_6]Cl_2$	
	hexacyanoferrate(II)

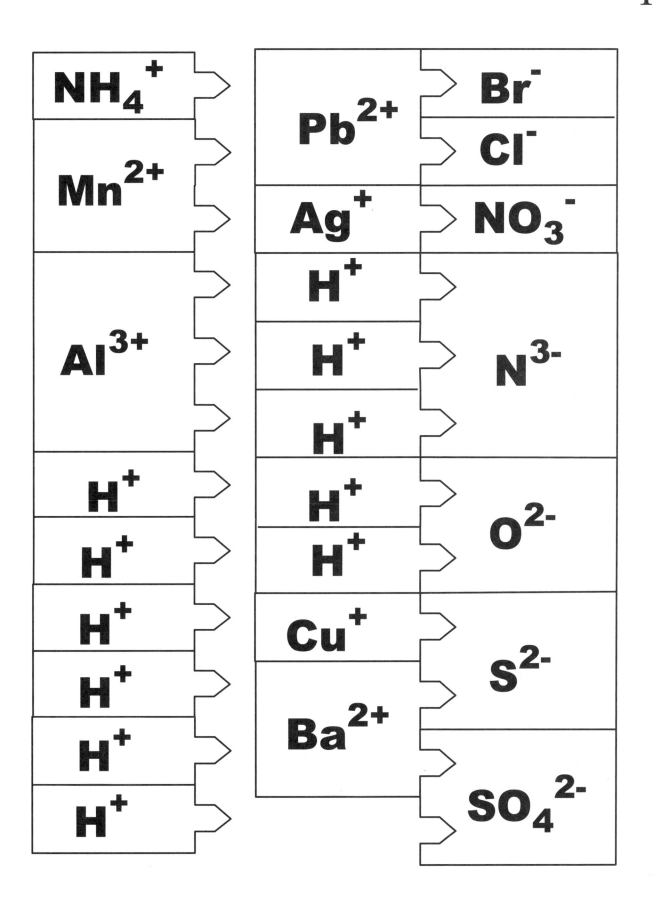

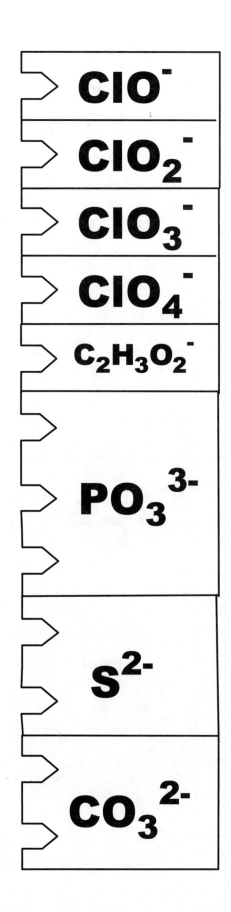

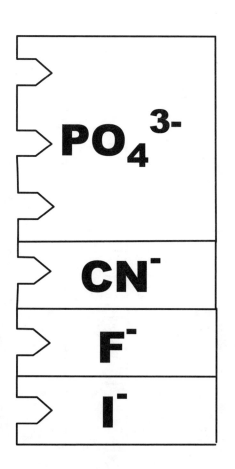

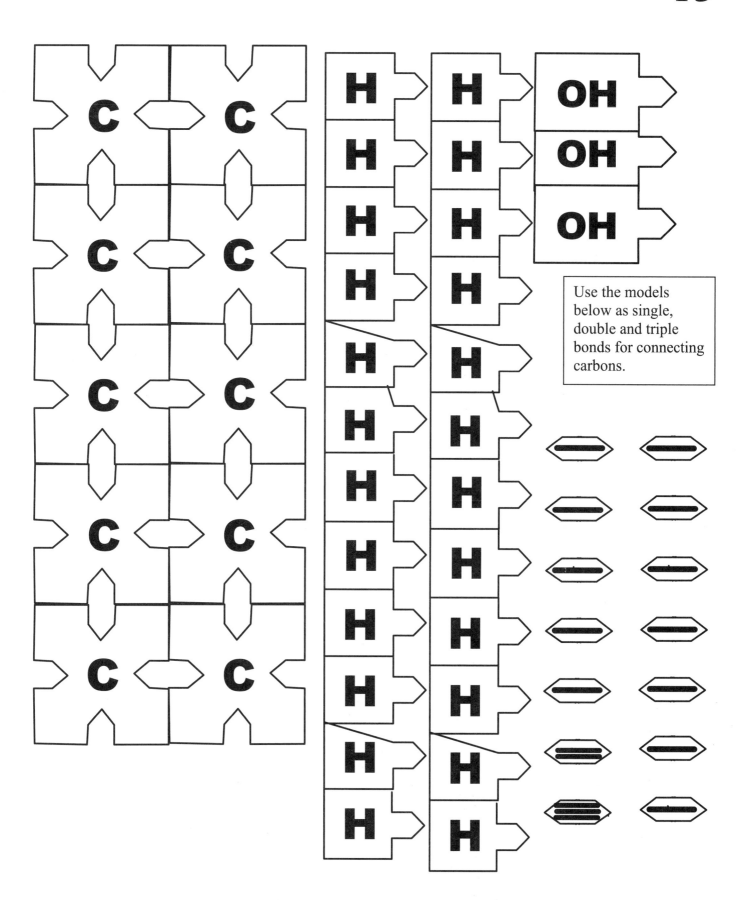

Use the models below as single, double and triple bonds for connecting carbons.

Net Ionic Equations
Making Sense of Chemical Reactions

OBJECTIVE

Students will be able to write net ionic equations from balanced molecular equations.

LEVEL

Chemistry

NATIONAL STANDARDS

UCP.1, UCP.2, B.3

TEKS

11(A), 11(B), 11(C), 12(B)

CONNECTIONS TO AP

AP Chemistry:
   III. Reactions A. Reaction Types 1. Acid-base reactions 2. Precipitation reactions 3. Oxidation-reduction reactions B. Stoichiometry 1. Ionic and molecular species present in chemical systems: net ionic equations

TIME FRAME

45 minutes

MATERIALS

note sheet   periodic table
solubility rules

TEACHER NOTES

This lesson should follow a study of the five main types of chemical reactions. Having students write net ionic equations will enhance their ability to understand "what is really happening in the beaker" type questions. The student lecture notes can provide the basis for your lecture over this material. Demonstrate the process one step at a time through several examples until the students feel confident. It often helps to demonstrate the reaction for the students by mixing the stated solutions in a test tube or beaker and asking them to explain in chemical terms what has taken place. Follow this lesson with *The Eight Solution Problem* lab to reinforce observing chemical reactions and give students more practice writing net ionic equations.

Train students to write the state symbols for each species. This will reduce the error of trying to cancel terms incorrectly identified as spectator ions. For example, if zinc metal reacts with hydrochloric acid, the following net ionic reaction results:

$$Zn_{(s)} + 2H^+_{(aq)} \rightarrow Zn^{2+}_{(aq)} + H_{2(g)}$$

Students often feel the need to cancel out the zinc term—by writing in state symbols, it is a bit more evident that the two terms are not identical and therefore do not cancel.

Students should memorize the strong acids and the strong bases. It is strongly recommended that your students begin memorizing the solubility rules during the first year of chemistry. The solubility rules must be memorized for the AP exam.

Continue writing net ionic reactions throughout the year in demonstrations and pre- and post-lab discussions to reinforce this concept.

Examples from the notes: Demonstrate any of these reactions for which you have materials as you work the examples. Recognition of state symbols is much easier when students physically see the reaction.

Example 1:
Solutions of iron (III) nitrate and potassium hydroxide are mixed.
- **Balanced formula equation:**
  $$Fe(NO_3)_{3(aq)} + 3KOH_{(aq)} \rightarrow 3KNO_{3(aq)} + Fe(OH)_{3(s)}$$

- **Total ionic equation:**
  $$Fe^{3+}_{(aq)} + 3NO_3^-{}_{(aq)} + 3K^+_{(aq)} + 3OH^-_{(aq)} \rightarrow 3K^+ + 3NO_3^-{}_{(aq)} + Fe(OH)_{3(s)}$$

- **Balanced net ionic equation:**
  $$Fe^{3+}_{(aq)} + 3OH^-_{(aq)} \rightarrow Fe(OH)_{3(s)}$$

Example 2:
Magnesium ribbon reacts with hydrochloric acid.
- **Balanced formula equation:**
  $$Mg_{(s)} + 2HCl_{(aq)} \rightarrow MgCl_{2(aq)} + H_{2(g)}$$

- **Total ionic equation:**
  $$Mg_{(s)} + 2H^+_{(aq)} + 2Cl^-_{(aq)} \rightarrow Mg^{2+}_{(aq)} + 2Cl^-_{(aq)} + H_{2(g)}$$

- **Balanced net ionic equation:**
  $$Mg_{(s)} + 2H^+_{(aq)} \rightarrow Mg^{2+}_{(aq)} + H_{2(g)}$$

Example 3:

Solutions of acetic acid and lithium bicarbonate are mixed.

- **Balanced formula equation:**

$$HC_2H_3O_{2(aq)} + LiHCO_{3(aq)} \rightarrow LiC_2H_3O_{2(aq)} + H_2CO_{3(aq)}$$

- **Total ionic equation:**

$$HC_2H_3O_{2(aq)} + Li^+_{(aq)} + HCO_3^-_{(aq)} \rightarrow Li^+_{(aq)} + C_2H_3O_2^-_{(aq)} + H_2O_{(l)} + CO_{2(g)}$$

- **Balanced net ionic equation:**

$$HC_2H_3O_{2(aq)}) + HCO_3^-_{(aq)} \rightarrow H_2O_{(l)} + CO_{2(g)} + C_2H_3O_2^-_{(aq)}$$

Example 4:

Solutions of magnesium chloride and calcium nitrate are mixed.

- **Balanced formula equation:**

$$MgCl_{2(aq)} + Ca(NO_3)_{2(aq)} \rightarrow CaCl_{2(aq)} + Mg(NO_3)_{2(aq)}$$

- **Total ionic equation:**

$$Mg^{2+}_{(aq)} + 2Cl^-_{(aq)} + Ca^{2+}_{(aq)} + 2NO_3^-_{(aq)} \rightarrow Ca^{2+}_{(aq)} + 2Cl^-_{(aq)} + Mg^{2+}_{(aq)} + 2NO_3^-_{(aq)}$$

- **Balanced net ionic equation:**

No net ionic equation—everything cancels so there is no driving force—this was just a physical combination of two different solutions.

ANSWERS TO THE CONCLUSION QUESTIONS

1. **Solutions of lead (II) nitrate and lithium chloride are mixed.**
   - $Pb(NO_3)_{2(aq)} + 2LiCl_{(aq)} \rightarrow 2LiNO_{3(aq)} + PbCl_{2(aq)}$
   - $Pb^{2+}_{(aq)} + 2NO_3^-_{(aq)} + 2Li^+_{(aq)} + 2Cl^-_{(aq)} \rightarrow 2Li^+_{(aq)} + 2NO_3^-_{(aq)} + PbCl_{2(s)}$
   - $Pb^{2+}_{(aq)} + 2Cl^-_{(aq)} \rightarrow PbCl_{2(s)}$

2. **Copper metal is placed into a solution of silver nitrate.**
   - $Cu_{(s)} + 2AgNO_{3(aq)} \rightarrow Cu(NO_3)_{2(aq)} + 2Ag_{(s)}$
   - $Cu_{(s)} + 2Ag^+_{(aq)} + 2NO_3^-_{(aq)} \rightarrow Cu^{2+}_{(aq)} + 2NO_3^-_{(aq)} + 2Ag_{(s)}$
   - $Cu_{(s)} + 2Ag^+_{(aq)} \rightarrow Cu^{2+}_{(aq)} + 2Ag_{(s)}$

3. **Solid potassium chlorate decomposes with heat.**
   - $2KClO_{3(s)} \rightarrow 2KCl_{(s)} + 3O_{2(g)}$
   - $2KClO_{3(s)} \rightarrow 2KCl_{(s)} + 3O_{2(g)}$

   *note—solids and gases do not ionize –the first reaction is the net ionic equation—students do not need to write three equations for full credit.

4. **Solid sodium metal is placed into distilled water.**
   - $2Na_{(s)} + 2H_2O_{(l)} \rightarrow 2NaOH_{(aq)} + H_{2(g)}$
   - $2Na_{(s)} + 2H_2O_{(l)} \rightarrow 2Na^+_{(aq)} + 2OH^-_{(aq)} + H_{2(g)}$

   * note—there are no spectator ions in this equation; therefore students could stop with the second equation for full credit.

5. **Chlorine gas is bubbled into a solution of magnesium bromide.**
   - $Cl_{2(g)} + MgBr_{2(aq)} \rightarrow MgCl_{2(aq)} + Br_{2(l)}$
   - $Cl_{2(g)} + Mg^{2+}_{(aq)} + 2Br^-_{(aq)} \rightarrow Mg^{2+}_{(aq)} + 2Cl^-_{(aq)} + Br_{2(l)}$
   - $Cl_{2(g)} + 2Br^-_{(aq)} \rightarrow 2Cl^-_{(aq)} + Br_{2(l)}$

6. **Methane gas is burned in the presence of oxygen gas.**
   - $CH_{4(g)} + 2O_{2(g)} \rightarrow CO_{2(g)} + 2H_2O_{(g)}$

   *note—all of these are gases or molecular so this is the net ionic equation.

7. **Solutions of silver acetate and barium chloride are mixed.**
   - $2AgC_2H_3O_{2(aq)} + BaCl_{2(aq)} \rightarrow 2AgCl_{(s)} + Ba(C_2H_3O_2)_{2(aq)}$
   - $2Ag^+_{(aq)} + 2C_2H_3O_2^-_{(aq)} + Ba^{2+}_{(aq)} + 2Cl^-_{(aq)} \rightarrow 2AgCl_{(s)} + Ba^{2+}_{(aq)} + 2C_2H_3O_2^-_{(aq)}$
   - $Ag^+_{(aq)} + Cl^-_{(aq)} \rightarrow AgCl_{(s)}$

8. **Solutions of sodium bicarbonate and hydrochloric acid are mixed.**
   - $NaHCO_{3(aq)} + HCl_{(aq)} \rightarrow NaCl_{(aq)} + H_2CO_{3(aq)}$
   - $Na^+_{(aq)} + HCO_3^-_{(aq)} + H^+_{(aq)} + Cl^-_{(aq)} \rightarrow Na^+_{(aq)} + Cl^-_{(aq)} + H_2O_{(l)} + CO_{2(g)}$
   - $HCO_3^-_{(aq)} + H^+_{(aq)} \rightarrow H_2O_{(l)} + CO_{2(g)}$

   * note—remember that carbonic acid quickly decomposes into water and carbon dioxide in an acid medium.

9. **Solutions of ammonium perchlorate and barium hydroxide are mixed.**
   - $2NH_4ClO_{4(aq)} + Ba(OH)_{2(aq)} \rightarrow 2NH_4OH_{(aq)} + Ba(ClO_4)_{2(aq)}$
   - $2NH_4^+_{(aq)} + 2ClO_4^-_{(aq)} + Ba^{2+}_{(aq)} + 2OH^-_{(aq)} \rightarrow 2NH_{3(g)} + 2H_2O_{(l)} + Ba^{2+}_{(aq)} + 2ClO_4^-_{(aq)}$
   - $2NH_4^+ + 2OH^- \rightarrow 2NH_3 + 2H_2O$

   * note—remember that ammonium hydroxide decomposes into ammonia and water.

10. **Solutions of tin (II) fluoride and lithium carbonate are mixed.**
    - $SnF_{2(aq)} + Li_2CO_{3(aq)} \rightarrow SnCO_{3(s)} + 2LiF_{(aq)}$
    - $Sn^{2+}_{(aq)} + 2F^-_{(aq)} + 2Li^+_{(aq)} + CO_3^{2-}_{(aq)} \rightarrow SnCO_{3(s)} + 2Li^+_{(aq)} + 2F^-_{(aq)}$
    - $Sn^{2+}_{(aq)} + CO_3^{2-}_{(aq)} \rightarrow SnCO_{3(s)}$

    * note—carbonate was not listed directly in the solubility rules so it is considered insoluble.

Net Ionic Equations
Making Sense of Chemical Reactions

Now that you have mastered writing balanced chemical equations it is time to take a deeper look at what is really taking place chemically in each reaction. There are many driving forces for chemical reactions. By writing the net ionic equation for a reaction it is often evident which driving force caused the reaction. The most common driving forces for chemical reactions are: formation of a precipitate, formation of a molecular compound such as water, and formation of a gas.

PURPOSE
In this activity you will write balanced chemical formulas, total ionic equations and net ionic equations to explain chemical reactions.

MATERIALS
note sheet   periodic table
solubility rules

CLASS NOTES
Net ionic equations represent only the species that are taking part in a chemical reaction. The parts of the equation that are not shown are known as the *spectator ions*. Spectator ions do just that—they "spectate"—they must be present in order for the reaction to occur since compounds are neutral, but they are not directly involved in the reaction. The type of equations that you have become familiar with are known as *balanced formula reactions*. In balanced formula equations all species are shown written as formulas with proper coefficients to balance the atoms. In a *total ionic equation,* substances that ionize extensively in solution are written as ions while all others are written as formulas.

How do you know when a substance will extensively ionize in solution? All strong electrolytes will ionize extensively in solution. Electrolytes are comprised of three classes of compounds—strong acids, strong bases and soluble salts. Using the information that follows you should be able to determine whether or not to ionize a particular substance.

1. **Strong Acids**: HCl, HBr, HI, H_2SO_4, HNO_3, $HClO_4$, $HClO_3$

2. **Strong Bases**: Hydroxides of group IA and IIA (Ba, Sr, Ca are marginal—the ions that do dissolve are in solution 100% and so are ionized when they are in dilute solutions. Be and Mg are weak and are not ionized.)

3. **Soluble Salts** (see Table 1): (ionic compounds: metal/nonmetal)

Table 1

Always Soluble if in a Compound	Except With
NO_3^-, Group IA, NH_4^+, $C_2H_3O_2^-$, ClO_4^-, ClO_3^-	No Exceptions
Cl^-, Br^-, I^-	Pb, Ag, Hg_2^{2+}
SO_4^{2-}	Pb, Ag, Hg_2^{2+} Ca, Sr, Ba

If the substance does not fit into one of the three rules listed in Table 1, assume that it is insoluble or it is a weak electrolyte and does not ionize in solution. (This won't always be correct, but will cover most of the situations you encounter.)

A few other important points:
- Gases, pure liquids, and solids are non-electrolytes.
- H_2CO_3 decomposes into $H_2O_{(l)}$ and $CO_{2(g)}$
- NH_4OH decomposes into $H_2O_{(l)}$ and $NH_{3(g)}$

Let's explore these terms using an example:
- Consider a reaction between solutions of sodium chloride and lead (II) nitrate. The three types of equations are:

- **Balanced formula equation**: (shows everything balanced)
  $2NaCl_{(aq)} + Pb(NO_3)_{2(aq)} \rightarrow 2NaNO_{3(aq)} + PbCl_{2(s)}$

- **Total ionic equation**: (ionize substances that are strong electrolytes)
  $2 Na^+_{(aq)} + 2 Cl^-_{(aq)} + Pb^{2+}_{(aq)} + 2 NO_3^-_{(aq)} \rightarrow 2 Na^+_{(aq)} + 2 NO_3^-_{(aq)} + PbCl_{2(s)}$

- **Balanced net ionic equation**: (what is left after canceling common terms—spectator ions)
  $2 Cl^-_{(aq)} + Pb^{2+}_{(aq)} \rightarrow PbCl_{2(s)}$

Identify the spectator ions in the equation above? Na^+ and NO_3^-

Example 1:
Solutions of iron (III) nitrate and potassium hydroxide are mixed.

Balanced formula equation:

Total ionic equation:

Balanced net ionic equation:

Example 2:
Magnesium ribbon reacts with hydrochloric acid.

Balanced formula equation:

Total ionic equation:

Balanced net ionic equation:

Example 3:
Solutions of acetic acid and lithium bicarbonate are mixed.

Balanced formula equation:

Total ionic equation:

Balanced net ionic equation:

Example 4:
Solutions of magnesium chloride and calcium nitrate are mixed.

Balanced formula equation:

Total ionic equation:

Balanced net ionic equation:

Name _____

Period _____

Net Ionic Equations
Making Sense of Chemical Reactions

CONCLUSION QUESTIONS

Write the net ionic reaction for each of the following chemical reactions on your own paper. Be sure to show the balanced formula equation, total ionic equation and net ionic equation for each of the following in order to receive full credit. Be sure to put in state symbols for each component and correct charges for ions where appropriate.

1. Solutions of barium nitrate and lithium chloride are mixed.

2. Copper metal is placed into a solution of silver nitrate.

3. Solid potassium chlorate decomposes with heat.

4. Solid sodium metal is placed into distilled water.

5. Chlorine gas is bubbled into a solution of magnesium bromide.

6. Methane gas is burned in the presence of oxygen gas.

7. Solutions of silver acetate and barium chloride are mixed.

8. Solutions of sodium bicarbonate and hydrochloric acid are mixed.

9. Solutions of ammonium perchlorate and barium hydroxide are mixed.

10. Solutions of tin (II) fluoride and lithium carbonate are mixed.

The Eight Solution Problem
Exploring Reactions of Aqueous Ionic Compounds

OBJECTIVE
Students will observe chemical reactions by mixing various ionic solutions in a grid-like fashion and then use these observations to identify eight unknowns.

LEVEL
Chemistry

NATIONAL STANDARDS
UCP.2, A.1, A.2, B.3,

TEKS
1(A), 2(B), 2(E), 11(A), 11(B), 11(C), 12(B)

CONNECTIONS TO AP
AP Chemistry:
  III. Reactions A. Reaction Types 1. Acid-base reactions 2. Precipitation reactions

TIME FRAME
45 minutes for Part I and 45 minutes for Part II.
45 minutes outside of class for completing equations and questions.

MATERIALS
(For a class of 28 working in pairs)

14 96 well plates
solutions of the following: $Pb(NO_3)_2$,
    Na_2CO_3, $CaCl_2$, HCl, H_2SO_4, $AgNO_3$,
    HNO_3, $NaOH$, KI, Na_3PO_4, $FeCl_3$, $CuSO_4$
28 lab aprons

1 box cotton swabs
14 sets of small scale pipets containing the
12 solutions
28 goggles
14 pieces of black paper or other dark surface

TEACHER NOTES

This lab is a good exercise in logical thinking, deductive reasoning, and problem solving. By having to rely on their observations from part I to identify unknowns in part II students will have to think carefully about which solutions to combine. This activity fits nicely with a study of chemical reactions, ionic equations, solubility rules, or precipitation reactions. Students will be asked to write net ionic equations so some instruction on proper procedure for this will need to have been presented.

Since this is a qualitative lab, the molarity of the above solutions is not a large concern. Any equimolar concentrations between 0.1M and 1.0M will suffice.

You can choose any of the solutions for your unknowns, but the following list tends to be easier for students to recognize: nitric acid, potassium iodide, sodium phosphate, silver nitrate, sodium hydroxide, lead (II) nitrate, and sulfuric acid. You will need to make a set of unknown pipets for each lab pair and label them with numbers. The wells the students are working with are very small, therefore, you may need to pull and trim the pipets to make them dispense less volume. To prevent students from figuring out they all have the same unknowns, create a numbering system. For example any number ending in zero is sulfuric acid, any number ending in 1 is silver nitrate, etc. Be sure to keep record of your scheme for refilling and grading students.

You may want to allow more than one night to complete the conclusion questions for this lab since the students will be writing equations for all the reactions and net ionic equations for all the applicable reactions.

Cotton swabs are included in the materials so that students can thoroughly clean the small chambers on the well plate. Some of the precipitates are sticky and cling to the plate.

	Pb(NO$_3$)$_2$	HCl	Na$_3$PO$_4$	KI	CuSO$_4$	H$_2$SO$_4$	NaOH	AgNO$_3$	CaCl$_2$	HNO$_3$	Na$_2$CO$_3$	FeCl$_3$
FeCl$_3$	① white ppt	② NR	③ NR	④ amber color	⑤ NR	⑥ NR	⑦ amber ppt	⑧ white ppt	⑨ NR	⑩ NR	⑪ orange milky ppt	
Na$_2$CO$_3$	⑫ white ppt	⑬ bubbles	⑭ NR	⑮ NR	⑯ thick orange green ppt	⑰ bubbles	⑱ NR	⑲ dirty white ppt	⑳ white ppt	㉑ bubbles		
HNO$_3$	㉒ NR	㉓ NR	㉔ NR	㉕ dark brown bubbles	㉖ NR	㉗ NR	㉘ NR	㉙ NR	㉚ NR			
CaCl$_2$	㉛ white ppt	㉜ NR	㉝ cloudy ppt	㉞ NR	㉟ green yellow ppt	㊱ ppt	㊲ white ppt	㊳ white ppt				
AgNO$_3$	㊴ NR	㊵ white ppt	㊶ milky light brown ppt	㊷ milky white ppt	㊸ white ppt	㊹ NR	㊺ orange ppt					
NaOH	㊻ clumpy white ppt	㊼ NR	㊽ NR	㊾ NR	㊿ blue ppt	51 NR						
H$_2$SO$_4$	52 white ppt	53 NR	54 NR	55 NR	56 NR							
CuSO$_4$	57 white ppt	58 NR	59 white ppt	60 brown ppt								
KI	61 orange ppt	62 NR	63 NR									
Na$_3$PO$_4$	64 white ppt	65 NR										
HCl	66 white ppt											

POSSIBLE ANSWERS TO THE CONCLUSION QUESTIONS AND SAMPLE DATA

1. On a sheet of paper write the full chemical equation, including state symbols for each of the reactions you observed. If there was no reaction place the symbol NR for no reaction in front of the reaction. Be careful here, just because no <u>precipitate</u> was formed doesn't mean there was no reaction; some reactions produce gases or other molecular compounds. For all reactions producing a precipitates, gases or other molecular compounds, write the net ionic equation. Be sure that all formulas are correct and that all equations are balanced. When writing the chemical equations start with the box labeled (1) on data table 1 and proceed in numerical order. Skip lines between equations. You may wish to type the equations for neatness.

1. $2FeCl_{3(aq)} + 3Pb(NO_3)_{2(aq)} \rightarrow 2Fe(NO_3)_{3(aq)} + 3PbCl_{2(s)}$
   $Pb^{2+}_{(aq)} + 2Cl^-_{(aq)} \rightarrow PbCl_{2\,(s)}$

NR 2. $FeCl_{3(aq)} + HCl_{(aq)} \rightarrow FeCl_{3(aq)} + HCl_{(aq)}$

3. $FeCl_{3(aq)} + Na_3PO_{4(aq)} \rightarrow FePO_{4(s)} + 3NaCl_{(aq)}$
   $Fe^{3+}_{(aq)} + PO_4^{3-}_{(aq)} \rightarrow FePO_{4(s)}$

NR 4. $FeCl_{3(aq)} + 3KI_{(aq)} \rightarrow FeI_{3(aq)} + 3KCl_{(aq)}$ * color change

   *Students will more than likely say NR—this reaction is actually a redox reaction

   $Fe^{3+}_{(aq)} + I^-_{(aq)} \rightarrow Fe^{2+}_{(aq)} + I_{2(s)}$

NR 5. $2FeCl_{3(aq)} + 3CuSO_{4(aq)} \rightarrow Fe_2(SO_4)_{3(aq)} + 3CuCl_{2(aq)}$

NR 6. $2FeCl_{3(aq)} + 3H_2SO_{4(aq)} \rightarrow Fe_2(SO_4)_{3(aq)} + 6HCl_{(aq)}$

7. $FeCl_{3(aq)} + 3NaOH_{(aq)} \rightarrow Fe(OH)_{3(s)} + 3NaCl_{(aq)}$
   $Fe^{3+}_{(aq)} + 3OH^-_{(aq)} \rightarrow Fe(OH)_{3(s)}$

8. $FeCl_{3(aq)} + 3AgNO_{3(aq)} \rightarrow Fe(NO_3)_{3(aq)} + 3AgCl_{(s)}$
   $Ag^+_{(aq)} + Cl^-_{(aq)} \rightarrow AgCl_{(s)}$

NR 9. $FeCl_{3(aq)} + CaCl_{2(aq)} \rightarrow FeCl_{3(aq)} + CaCl_{2(aq)}$

NR 10. $2FeCl_{3(aq)} + 3HNO_{3(aq)} \rightarrow Fe(NO_3)_{3(aq)} + 3HCl_{(aq)}$

11. $2FeCl_{3(aq)} + 3Na_2CO_{3(aq)} \rightarrow Fe_2(CO_3)_{3(s)} + 6NaCl_{(aq)}$
    $2Fe^{3+}_{(aq)} + 3CO_3^{2-}_{(aq)} \rightarrow Fe_2(CO_3)_{3(s)}$

12. $Na_2CO_{3(aq)} + Pb(NO_3)_{2(aq)} \rightarrow 2NaNO_{3(aq)} + PbCO_{3(s)}$
    $Pb^{2+}_{(aq)} + CO_3^{2-}_{(aq)} \rightarrow PbCO_{3(s)}$

13. $Na_2CO_{3(aq)} + 2HCl_{(aq)} \rightarrow 2NaCl_{(aq)} + \cancel{H_2CO_{3(aq)}}\ H_2O_{(l)} + CO_{2(g)}$
    $2H^+_{(aq)} + CO_3^{2-}_{(aq)} \rightarrow H_2O_{(l)} + CO_{2(g)}$

NR 14. $Na_2CO_{3(aq)} + Na_3PO_{4(aq)} \rightarrow Na_2CO_{3(aq)} + Na_3PO_{4(aq)}$

NR 15. $Na_2CO_{3(aq)} + 2KI_{(aq)} \rightarrow 2NaI_{(aq)} + K_2CO_{3(aq)}$

16. $Na_2CO_{3(aq)} + CuSO_{4(aq)} \rightarrow CuCO_{3(s)} + Na_2SO_{4(aq)}$
$Cu^{2+}_{(aq)} + CO_{3(aq)} \rightarrow CuCO_{3(s)}$

17. $Na_2CO_{3(aq)} + H_2SO_{4(aq)} \rightarrow Na_2SO_{4(aq)} + \sout{H_2CO_{3(aq)}}\, H_2O_{(l)} + CO_{2(g)}$
$2\,H^+_{(aq)} + CO_3^{2-}_{(aq)} \rightarrow H_2O_{(l)} + CO_{2(g)}$

NR 18. $Na_2CO_{3(aq)} + NaOH_{(aq)} \rightarrow Na_2CO_{3(aq)} + NaOH_{(aq)}$

19. $Na_2CO_{3(aq)} + 2AgNO_{3(aq)} \rightarrow 2NaNO_{3(aq)} + Ag_2CO_{3(s)}$
$2Ag^+_{(aq)} + CO_3^{2-}_{(aq)} \rightarrow Ag_2CO_{3(s)}$

20. $Na_2CO_{3(aq)} + CaCl_{2(aq)} \rightarrow 2NaCl_{(aq)} + CaCO_{3(s)}$
$Ca^{2+}_{(aq)} + CO_3^{2-}_{(aq)} \rightarrow CaCO_{3(s)}$

21. $Na_2CO_{3(aq)} + 2HNO_{3(aq)} \rightarrow 2NaNO_{3(aq)} + \sout{H_2CO_{3(aq)}}\, H_2O_{(l)} + CO_{2(g)}$
$2\,H^+_{(aq)} + CO_3^{2-}_{(aq)} \rightarrow H_2O_{(l)} + CO_{2(g)}$

NR 22. $HNO_{3(aq)} + Pb(NO_3)_{2(aq)} \rightarrow HNO_{3(aq)} + Pb(NO_3)_{2(aq)}$

NR 23. $HNO_{3(aq)} + HCl_{(aq)} \rightarrow HNO_{3(aq)} + HCl_{(aq)}$

24. $3HNO_{3(aq)} + Na_3PO_{4(aq)} \rightarrow H_3PO_{4(aq)} + 3NaNO_{3(aq)}$

$3H^+_{(aq)} + PO_4^{3+}_{(aq)} \rightarrow H_3PO_{4\,(aq)}$ *weak acid only partial dissociation

25. $HNO_{3(aq)} + KI_{(aq)} \rightarrow HI_{(aq)} + KNO_{3(aq)}$ *really dark bubbles, brown

*Students will more than likely say NR—this reaction is actually a redox reaction

$H^+_{(aq)} + NO_3^-{}_{(aq)} + I^-_{(aq)} \rightarrow NO_{(g)} + I_{2(s)} + H_2O_{(l)}$

NR 26. $2HNO_{3(aq)} + CuSO_{4(aq)} \rightarrow H_2SO_{4(aq)} + Cu(NO_3)_{2(aq)}$

NR 27. $HNO_{3(aq)} + H_2SO_{4(aq)} \rightarrow HNO_{3(aq)} + H_2SO_{4(aq)}$

28. $HNO_{3(aq)} + NaOH_{(aq)} \rightarrow NaNO_{3(aq)} + H_2O_{(l)}$
$H^+_{(aq)} + OH^-_{(aq)} \rightarrow H_2O_{(l)}$

NR 29. $HNO_{3(aq)} + AgNO_{3(aq)} \rightarrow HNO_{3(aq)} + AgNO_{3(aq)}$

NR 30. $2HNO_{3(aq)} + CaCl_{2(aq)} \rightarrow 2HCl_{(aq)} + Ca(NO_3)_{2(aq)}$

31. $CaCl_{2(aq)} + Pb(NO_3)_{2(aq)} \rightarrow Ca(NO_3)_{2(aq)} + PbCl_{2(s)}$
$Pb^{2+}_{(aq)} + 2Cl^-_{(aq)} \rightarrow PbCl_{2(s)}$

NR 32. $CaCl_{2(aq)} + HCl_{(aq)} \rightarrow CaCl_{2(aq)} + HCl_{(aq)}$

33. $3CaCl_{2(aq)} + 2Na_3PO_{4(aq)} \rightarrow Ca_3(PO_4)_{2(s)} + 6NaCl_{(aq)}$
    $3Ca^{2+}_{(aq)} + 2PO_4^{3-}_{(aq)} \rightarrow Ca_3(PO_4)_{2(s)}$

NR 34. $CaCl_{2(aq)} + 2KI_{(aq)} \rightarrow CaI_{2(aq)} + KCl_{(aq)}$

35. $CaCl_{2(aq)} + CuSO_{4(aq)} \rightarrow CaSO_{4(s)} + CuCl_{2(aq)}$
    $Ca^{2+}_{(aq)} + SO_4^{2-}_{(aq)} \rightarrow CaSO_{4(s)}$

    * students may actually mark this reaction NR—calcium sulfate is moderately soluble and therefore, depends on concentration

36. $CaCl_{2(aq)} + H_2SO_{4(aq)} \rightarrow CaSO_{4(s)} + 2HCl_{(aq)}$
    $Ca^{2+}_{(aq)} + SO_4^{2-}_{(aq)} \rightarrow CaSO_{4(s)}$

    * students may actually mark this reaction NR—calcium sulfate is moderately soluble and therefore, depends on concentration

NR 37. $CaCl_{2(aq)} + 2NaOH_{(aq)} \rightarrow Ca(OH)_{2(aq)} + 2NaCl_{(aq)}$

38. $CaCl_{2(aq)} + 2AgNO_{3(aq)} \rightarrow Ca(NO_3)_{2(aq)} + 2AgCl_{(s)}$
    $Ag^+_{(aq)} + Cl^-_{(aq)} \rightarrow 2AgCl_{(s)}$

NR 39. $AgNO_{3(aq)} + Pb(NO_3)_{2(aq)} \rightarrow AgNO_{3(aq)} + Pb(NO_3)_{2(aq)}$

40. $AgNO_{3(aq)} + HCl_{(aq)} \leftrightarrow AgCl_{(s)} + HNO_{3(aq)}$
    $Ag^+_{(aq)} + Cl^-_{(aq)} \rightarrow AgCl_{(s)}$

41. $3AgNO_{3(aq)} + Na_3PO_{4(aq)} \rightarrow Ag_3PO_{4(s)} + 3NaNO_{3(aq)}$
    $3Ag^+_{(aq)} + PO_4^{3-}_{(aq)} \rightarrow Ag_3PO_{4(s)}$

42. $AgNO_{3(aq)} + KI_{(aq)} \rightarrow AgI_{(s)} + KNO_{3(aq)}$
    $Ag^+_{(aq)} + I^-_{(aq)} \rightarrow AgI_{(s)}$

43. $2AgNO_{3(aq)} + CuSO_{4(aq)} \rightarrow Ag_2SO_{4(aq)} + Cu(NO_3)_{2(aq)}$
    $2Ag^+_{(aq)} + SO_4^{2-}_{(aq)} \rightarrow Ag_2SO_{4(s)}$

    * students may actually mark this reaction NR—silver sulfate is moderately soluble and therefore, depends on concentration

44. $2AgNO_{3(aq)} + H_2SO_{4(aq)} \rightarrow Ag_2SO_{4(aq)} + 2HNO_{3(aq)}$
    $2Ag^+_{(aq)} + SO_4^{2-}_{(aq)} \rightarrow Ag_2SO_{4(s)}$

    * students may actually mark this reaction NR—silver sulfate is moderately soluble and therefore, depends on concentration

45. $AgNO_{3(aq)} + NaOH_{(aq)} \rightarrow AgOH_{(s)} + NaNO_{3(aq)}$
    $Ag^+_{(aq)} + OH^-_{(aq)} \rightarrow AgOH_{(s)}$

46. $2NaOH_{(aq)} + Pb(NO_3)_{2(aq)} \rightarrow 2NaNO_{3(aq)} + Pb(OH)_{2(s)}$
$Pb^{2+}_{(aq)} + 2OH^-_{(aq)} \rightarrow Pb(OH)_{2(s)}$

47. $NaOH_{(aq)} + HCl_{(aq)} \rightarrow NaCl_{(aq)} + H_2O_{(l)}$
$H^+_{(aq)} + OH^-_{(aq)} \rightarrow H_2O_{(l)}$

NR 48. $NaOH_{(aq)} + Na_3PO_{4(aq)} \rightarrow NaOH_{(aq)} + Na_3PO_{4(aq)}$

NR 49. $NaOH(aq) + KI(aq) \rightarrow NaI(aq) + KOH(aq)$

50. $2NaOH_{(aq)} + CuSO_{4(aq)} \rightarrow Na_2SO_{4(aq)} + Cu(OH)_{2(s)}$
$Cu^{2+}_{(aq)} + 2OH^-_{(aq)} \rightarrow Cu(OH)_{2(s)}$

51. $2NaOH_{(aq)} + H_2SO_{4\ (aq)} \rightarrow Na_2SO_{4\ (aq)} + 2H_2O_{(l)}$
$2H^+_{(aq)} + 2OH^-_{(aq)} \rightarrow 2H_2O_{(l)}$

52. $H_2SO_{4\ (aq)} + Pb(NO_3)_{2(aq)} \rightarrow 2HNO_{3(aq)} + PbSO_{4(s)}$
$Pb^{2+}_{(aq)} + SO_4^{2-}_{(aq)} \rightarrow PbSO_{4(s)}$

N NR 53. $H_2SO_{4\ (aq)} + HCl_{(aq)} \rightarrow HCl_{(aq)} + H_2SO_{4(aq)}$

54. $3H_2SO_{4\ (aq)} + 2Na_3PO_{4(aq)} \rightarrow 2H_3PO_{4(aq)} + 3Na_2SO_{4(aq)}$
$6H^+_{(aq)} + 3PO_4^{3-}_{(aq)} \rightarrow 2H_3PO_{4\ (aq)}$

N NR 55. $H_2SO_{4\ (aq)} + KI_{(aq)} \rightarrow HI_{(aq)} + K_2SO_{4(aq)}$

N NR 56. $H_2SO_{4\ (aq)} + CuSO_{4(aq)} \rightarrow CuSO_{4(aq)} + H_2SO_{4\ (aq)}$

57. $CuSO_{4(aq)} + Pb(NO_3)_{2(aq)} \rightarrow Cu(NO_3)_{2(aq)} + PbSO_{4(s)}$
$Pb^{2+}_{(aq)} + SO_4^{2-}_{(aq)} \rightarrow PbSO_{4(s)}$

N NR 58. $CuSO_{4(aq)} + 2HCl_{(aq)} \rightarrow CuCl_{2(aq)} + H_2SO_{4(aq)}$

59. $3CuSO_{4(aq)} + 2Na_3PO_{4\ (aq)} \rightarrow Cu_3(PO_4)_{2\ (s)} + 3Na_2SO_{4(aq)}$
$3Cu^{2+}_{(aq)} + 2PO_4^{3-}_{(aq)} \rightarrow Cu_3(PO_4)_{2\ (s)}$

60. $CuSO_{4(aq)} + 2KI_{(aq)} \rightarrow CuI_{2(s)} + K_2SO_{4(aq)}$
$Cu^{2+}_{(aq)} + 2I^-_{(aq)} \rightarrow CuI_{2(s)}$

61. $2KI_{(aq)} + Pb(NO_3)_{2(aq)} \rightarrow 2KNO_{3(aq)} + PbI_{2(s)}$
$Pb^{2+}_{(aq)} + 2I^-_{(aq)} \rightarrow PbI_{2(s)}$

N NR 62. $KI_{(aq)} + HCl_{(aq)} \rightarrow KCl_{(aq)} + HI_{(aq)}$

N NR 63. $3KI_{(aq)} + Na_3PO_{4\ (aq)} \rightarrow K_3PO_{4(aq)} + 3\ NaI_{(aq)}$

64. $2Na_3PO_{4\ (aq)} + 3Pb(NO_3)_{2(aq)} \rightarrow 6NaNO_{3(aq)} + Pb_3(PO_4)_{2(s)}$
$3Pb^{2+}_{(aq)} + 2PO_4^{3-}_{(aq)} \rightarrow Pb_3(PO_4)_{2(s)}$

65. $Na_3PO_{4\,(aq)} + 3HCl_{(aq)} \rightarrow 3NaCl_{(aq)} + H_3PO_{4(aq)}$
    $3H^+_{\,(aq)} + PO_4^{\,3+}_{\,(aq)} \rightarrow H_3PO_{4\,(aq)}$

66. $2HCl_{(aq)} + Pb(NO_3)_{2(aq)} \rightarrow 2HNO_{3(aq)} + PbCl_{2(s)}$
    $Pb^{2+}_{\,(aq)} + 2Cl^-_{\,(aq)} \rightarrow PbCl_{2(s)}$

2. Which reactions formed gas? What do they all have in common? Write a net ionic equation for one of the reactions.
   - Hydrochloric acid and sodium carbonate
     $2HCl + Na_2CO_3 \rightarrow \cancel{H_2CO_3}\,H_2O + CO_2 + 2NaCl$
     $2H^+ + CO_3^{\,2-} \rightarrow H_2O + CO_2$

   - Sulfuric acid and sodium carbonate
     $H_2SO_4 + Na_2CO_3 \rightarrow \cancel{H_2CO_3}\,H_2O + CO_2 + Na_2SO_4$
     $2H^+ + CO_3^{\,2-} \rightarrow H_2O + CO_2$

   - Nitric acid and sodium carbonate
     $2HNO_3 + Na_2CO_3 \rightarrow \cancel{H_2CO_3}\,H_2O + CO_2 + 2NaNO_3$
     $2H^+ + CO_3^{\,2-} \rightarrow H_2O + CO_2$

   - They are all strong acids reacting with sodium carbonate

3. Some of the reactions produced distinct color changes but did not produce a precipitate. Does this mean that a reaction did not take place? Explain your answer.
   - No, a reaction did take place due to the fact there is a color change.

4. Were there any neutralization reactions? If so, write them below. What do they all have in common?
   - Yes, nitric acid plus sodium hydroxide, sodium hydroxide plus sulfuric acid, sodium hydroxide plus hydrochloric acid.
   - Neutralization reactions all involve an acid plus a base and they all produce water and an ionic salt.

5. Many of the equations you wrote earlier were labeled NR for no reaction. Choose one of the reactions and explain the NR by showing ion cancellation leaving no net reaction.

   58. $CuSO_{4(aq)} + 2HCl_{(aq)} \rightarrow CuCl_{2(aq)} + H_2SO_{4(aq)}$

   $\cancel{Cu^{2+}} + \cancel{SO_4^{\,2-}} + \cancel{2H^+} + \cancel{2Cl^-} \rightarrow \cancel{Cu^{2+}} + \cancel{2Cl^-} + \cancel{2H^+} + \cancel{SO_4^{\,2-}}$

The Eight Solution Problem
Exploring Reactions of Aqueous Ionic Compounds

INTRODUCTION
Your goal in this lab is to identify eight "unknown" solutions. You and your partner will first collect data by observing reactions between various known ionic compounds. You will then use this data to determine the identity of the unknowns. This type of lab requires organization, clear logical thinking, and excellent observation/data-taking skills.

PURPOSE
In this activity you will observe a variety of double replacement reactions, identify their solubilities and write net ionic equations for all reactions producing precipitates. You will also use your observation and data-taking skills to identify 8 unknown solutions.

MATERIALS
96 well plate

cotton swabs

solutions of the following: $Pb(NO_3)_2$, Na_2CO_3, $CaCl_2$, HCl, H_2SO_4, $AgNO_3$, HNO_3, NaOH, KI, Na_3PO_4, $FeCl_3$, $CuSO_4$

small scale pipets of all solutions

goggles

black paper or surface

lab aprons

Safety Alert
1. Wear your safety goggles and lab aprons.
2. If your hands come in contact with any of the chemicals make sure to rinse them under water and dry.
3. If the chemicals come in contact with your eyes, please use the eye wash according to your teacher's instructions.

PROCEDURE
PART I

1. Obtain your 96-well plate and orient it on the table so that the long edge is at the top. You might find it helpful to put piece of black or dark paper behind the plate to make any precipitates more visible.

2. You will place 2 drops of each horizontal and vertical reactant listed on data table 1 into each well. To begin, place 2 drops of iron (III) chloride in each of 11 wells across the first row. As you begin to add the second solution, make sure not to touch the pipet tip to the well or its contents. Any contamination of the pipet will cause error in subsequent reactions.

3. To the first well add two drops of lead (II) nitrate. Use a toothpick and swirl the solutions together until mixed. Take care not to scratch the bottom of the well plate. Record your observations in data table 1 on the student answer page.

4. Continue in the same fashion by observing all of the combinations listed on your data table.

5. When you have finished, carefully rinse the microplate with water and use cotton swabs to clean any wells that have remaining precipitate.

PART II

1. Obtain the eight unknown solutions as directed by your teacher. These 8 unknown solutions came from the original 12 you have already tested. Write the numbers for each of your unknowns horizontally on data table 2. Reverse the order of the numbers as you write them down the vertical column. Mix two drops of each solution in a well just as you did in part I. Stir them with a toothpick and then record your observations in data table 2. Take care not to scratch the bottom of the well plate.

2. Analyze the results of your unknown mixings by comparing them with the results recorded in part I. Identify all of the unknowns which had distinct reactions.

3. You may need to mix a few of the original 12 known solutions to correctly identify all of your unknowns. Make a plan before mixing.

4. Once you have identified the eight unknown solutions, write the number, name, and formula in data table 2 in the spaces provided on the student answer page.

Name ______________________________

Period ______________________________

The Eight Solution Problem
Reactions of Aqueous Ionic Compounds Page

DATA AND OBSERVATIONS

	$Pb(NO_3)_2$	HCl	Na_3PO_4	KI	$CuSO_4$	H_2SO_4	NaOH	$AgNO_3$	$CaCl_2$	HNO_3	Na_2CO_3	$FeCl_3$
$FeCl_3$	①	②	③	④	⑤	⑥	⑦	⑧	⑨	⑩	⑪	
Na_2CO_3	⑫	⑬	⑭	⑮	⑯	⑰	⑱	⑲	⑳	㉑		
HNO_3	㉒	㉓	㉔	㉕	㉖	㉗	㉘	㉙	㉚			
$CaCl_2$	㉛	㉜	㉝	㉞	㉟	㊱	㊲	㊳				
$AgNO_3$	㊴	㊵	㊶	㊷	㊸	㊹	㊺					
NaOH	㊻	㊼	㊽	㊾	㊿	(51)						
H_2SO_4	(52)	(53)	(54)	(55)	(56)							
$CuSO_4$	(57)	(58)	(59)	(60)								
KI	(61)	(62)	(63)									
Na_3PO_4	(64)	(65)										
HCl	(66)											

Unknown
Numbers

Unknown Name and Formula

Number	Name and Formula
_____	_____
_____	_____
_____	_____
_____	_____
_____	_____
_____	_____
_____	_____
_____	_____

CONCLUSION QUESTIONS

1. On a sheet of paper write the full chemical equation, including state symbols for each of the reactions you observed. If there was no reaction place the symbol NR for no reaction in front of the reaction. Be careful here, just because no <u>precipitate</u> was formed doesn't mean there was no reaction; some reactions produce gases or other molecular compounds. For all reactions producing a precipitates, gases or other molecular compounds, write the net ionic equation. Be sure that all formulas are correct and that all equations are balanced. When writing the chemical equations start with the box labeled (1) on data table 1 and proceed in numerical order. Skip lines between equations. You may wish to type the equations for neatness.

2. Which reactions formed gas? What do they all have in common? Write a net ionic equation for one of the reactions.

3. Some of the reactions produced distinct color changes but did not produce a precipitate. Does this mean that a reaction did not take place? Explain your answer.

4. Were there any neutralization reactions? If so, write them below. What do they all have in common?

5. Many of the equations you wrote earlier were labeled NR for no reaction. Choose one of the reactions and explain the NR by showing ion cancellation leaving no net reaction.

Stoichiometry
Exploring a Student-Friendly Method of Problem Solving

OBJECTIVE
Students will be introduced to an alternative and student-friendly method of stoichiometry. This method will enhance their readiness for solving equilibrium problems using a RICE table.

LEVEL
Chemistry

NATIONAL STANDARDS
UCP.1, UCP.2, UCP.3, B.2, B.3

TEKS
2(C)

CONNECTIONS TO AP
AP Chemistry:
   III. Reactions B. Stoichiometry 2. Balancing equations including those for redox reactions 3. Mass and volume relations with emphasis on the mole concept, including empirical formulas and limiting reactants.

TIME FRAME
45 minutes

MATERIALS
   calculator   paper and pencil

TEACHER NOTES
This lesson is designed as a direct teach and designed to give students an alternative method of solving problems involving stoichiometry—especially students who are not demonstrating mastery. This lesson should follow direct instruction on how to properly write and name chemical formulas, calculate molar masses, and balance chemical equations. It is a good idea to hold an out-of-class tutorial on each of these skills at the first sign of trouble once beginning this unit. Even good students generally find this unit among the most difficult in the first-year course and frequently struggle with the dimensional analysis.

The mole concept is introduced using a succinct visual tool called the "mole map". Instead of lengthy dimensional analysis conversion, this reaction stoichiometry method uses a table format, not unlike the RICE table format you will introduce to them later in the course when solving equilibrium problems. This method of problem-solving is also very student-friendly since it capitalizes on their ability to solve simple proportion problems. Although it appears labor intensive at first, many students find it is not as time intensive as the dimensional analysis approach, especially when answering multi-stepped problems like those on the AP Chemistry exam.

If it has been your professional experience that stoichiometry is a difficult unit to teach, or a difficult one for students to master, you are encouraged to try this lesson with your students. Try presenting dimensional analysis as one approach and on a subsequent class day, introduce this method. Allow students to choose whichever method will allow them to work swiftly and accurately since that is essential to their success on the AP Chemistry exam. You may also want to use the method on sample problems from your textbook prior to tackling the more intricate problems included on the student answer pages. It is not recommended that you skip the conclusion questions included in this lesson. They are structured much like an AP stoichiometry problem and are designed to teach some descriptive chemistry along the way.

To get the students warmed up and to assess their readiness for stoichiometry problems, a table of simple mole conversions is included in the student introduction. Once students have completed the table you should provide them with the answers to check their work.

Example 1:

Substance	MM	Number of moles	Mass in grams	Number of Particles	Liters of gas @ STP
carbon dioxide, CO_2	44.01	3.00	132	1.81×10^{24}	67.2
oxygen, O_2	32.00	2.00	64.0	1.20×10^{24}	44.8
methane, CH_4	16.05	0.279	4.48	1.68×10^{23}	6.25
nitrogen, N_2	28.02	158	4430	9.50×10^{25}	3540

A note about significant digits: Three significant digits were used throughout this example with the exception of molar masses where two decimal places were used. The absolutely correct way to determine the number of significant digits for a molar mass is to survey the stoichiometry problem, determine the quantity given with the least number of significant digits and round each atomic mass from the periodic table to that number of digits. Next, multiply each mass by its subscript and add the masses respecting the rules of significant digit addition. This rule dictates that the sum will be reported to the least number of *decimal places*. Most teachers surrender and use two decimal places as a rule of thumb when reporting atomic or molecular masses.

ANSWERS TO THE CONCLUSION QUESTIONS

NOTE: There will be slight variations in the answers below depending on whether or not you left all of the numbers in the calculator and rounded only at the end, or started over from moles each time.

1. Automotive air bags inflate when a sample of NaN_3 is very rapidly decomposed. (a) What mass of NaN_3 is required to produce 375 L of nitrogen gas at STP? (b) How many sodium atoms are produced when 375 L of nitrogen gas at STP is produced? (c) What is the mass of the sodium atoms produced?

Molar Mass:	(65.02)		(22.99)		(28.02)	
Balanced Eq'n	$2NaN_3$	$\rightarrow$	2Na	+	$3N_2$	
mole:mole	**2**		**2**		**3**	
# moles	11.2 moles		11.2 moles		16.7 moles	
amount	(a) 728 g NaN_3 formed		(b) 6.74×10^{24} atoms Na (c) 257 g Na		375 L $\div 22.4$L $= 16.7$ moles	

2. When ignited or subjected to sudden impact, nitroglycerine decomposes very rapidly and exothermically:

$$4C_3H_5N_3O_9 \text{ (l)} \rightarrow 6N_2(g) + 12CO_2(g) + 10H_2O \text{ (g)} + O_2(g) + \text{energy}$$

Pure nitroglycerine is quite dangerous. Alfred Nobel discovered in 1867 that if nitroglycerine was absorbed into porous silica, it can be handled quite safely. He named this new mixture dynamite. This brought him a great fortune which he used to establish the Nobel Prizes given today. (a) The density of nitroglycerine is 1.60 g/ml. What is the volume in liters of a mole of liquid nitroglycerine? (b) How many moles of water vapor are formed when one mole of nitroglycerine detonates and decomposes completely? (c) What is the total volume of the gases produced when one mole of liquid nitroglycerine is completely vaporized? (d) By what factor does the volume increase once the liquid nitrogen sample has been completely vaporized?

Molar Mass:	(227.11)		(28.02)		(44.01)	(18.02)		(32.00)
Balanced Eq'n	$4C_3H_5N_3O_9$ (l)	$\rightarrow$	$6N_2(g)$	+	$12CO_2(g)$ +	$10H_2O$ (g)	+	$O_2(g)$
mole:mole	**4**		**6**		**12**	**10**		**1**
# moles	1 mole		1.5 mole		3 mole	2.5 mole		0.25 mole
amount								

(a) 1 mole $= 227.11\text{g} \times \dfrac{1 \text{ mL}}{1.60 \text{ g}} = 141.9$ mL $= 0.142$ L of liquid nitroglycerine

(b) Total moles of gas formed $= 1.5 + 3 + 2.5 + 0.25 = 7.25$ moles

(c) $7.25 \text{ moles} \times 22.4 \dfrac{L}{\text{mole}} = 162L$ of gas formed from 0.762 L of liquid nitroglycerine.

(d) $\dfrac{162 \text{ L}}{0.142 \text{ L}}$ = by 1,140 times its original volume.

3. On April 6, 1938, at DuPont's Jackson Laboratory in New Jersey, DuPont chemist, Dr. Roy J. Plunkett, was working with gases related to Freon® refrigerants, another DuPont product. Upon checking a frozen, compressed sample of tetrafluoroethylene, he and his associates discovered that the sample had polymerized spontaneously into a white, waxy solid to form polytetrafluoroethylene (PTFE). This was not expected and a bit of a disaster. It would set their work back by many days. As it turns out PTFE is inert to virtually all chemicals and is considered the most slippery material in existence. Thus, Teflon® was born. Teflon® is a polymer. The synthesis reaction for a fragment of Teflon® is given below:

$$C_2F_4 \;\rightarrow\; C_{50}F_{200}$$

(a) What mass of reactant is required to form 575 grams of Teflon®? (b) A skillet is massed before the application of a Teflon® coating. Its initial mass is 235 g. The coating is applied and its final mass is determined to be 269g. How many C_2F_4 particles were contained in the Teflon® coating applied to the skillet? (c) How many fluorine atoms were contained in the coating?

Molar Mass:	(100.02)		(4,400.50)
Balanced Eq'n	25 C_2F_4	$\rightarrow$	$C_{50}F_{200}$
mole:mole	**25**		**1**
# moles	3.27 moles		0.131 moles
amount	(a) 327 g		575 g

For Parts (b) and (c)

	25 C_2F_4	$\rightarrow$	$C_{50}F_{200}$
mole:mole	**25**		**1**
# moles	0.19 moles		7.7×10^{-3} moles
amount	(b) 1.2×10^{23} particles		269g after coating – 235g
	(c) 4.7×10^{23} F atoms		before coating = 34g of
	(4 F atoms per particle)		Teflon® coating

4. NASA has engineered reusable booster rockets for the U.S. space shuttle program. These booster rockets use a mixture of powdered aluminum metal and solid ammonium perchlorate for fuel. Solid aluminum oxide is formed as a product along with solid ammonium chloride. It is the formation of nitrogen monoxide gas and water vapor from the two solid reactants that gives the shuttle its boost. (a) What mass of ammonium perchlorate should be used in the fuel mixture for every kilogram of aluminum? (b) What is the *total* number of gas moles produced? (c) How many liters of water vapor are produced per kilogram of aluminum in the fuel mixture? (d) How many total liters of gas at STP are produced when 100.0 grams of powdered aluminum reacts with 100.0 grams of ammonium chloride.

Molar Mass:	(26.98)	(117.50)	(101.96)	(53.50)	(30.01)	(18.02)
Balanced Eq'n	$3Al(s)$ +	$3NH_4ClO_4(s)$ →	$Al_2O_3(s)$ +	$NH_4Cl(s)$ +	$3NO(g)$ +	$6H_2O(g)$
mole:mole	3	3	1	1	3	6
# moles	37.1 moles	37.1	12.4	12.4	37.1	74.2
amount	1.00 kg*	(a) 4,360 g = 4.36 kg				(c) 1,660 L

For part (d) limiting reactant

mole:mole	3	3	1	1	3	6
# moles	3.706 moles ∴ 0.8510 moles used	0.8510 moles	0.2837	0.2837	0.8510	1.702
amount	100.0 g	100.0g				Total = 0.8510 + 1.702 = 2.553 × 22.4 L/mol = 57.19 L at STP

For part (a) *3 significant digits chosen arbitrarily.

(b) Total moles of gas = 37.1 + 74.2 = 111.3 moles of gas formed

Molar Mass:	(12.01)	(2.02)		(28.02)	(32.00)		(238.26)
Balanced Eq'n	(a) 14 C +	5 H$_2$ +		N$_2$ +	O$_2$	→	C$_{14}$H$_{10}$N$_2$O$_2$
mole:mole	14	5		1	1		1
# moles	249.8	~~247.5~~ (c) ∴ 89.22 moles used AND 247.5 – 89.22 = 158.3 moles remain ∴ 319.7 g excess H$_2$		17.84	17.84		17.84
amount		~~500.0 g~~		500.0 g			(b) 4.251 kg (d)1.074 × 10^{25}

5. Another DuPont product, Kevlar®, was originally developed as a replacement for steel in radial tires. Kevlar® is now used in a wide range of applications including bullet proof vests since it is twenty times stronger than steel. Kevlar® was first synthesized in 1964 by Stephanie Kwolek at the DuPont laboratories in Wilmington, Delaware. A monomer of Kevlar is C$_{14}$H$_{10}$N$_2$O$_2$. (a) Write the chemical reaction that would synthesize a monomer of Kevlar® from its elements. (b) If 500.0 g of both nitrogen gas and hydrogen gas at STP were reacted with excess carbon and oxygen gas, how many kilograms of Kevlar® could be produced? (c) What mass of excess reagent remains unreacted? (d) How many Kevlar® monomer units are produced as a result?

Stoichiometry
Exploring a Student-Friendly Method of Problem Solving

Stoichiometry comes in two forms: composition and reaction. If the relationship is between the quantities of each element in a compound it is called composition stoichiometry. If the relationship is between quantities of substances as they undergo a chemical reaction it is called reaction stoichiometry. In plain English, if you have to calculate just about anything to do with moles or chemical quantities, we collectively call those computations *stoichiometry*.

THE MOLE CONCEPT
To be successful in solving stoichiometry problems you must be familiar with the following terms and their meaning:
- **mole**—the number of C atoms in exactly 12.0 grams of ^{12}C; also a number, 6.02×10^{23} just as the word "dozen" means 12 and "couple" means 2.
- **Avogadro's number**, n—6.02×10^{23}, the number of particles in a mole of anything
- **molar mass**—the mass of one mole of particles in grams; the sum of the atomic masses in a chemical formula. When calculating be sure to multiply the atomic mass of each element by the subscript following that element. The molar mass for H_2O is 2(1.01 for hydrogen) + 16.00 for oxygen = 18.02 grams for a mole of water molecules.

THE MOLE MAP
The mole map is a quick way to calculate any quantity you desire that is related to moles. Learn to reproduce this map (in other words, memorize it) so that you may use it as a problem-solving tool when faced with stoichiometry problems.

Notice the number of moles is *always* in the center of the map which means when converting moles into any other unit you must multiply the number of moles by the conversion factor found next to the arrow. The conversion factors included on this map will allow you to convert the number of moles to either the number of particles, mass of matter in grams, or liters of a gas at standard temperature and pressure, STP (1 atm pressure and 273 K).

If you have one of the above quantities and wish to find the number of moles simply start at the spot on the map that corresponds to the quantity "given" in your problem and divide by the appropriate conversion factor. Once there, you can multiply by the appropriate conversion factor to solve for a different quantity.

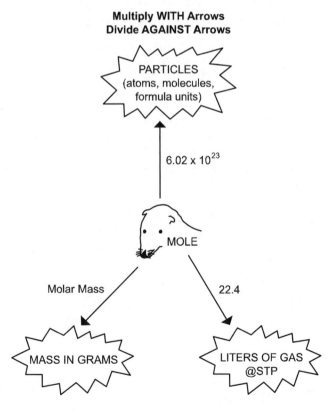

Multiply WITH Arrows
Divide AGAINST Arrows

PARTICLES (atoms, molecules, formula units)

6.02×10^{23}

MOLE

Molar Mass

22.4

MASS IN GRAMS

LITERS OF GAS @STP

Substance	MM	Number of moles	Mass in grams	Number of Particles	Liters of gas @ STP
carbon dioxide		3.0			
oxygen			64.0		
methane					6.25
nitrogen				9.50×10^{25}	

Example: Complete the following table using the mole map and a periodic table. Your teacher has the correct answers.

Now that you understand how to use the mole map, you must realize that being successful in solving stoichiometry problems requires that you have demonstrated proficiency in the following individual skills:

- *Correctly* writing chemical formulas—this requires knowledge of your polyatomic ions and being able to use the periodic table to deduce what you have not had to memorize.
- *Correctly* calculating molar masses from a chemical formula.
- *Correctly* balancing chemical equations.

SOLVING STOICHIOMETRY PROBLEMS

This method will use a table to help you organize your thoughts and focus on the task at hand. Each time you embark upon solving a reaction stoichiometry problem, reproduce the row headings found below and place them on the left side of your paper as shown:

Molar Mass:
Balanced Eq'n:
mole:mole
 # moles:
amount:

Once you have a clean template written on your paper, work through Example 2 and add the appropriate information to you table.

Example 2: What mass of oxygen will react with 96.1 grams of propane? (Notice the balanced equation was not provided, you will have to supply the chemical formulas!)

1. Write a chemical equation. Double check to be sure you have correct chemical formulas!

Molar Mass:
Balanced Eq'n C_3H_8 + O_2 → CO_2 + H_2O
mole:mole
 # moles
amount

2. Calculate the molar masses and write them in parentheses above the formulas—soon you'll figure out you don't have to do this for every reactant and product, just those your question uses.

Molar Mass:	(44.11)	(32.00)	(44.01)	(18.02)
Balanced Eq'n	C_3H_8 +	O_2 →	CO_2 +	H_2O
mole:mole				
# moles				
amount				

3. Balance the chemical equation using coefficients. Do not attempt to change the subscripts!

Molar Mass:	(44.11)	(32.00)	(44.01)	(18.02)
Balanced Eq'n	C_3H_8 +	5 O_2 →	3 CO_2 +	4 H_2O
mole:mole				
# moles				
amount				

4. Look at the coefficients in the balanced equation. These numbers ARE the **mole:mole** ratio. Re-write them under the balanced equation so that they are easy to see.

Molar Mass:	(44.11)	(32.00)	(44.01)	(18.02)
Balanced Eq'n	C_3H_8 +	5 O_2 →	3 CO_2 +	4 H_2O
mole:mole	1	5	3	4
# moles				
amount				

5. Next, re-read the problem and look for some information about an amount of one of the substances—in this example it is 96.1 g of propane.

Molar Mass:	(44.11)	(32.00)	(44.01)	(18.02)
Balanced Eq'n	C_3H_8 +	5 O_2 →	3 CO_2 +	4 H_2O
mole:mole	1	5	3	4
# moles				
amount	96.1 grams			

6. Use the mole map to calculate the number of moles of something, anything! Start at 96.1 grams, divide (against the arrow) by molar mass to get the # moles of propane.

Molar Mass:	(44.11)	(32.00)	(44.01)	(18.02)
Balanced Eq'n	C_3H_8 +	5 O_2 →	3 CO_2 +	4 H_2O
mole:mole	1	5	3	4
# moles	2.18	10.9	6.54	8.72
amount	$\frac{96.1g}{44.11g/mol}=2.18mol$			

7. USE the mole: mole ratio to find the moles of EVERYTHING! If 1 = 2.18 mol propane, then moles oxygen equals 5(2.18) mol or 10.9 moles etc…. (IF you found the number of moles for something that has a coefficient other than "1", just divide by the coefficient it has and multiply by the new coefficient.) Leave everything in your calculator—we only rounded to save space!

Molar Mass:	(44.11)	(32.00)	(44.01)	(18.02)
Balanced Eq'n	C_3H_8 +	5 O_2 →	3 CO_2 +	4 H_2O
mole:mole	1	5	3	4
# moles	2.18	10.9	6.54	8.72
amount	9.16 g			

8. Re-read the problem to determine which amount you are seeking. First, we'll find the mass of oxygen required since that's what the problem asked. 10.9 moles x 32.00 g/mol = 349 g of oxygen.

Molar Mass:	(44.11)	(32.00)	(44.01)	(18.02)
Balanced Eq'n	C_3H_8 +	5 O_2 →	3 CO_2 +	4 H_2O
mole:mole	1	5	3	4
# moles	2.18	10.9	6.54	8.72
amount	9.16 g	349 g	146 L	

What if part (b) of Example 2 asked for liters of CO_2 at STP? Simply take the number of moles of CO_2 from the table and use the mole map to solve for the liters of gas at STP. Start in the middle of the map with 6.54 moles and multiply by 22.4 L/mol to determine that 146 L of CO_2 gas is produced at STP.

Molar Mass:	(44.11)	(32.00)	(44.01)	(18.02)
Balanced Eq'n	C_3H_8 +	5 O_2 →	3 CO_2 +	4 H_2O
mole:mole	1	5	3	4
# moles	2.18 mol	10.9	6.54	8.72
amount	9.16 g	349 g		

What if part (c) asked how many water molecules are produced? Simply multiply the number of moles of water 8.72 mol by 6.02 X 10^{23} to determine that 5.25 x 10^{24} molecules of water were produced.

Molar Mass:	(44.11)	(32.00)	(44.01)	(18.02)
Balanced Eq'n	C_3H_8 +	5 O_2 →	3 CO_2 +	4 H_2O
mole:mole	1	5	3	4
# moles	2.18	10.9	6.54	8.72
amount	96.1 g			5.25 x 10^{24}

CALCULATIONS INVOLVING A LIMITING REACTANT

Ever notice how hot dogs are sold in packages of 10 and the buns come in packages of 8? The bun is the limiting reactant and limits hot dog production to 8 when one package of each is purchased. The limiting reactant (or reagent) is the one that runs out first in the chemical reaction.

Plan of attack: First, you need to recognize that you ***need*** a plan of attack! Your clue is that you are given TWO amounts of matter that react.

Next, when in doubt…find the number of moles. Set up your table like before, only now you will have TWO amounts and thus TWO numbers of moles to get you started. Simply divide each number of moles by its coefficient in the balanced equation and select the smaller number of moles since it is the limiting reagent. Be sure to cross out the larger value of moles and recalculate its value based on the limiting reagent.

The Haber process synthesizes ammonia, often used for fertilizer production, from the reaction of nitrogen and hydrogen gases. The nitrogen gas is drawn from the atmosphere while the hydrogen gas is obtained from the reaction of methane with water vapor. It is carried out at 250 atm and 400°C in the presence of a catalyst. This process has ultimately saved millions from starvation. The reaction is below:

Molar Mass:	(28.02)		(2.02)		(17.04)
Balanced Eq'n	N_2	+	3 H_2 →		2 NH_3
mole:mole	1		3		2
# moles					
amount					

Suppose 25.0 kg of nitrogen reacts with 5.00 kg of hydrogen to form ammonia. (a) What mass of ammonia can be produced? (b) Which reactant is the limiting reactant? (c) What is the mass of the reactant that is in excess?

Insert the masses in the amount row and find the number of moles of BOTH!

Molar Mass:	(28.02)		(2.02)		(17.04)
Balanced Eq'n	N_2	+	3 H_2 →		2 NH_3
mole:mole	1		3		2
# moles	892 moles		2,475 moles		
amount	25,000 g		5,000 g		

Next, divide each number of moles by its coefficient. Compare these mole values and select the smaller number of moles as your limiting reactant. N_2 has a coefficient of one already, so just focus on the 892 moles present. For H_2, your focus becomes 2,475 ÷ 3 = 825 moles, which is smaller than the number of N_2 moles, so H_2 is your limiting reagent.

Now you are ready to modify the table: Cross out 25,000 g and 892 moles of N_2 since they will not limit the reaction and replace it with the 825 moles of N_2 consumed (since nitrogen has a coefficient of one). If needed, you could then calculate the actual mass of N_2 that reacts.

Molar Mass:	(28.02)	(2.02)	(17.04)
Balanced Eq'n	N₂ +	3 H₂ →	2 NH₃
mole:mole	**1**	**3**	**2**
# moles	825 mol used	∴ 2,475 moles used	∴ 1,650 mol produced
	↑ 892 moles		
amount	∴ 825 mol x 28.02 = 23,117 g used		1,650 mol (17.04) = 28,116 g produced
		5,000 g	
	25,000 g 1,883 g excess!!		

Here's the question again, let's clean up any significant digit issues:

Suppose 25.0 kg of nitrogen reacts with 5.00 kg of hydrogen to form ammonia. (3 sig. dig. limit)

(a) What mass of ammonia can be produced? 28,100 g produced = 28.1 kg (it is always polite to respond in the mass unit given).

(b) Which reactant is the limiting reactant? hydrogen—once that's established, forget about how much nitrogen you were given and base the calculation on hydrogen's amount.

(c) What is the mass of the reactant that is in excess? 1,883 g = 1.88 kg excess nitrogen.

PURPOSE

In this activity you will explore a different method of solving stoichiometry problems. At the end of this activity you should be able to determine which method of problem solving allows you to work the fastest and most accurately.

MATERIALS

calculator   paper and pencil

PROCEDURE

1. Solve the problems found on your student answer pages. Use your own paper but staple your student answer page to the front of your notebook paper to turn in to your teacher.

2. Be sure to show all work and pay close attention to the proper use of significant digits and units.

Name _____

Period _____

Stoichiometry
Exploring a Student-Friendly Method of Problem Solving

CONCLUSION QUESTIONS

1. Automotive air bags inflate when a sample of NaN_3 is very rapidly decomposed. (a) What mass of NaN_3 is required to produce 375 L of nitrogen gas at STP? (b) How many sodium atoms are produced when 375 L of nitrogen gas at STP is produced? (c) What is the mass of the sodium atoms produced?

2. When ignited or subjected to sudden impact, nitroglycerine decomposes very rapidly and exothermically:

$$4C_3H_5N_3O_9 \text{ (l)} \rightarrow 6N_2\text{(g)} + 12CO_2\text{(g)} + 10H_2O \text{ (g)} + O_2\text{(g)} + \text{energy}$$

Pure nitroglycerine is quite dangerous. Alfred Nobel discovered in 1867 that if nitroglycerine was absorbed into porous silica, it can be handled quite safely. He named this new mixture dynamite. This brought him a great fortune which he used to establish the Nobel Prizes given today. (a) The density of nitroglycerine is 1.60 g/ml. What is the volume in liters of a mole of liquid nitroglycerine? (b) How many moles of water vapor are formed when one mole of nitroglycerine detonates and decomposes completely? (c) What is the total volume of the gases produced when one mole of liquid nitroglycerine is completely vaporized? (d) By what factor does the volume increase once the liquid nitrogen sample has been completely vaporized?

3. On April 6, 1938, at DuPont's Jackson Laboratory in New Jersey, DuPont chemist, Dr. Roy J. Plunkett, was working with gases related to Freon® refrigerants, another DuPont product. Upon checking a frozen, compressed sample of tetrafluoroethylene, he and his associates discovered that the sample had polymerized spontaneously into a white, waxy solid to form polytetrafluoroethylene (PTFE). This was not expected and a bit of a disaster. It would set their work back by many days. As it turns out PTFE is inert to virtually all chemicals and is considered the most slippery material in existence. Thus, Teflon® was born. Teflon® is a polymer. The synthesis reaction for a fragment of Teflon® is given below:

$$C_2F_4 \rightarrow C_{50}F_{200}$$

(a) What mass of reactant is required to form 575 grams of Teflon®? (b) A skillet is massed before the application of a Teflon® coating. Its initial mass is 235 g. The coating is applied and its final mass is determined to be 269g. How many C_2F_4 particles were contained in the Teflon® coating applied to the skillet? (c) How many fluorine atoms were contained in the coating?

4. NASA has engineered reusable booster rockets for the U.S. space shuttle program. These booster rockets use a mixture of powdered aluminum metal and solid ammonium perchlorate for fuel. Solid aluminum oxide is formed as a product along with solid ammonium chloride. It is the formation of nitrogen monoxide gas and water vapor from the two solid reactants that gives the shuttle its boost. (a) What mass of ammonium perchlorate should be used in the fuel mixture for every kilogram of aluminum? (b) What is the total number of gas moles produced? (c) How many liters of water vapor are produced per kilogram of aluminum in the fuel mixture? (d) How many total liters of gas at STP are produced when 100.0 grams of powdered aluminum reacts with 100.0 grams of ammonium chloride.

5. Another DuPont product, Kevlar®, was originally developed as a replacement for steel in radial tires. Kevlar® is now used in a wide range of applications including bullet proof vests since it is twenty times stronger than steel. Kevlar® was first synthesized in 1964 by Stephanie Kwolek at the DuPont laboratories in Wilmington, Delaware. A monomer of Kevlar is $C_{14}H_{10}N_2O_2$. (a) Write the chemical reaction that would synthesize a monomer of Kevlar® from its elements. (b) If 500.0 g of both nitrogen gas and hydrogen gas at STP were reacted with excess carbon and oxygen gas, how many kilograms of Kevlar® could be produced? (c) What mass of excess reagent remains unreacted? (d) How many Kevlar® monomer units are produced as a result?

Simple vs. True
Calculating Empirical and Molecular Formulas

OBJECTIVE
Students will learn to calculate empirical and molecular formulas using logical problem-solving steps.

LEVEL
Chemistry

NATIONAL STANDARDS
UCP.1, UCP.2, UCP.3, B.2

TEKS
2(C), 11(A)

CONNECTIONS TO AP
AP Chemistry:
  III. Reactions B. Stoichiometry 3. Mass and volume relations with emphasis on the mole concept, including empirical formulas

TIME FRAME
45 minutes

MATERIALS
  calculator   periodic table

TEACHER NOTES
This lesson is appropriate for inclusion in your unit on the mole concept. These calculations logically follow mastery of percent composition calculations. The student notes can provide the basis for your lecture over this material. Work the example problems with the students either on an overhead projector or chalkboard. Be sure to explain each step as you work through the problems. The example problems are worked out below.

EXAMPLE PROBLEM 1:

Many of the chemicals in our body consist of the elements carbon, hydrogen, oxygen and nitrogen. One of these chemicals, norepinephrine, is often released during stressful times and serves to increase our metabolic rate. The percent composition of this hormone is 56.8% C, 6.56% H, 28.4% O, and 8.28% N. Calculate the simplest formula for this biological compound.

Step 1: Convert to moles (Assume a 100. gram sample)

$$56.8 \text{ g C} \times \frac{1\,\text{mol C}}{12.01\,\text{g C}} = 4.729 \text{ mol C}$$

$$6.56 \text{ g H} \times \frac{1\,\text{mol H}}{1.01\,\text{g H}} = 6.495 \text{ mol H}$$

$$28.4 \text{ g O} \times \frac{1\,\text{mol O}}{16.00\,\text{g O}} = 1.775\,\text{mol O}$$

$$8.28 \text{ g N} \times \frac{1\,\text{mol N}}{14.01\,\text{g N}} = 0.5910\,\text{mol N}$$

Step 2: Mole ratio

* N is the smallest number of moles – divide all moles by this –

$$\frac{4.729\,\text{mol C}}{0.5910\,\text{mol N}} = 8$$

$$\frac{6.495\,\text{mol H}}{0.5910\,\text{mol N}} = 11 \qquad \text{Empirical Formula} = C_8H_{11}O_3N$$

$$\frac{1.775\,\text{mol O}}{0.5910\,\text{mol N}} = 3$$

$$\frac{0.5910\,\text{mol N}}{0.5910\,\text{mol N}} = 1$$

EXAMPLE PROBLEM 2:

A sample of a white, granular ionic compound weighing 41.764 grams was found in the photography lab. Analysis of this compound revealed that the compound was composed of 12.144 grams of sodium, 16.948 grams of sulfur, and the rest of the compound was oxygen. Calculate the formula for this compound and provide its name.

Step 1: Convert to moles

$$12.144 g\, Na\ x\ \frac{1\,mol\,Na}{22.99 g\,Na} = 0.5282\ mol\,Na$$

$$16.948 g\, S\ x\ \frac{1\,mol\,S}{32.06 g\,S} = 0.5286\,mol\,S$$

Find grams of oxygen → total mass = 41.764

$$41.764 - (12.144 + 16.948) = 12.672 g\,O$$

$$12.672 g\,O\ x\ \frac{1\,mol\,O}{16.00 g\,O} = 0.792\,mol\,O$$

Step 2: Mole ratio   *moles of sodium is smallest

$$\frac{0.5282\,mol\,Na}{0.5282\,mol\,Na} = 1$$

$$\frac{0.5286\,mol\,S}{0.5282\,mol\,Na} = 1$$

$$\frac{0.792\,mol\,O}{0.5282\,mol\,Na} = 1.499 = 1.5$$

*Multiply all by 2 to obtain whole numbers –
Empirical formula = $Na_2S_2O_3$ = Sodium thiosulfate

EXAMPLE PROBLEM 3:

Calculate the molecular formula for an organic compound whose molecular mass is 180. grams/mole and has an empirical formula of CH_2O. Name this compound.

Step 1: Molecular formula $= \dfrac{\text{molecular mass}}{\text{empirical mass}} = \dfrac{180.}{30.0} = 6$

Find empirical mass for $CH_2O = 30.0 \text{g/mol}$

Step 2: Multiply to get new subscripts.

$$6(CH_2O) = C_6H_{12}O_6 = \text{glucose}$$

EXAMPLE PROBLEM 4:

An organic alcohol was quantitatively found to contain the following elements in the given proportions:
C = 64.81% ; H = 13.60%; O = 21.59%

Given that the molecular weight of this alcohol is 74 g/mol, find the molecular formula and name this alcohol.

Step 1: Convert to moles

$$64.81\text{g C} \times \frac{1\,\text{mol C}}{12.01\,\text{g C}} = 5.396\,\text{mol C}$$

$$13.60\text{g H} \times \frac{1\,\text{mol H}}{1.01\,\text{g H}} = 13.465\,\text{mol H}$$

$$21.59\text{g O} \times \frac{1\,\text{mol O}}{16.00\,\text{g O}} = 1.349\,\text{mol O}$$

Step 2: Mole ratio   *Oxygen has smallest number of moles

$$\frac{5.385\,\text{mol C}}{1.349\,\text{mol O}} = 4$$

$$\frac{13.465\,\text{mol H}}{1.349\,\text{mol O}} = 10 \qquad\qquad C_4H_{10}O = C_4H_9OH$$

butanol = empirical formula

$$\frac{1.349\,\text{mol O}}{1.349\,\text{mol O}} = 1$$

Step 3: $\dfrac{\text{molecular mass}}{\text{emp. mass}} = \dfrac{74}{74} = 1$ The empirical and molecular formulas are the same

ANSWERS TO THE ANALYSIS QUESTIONS

1. A 2.676 gram sample of an unknown compound was found to contain 0.730 g of sodium, 0.442 g of nitrogen and 1.504 g of oxygen. Calculate the empirical formula for this compound and name it.

Step 1:

$$.730g \text{ Na} \times \frac{1 \text{mol Na}}{22.99g \text{Na}} = 0.03175 \text{ mol Na}$$

$$.442g \text{ N} \times \frac{1 \text{mol N}}{14.01g \text{N}} = 0.03155 \text{ mol N}$$

$$1.504g \text{ O} \times \frac{1 \text{mol O}}{16.00g \text{ O}} = 0.9400 \text{ mol O}$$

Step 2:

$$\frac{0.03175 \text{ mol Na}}{0.03155 \text{ mol N}} = 1$$

$$\frac{0.03155 \text{ mol N}}{0.03155 \text{ mol N}} = 1$$

$$\frac{0.09400 \text{ mol O}}{0.03155 \text{ mol N}} = 3$$

Empirical formula $= \text{NaNO}_3$ sodium nitrate

2. A mysterious white powder was found on the kitchen counter. It was found quantitatively to contain 27.37% sodium, 1.20% hydrogen, 14.30% carbon, and the rest was found to be oxygen. Calculate the empirical formula for this compound and name it.

Step 1:

$$27.37 \text{g Na} \times \frac{1 \text{mol Na}}{22.99 \text{g Na}} = 1.191 \text{mol Na}$$

$$1.20 \text{g H} \times \frac{1 \text{mol H}}{1.01 \text{g H}} = 1.188 \text{mol H}$$

$$14.30 \text{g C} \times \frac{1 \text{mol C}}{12.01 \text{g C}} = 1.191 \text{mol C}$$

$$\text{Oxygen} = 100 - (27.37 + 1.20 + 14.30) = 57.13 \% \text{ O}$$

$$57.13 \text{g O} \times \frac{1 \text{mol O}}{16.00 \text{g O}} = 3.571 \text{mol O}$$

Step 2:

$$\frac{1.191 \text{mol Na}}{1.188 \text{mol H}} = 1$$

$$\frac{1.188 \text{mol H}}{1.188 \text{mol H}} = 1$$

$$\frac{1.191 \text{mol C}}{1.188 \text{mol H}} = 1$$

$$\frac{3.571 \text{mol O}}{1.188 \text{mol H}} = 3$$

Empirical formula = $NaHCO_3$ = sodium bicarbonate

3. A common organic solvent has an empirical formula of CH and a molecular mass of 78 g/mole. Calculate the molecular formula for this compound and name it.

$$\text{Molecular formula} = \frac{\text{molar mass}}{\text{empirical mass}} = \frac{78}{13} = 6$$

$CH = 13\text{g/mol}$ $\qquad 6(CH) = C_6H_6$ benzene

Students may have to look up this name

4. A gas was qualitatively analyzed and found to contain only the elements nitrogen and oxygen. The compound was further analyzed to discover that the gas was composed of 30.43% nitrogen. Given that the molecular mass of the compound is 92.0 g/mole, calculate the molecular formula.

Step 1:

$$30.43 \text{g N} \times \frac{1 \text{mol N}}{14.01 \text{g N}} = 2.172 \text{mol N}$$

$$\text{Oxygen} = 100 - 30.43 = 69.57\%$$

$$69.57 \text{g O} \times \frac{1 \text{mol O}}{16.00 \text{g O}} = 4.348 \text{mol O}$$

Step 2:

$$\frac{2.172 \text{mol N}}{2.172 \text{mol N}} = 1$$

$$\frac{4.348 \text{mol O}}{2.172 \text{mol N}} = 2$$

Empirical formula = NO_2 = 46g/mol   nitrogen dioxide

Step 3:

$$\frac{\text{molecular mass}}{\text{empirical mass}} = \frac{92.0}{46} = 2$$

Step 4:

$$2(NO_2) = N_2O_4 \quad \text{dinitrogen tetraoxide}$$

Simple vs. True
Calculating Empirical and Molecular Formulas

Formula writing is a key component for success in chemistry. How do scientists really know what the "true" formula for a compound might be? In this lesson we will explore some of the mathematics supporting many of the chemical formulas that you have encountered.

PURPOSE
In this lesson you will learn problem-solving strategies that will enable you to calculate empirical and molecular formulas given experimental data.

MATERIALS
calculator   periodic table

CLASS NOTES
The simplest formula, *empirical formula*, for a compound is the smallest whole number ratio of atoms present in a given formula. The *molecular formula* represents the true ratio of atoms in a molecular compound. Sometimes the empirical formula and the molecular formula are the same. For example, the formula for water, H_2O, is both the simplest ratio of elements and the true ratio of elements. In other instances, the molecular formula is a whole number multiple of the empirical formula. For example, the formula for butane is C_4H_{10}. This formula represents the molecular formula, the true ratio of atoms. The empirical formula for this compound is easily found by reducing the subscripts (dividing by the greatest common factor) to end with C_2H_5. For ionically-bonded substances, the empirical formula is the representation that is commonly written as the formula. For example, in the formula NaCl, Na and Cl are in a 1:1 ratio in this empirical formula, however, sodium chloride crystals are actually face-centered cubic in shape and one unit cell requires many more than 2 ions while maintaining the 1:1 ratio.

When a new substance is found or discovered, the formula is unknown until some qualitative and quantitative analyses are performed on the compound. First, qualitative analysis might reveal which elements are in the compound and then quantitatively, the amounts of those elements in the compound must be found. From this type of experimental data, the empirical formula may be calculated. Use the following strategies to make calculating empirical and molecular formulas a breeze!

EMPIRICAL FORMULA

1. Convert the percentage or grams of each element into moles. (Remember that the sum of the percentages must total 100% so by assuming a 100. gram sample, you can simply drop the % and replace it with grams. For example, if a compound has 20.0% Na, then convert this to 20.0 grams of sodium.) Leave answers to at least 4 significant figures in this step. Rounding early could yield incorrect answers later.

2. Set up a mole ratio for each element. Divide each of the moles calculated by the element with the smallest number of moles. This step may yield whole numbers. If so, these whole numbers are the subscripts for the formula.

   a. If the numbers are not whole numbers you may have to multiply by some factor (try multiplying by 2 first, then by 3, etc.) to make them whole numbers. Don't just round!

   b. For example: If the ratio comes out 1: 2.5: 1—multiplying each number by 2 will yield the same proportion with no decimals. The ratio becomes 2: 5: 2. Remember that subscripts must be whole numbers.

   c. Watch for numbers that have the following terminal decimals:
      - 0.20 (could be multiplied by 5 to yield whole numbers)
      - 0.25 (could be multiplied by 4 to yield whole numbers)
      - 0.33 (could be multiplied by 3 to yield whole numbers)
      - 0.50 (could be multiplied by 2 to yield whole numbers)

3. Write the formula with proper subscripts and name the compound if asked. (Usually the elements will be listed in order within the problem.)

EXAMPLE PROBLEM 1:

Many of the chemicals in our body consist of the elements carbon, hydrogen, oxygen and nitrogen. One of these chemicals, norepinephrine, is often released during stressful times and serves to increase our metabolic rate. The percent composition of this hormone is 56.8% C, 6.56% H, 28.4% O, and 8.28% N. Calculate the simplest formula for this biological compound.

EXAMPLE PROBLEM 2:
A sample of a white, granular ionic compound weighing 41.764 grams was found in the photography lab. Analysis of this compound revealed that the compound was composed of 12.144 grams of sodium, 16.948 grams of sulfur, and the rest of the compound was oxygen. Calculate the formula for this compound and name it!

MOLECULAR FORMULA
It may be necessary for you to calculate the empirical formula first in order to determine the molecular formula.

1. Divide the molecular mass by the empirical formula mass. This should yield a whole number.

2. Multiply all of the subscripts in the empirical formula by the whole number obtained from the previous step to get the true ratio of atoms in the molecular formula.

EXAMPLE PROBLEM 3:
Calculate the molecular formula for an organic compound whose molecular mass is 180. grams/mole and has an empirical formula of CH_2O. Name this compound.

EXAMPLE PROBLEM 4:
An organic alcohol was quantitatively found to be contain the following elements in the given proportions: C = 64.81% ; H = 13.60%; O = 21.59%

Given that the molecular weight of this alcohol is 74 g/mole, find the molecular formula and name this alcohol.

Name _____

Period _____

Simple vs. True
Calculating Empirical and Molecular Formulas

ANALYSIS

Solve the following problems on this answer page. Be sure to show all work for full credit paying special attention to significant figures and units.

1. A 2.676 gram sample of an unknown compound was found to contain 0.730 g of sodium, 0.442 g of nitrogen and 1.504 g of oxygen. Calculate the empirical formula for this compound and name it.

2. A mysterious white powder was found on the kitchen counter. It was found quantitatively to contain 27.37% sodium, 1.20% hydrogen, 14.30% carbon, and the rest was found to be oxygen. Calculate the empirical formula for this compound and name it.

3. A common organic solvent has an empirical formula of CH and a molecular mass of 78 g/mole. Calculate the molecular formula for this compound and name it.

4. A gas was qualitatively analyzed and found to contain only the elements nitrogen and oxygen. The compound was further analyzed to discover that the gas was composed of 30.43% nitrogen. Given that the molecular mass of the compound is 92.0 g/mole, calculate the molecular formula.

Limiting Reactant
Exploring Molar Relationships

OBJECTIVE
Students will react aluminum with copper (II) chloride to determine the limiting reactant. They will present their results in a formal lab report.

LEVEL
Chemistry

NATIONAL STANDARDS
UCP.2, A.1, A.2, B.3

TEKS
1(A), 2(A), 2(B), 2(C), 2(D), 11(A), 11(B), 11(C)

CONNECTIONS TO AP
AP Chemistry:
    III. Reactions B. Stoichiometry 2. Balancing of equations including those for redox reactions
    3. Mass and volume relations with emphasis on the mole concept, including empirical formulas and limiting reactant

TIME FRAME
45 minutes for experiment
Lab report prepared outside of class

MATERIALS
(For a class of 28 working in pairs)

2 L 1.0 M copper (II) chloride   14 tweezers or tongs
14 strips of aluminum metal   paper towels
14 250 mL beakers   balances
14 100 mL graduated cylinders

TEACHER NOTES

This laboratory activity gives students the opportunity to design a simple experiment to determine whether a strip of aluminum or a measured quantity of copper (II) chloride solution acts as the limiting reactant in the following reaction:

$$2 \text{ Al}_{(s)} + 3 \text{ CuCl}_{2(aq)} \rightarrow 2 \text{ AlCl}_{3(aq)} + 3\text{Cu}_{(s)}$$

This activity should be used during a unit on stoichiometry. The 2:3:2:3 mole ratio requires students to have a clear understanding of mole ratios and their effect on the quantities of a reaction. Students will also need to be able to predict the products of this single replacement reaction and calculate the molarity of the $CuCl_2$ solution you provide. If you have not covered molarity calculations prior to this activity you may wish to provide students with the molarity formula for their analysis.

To prepare a 0.50 M $CuCl_2$ solution:

Place 85.24 g of $CuCl_2 \bullet 2 \text{ H}_2O$ (copper(II) chloride dihydrate) in a 1.00 L volumetric flask. Dilute to the mark with distilled water. Repeat this procedure to make a second 1.00 L batch of $CuCl_2$. Combine the two batches in one container and mix thoroughly before dispensing to students.

You can structure the lab so that either of the reactants is the limiting one, however, it is recommended for both time and simplicity to let $CuCl_2$ be the limiting reactant. By letting the $CuCl_2$ be the limiting reactant students will only need to measure the initial and final mass of the aluminum strip to calculate the mass of aluminum consumed. By knowing the mass of aluminum consumed and the $Al:CuCl_2$ ratio from the balanced equation, students can calculate the moles of $CuCl_2$ that were consumed by the reaction and ultimately the molarity of the 100 mL $CuCl_2$ sample.

Aluminum strips can be cut from a sheet of aluminum metal or may be purchased in pre-cut strips from scientific supply companies such as Flinn (www.flinnsci.com) or Science Kit (www.sciencekit.com). Be sure that the mass of your strips is such that it will be the excess reagent when combined with 100 mL of the $CuCl_2$ solution. To do this for a 0.50 M $CuCl_2$ solution:

$$100 \text{ mL} \times \frac{1 \text{ L}}{1000 \text{ mL}} \times \frac{0.50 \text{ mol CuCl}_2}{\text{L}} \times \frac{2 \text{ mol Al}}{3 \text{ mol CuCl}_2} \times \frac{26.98 \text{ g}}{1 \text{ mol Al}} = 0.899 \text{ g Al consumed}$$

Use a piece of aluminum that is at least double the required mass so that a sizeable piece will remain after the reaction.

Although this reaction appears to be a simple single replacement reaction, you will notice that bubbles are produced during this reaction. Actually, the bubbling is caused by the production of hydrogen gas. As it is too difficult for students to collect the gas and test it with a glowing splint, it is suggested that the collection and testing of the gas be a teacher demonstration if desired. It would be necessary to repeat the reaction using a gas collecting apparatus.

Technical note: The reaction is very complex.

Students are asked to produce a formal lab report from their findings. A handout entitled Formal Lab Report Guidelines is included for student reference. A suggested scoring rubric is included below.

Report Item	Possible Points	Earned Points/Comments
Introduction	**(10)**	
Background information	5	
Purpose of the experiment	5	
Hypothesis (if-then)	**(10)**	
Written in if-then format	5	
Testable	5	
Materials	**(5)**	
Materials listed	5	
Procedures	**(20)**	
Procedures stated clearly	10	
Measurement procedure indicated	10	
Data Table/Observations	**(10)**	
Organized table	5	
Clear observation statements	5	
Analysis	**(10)**	
Clearly labeled work	5	
Correct answer	5	
Conclusion	**(30)**	
Summarizes results	6	
Cites data or observations	6	
Evaluates errors	6	
Answers conclusion questions	12	
Student answer page attached	**(5)**	
TOTAL POINTS	**(100)**	

SAMPLE DATA AND POSSIBLE ANSWERS TO THE CONCLUSION QUESTIONS

HYPOTHESIS

If copper (II) chloride is the limiting reactant then the reaction will stop before all of the aluminum is consumed.

DATA AND OBSERVATIONS

	Mass of Aluminum (g)
Initial	1.40 g
Final	0.54 g
Mass of aluminum consumed	0.86 g

During the reaction the beaker got very warm and bubbles were produced. As the reaction occurred the blue color of the solution faded to colorless. Rust colored flakes of solid collected on the aluminum strip.

ANALYSIS

$$Molarity = \frac{moles}{L}$$

$$0.86 \ g \ Al \times \frac{1 \ mol \ Al}{26.98 \ g} = 0.032 \ mol \ Al$$

Balanced equation shows 2 mol Al : 3 mol $CuCl_2$

$$0.032 \ mol \ Al \times \frac{3 \ mol \ CuCl_2}{2 \ mol \ Al} = 0.048 \ mol \ CuCl_2$$

$$Molarity = \frac{0.48 \ mol \ CuCl_2}{0.100 \ L} = 0.048 \ M$$

$$\% \ error = \frac{\left| theoretical - actual \right|}{theoretical} \times 100$$

$$\% \ error = \frac{\left| 0.50 - 0.48 \right|}{0.50} \times 100 = 4.0\%$$

CONCLUSION QUESTIONS

1. Write the balanced chemical equation for this reaction.
   - $2\ Al_{(s)} + 3\ CuCl_{2(aq)} \rightarrow 2\ AlCl_{3(aq)} + 3Cu_{(s)}$

2. Which reactant was the limiting reactant? What evidence do you have to support this?
   - $CuCl_2$ was the limiting reactant because there was still aluminum remaining after the reaction had stopped.
   - The blue color of the $CuCl_2$ disappeared indicating that the Cu^{2+} ion was completely consumed.

3. A student forgets to dry the piece of aluminum before taking the final mass. Explain how this error will affect the calculated molarity of the copper (II) chloride solution. Clearly state whether the calculated molarity will be too high, too low, or unchanged by this error and include mathematical justification for your answer.
   - Forgetting to dry the aluminum will result in a final mass of aluminum that is too high. When subtracted from the initial mass the apparent mass of aluminum consumed will be too low.
   - If the mass of Al is too low then the moles Al will be too low.
   - If the moles of Al is too low then the moles $CuCl_2$ will be too low.
   - Since moles of $CuCl_2$ is too low and in the numerator the molarity will be too low.

4. Propose a procedure that would allow you to experimentally determine the ratio of moles of Al consumed to moles of Cu produced in this reaction.
   - Students can collect and dry the copper produced. After finding the mass of the dry copper this quantity can be converted to moles of copper.
   - The moles of aluminum can be found by subtracting the initial and final masses of the strip and converting this difference to moles of aluminum.
   - To calculate the mole ratio divide the moles of aluminum by the moles of copper. The balanced equation shows a 2:3 ratio.

Limiting Reactant
Exploring Molar Relationships

When two substances react in a chemical reaction there is generally a limiting reactant. This reactant is the one that is consumed entirely and limits how far the reaction can proceed. The amount of product produced is dependant on the number of moles of this limiting reactant. The reactant that is not entirely consumed is called the excess reactant.

PURPOSE

In this activity you will react solid aluminum and a measured quantity of copper (II) chloride solution to determine which one is the limiting reactant. You will present your results in a formal written lab report.

MATERIALS

1 strip of aluminum metal   tweezers or tongs
100 mL of copper (II) chloride   balance
250 mL beaker   paper towels
100 mL graduated cylinder

Safety Alert
1. Wear eye protection.
2. Wash your hands before leaving the lab.

PROCEDURE

1. Your task is to design an experiment that will determine whether a strip of aluminum or 100 mL of copper (II) chloride is the limiting reactant in this reaction. You will also determine the molarity of the copper (II) chloride solution from your calculations.

2. In the space marked HYPOTHESIS on your student answer page, write a testable if-then hypothesis statement regarding the limiting reactant for this reaction.

3. In the space marked PROCEDURE on your student answer page, record the exact steps you use as you perform your experiment.

4. In the space marked DATA AND OBSERVATIONS on your student answer page, record your observations and relevant measurements.

5. When you have completed the experiment, clean up your lab area and wash your hands.

6. Refer to the handout entitled Formal Lab Report Guidelines for the format of your lab report, Be sure to answer the conclusion questions in your report.

Name _____

Period _____

Limiting Reactant
Exploring Molar Relationships

HYPOTHESIS _____

PROECEDURE _____

DATA AND OBSERVATIONS _____

Use this space to record your observations as the reaction takes place. You will also need to record any measurements you made.

ANALYSIS

What is the molarity of the copper (II) chloride solution your teacher provided? Show calculations below. Obtain the true molarity of the copper (II) chloride solution and calculate percent error.

CONCLUSION QUESTIONS

1. Write the balanced chemical equation for this reaction.

2. Which reactant was the limiting reactant? What evidence do you have to support this?

3. A student forgets to dry the piece of aluminum after before taking the final mass. Explain how this error will affect the calculated molarity of the copper (II) chloride solution. Clearly state whether the calculated molarity will be too high, too low, or unchanged by this error and include mathematical justification for your answer.

4. Propose a procedure that would allow you to experimentally determine the ratio of moles of Al consumed to moles of Cu produced in this reaction.

Formal Lab Report Guidelines

Prepare a written report of your experiment which includes the section titles listed below. These section titles should be used to label each section of your report.

   I.   Introduction
   II.   Hypothesis
   III.   Materials
   IV.   Procedures
   V.   Observations/Data Collection
   VI.   Analysis
   VII.   Conclusions

The following information should be included in each section of the lab report.

I.   **Introduction** – In this section of the report you should give the reader background information that will help them understand the experiment that you have conducted. Important terms, equations, and reactions should be presented in the section. Additionally the purpose of the lab should be clearly stated in the introduction.

II.   **Hypothesis** – This relatively short section should include a testable hypothesis written in an if-then format.

III.   **Materials** – A complete listing of the materials and supplies that were used to conduct the experiment should be included in this portion of the report.

IV.   **Procedures** – In this section of the report you should present the exact steps that were followed in your experiment. If applicable, clearly identify the control, variables and the measurement techniques used.

V.   **Observations/Data Collection** – All of the data and observations that were collected during the experiment should be presented in a data table or tables. If applicable, a graph of the data should be included in this section. Make sure that the graph is appropriately titled and labeled. Include a legend if necessary.

VI.   **Analysis** – This section should show all mathematical calculations that were performed on the collected data. Equations and thorough work should support your calculations.

VII.   **Conclusions** – This portion of the report is used to clearly explain whether the results support or refute the hypothesis being tested by citing any data or evidence from the experiment. You should also answer any conclusion questions posed in the lab and discuss sources of error that were present.

Charles' Law
Investigating the Relationship Between Temperature and Volume of a Gas

OBJECTIVE
Students will investigate the relationship between the volume of a gas and the temperature. In the process they will extrapolate the data to determine a value for absolute zero.

LEVEL
Chemistry

NATIONAL STANDARDS
UCP.1, UCP.2, B.2, B.6, E.1, E.2

TEKS
2(A), 2(B), 2(C), 2(D), 2(E), 4(B), 7(B)

CONNECTIONS TO AP
AP Chemistry:
    II. States of Matter A. Gases 1. Laws of ideal gases

TIME FRAME
45 minutes

MATERIALS
(For a class of 28 working in pairs)

7 hot plates	7 tall form 600-mL beakers
3 L of corn oil	14 graphing calculators-one for each pair of students
14 rulers	
14 data collection devices — CBL2, or LabPro	14 temperature probes
latex tubing with 0.5 cm inner diameter	100-mm capillary melting point tubes
paper towels	computer graphing software

TEACHER NOTES
Set-up: The hot plates should be positioned about the room in the safest possible places. Each beaker should be approximately 2/3 filled with corn oil. The beakers should be clamped so that they may not be tipped over. Students will secure the capillary tube to the thermometer and will completely immerse it in the hot oil, so the oil needs to be at least 100 mm deep. Remember that oil has a considerable coefficient of volume expansion. Do not fill the beaker to the top. The corn oil may be saved and re-used for several years.

Cut the 0.5 mm latex tubing into small bands to affix the capillary tubing to the temperature probe. Orthodontic rubber bands will also work. Each pair will need at least two bands.

The original CBLs will work with this lab as well if you use the brass temperature probes. Other than that, the protocol will be the same.

POSSIBLE ANSWERS TO THE CONCLUSION QUESTION AND SAMPLE DATA

Length of trapped gas sample (mm)	Temperature (C°)
100. mm	108
95	79
90	55
85	38
80	30

1. At what temperature does your extended graph line intersect the *y*-axis? What does this number represent?
   - The y-intercept should be between -250 to -300. It represents the temperature when the volume is theoretically zero. The y-intercept for the sample data is -293

2. Realizing that this value should be -273° C, calculate a percent error for your lab.
   - $\%Error = \dfrac{|\,Observed\ value - Theoretical\ value\,(-273)\,|}{|\,Theoretical\ value\,(-273)\,|}$
   - The sample data has $\%Error = \dfrac{|\,(-293) - (-273)\,|}{|\,(-273)\,|} = 7.33\%$

3. What is the equation for the regression line?
   - Student answers will vary. The equation for the data shown above is y = 3.94x + -293. Anytime data is analyzed graphically, students should report the regression equation.

4. What would be the volume of your gas sample at the temperature of the y-intercept?
   - The volume of the gas would be zero.

5. Why is this volume only theoretical?
   - Real gases actually have molecular volume and therefore would condense into a liquid and then a solid at this temperature.

Re-number the temperature scale on your graph, assigning the value zero the temperature at which the graph intersects the y-axis. Do this through the STAT Edit (ENTER) function with a batch transform. To do a batch transform on the temperature, first determine the y-intercept. Move the cursor to L2 and arrow up so that it actually sits on the L2. Type 2nd 2 (L2) + the absolute value of the y-intercept. ENTER. The absolute value of the y-intercept will be added to all of the values in L2. Return to STAT ► CALC to run a new linear regression and paste the function into Y= as you did before. Adjust your window appropriately. The new scale expresses temperature in Kelvins (K).

One Kelvin is the same size as on degree Celsius. However, unlike zero degrees Celsius, zero Kelvin is the lowest temperature theoretically possible—absolute zero.

- In all likelihood, you will have to walk the students through this step. This is particularly true if this is the first time that they have done a transformation that is similar to this one.

6. Based on your new graph, what Kelvin (K) temperature would correspond to 0° C? What Kelvin (K) temperature would correspond to 100° C?
   - Student answers will vary. The sample data provides an answer of 252 K.

7. After adjusting your equation to the absolute scale, download your data to Graphical Analysis™, LoggerPro, or Excel. Appropriately label your axes and title the graph. Print a copy of the graph to turn in with this lab report.
   - If computers are not available, students may produce a hand-drawn graph to turn in with this lab.

Charles' Law
Investigating the Relationship Between Temperature and Volume of a Gas

Most people realize that gases expand when heated and contract when cooled. However, in this lab you will investigate this relationship quantitatively.

PURPOSE
In this activity you will investigate the relationship between temperature and volume of a gas. You will also extrapolate the data to determine a value for absolute zero.

MATERIALS

hot plate
graphing calculator with CHEMBIO or
    DATAMATE loaded into the APPS
data collection devices — CBL2, or LabPro
ruler
paper towels — 2 sheets
computer graphing software (optional)

tall form 600-mL beaker filled 2/3 full of
    corn oil (beaker should be securely
    clamped)
temperature probe
2 latex tubing "rubber bands"
pencil
100-mm capillary melting point tubes

Safety Alert
1. The oil is very hot. Use caution around the hot plates.
2. Take care that the cords from the data collection devices do not come in contact with the hot plate.
3. Dispose of glass capillary tubes in a glass disposal box or as instructed by your teacher.

PROCEDURE

1. In the space provided on your student answer page, write an if-then hypothesis examining the relationship between temperature and volume of a gas.

2. Set up the data collection device with the temperature probe in channel 1. Connect the calculator firmly to the data collection device.

3. Fasten a capillary (melting point tube closed at one end) tube to the lower end of the temperature probe using two rubber bands (0.5 cm. inner diameter). Both bands should be within 60 mm from the top (closed end) of the tube. The open end of the tube should be at the terminal end of the temperature probe.

4. Place the temperature probe with capillary tube on a clean paper towel. With a pencil, mark and label the top and bottom of the capillary tube. The paper towel will be marked in 5 mm increments to serve as a ruler for measuring the length of the trapped gas in the tube. It will only be necessary to mark the region between 60 and 100 mm as the gas bubble will not exceed this range. The closed end of the tube will be considered 0mm and the open end will be considered 100 mm. Use your pencil on the paper towel to mark 5 mm increments between 60 and 100 mm.

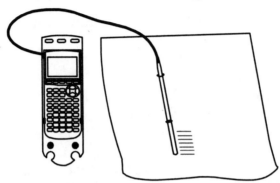

5. Setting up the program.
   - Turn on the calculator. Access CHEMBIO from the APPS menu.
   - Select SET UP Probes from Main Menu. [ENTER]
   - # of probes: 1[ENTER]
   - Select Channel 1. [ENTER]
   - Select (2) COLLECT DATA [ENTER]
   - Select (4) TRIGGER. [ENTER]

You are now ready to collect data.

6. Take the temperature probe and capillary tube assembly to the hot oil. Immerse the tube and the temperature probe in the hot oil. Be sure the entire capillary tube is immersed in the oil. Wait for your tube and temperature probe to reach the temperature of the oil (about 125 °C). **Be sure not to go above 125 °C.**

7. When the temperature reaches 124-125 °C, push (+) on the calculator. This will be your first data point for the full length of the capillary tube (100 mm). Then push (1) MORE DATA on the calculator. You are now ready to collect your second point.

8. After you have recorded your first data point, lift the temperature probe so that about 1-cm of the capillary tube is remaining in the oil. Pause here for about 3 seconds to allow some oil to rise into the tube. Then quickly place the apparatus on the marked paper towel.

9. Place the temperature probe with the capillary tube on the marked paper towel you made in step 4, carefully aligning the apparatus with your marks. Do this quickly as the longer you take, the less accurate your values will be.

10. When the **top** of the oil drop reaches one of the marked values (85mm, 80mm, 75mm, etc.) push (4) TRIGGER, then (1) MORE DATA. Collect at least four data points including the 100-mm value. After the last data point, enter STOP. Since the capillary tube has a uniform diameter, length serves as a relative measure of the gas volume.

11. Discard the rubber bands and capillary tubes according to your teacher's instructions. Wipe the thermometer probe dry with a paper towel. Clean up any drops of oil that may be on your lab table.

Name _____

Period _____

Charles' Law
Investigating the Relationship Between Temperature and Volume of a Gas

HYPOTHESIS

DATA AND OBSERVATIONS

Length of trapped gas sample (mm)	Temperature (C°)
100. mm	

ANALYSIS

The data that you collected will be in L_1. Press [STAT] [ENTER] to view this data. Record the data in the table above. To transfer the data from L_1 to L_2 move the cursor to the L_2 column and arrow up so that it sits on L_2. Press [CLEAR] and [ENTER]. Now move the cursor back up to L_2 and type [2nd][1] [ENTER]. Next, arrow over to L_1 and clear that list as you did with L_2. Enter the values for the column length in L_1.

To be sure that your correlation coefficient will appear, go to the CATALOG. To do this, hit [2nd] [0] [x⁻¹]. This will bring you into the "d's" of the CATALOG. Select DIAGNOSTICS ON. [ENTER]. [ENTER].

Run a linear regression and paste the equation into [Y=]. If you have forgotten how to do this, the key strokes are: [STAT][▶] CALC 4:LinReg [VARS] [▶] Y-VARS [ENTER][ENTER][ENTER].

Set up your graph with STAT PLOT ([2nd][Y=]) so that it looks like the screen below.

Extend the window so that we can extrapolate the data. In order to do this, go to [WINDOW]. Set your window as shown below.

```
WINDOW
 Xmin=-15
 Xmax=100
 Xscl=1
 Ymin=-350
 Ymax=150
 Yscl=50
 Xres=1
```

Because we have manually set the window, we must now press [GRAPH], rather than our usual method of going to [ZOOM][9]. To use the ZOOM function would erase our pre-set window. If you find that you do not have axes, go to CATALOG [2nd][0] AXES ON and select and execute that function.

CONCLUSION QUESTIONS

1. At what temperature does your extended graph line intersect the *y*-axis? What does this number represent?

2. Realizing that this value should be -273° C, calculate a percent error for your lab.

3. What is the equation for the regression line?

4. What would be the volume of your gas sample at the temperature of the y-intercept?

5. Why is this volume only theoretical?

   Re-number the temperature scale on your graph, assigning the value zero the temperature at which the graph intersects the y-axis. Do this through the [STAT] Edit ([ENTER]) function with a batch transform. To do a batch transform on the temperature, first determine the y-intercept from your regression equation. Move the cursor to L_2 and arrow up so that it actually sits on the L_2. Type [2nd] [2] (L_2) [+] the absolute value of the y-intercept. [ENTER]. The absolute value of the y-intercept will be added to all of the values in L_2. Return to [STAT][▶] CALC to run a new linear regression and paste the function into [Y=] as you did before. Adjust your window appropriately. The new scale expresses temperature in Kelvins (K). One Kelvin is the same size as one degree Celsius. However, unlike zero degrees Celsius, zero Kelvin is the lowest temperature theoretically possible—**absolute zero.**

6. Based on your new graph, what Kelvin (K) temperature would correspond to 0° C? What Kelvin (K) temperature would correspond to 100° C?

7. After adjusting your equation to the absolute scale, download your data to Graphical Analysis™, LoggerPro, or Excel. Appropriately label your axes and title the graph. Print a copy of the graph to turn in with this lab report.

Airbags
Designing a Lab with Gas Laws

OBJECTIVE
Students will design an airbag with baking soda, hydrochloric acid, and a resealable plastic bag. They will perform calculations to determine optimum quantities to just fill the bag with gas.

LEVEL
Chemistry

NATIONAL STANDARDS
UCP.2, UCP.3, A.1, A.2, B.2, B.3, E.1, E.2, F.1, G.1, G.2

TEKS
2(A), 2(B), 2(C), 2(D) 4(B), 7(A), 7(B), 11(B)

CONNECTIONS TO AP
All AP Science Courses—Lab design

AP Chemistry:
    II. States of Matter A. Gases 1. Laws of ideal gases

TIME FRAME
45 minutes

MATERIALS
(For a class of 28 students working in pairs.)

28 resealable plastic bags (pint or quart)	14 small boxes of baking soda
14 bottles containing 250 mL 6 M HCl each	56 disposable pipets
several balances	A number of graduated cylinders. Some
calculators	250 mL and 500 mL are helpful.
small condiment cups for massing $NaHCO_3$	14 scoops
assorted beakers	14 thermometers
paper towels	

TEACHER NOTES
This is an excellent opportunity to connect real-life chemistry to the classroom. Not only does this activity give the students a chance to design a lab, but it also provides an entry to discuss consumer chemistry, as well as safety in and out of the lab. There are many excellent resources on the Internet. This one is particularly useful: http://wunmr.wustl.edu/EduDev/LabTutorials/Airbags/airbags.html. If you do not have classroom access to computers, you might want to print several copies for the students to use during the activity.

Students will need to know the barometric pressure. Post in a clear, obvious place in the room, but do not call deliberate attention to it. Students will need some background with gas laws to use this lesson.

POSSIBLE ANSWERS TO THE CONCLUSION QUESTIONS AND SAMPLE DATA

Students will develop a number of different solutions to the problem. Several points that need to be considered are:

- Students need to determine the volume of the baggie. The easiest way to accomplish this is for students to fill the baggie with water and then pour the water into the graduated cylinder to measure the volume.

- Students need to decide which reactant will be the limiting reagent. Using $NaHCO_3$ as the limiting reagent is the easiest, but not the only solution. If they use baking soda as the limiting reagent, they will need to determine the mass of baking soda required to fill the bag with gas when the reaction is complete.

- Students need to develop a way to place both reagents into the bag without mixing them before the bag is zipped close. Several workable solutions include:
  - put the HCl in the pipets and discharge once the bag is closed
  - put the HCl in a condiment cup and carefully place it in the bag on top of the massed baking soda and tip after closing the bag
  - fold the bag in such a way that the reagents can not mix.

SAMPLE DATA

Volume of bag (volume of bag will vary with size and brand)	0.950 L
Temperature of the room	23° C
Barometric pressure	763 torr
Molar mass of $NaHCO_3$	84 g/mol
Mass of baking soda required	3.3 g

1. Write a **balanced** chemical equation for the reaction used to produce the gas in your simulated air bag.
   - $NaHCO_3 + HCl \rightarrow NaCl + CO_2 + H_2O$

2. The reaction in this lab is only a **simulation** for that in an automobile airbag. Write a balanced chemical equation for the reaction that actually takes place in an automobile airbag. You may need to do some research. The primary reaction involves the decomposition of sodium azide, NaN_3. A second reaction removes the metallic sodium by reacting it with potassium nitrate. A third reaction removes the sodium oxide and potassium oxide by reacting it with silicon dioxide to produce harmless silicates.
   - $2NaN_3 \rightarrow 2Na + 3 N_2$

   - $10Na + 2KNO_3 \rightarrow K_2O + 5Na_2O + N_2$

   - $K_2O + SiO_2 \rightarrow K_2SiO_3$ and $Na_2O + SiO_2 \rightarrow Na_2SiO_3$

3. What was the limiting reagent in your experiment? Justify your answer with data.
    *   These answers will vary with the students' lab designs.

Airbags
Designing a Lab with Gas Laws

Since model year 1998, all new cars have been required to have air bags on both driver and passenger sides. To date, statistics show that air bags reduce the risk of dying in a direct frontal crash by about 30 percent. Many people believe that the gas used to inflate the bag comes from a compressed air tank. However, the airbags are filled with a gas as the result of a rapid chemical reaction. In this activity, you will simulate this process using a zip-type plastic bag and a different chemical reaction.

This exercise tests your ability to design and carry out laboratory experiments. You and your partner will be graded on experimental design, data collection skills, and on the accuracy and precision of your results. Clear thought processes and well written responses will contribute to your success on this assignment. Your written responses need to be as clear and concise as possible. You must follow proper safety procedures.

PURPOSE
The task is to generate a gas that will **just** fill a small resealable plastic bag using baking soda and 6-M HCl. The ideal result is to fill the bag to plumpness, yet not to over-inflate or under-inflate the bag; the bag may also contain unreacted chemicals and/or other products of the reaction. You will be asked to describe the method you develop to solve the problem. You must complete this assignment (including the report) during the assigned period. You may not share information between groups.

MATERIALS
2 resealable plastic bags (pint or quart)   1 small box of baking soda
bottles containing 250 mL 6 M HCl each   4 disposable pipets
several balances   graduated cylinders
small condiment cups for massing $NaHCO_3$ other equipment as allowed by your teacher
assorted beakers   scoops
thermometer   calculator

Safety Alert
1. Goggles are required throughout this entire lab.
2. 6-M HCl is a concentrated acid. Clean up spills by reacting with excess baking soda. If the acid is splashed on skin, flush with water for 15 minutes. Notify the teacher.
3. Wash hands at the end of the lab.

PROCEDURE

1. State your hypothesis in an if-then format.

2. With your partner, plan a design for your project and record your design on the student answer page. Have your teacher initial your design before beginning work.

3. Carry out your plan. Record your observations.

4. Revise your design. Repeat and show your teacher the filled bag.

Name _____

Period _____

Airbags
Designing a Lab with Gas Laws

HYPOTHESIS _____

DESCRIPTION OF PLAN, INCLUDING CALCULATIONS _____

TEACHER APPROVAL OF FILLED BAG _____

OBSERVATIONS

CONCLUSION QUESTIONS

1. Write a **balanced** chemical equation for the reaction used to produce the gas in your simulated air bag.

2. The reaction in this lab is only a **simulation** for that in an automobile airbag. Write a balanced chemical equation for the reaction that actually takes place in an automobile airbag. You may need to do some research. The Internet is a resource. One possible website is: http://wunmr.wustl.edu/EduDev/LabTutorials/Airbags/airbags.html

3. What was the limiting reagent in your experiment? Justify your answer with data.

Heating Curves and Phase Diagrams
Investigating Changes of State

OBJECTIVE

The students will participate in several activities to develop and enrich their understanding of phase changes. They will investigate equilibrium vapor pressure curves, phase diagrams and will develop a heating curve of water. They will use the various graphs to solve problems and will apply the information to enrich their understanding of the underlying physical phenomena.

LEVEL

Chemistry

NATIONAL STANDARDS

UCP.1, UCP.2, UCP.3, A.1, A.2, B.2, B.6, E.1, E.2

TEKS

2(B), 2(C), 3(A), 5(A), 5(C)

CONNECTIONS TO AP

AP Chemistry:

    II. States of Matter B. Liquids and solids 2. Phase diagrams of one-component systems

    3. Changes of state, including critical points and triple points

    III. Reactions E. Thermodynamics 2. First law: change in enthalpy; heat of formation; heat of reaction; Hess's law; heats of vaporization and fusion; calorimetry

TIME FRAME

135 minutes: 20 minutes for equilibrium vapor pressure graphing exercise; 25 minutes for teacher-directed phase diagram lesson and questions; 45 minutes for development of heating curve; 45 minutes for problem solving and lesson on phase diagrams.

MATERIALS

(For a class of 28 working in pairs)

14 15 mL polyethylene centrifuge tubes	14 600-mL beakers
14 ring stands	14 rings large enough to support the beakers
1 1000-mL beaker	to avoid spillage
1 hammer	1 quart acetone
28 TI-83+ calculators	wooden paint stirrer
28 stainless steel temperature probes	14 CBLs or LabPros with link cords
computers with TI Connect and GraphLinks	computer graphing software
highlighters or markers of various colors	printers for computers
	14 rulers

TEACHER NOTES: ACTIVITY ONE

Students will be graphing the equilibrium vapor pressure of three different substances on their graphing calculators. If calculators are not available, this could be a paper/pencil exercise; however, it will take considerably more time.

When the students start to determine the proper regression equation that will fit the data, it is a good idea to tie it to the families of functions that they may be studying in their math classes. The proper regression for this data is exponential. If the mathematical ability of the students is not yet at the level where they can look at a graph and make an intelligent "guess" about the proper function, you might narrow the field for them by giving them three choices out of the many that are available on their calculators.

Ideas for this activity were taken from David W. Brooks, University of Nebraska-Lincoln web site.

DATA AND OBSERVATIONS: ACTIVITY ONE

Draw a rough sketch of the graph that is on your calculator.

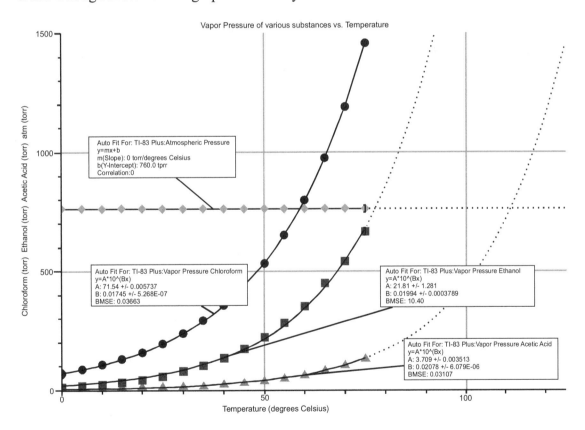

POSSIBLE ANSWERS TO QUESTIONS: ACTIVITY ONE

1. What does the horizontal line in Y_4 represent?
    *   Y=760 represents the pressure of one atmosphere.

2. Why do you think that the vapor pressure vs. temperature graph is not a linear relationship?
   - This is a very difficult question, and one that the students should be encouraged to think about. Temperature is related to kinetic energy. $(KE = \frac{1}{2} mv^2)$ Molecules have to be moving fast enough to escape the intermolecular forces and vaporize. Once students see the velocity squared term, they usually realize that the function cannot be linear.

3. Which substance has the strongest intermolecular forces? Explain your answer.
   - Acetic acid has the strongest intermolecular forces because it has the lowest vapor pressure.
   - Additional information (although not explicitly required by the question): Chloroform is slightly polar and has relatively strong London forces. Both ethanol and acetic acid are able to form hydrogen bonds. Acetic acid forms dimers with two hydrogen bonds/molecule.

4. Determine the normal boiling point for
   a. Chloroform:
      - 58.8°C

   b. Ethanol
      - 74.4°C

   c. Acetic acid
      - 111°C

5. At 60°C, which substance has the highest vapor pressure?
   - Chloroform

6. If the pressure were dropped to 600 torr, what would be the boiling temperature of ethanol?
   - 69.9°C

7. If the pressure were dropped to 275 torr, which substance(s) would be boiling at or below 55°C?
   - Both chloroform and ethanol would be boiling at 275 torr and 55°C.

TEACHER NOTES: ACTIVITY TWO

Use a copier to increase the size of the two phase diagrams and make overhead transparencies of each one. The first diagram could be a generic diagram for most substances. There are no numbers on the axes, so it is impossible to tell what it is. Use this diagram to point out the important features of this graph. It is important that students know the following:

1. That the y-axis is pressure and the x-axis is temperature.
2. Where the regions of solid, liquid, and gas (vapor) are located.
3. The meaning of the triple point.
4. The meaning of the critical point.
5. The fact that the lines that separate the different phases are equilibrium points.

Pick a pressure—any pressure—and increase the temperature. Explain what happens to the substance when the temperature increases. Then pick a temperature and increase the pressure. What happens to the substance?

Point out the difference between the slopes of the liquid/solid equilibrium lines of the two graphs. Increasing the pressure will always increase the density. Does increasing the pressure favor the liquid phase or the solid phase? If the liquid phase is favored, then the liquid will be more dense than the solid and the solid will float in the liquid. However, if the solid phase is favored, then the solid is more dense and will sink in the liquid. Water is the most notable of substances that has a liquid phase that is more dense than the solid, but it is not the only substance that does this.

POSSIBLE ANSWERS TO QUESTIONS: ACTIVITY TWO

1. Use the following information about ammonia to draw a phase diagram on the student answer sheet. The triple point of ammonia is 195.42 K and 0.05997 atm. The critical point is 405.38 K and 111.5 atm. The normal freezing point is 195.45 K and the normal boiling point is 239.8 K.

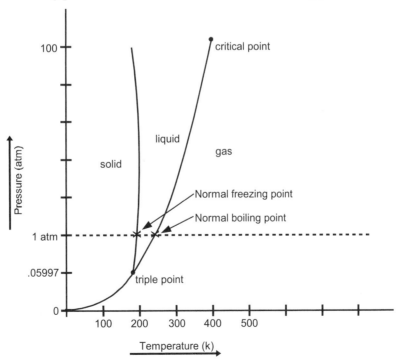

2. Describe what would happen when you make each of the following changes:
   a. While the system is in an enclosed vessel, heat is constantly added. The temperature is increased from 150 K to 250 K while the pressure is kept constant at 0.75 atm.
      - The solid heats until it reaches the equilibrium line between solid and liquid. It stays at a constant temperature until all of the solid melts. Then the liquid will heat until the temperature reaches the equilibrium between liquid and vapor. The substance boils. Once the entire substance is in the vapor state, the vapor heats to 250 K.
   b. While the temperature is kept constant at 225 K, the pressure is increased from 0.1 atm to 0.95 atm.
      - The ammonia is in the gas phase. As the pressure is increased, the ammonia reaches the point where it is in equilibrium between gas and liquid, where it condenses.

3. Describe the system at 407 K and 112 atm.
   - The system is beyond the critical point and is a one phase system. At this temperature and pressure, it would be a supercritical fluid—one phase but with characteristics unique to that system. You might want students to do some research on supercritical fluids. Water and carbon dioxide are particularly interesting.

TEACHER NOTES: ACTIVITY THREE
Acetone may be obtained at a local hardware or paint store. Many grocery stores now have dry ice available for purchase.

To prepare the frozen temperature probes:
The probes need to be frozen in the middle of the centrifuge tubes which have been ½ filled with distilled water. While wearing goggles, pound the dry ice into a powder. Put the powdered dry ice in a 1000-mL beaker until ¾ full. Slowly add acetone while stirring with the paint stick to make a dry ice/acetone slurry. This mixture will bubble considerably when you begin, so it is advisable that you perform this step in a sink. Once the slurry has been prepared, it can be replenished all day simply by adding more dry ice.

Place the temperature probe into the plastic centrifuge tube that is ½ filled with distilled water and place the test tube into the dry ice/acetone slurry. It is very important that the probe be frozen in the middle of the ice. If the tip of the probe rests on the bottom or side of the test tube, the student data will not be accurate. Hold the probe up so it does not touch the bottom or sides. Once a layer of ice has formed all around the sides of the test tube, you can allow the probe to rest on this layer and start the next probe. Test tubes can remain in the slurry until the students are ready to use them. While the students are collecting data, you can prepare the probes for the next class.

While CBLs or LabPros can be run with batteries, using a power adapter saves on battery cost and sometimes gives more consistent results.

Graphing for this lab may be done with paper and pencil; however, because of the number of data points, it is quite time consuming.

PRE-LAB: ACTIVITY THREE

1. Define the following terms:
   a. Endothermic
      - A process which takes in energy (usually in the form of heat) to occur.
   b. Exothermic
      - A process which releases energy to occur.
   c. Potential energy
      - Stored energy. In chemistry, the energy is stored in the chemical bonds. Students have probably talked about gravitational or positional potential energy. Be sure they understand that potential energy is "stored" energy and that it can be stored in many ways.
   d. Kinetic energy
      - Energy of motion. $E_k = \frac{1}{2} mv^2$ where E_k is kinetic energy (in joules), m is mass (in kilograms) and v is velocity (in meters/second). Kinetic energy is directly proportional to temperature, so when the temperature is rising, kinetic energy is being added.
   e. Specific heat
      - The amount of heat it takes to increase the Celsius temperature of one gram of substance by one degree.
   f. Latent heat of fusion
      - The amount of heat it takes to melt one gram of a substance.
   g. Fusion
      - Melting
   h. Latent heat of vaporization
      - The amount of heat it takes to vaporize one gram of a substance.
   i. Vaporizing
      - Turning a liquid into a gas phase
   j. Melting
      - Turning a solid into a liquid
   k. Freezing
      - Turning a liquid into a solid
   l. Boiling
      - Rapidly turning a liquid into a vapor so that the entire liquid is disturbed with vapor bubbles
   m. Condensing
      - Turning a vapor into a liquid

2. Since we are continuously adding heat energy to this system, what is happening to that heat when the ice is melting? What type of energy is this?
   - The energy is being used to break some of the hydrogen bonds. These intermolecular attractive forces between the water molecules are responsible for holding the water molecules in its crystal lattice.
   - This is potential energy or the energy stored in the chemical bonds.

3. In a cooking class, a teacher told the class to bring the water to a boil and allow it to boil for 5 minutes so that it would be "good and hot." What is wrong with this statement?
   - Boiling occurs at a constant temperature and will not get hotter than $100^\circ C$ at one atmosphere of pressure.

4. You have been given a picture of a theoretical heating curve for water. After reading the lab, sketch a picture of the heating curve which you expect to produce with this lab.

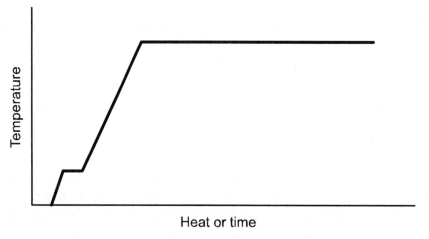

DATA AND OBSERVATIONS: ACTIVITY THREE

1. Download your data from the calculator. Print a copy of the data table and turn it in with the lab. Be sure to print a copy for everyone in your group.

2. Graph temperature vs. time. Title your graph and label the axes. Be sure to include units. Print the graph and turn it in with the lab. Be sure to print a copy for everyone in your group.

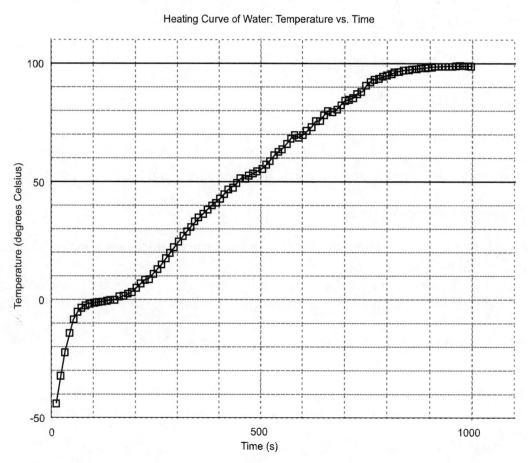

ANALYSIS: ACTIVITY THREE

1. Use a highlighter to indicate the region of the graph where melting occurred. Use a different colored highlighter to indicate the region where boiling occurred. Clearly label each one.

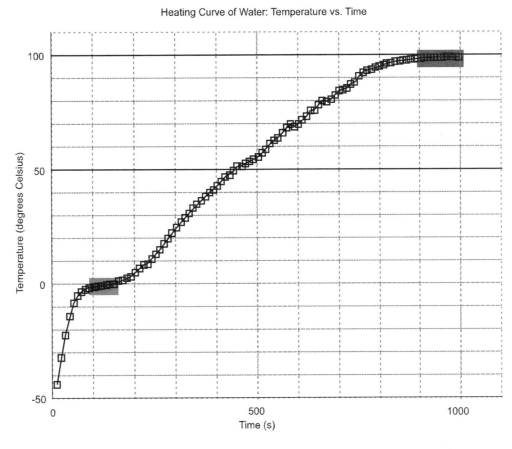

Heating Curve of Water: Temperature vs. Time

2. Look at the region where the ice was heating. Compare it to the region where the water was heating. Suggest a reason why the slopes are not the same.
   - The specific heat of ice is not as great as the specific heat of water. It should take less heat to raise the temperature of the ice than the water so the slope should be steeper.

3. Citrus growers in South Texas often spray their crops with water if the weather forecast indicates that the temperature will drop below 0.0°C. Explain why this would protect the crop.
   - As the water freezes it holds the temperature constant at 0°C and freezing is an exothermic process, transferring heat energy to the crop.

4. This is a qualitative lab, rather than a quantitative lab. What is the difference?
   - A quantitative lab measures actual amounts. We have no idea how much heat energy is actually being transferred to our test tube. Therefore it is only a general shape of the graph of temperature vs. time. Qualitative labs tell us generally what is happening without worrying about "how much".

TEACHER PAGES

5. Explain any differences between your graph and the standard graph shown in the introduction of this lab.
   - Student answers will vary. None of the graphs will have the portion of the graph showing heating of the vapor phase.

CONCLUSION QUESTIONS: ACTIVITY THREE

1. Indicate whether each of the following is exothermic or endothermic.
   a. Boiling: <u>Endothermic</u>

   b. Melting: <u>Endothermic</u>

   c. Condensing: <u>Exothermic</u>

   d. Freezing: <u>Exothermic</u>

Use this table to work the following problems:

Specific heat of ice	2.1 J/g $^\circ$C
Specific heat of water (liquid)	4.2 J/g $^\circ$C
Specific heat of steam	2.1 J/g $^\circ$C
Latent heat of fusion of ice	336 J/g
Latent heat of vaporization of water	2264 J/g

2. How much heat is given off by 75.0 g of water freezing?

   $$q = L_f m$$

   - $$\frac{336\,J}{\cancel{g}} \times \frac{75.0\,\cancel{g}}{} = 25\,200\,J$$

3. If I start with 20.0 g of ice at -5.0°C and heat it until it is liquid water at 60.°C, how much heat (in kJ) does it take?

   $$q = m_{ice}c_{ice}\Delta T_{ice} + L_{f_{ice}} m_{ice} + m_{water}c_{water}\Delta T_{water}$$

   -
   $$(\frac{20.0\cancel{g}}{} \times \frac{2.1\cancel{J}}{\cancel{g}\,^\circ\cancel{C}} \times \frac{5\,^\circ\cancel{C}}{} + \frac{336\cancel{J}}{\cancel{g}} \times \frac{20.0\,\cancel{g}}{} + \frac{20.0\cancel{g}}{} \times \frac{4.2\cancel{J}}{\cancel{g}\,^\circ\cancel{C}} \times \frac{60\,^\circ\cancel{C}}{}) \times \frac{kJ}{1000\cancel{J}} = 12\,kJ$$

   - Notice that there are three terms in this equation. Students may want to break the problem into three separate steps and then add them together at the end. Referring to the phase diagram when working these problems is very helpful.

4. A mixture of 20. g of ice and 100. g of liquid water at $0^\circ C$ are heated until all the ice melts and the combined water is at $30.^\circ C$. How much heat (in kJ) is added to this system to cause this to happen?

$$q = L_{f_{ice}} m_{ice} + m_{water} c_{water} \Delta T_{water}$$

* $$(\frac{336 \cancel{J}}{\cancel{g}} \times \frac{20. \cancel{g}}{} + \frac{120. \cancel{g}}{} \times \frac{4.2 \cancel{J}}{\cancel{g} \, ^\circ \cancel{C}} \times \frac{30. \, ^\circ \cancel{C}}{}) \times \frac{kJ}{1000 \cancel{J}} = 22 \, kJ$$

Heating Curves and Phase Diagrams
Investigating Changes of State

ACTIVITY ONE: EQUILIBRIUM VAPOR PRESSURES

Everyone knows that if you leave an open beaker of water out in the room for a period of time, the water will evaporate. However, if you place a tight cover on the beaker, the water will still be there several weeks later. Is there nothing happening in the covered beaker? Actually, some of the water will evaporate and some will condense. After putting the lid on the system, the system will reach equilibrium between the number of water molecules that are in the liquid state, and the number of water molecules that are in the vapor state. Chemists would express what is occurring like this:

$$\text{water (l)} \rightleftharpoons \text{water (g)}$$

The double arrow indicates that the reaction is going forward and backward at the same rate. It does not indicate that there is the same number of molecules in the liquid state as there are in the vapor state. This is called equilibrium and it is referred to as dynamic equilibrium because the molecules are constantly changing places even though there is no net change in the number of molecules in each phase.

Because some of the molecules are in the gas phase, they exert a vapor pressure. This is called the equilibrium vapor pressure and it is temperature dependent. That is to say that, if we increase the temperature, more of the molecules will go into the vapor phase and a new equilibrium will be established at a new temperature. All liquids (and some solids) have a vapor pressure. During this first activity, you will use your calculator to graph the equilibrium vapor pressures of some substances other than water and then use your graph to answer questions about the liquids.

PURPOSE: ACTIVITY ONE

In this activity you will graph equilibrium vapor pressure versus temperature and interpret the graph.

MATERIALS

graphing calculator   paper/pencil
student answer page

PROCEDURE: ACTIVITY ONE

1. Using your graphing calculator, enter the following data:

L₁	L₂	L₃	L₄
0	71.5	12.2	3.7
5	87.5	17.3	4.7
10	106.9	23.6	6
15	130.7	32.2	7.6
20	159.8	43.9	9.7
25	195.4	59	12.3
30	238.8	78.8	15.6
35	292	103.7	19.8
40	356.9	135.3	25.1
45	436.4	174	31.9
50	533.4	222.2	40.6
55	652.1	280.6	51.5
60	797.3	352.7	65.5
65	974.6	448.8	83.2
70	1191.5	542.5	105.6
75	1456.6	666.1	134.2

   a. Temperature: L_1

   b. Chloroform: L_2

   c. Ethanol, L_3

   d. Acetic acid, L_4

2. Go to [2nd][Y=] to get to STATPLOT. Turn on all three plots. Plot 1 should be XList: L1; YList: L2. Plot 2 should be XList: L1; YList: L3. Plot 3 should be XList: L1; YList: L3. Choose a different Mark for each list. Go to [ZOOM][9] to look at the graph.

3. Be sure that your diagnostics are turned on. ([2nd][0] to get to the catalog; [x⁻¹] will bring you into the "d's". Arrow down [▼][▼][▼] several times to DiagnosticON and [ENTER][ENTER].)

4. You will want to run curve fits to determine the functions for the data. They will all use the same regression, so once you determine one of them, you may run the same regression on each. Go to [STAT][▸] to CALC. After looking at the graphs, you may have an idea as to which regression to try. The goal is to try and get the regression coefficient, "r", as close to 1.000 as possible. Once you determine which regression to run, paste it into [Y=] as Y1. Run the same regression on L1, L3 and L1, L4. Paste those regressions into Y2 and Y3 respectively. (Refresher instructions: Choose regression [ENTER]. [2nd][1][,][2nd][3][,][VARS][▸] to Y-VARS [ENTER][▼] to Y2 [ENTER][ENTER]. Be sure that you read the screen carefully as you enter each key stroke so that you understand what you are doing and can remember it for future use.) Repeat for L1, L4 and paste that function into Y3. To view your graphs, go to [ZOOM][9].

5. Go to [Y=] and arrow down to Y4. Enter 760.

6. You will need to extend the window for this exercise. Go to WINDOW and change Xmax to 125. This is the only parameter that you will need to change. After this step, you will need to go to GRAPH instead of ZOOM 9 to view your graph.

7. Use the graph to answer the questions for Activity One on the student answer page.

ACTIVITY TWO: PHASE DIAGRAMS

The following diagram (Figure 1) is a phase diagram for a pure substance. You have already seen a similar portion of this graph in Activity One. The line that shows the equilibrium between liquid and gas is the equilibrium vapor pressure. A phase diagram extends both the temperature and pressure axes so that you can see the solid phase and more of the gas phase. There are several important points or areas to note in these diagrams.

A. The point where all three phases exist in equilibrium is called the *triple point*.

B. If you draw a horizontal line at 1 atm (760 torr), the point where it crosses the equilibrium line between solid and liquid is the normal freezing point (melting point). The point when the 1 atm line crosses the equilibrium line between liquid and gas is the normal boiling point.

C. The critical point is the temperature beyond which a gas cannot be pressurized enough to become a liquid. The phase boundary between the liquid and gas is very indistinct. Substances in this region have characteristics of both gas and liquid as well as having distinct characteristics specific to that phase. They are referred to as supercritical fluids.

D. If the atmospheric pressure is below the triple point, the substance will sublime rather than vaporize from the liquid phase.

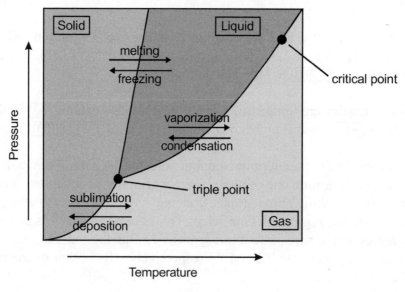

Fig. 1

Below is shown the phase diagram for water (Figure 2). Compare the two phase diagrams. What differences do you see between the two phase diagrams? Did you notice that the equilibrium line between liquid and solid has a negative slope rather than a positive slope? This means that increasing pressure will favor the liquid phase and the liquid will be more dense than the solid. Remember that this diagram is for water. Everyone knows that ice is less dense than liquid water because ice floats. Did you also know that you can melt ice by increasing the pressure? Just ask any ice skater!

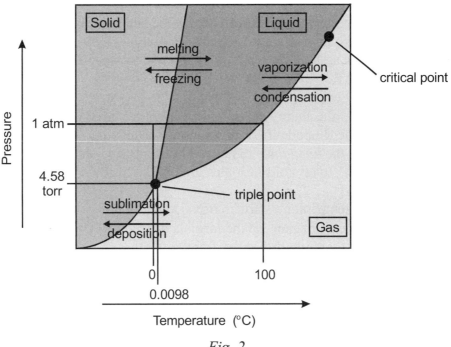

Fig. 2

PURPOSE: ACTIVITY TWO

In this activity you will use various data points to recreate a phase diagram and then interpret the graph.

MATERIALS: ACTIVITY TWO

student answer page   pencil
ruler

PROCEDURE: ACTIVITY TWO

1. Use the following information about ammonia to draw a phase diagram on the student answer page. The triple point of ammonia is 195.42 K and 0.05997 atm. The critical point is 405.38 K and 111.5 atm. The normal freezing point is 195.45 K and the normal boiling point is 239.8 K. Be sure to label the following:

   a. Both axes

   b. Regions containing solid, liquid, and gas

   c. Normal boiling point

   d. Normal freezing point

   e. Triple point

   f. Critical point

2. Answer the Activity Two questions on you student answer page.

ACTIVITY THREE: HEATING CURVES

Most pure substances exist in three phases: solid, liquid, and gas. However, the temperature ranges necessary to view all three phases is often very great. Water exists in all three states at temperature ranges that are common to our everyday experience.

If we start with water that has been frozen and is well below $0^\circ C$ and steadily add heat to it, we will find that the temperature rises at a constant rate until it reaches the melting point. This process is controlled by the amount of heat put into the system, the mass of the ice, and the specific heat of the ice. Remember the formula: $q = mc\Delta T$.

When the ice begins to melt, all of the energy going into the system is used to break hydrogen bonds between the water molecules. The temperature does not rise. The amount of heat that it takes to melt one gram of a substance is called the *latent heat of fusion*. Fusion is a scientific term that means "melting". It is symbolized by L_f. It has units of joules/gram. The formula that we use is $q = L_f m$.

Once the water is all melted, the temperature starts to rise again until the water reaches the boiling point. During this region of the graph we will again use the formula $q = mc\Delta T$. Once the water begins to boil, the heat is being used to separate the water molecules from each other so that they go into a vapor state. The temperature does not rise. The amount of heat that it takes to vaporize one gram of a substance is called the *latent heat of vaporization* and is symbolized by L_v. It also has units of joules/gram, so the formula that we use is $q = L_v m$.

It takes special equipment to collect and continue heating a substance in the gas phase. We will not perform that part of the procedure in this lab. However, you can look at the heating curve below to see what happens to a system when you are able to heat the vapor state.

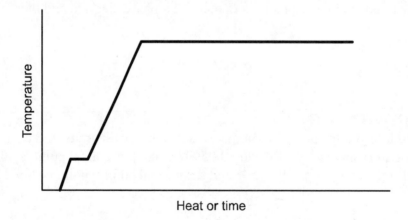

Heat or time

PURPOSE: ACTIVITY THREE

In this activity you will start with ice that is well below the freezing point and continuously add heat to the system while collecting data with a temperature probe. The resulting heating curve will be downloaded to a computer to be analyzed and printed.

MATERIALS

1 temperature probe frozen into a polyethylene centrifuge tube of water

ring stand

600-mL beaker ¾ filled with distilled water

CBL or LabPro

2 different colored highlighters or markers

hot plate

ring

TI-83+ graphing calculator

computer with software and link cord

Safety Alert

1. Wear your goggles.
2. The test tubes containing the probes are very cold. Do not handle with your bare hands.
3. Be very careful that the temperature probe cords do not touch the hot plates.

PRE-LAB: ACTIVITY THREE

1. Define the terms found on your student answer page.

2. Carefully read the lab procedure. Examine the heating curve included.

3. Answer the pre-lab questions.

PROCEDURE: ACTIVITY THREE

1. Set up a hot plate with a ring stand and ring. Fill the 600-mL beaker ¾ full of distilled water. Use the ring to steady the beaker so that it cannot tip over.

2. Set up the calculator and CBL for one temperature probe. Do not bring the temperature probe to your station until everything else is ready.
   - Turn on the CBL unit and calculator. Start the CHEMBIO or PHYSICS program from the APPS menu.
   - Select SET UP PROBES from the MAIN MENU.
   - Enter "1" as the number of probes.
   - Select TEMPERATURE from the SELECT PROBE menu.
   - Enter "1" as the channel number.
   - Select USE STORED from the CALIBRATION menu.

3. Set up the calculator and CBL for data collection.
   - Select COLLECT DATA from the MAIN MENU.
   - Select TIME GRAPH from the DATA COLLECTION menu.
   - Enter "10" as the time between samples, in seconds.
   - Enter "100" as the number of samples.
   - Press ENTER. Select USE TIME SETUP to continue. If you want to MODIFY SETUP you may change the time or sample numbers.
   - Enter "-20" as the Ymin
   - Enter "110" as the Ymax
   - Enter "5" as the Yscl
   - Do not begin colleting data until instructed to do so in step 6.

4. Obtain the temperature probe that has been frozen in a test tube of water from your teacher. *Do not touch the test tube with your bare hands.* It has been sitting in a slurry of dry ice and acetone and will begin at a temperature of -78°C. This temperature is cold enough to give you frostbite.

5. Place the test tube with frozen probe into the beaker of water and turn the hot plate to its highest setting. Start collecting data by pressing ENTER on your calculator.

6. Be very careful not to allow the cords to touch the hot plate. After all the ice has melted, you may use the probe to stir the water in the test tube. Continue taking data until the program stops and displays "DONE" on your calculator screen.

7. Press ENTER to display a graph of temperature vs. time.

8. Remove the probe from the hot water. Turn off the hot plate and allow it to cool.

9. Download your data to a computer graphing program such as Graphical Analysis, Logger Pro or Excel as instructed by your teacher.

10. Return the probe and plastic test tube to your teacher. Clean up your lab area.

Name _____

Period _____

Heating Curves and Phase Diagrams
Investigating Changes of State

DATA AND OBSERVATIONS: ACTIVITY ONE

Draw a rough sketch of the graph that is on your calculator.

QUESTIONS: ACTIVITY ONE

1. What does the horizontal line in Y4 represent?

2. Why do you think that the vapor pressure vs. temperature is not a linear relationship?

3. Which substance has the strongest intermolecular forces? Explain your answer.

4. Determine the normal boiling point for
   a. Chloroform:

   b. Ethanol:

   c. Acetic acid:

5. At 60°C, which substance has the highest vapor pressure?

6. If the pressure were dropped to 600 torr, what would be the boiling temperature of ethanol?

7. If the pressure were dropped to 275 torr, which substance(s) would be boiling at or below 55°C?

QUESTIONS: ACTIVITY TWO

1. Phase diagram

2. Describe what would happen when you make each of the following changes:
    a. While the system is in an enclosed vessel, heat is constantly added. The temperature is increased from 150 K to 250 K while the pressure is kept constant at 0.75 atm.

    b. While the temperature is kept constant at 225 K, the pressure is increased from 0.1 atm to 0.95 atm.

3. Describe the system at 407 K and 112 atm.

PRE-LAB: ACTIVITY THREE

1. Define the following terms:
    a. Endothermic

    b. Exothermic

    c. Potential energy

   d. Kinetic energy

   e. Specific heat

   f. Latent heat of fusion

   g. Fusion

   h. Latent heat of vaporization

   i. Vaporization

   j. Melting

   k. Freezing

   l. Boiling

   m. Condensing

2. Since we are continuously adding heat energy to this system, what is happening to that heat when the ice is melting? What type of energy is this?

3. In a cooking class, a teacher told the class to bring the water to a boil and allow it to boil for 5 minutes so that it would be "good and hot." What is wrong with this statement?

4. You have been given a picture of a theoretical heating curve for water. After reading the lab, sketch a picture of the heating curve which you expect to get with this lab.

DATA AND OBSERVATIONS: ACTIVITY THREE

1. Download your data from the calculator. Print a copy of the data table and turn it in with the lab. Be sure to print a copy for everyone in your group.

2. Graph temperature vs. time. Title your graph and label the axes. Be sure to include units. Print the graph and turn it in with the lab. Be sure to print a copy for everyone in your group.

ANALYSIS: ACTIVITY THREE

1. Use a highlighter to indicate the region of the graph where melting occurred. Use a different colored highlighter to indicate the region where boiling occurred. Clearly label each one.

2. Look at the region where the ice was heating. Compare it to the region where the water was heating. Suggest a reason why the slopes are not the same.

3. Citrus growers in South Texas often spray their crops with water if the weather forecast indicates that the temperature will drop below 0.0°C. Explain why this would protect the crop.

4. This is a qualitative lab, rather than a quantitative lab. What is the difference?

5. Explain any differences between your graph and the standard graph shown in the introduction of this lab.

CONCLUSION QUESTIONS: ACTIVITY THREE

1. Indicate whether each of the following is exothermic or endothermic.

   a. Boiling _____

   b. Melting _____

   c. Condensing _____

   d. Freezing _____

Use this table to work the following problems:

Specific heat of ice	2.1 J/g °C
Specific heat of water (liquid)	4.2 J/g °C
Specific heat of steam	2.1 J/g °C
Latent heat of fusion of ice	336 J/g
Latent heat of vaporization of water	2264 J/g

2. How much heat is given off by 75.0 g of water freezing?

3. If I start with 20.0 g of ice at -5.0°C and heat it until it is liquid water at 60.°C, how much heat (in kJ) does it take?

4. A mixture of 20. g of ice and 100. g of liquid water at 0°C are heated until all the ice melts and the combined water is at 30.°C. How much heat (in kJ) is added to this system to cause this to happen?

Conductivity of Ionic Solutions
Exploring Ions in Solutions

OBJECTIVE
Students will investigate the electrical conductivity of several ionic solutions and relate this do the number of ions formed during dissociation.

LEVEL
Chemistry

NATIONAL STANDARDS
UCP.1, UCP.2, UCP.3, B.2, E.1, E.2

TEKS
13(B)

CONNECTIONS TO AP
AP Chemistry:
   II. States of Matter   C. Solutions   3. Raoult's Law and colligative properties

TIME FRAME
45 minutes

MATERIALS
(For a class of 28 working in pairs)

14 ring stands	14 utility clamps
14 large test tubes. Test tubes must hold at least 50.0 mL (25 mm x 150 mm size.)	14 CBL2s or LabPros
14 TI-83+ graphing calculators	14 link cords
14 Vernier AC adapters*	computer graphing software
14 conductivity probes**	each group will need a set of 3 dropping bottles containing 0.1 M NaCl, 0.1 M $CaCl_2$, and 0.1 M $AlCl_3$

TEACHER NOTES
* You may choose to use batteries in your CBL2 or LabPro, but over time, use of an AC adapter will save money.

** Depending upon when the conductivity probes were purchased, a DIN adapter may be necessary.

Electrical conductivity is dependent upon charged particles that are free to move. In dilute ionic solutions, the degree of electrical conductivity is a measure of the concentration of the ions in solution. Because of this, it is important to be careful when making the solutions. The solutions will have a long shelf life if kept in air-tight containers at room temperature.

Instructions for making 500mL of each of the three solutions:

A. Place 2.93 g of NaCl in a 500 mL volumetric flask. Dilute to the mark with distilled water.

B. Place 7.35 g of $CaCl_2 \cdot 2H_2O$ in a 500 mL volumetric flask. Dilute to the mark with distilled water.

C. Place 12.1 g of $AlCl_3 \cdot 6H_2O$ in a 500 mL volumetric flask. Dilute to the mark with distilled water.

POSSIBLE ANSWERS TO THE CONCLUSION QUESTIONS AND SAMPLE DATA

NaCl	
Drops of NaCl	Conductivity (μS)
0	19.094
20	591.92
30	868.78
40	1126.6
50	1374.8

Equation of line: $y = 27.2x + 35.1$

Correlation coefficient: 0.9995

CaCl$_2$	
Drops of CaCl$_2$	Conductivity (μS)
0	19.094
20	887.87
30	1298.4
40	1689.8
50	2024.0

Equation of line: $y = 40.4x + 53$

Correlation coefficient: 0.9988

AlCl₃	
Drops of AlCl₃	Conductivity (µS)
0	9.547
20	1202.9
30	1804.4
40	2329.5
50	2864.0

Equation of line: $y = 57.2x + 40.0$

Correlation coefficient: 0.9995

CONCLUSION QUESTIONS

1. Find the ratios of the slopes of each line. To do this, divide each slope by the smallest slope. If you do not get numbers that are integers, convert each number to an improper fraction and use the numerators as the ratios.
    *   The ratios should be close to 1:1.5:2. That would be the same as 2:3:4.
    *   The ratios of the sample data are 1:1.49:2.1. That is very close to the expected values.

2. Plot all three data sets on the same graph. Title the graph and label the axes. The units for the x-axis should be drops/25 mL. Print a copy of the graph for your lab.
    *   The graphs should be linear with increasingly steep slopes

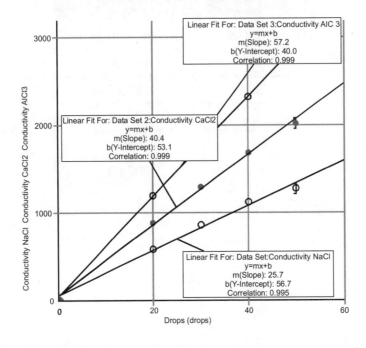

3. Write an equation for the dissociation of each of the three compounds.
    - $NaCl_{(s)} \rightarrow Na^+_{(aq)} + Cl^-_{(aq)}$
    - $CaCl_{2(s)} \rightarrow Ca^{2+}_{(aq)} + 2\ Cl^-_{(aq)}$
    - $AlCl_{3(s)} \rightarrow Al^{3+}_{(aq)} + +3\ Cl^-_{(aq)}$

4. Is there a relationship between the ratios of slopes and the equations from question 3? Explain your answer.
    - The ratios of the slopes should be 2:3:4. The number of ions formed in the successive equations are 2:3:4 as well.

5. Are your correlation coefficients similar? Do they all suggest a linear relationships between conductivity and concentration?
    - If the students are careful in counting the drops, the relationships should be very linear with correlation coefficients very close to 1.0000.

6. Are there any anomalies in your data? If so, can you suggest an explanation?
    - If the student does not measure the initial water carefully, or fails to accurately count the drops, the data will not be accurate.

7. The student lab aide who mixed the 0.1 M solutions failed to realize that aluminum chloride really was aluminum chloride hexahydrate. What effect would this mistake have on the value of the slope?
    - The calculated molar mass was much lower than the actual molar mass, so the solution was less concentrated than the other solutions. Therefore the slope was less than it should have been.

Conductivity of Ionic Solutions
Exploring Ions in Solutions

In order to conduct an electric current, there must be charged particles such as electrons or ions present, and those particles must be free to move. Materials that tend to conduct electricity are metals with mobile electrons, fused (melted) ionic substances, or solutions containing ions.

PURPOSE

In this activity you will quantitatively investigate the electrical conductivity of three different ionic solutions. The data will be graphed and the slopes of the graphs will be compared to assess the ratios of the ions in each solution.

MATERIALS

ring stand
large test tube (must contain at least 50 mL)
TI-83+ graphing calculator
Vernier AC adapter
conductivity probe
utility clamps

CBL2 or LabPro
link cord/DIN adapter if necessary
computer graphing software
each group will need a set of 3 dropping
    bottles containing 0.1 M NaCl, 0.1 M
    $CaCl_2$, and 0.1 M $AlCl_3$

Safety Alert
1. Be sure to wear your goggles.

SET UP AND PROCEDURE

1. Attach the utility clamp to the ring stand and clamp the test tube vertically in the utility clamp.

2. Carefully measure 25.0 mL distilled water and place it in the test tube. Place the conductivity probe in the test tube with the water.

3. Connect the CBL2 or LabPro to the calculator with the link cord. If you have LabPros, you may use Logger Pro directly with a computer.

4. Plug the conductivity probe into Channel 1. The switch on the probe box should be set in the up position (0-20000).

5. Go to APPS and choose CHEMBIO. Set up CONDUCTIVITY probe. Use the stored calibration: Conductivity:5:0-20000.

6. COLLECT DATA: TRIGGER PROMPT.

7. Take the first point with the probe in the distilled water. Follow the instructions on the screen and enter 0 when asked for a response. Collect more data.

8. Add 20 drops NaCl solution. Stir thoroughly and allow at least 10 seconds to equilibrate. When the voltage settles, press ⊞. Enter 20 to represent the total drops of NaCl in solution.

9. Add 10 more drops of NaCl and record the conductivity and number of drops. Continue this procedure until you have added a total of 50 drops. You should have entered 0, 20, 30, 40, 50. Then select Stop and Graph. Download data to Logger Pro or other computer graphing software program. If you have Graphical Analysis or Logger Pro with a Graph Link cable, the data can be downloaded directly from the calculator by selecting the "import data from TI-device" command from the file menu. If you do not have this equipment, you may enter the data manually. Run a linear regression on the data and record the equation for the line and the correlation coefficient. If you need help doing this, refer to the Foundation Lessons on Graphing Calculators found in this manual.

10. Thoroughly wash the test tube with distilled water. Rinse the conductivity probe thoroughly with distilled water. Set up again as before. Repeat the experiment with $CaCl_2$. When finished, download your data to the same file as before as a "New Data Set".

11. Clean your equipment and repeat the entire experiment with $AlCl_3$. Download as a new data set.

Name _____

Period _____

Conductivity of Ionic Solutions
Exploring Ions in Solutions

DATA AND OBSERVATIONS

NaCl	
Drops of NaCl	Conductivity (μS)
0	
20	
30	
40	
50	

Equation of line:

Correlation coefficient:

CaCl$_2$	
Drops of CaCl$_2$	Conductivity (μS)
0	
20	
30	
40	
50	

Equation of line:

Correlation coefficient:

AlCl₃	
Drops of AlCl₃	Conductivity (µS)
0	
20	
30	
40	
50	

Equation of line:

Correlation coefficient:

CONCLUSION QUESTIONS

1. Find the ratios of the slopes of each line. To do this, divide each slope by the smallest slope. If you do not get numbers that are integers, convert each number to an improper fraction and use the numerators as the ratios.

2. Plot all three data sets on the same graph. Title the graph and label the axes. The units for the x-axis should be drops/25 mL. Print a copy of the graph for your lab.

3. Write an equation for the dissociation of each of the three compounds.

   • NaCl(s) →

   • CaCl₂(s) →

   • AlCl₃(s) →

4. Is there a relationship between the ratios of slopes and the equations from question 3? Explain your answer.

5. Are your correlation coefficients similar? Do they all suggest a linear relationships between conductivity and concentration?

6. Are there any anomalies in your data? If so, can you suggest an explanation?

7. The student lab aide who mixed the 0.1 M solutions failed to realize that aluminum chloride really was aluminum chloride hexahydrate. What effect would this mistake have on the value of the slope?

Colligative Properties
Calculating Freezing Point and Boiling Point Changes

OBJECTIVE
Students will explore the nature of colligative properties and be able to calculate changes in boiling point and freezing point of various solutions. They will also determine the molecular mass of an unknown solute using freezing point data.

LEVEL
Chemistry

NATIONAL STANDARDS
UCP.2, UCP.3, B.2, G.1, G.2

TEKS
4(A), 4(B), 4(C), 8(B), 8(D), 13(B)

CONNECTIONS TO AP
AP Chemistry:
   II. States of Matter C. Solutions 2. Methods of expressing concentration 3. Raoult's law and colligative properties (nonvolatile solutes)

TIME FRAME
45 minutes class; 30 minutes homework

MATERIALS
   calculator

TEACHER NOTES
The discussion of colligative properties should be included in a unit on liquids and solutions. Many textbooks qualitatively mention colligative properties but do not have students perform any calculations using this concept. Sharpening students' math skills in your first year course will only better prepare them for the advanced coursework that follows.

This lesson is designed to be a teacher led lecture and discussion. Modeling how to solve the problems and "thinking out loud" (metacognition) as you work through the examples will ensure that students understand the "why's" behind each step.

A popular lab activity to include during a unit on colligative properties is to have students make ice cream. There are many versions of this experiment that range from simple re-sealable plastic bag or coffee can freezers to more elaborate real ice cream freezers. Students rarely have the opportunity to experience such fun applications of their scientific learning. Whichever version you choose, be sure to have students make quantitative measurements throughout the lab activity such as monitoring the initial temperature of the ice and then the ice/salt mixture during the freezing process. The temperature of the ice cream mixture can also be obtained before and after the experiment. One version of this experiment can be found at www.tealighthouse.org listed in the chemistry activities at the time of this publication.

Example Problems in Student Notes:

EXAMPLE PROBLEM #1:
Calculate the new boiling point and the new freezing point for a 0.50 m solution of sucrose, $C_{12}H_{22}O_{11}$. We have everything that we need to solve this problem, substitute values into the formulas:

$$\Delta T_b = K_b \; m_{solution} \; i \qquad\qquad \Delta T_f = K_f \; m_{solution} \; i$$

Note that sucrose is a nonelectrolyte, so the van't Hoff factor is = 1.
Therefore,
New boiling point: $\Delta T_b = (0.50 \text{ m})(0.512 \text{ °C/molal}) = 0.26 \text{ °C}$
Therefore, the new b.p. = 100.00 + 0.26 = 100.26 °C
New freezing point: $\Delta T_f = (0.50 \text{ m})(1.86 \text{ °C/molal}) = 0.93 \text{ °C}$
Therefore, the new f.p. = 0.00 – 0.93 = -0.93 °C

EXAMPLE PROBLEM #2:
Calculate the new boiling point and the new freezing point for a solution composed of 45.00 grams of sodium chloride dissolved in 600.0 grams of water.

This time we have an electrolyte: $NaCl_{(s)} \rightarrow Na^+_{(aq)} + Cl^-_{(aq)}$ (2 moles of ions)
We also must calculate the molality:

$$m = \frac{mol \, solute}{kg \, solvent}$$

$$45.00 \, g \, NaCl \times \frac{1 \, mol \, NaCl}{58.5 \, g \, NaCl} = 0.7692 \, mol \, NaCl$$

$$m = \frac{0.7692 \, mol \, NaCl}{0.6000 \, kg \, water} = 1.282 \, m$$

New boiling point: $\Delta T_b = (1.282 \text{ m}) (0.512 \text{ °C/molal}) (2) = 1.313 \text{ °C}$
Therefore, the new b.p. = 100.00 + 1.313 = 101.313 °C
New freezing point: $\Delta T_f = (1.282 \text{ m}) (1.86 \text{ °C/molal}) (2) = 4.769 \text{ °C}$
Therefore, the new f.p. = 0.00 – 4.769 = -4.769 °C

EXAMPLE PROBLEM #3:

Another really useful property of freezing point depression and boiling point elevation is the ability to calculate the molecular weight of an unknown nonvolatile solute.

A 1.20 gram sample of an unknown molecular compound is dissolved in 50.00 grams of benzene. The solution freezes at a temperature of 4.93 °C. Calculate the molecular weight of the compound.

(The normal freezing point for benzene is 5.48 °C and the K_f for benzene is 5.12 °C/m)

$$\Delta T_f = K_f \; m_{solution} \; i$$

Remember that we are looking for molecular weight: grams/mol; we know the grams of solute—we just need to know the moles of solute. We must work backwards with our formula and solve for molality.

Since the solute is a molecular compound, $i = 1$.
$$\Delta T_f = 5.48 - 4.93 = 0.55 \; °C$$
$$m = \frac{\Delta T_f}{k_f} = \frac{0.55 \, °C}{5.12 \, °C/m} = 0.107 \, m$$

Now, use the molality formula to find moles of solute: $m = mol /kg$
moles solute = (molality)(kg solvent)
moles solute = (0.107m) (0.05000 kg) = 0.00535 mole solute

$$\text{Molecular weight} = \frac{1.20 \, g}{0.00535 \, mol} = 224 \, g/mol$$

ANSWERS TO THE CONCLUSION QUESTIONS

1. Glucose is often used as a form of nourishment to hospital patients through intravenous feeding. Calculate the new boiling point and the new freezing point for a 0.70 m solution of glucose, $C_6H_{12}O_6$.
   - We have everything that we need to solve this problem, substitute values into the formulas:
     $$\Delta T_b = K_b \; m_{solution} \; i \qquad\qquad \Delta T_f = K_f \; m_{solution} \; i$$

     Note that sucrose is a nonelectrolyte, so the van't Hoff factor is $= 1$.
     Therefore,
     New boiling point: $\Delta T_b = (0.70 \, m)(0.512 \, °C/molal) = 0.36 \, °C$
     Therefore, the new b.p. $= 100.00 + 0.36 = 100.36 \, °C$
     New freezing point: $\Delta T_f = (0.70 \, m)(1.86 \, °C/molal) = 1.30 \, °C$
     Therefore, the new f.p. $= 0.00 - 1.30 = -1.30 \, °C$

2. Calcium chloride is a common drying agent and is very exothermic when dissolved in water. If 47.50 grams of calcium chloride is dissolved in 395 mL of water, calculate the boiling point and freezing point of the solution that is formed.
   - Remember that 1 mL = 1 g of water since the density of water is 1.00 g/mL
   - $CaCl_2$ is an electrolyte that decomposes as follows: $CaCl_2 \rightarrow Ca^{2+} + 2Cl^-$ (3 ions in solution)

- We must calculate molality first.

$$m = \frac{mol\,solute}{kg\,solvent}$$

$$47.50\,g\,CaCl_2 \times \frac{1\,mol\,CaCl_2}{110.98\,g\,CaCl_2} = 0.4280\,mol\,CaCl_2$$

$$m = \frac{0.4280\,mol\,CaCl_2}{0.395\,\,kg\,water} = 1.08\,m$$

New boiling point: $\Delta T_b = (1.08\,m)\,(0.512\,°C/molal)\,(3) = 1.66\,°C$
Therefore, the new b.p. $= 100.00 + 1.66 = 101.66\,°C$
New freezing point: $\Delta T_f = (1.08\,m)\,(1.86\,°C/molal)\,(3) = 6.03\,°C$
Therefore, the new f.p. $= 0.00 - 6.03 = -6.03\,°C$

3. Moth balls are often composed of pure naphthalene, $C_{10}H_8$. The normal freezing point of naphthalene is 80.2 °C. When 5.25 grams of an unknown nonvolatile substance was mixed with 35.00 grams of the naphthalene, the freezing point was found to be 78.8 °C. Calculate the molecular weight of the unknown substance. (the K_f for $C_{10}H_8$ is 6.9 °C/m)

    $\Delta T_f = K_f\,m_{solution}\,\,i$

- Remember that we are looking for molecular weight: grams/mol; we know the grams of solute—we just need to know the moles of solute. We must work backwards with our formula and solve for molality.

    Since the solute is a molecular compound, i = 1.
    $\Delta T_f = 80.2 - 78.8 = 1.4\,°C$

    $$m = \frac{\Delta T_f}{k_f} = \frac{1.4°\,C}{6.9\,°C/m} = 0.2029\,m$$

    Now, use the molality formula to find moles of solute: m = mol /kg
    moles solute = molality x kg solvent
    moles solute = (0.2029m) (0.03500 kg) = 0.007102 mole solute

    $$Molecular\,weight = \frac{5.25\,g}{0.007102\,mol} = 739\,g/mol$$

4. Which solution freezes at a lower temperature, 0.20m sodium sulfate or 0.20m magnesium sulfate. Explain.
   - Sodium sulfate should freeze at a lower temperature.
   - Both solutions have the same molality; both are dissolved in water and thus have the same K_f value.
   - The difference between the solutions is the number of ions that are released into solution. Sodium sulfate produces 3 ions in solution while magnesium sulfate only produces 2 ions per mole in solution.
   - $Na_2SO_4 \rightarrow 2\,Na^+ + SO_4^{2-}$ (3 ions)
   - $MgSO_4 \rightarrow Mg^{2+} + SO_4^{2-}$ (2 ions)

5. Automobiles need antifreeze in the winter to prevent water from freezing in the radiator. How many grams of antifreeze, ethylene glycol, $C_2H_4(OH)_2$, are needed for every 250. grams of water to maintain a freezing temperature of -25.00 °C?

   $\Delta T_f = K_f\ m_{solution}\ i$
   - Solve for molality.
   - $\Delta T = 25.00$ °C since the freezing point of water is 0.00 °C.
   - $i = 1$ since we have a nonelectrolyte
   - $m = \dfrac{\Delta T_f}{k_f} = \dfrac{25.00°C}{1.86°C/m} = 13.44\,m$

   Now, use the molality formula to find moles of solute: m = mol /kg
   moles solute = molality x kg solvent
   moles solute = (13.44m) (0.250 kg) = 3.36 mole solute

   $$3.36\,mol\,C_2H_4(OH)_2 \times \frac{62.08\,g\,C_2H_4(OH)_2}{1\,mol\,C_2H_4(OH)_2} = 209\,g\,C_2H_4(OH)_2$$

Colligative Properties
Calculating Freezing Point and Boiling Point Changes

Have you ever given much thought as to why a city spreads sand or salt on an icy road? Sure, it makes the ice melt—but exactly how and why this happens will be explored in this lesson.

PURPOSE
Building on your knowledge of solutions we will explore some colligative properties of solutions. We will calculate changes in freezing and boiling points and determine the molar mass of an unknown solute.

MATERIALS
    calculator

CLASS NOTES: COLLIGATIVE PROPERTIES
Properties that depend upon the **number** of solute particles present per solvent molecule and **not** the identity of the solute are termed *colligative properties*. These properties include; freezing point, boiling point, vapor pressure and osmotic pressure. In this lesson we will focus only on freezing point depression and boiling point elevation. You will study the others in science courses that follow.

In order to fully understand the nature of these colligative properties we must understand the concept of vapor pressure. The vapor pressure is the partial pressure of a vapor in equilibrium with its parent liquid or solid. In other words, vapor pressure is a measure of the tendency for molecules to escape from the surface of the parent liquid or solid. The less energy that it takes to overcome the intermolecular attractive forces between molecules in the liquid or the solid, the higher the vapor pressure. Substances with very high vapor pressures, like gasoline, are often referred to as volatile. Substances, like water, that have relatively low vapor pressures are termed non-volatile. In general, the stronger the intermolecular forces of attraction, the more energy required for molecules to leave the parental surface, the lower the vapor pressure.

So how do solvent molecules escape the surface of a parent liquid? They must gain enough energy through collisions with other particles to overcome the attractive forces holding the molecules together. It seems logical then, remembering intermolecular forces, that molecules with strong intermolecular forces, like hydrogen bonding in water, take much greater energy to overcome than the weak London dispersion forces holding gasoline molecules to each other.

By definition, boiling point is the temperature at which the equilibrium vapor pressure of a liquid is equal to the atmospheric pressure. *Normal* boiling is defined as the temperature at which the vapor pressure is equal to standard sea-level pressure, 1 atmosphere. When solute particles interfere with molecules being able to escape the surface, the vapor pressure is decreased. This in turn, will prevent the liquid from boiling until enough gas particles move into the vapor phase and allow equilibrium to be established. Solute particles dissolved in a pure solvent do just this. They "get in the way" of escaping particles — think of it like a net over the solution where particles can only escape at certain places.

Much more energy will be needed for the solvent molecules to escape, and thus, the boiling point is now elevated to a higher temperature.

Freezing point is defined as the temperature at which the vapor pressure of the solid and the liquid are in equilibrium with each other. *Normal* freezing is defined as the temperature at which the vapor pressure is equal to standard sea-level pressure, 1 atmosphere. Using the same argument from the previous paragraph, if the vapor pressure of the liquid is lowered, the freezing point will also be lowered. This change in freezing temperature is why $NaCl$ and $MgCl_2$ are used on icy roads and sidewalks.

What is happening when a solution's freezing point is depressed? The answer lies in the fact that molecules need to cluster in order to freeze. They must be attracted to one another with greater energy than it takes for them to overcome their kinetic energy and they must have a spot in which to cluster. When solute molecules get in the way of this solvent-solvent attraction the only way for freezing to occur is for the temperature to be lower so that intermolecular attractive forces can dominate.

Let's take a look at how drastically the temperature is really changed by performing some calculations with these concepts.

BOILING POINT ELEVATION:

$$\Delta T_b = K_b \, m_{solution} \, i$$

ΔT_b = change in the temperature of the boiling point — this amount must be added to the normal boiling point of the solvent (the normal boiling point for water is 100 °C)

K_b for water = 0.512 °C/molal = This is called the molal boiling point elevation constant (this value will change depending on the solvent)

m = the concentration in molality (moles solute / kg solvent)

i = van't Hoff factor

The van't Hoff factor is a measure of the extent of ionization of an electrolyte in solution. For our purposes, it will be equal to the number of ions that an electrolyte dissociates into. For instance, if $NaCl$ is the solute, we would say that i = 2. If $MgCl_2$ is the solute, then i = 3.

For nonelectrolytes, i = 1. If sucrose, $C_{12}H_{22}O_{11}$, is the solute then i = 1. Remember that nonelectrolytes do not ionize in solution but stay together as a molecule.

The big idea is that the more ions in solution, the greater the effect on the colligative properties because there are more particles to interfere with the solvent-solvent interactions.

FREEZING POINT DEPRESSION:

$\Delta T_f = K_f \, m_{solution} \, i$

ΔT_f = change in freezing temperature — this amount must be subtracted from the normal freezing point of the solvent (the normal freezing point for water is 0 °C)

K_f for water = 1.86°C/molal = This is called the molal freezing point depression constant (this value will differ depending on the solvent)

m = molality

i = van't Hoff factor

EXAMPLE PROBLEM #1:

Calculate the new boiling point and the new freezing point for a 0.50 m solution of sucrose, $C_{12}H_{22}O_{11}$.

EXAMPLE PROBLEM #2:

Calculate the new boiling point and the new freezing point for a solution composed of 45.00 grams of sodium chloride dissolved in 600.0 grams of water.

EXAMPLE PROBLEM #3:

Another really useful property of freezing point depression and boiling point elevation is the ability to calculate the molecular weight of an unknown nonvolatile solute.

A 1.20 gram sample of an unknown molecular compound is dissolved in 50.00 grams of benzene. The solution freezes at a temperature of 4.93 °C. Calculate the molecular weight of the compound.

(The normal freezing point for benzene is 5.48 °C and the K_f for benzene is 5.12 °C/m)

Show all work for the problems shown on the student answer page.

Name _____

Period _____

Colligative Properties
Calculating Freezing Point and Boiling Point Changes

CONCLUSION QUESTIONS

1. Glucose is often used as a form of nourishment to hospital patients through intravenous feeding. Calculate the new boiling point and the new freezing point for a 0.70 m solution of glucose, $C_6H_{12}O_6$.

2. Calcium chloride is a common drying agent and is very exothermic when dissolved in water. If 47.50 grams of calcium chloride is dissolved in 395 mL of water, calculate the boiling point and freezing point of the solution that is formed.

3. Moth balls are often composed of pure naphthalene, $C_{10}H_8$. The normal freezing point of naphthalene is 80.2 °C. When 5.25 grams of an unknown nonvolatile substance was mixed with 35.00 grams of the naphthalene, the freezing point was found to be 78.8 °C. Calculate the molecular weight of the unknown substance. (the K_f for $C_{10}H_8$ is 6.9°C/m)

4. Which solution freezes at a lower temperature, 0.20m sodium sulfate or 0.20m magnesium sulfate. Explain.

5. Automobiles need antifreeze in the winter to prevent water from freezing in the radiator. How many grams of antifreeze, ethylene glycol, $C_2H_4(OH)_2$, are needed for every 250. grams of water to maintain a freezing temperature of -25.00 °C?

Chemical Reaction Rates I
Solving Kinetics Problems Involving Differential Rate Law

OBJECTIVE

Students will determine the order of a reactant from concentration-rate data using the methods of differential rate law.

LEVEL

Chemistry

NATIONAL STANDARDS

UCP.1, UCP.2, UCP.3, B.3, B.6

TEKS

15(B)

CONNECTIONS TO AP

AP Chemistry:
    III. Reaction types D. Kinetics 1. Concept of rate of reaction 2. Use of experimental data and graphical analysis to determine reactant order, rate constants, and reaction rate laws 3. Effect of temperature changes on rates 4. Energy of activation; the role of catalysts

TIME FRAME

two 45 minute class periods

MATERIALS

    graphing calculator   pencil

TEACHER NOTES

This activity is designed to provide an introduction to the types of problems students will face on the AP Chemistry exam. It provides only an introduction, one that is usually absent from a first-year Chemistry textbook. The mathematical skills and graphing calculator skills are those students should have mastered in their mathematics class since they are required of Algebra I students on the TAKS test. Students should complete the conclusion questions for homework. The kinetics topics involving mechanisms and the Arrhenius equation for calculating the activation energy are not addressed in this lesson.

Prior to this lesson, students should have an understanding of basic collision theory, especially reaction coordinate diagrams (a.k.a. potential energy diagrams). From these diagrams they should be able to compare the energy of the reactants, the energy of the products, and the activation energy. Emphasize that collisions between molecules must occur for chemical reactions to take place. Also emphasize that a catalyst lowers the activation energy by providing an alternate collision pathway, one that requires less energy. Remind students that a catalyst is neither a reactant nor a product, so it can participate in chemical reactions over and over again. Be sure students also have an understanding of a reaction intermediate and that it is a highly unstable species that is formed at the peak of the potential energy diagram.

ANSWERS TO THE CONCLUSION QUESTIONS

1. The rate of the reaction between CO and NO_2

$$CO(g) + NO_2(g) \rightarrow CO_2(g) + NO(g)$$

was studied at 650 K starting with various concentrations of CO and NO_2, and the data in the table were collected.

Experiment Number	Initial Rate mol/(Lx h)	Initial concentration [CO]	Initial concentration [NO₂]
1	3.4×10^{-4}	0.510	0.0350
2	6.8×10^{-4}	0.510	0.0700
3	1.7×10^{-4}	0.510	0.0175
4	6.8×10^{-4}	0.102	0.0350
5	10.2×10^{-4}	0.153	0.0350

a. Write the rate law that is consistent with the data.
   - In experiments 1 & 2 [CO] is held constant and [NO₂] is doubled, the rate also doubles in response ∴ the reaction is first order with respect to NO₂.
   - In experiments 1 & 4 [NO₂] is held constant and [CO] is doubled, the rate also doubles in response ∴ the reaction is first order with respect to CO.
   - The rate law is now written as Rate = k[CO][NO₂].

b. Calculate the value of the specific rate constant, k, and specify units.
   - Using experiment 1,
     rate = k[CO][NO$_2$]
   - $$\frac{rate}{[CO][NO_2]} = k = \frac{3.4 \times 10^{-4}}{[0.510][0.0350]} = \frac{0.019}{M \bullet h} ; \text{ where M is } \frac{mol}{L}.$$
     The answer may also be expressed as 0.019 $M^{-1}h^{-1}$

c. What is the overall order for the reaction?
   - Second order overall.

2. The initial rate of the reaction of nitrogen monoxide and oxygen was measured at 35°C for various initial concentrations of NO and O_2.

$2 \, NO(g) \; + \; O_2(g) \; \rightarrow 2 \, NO_2(g)$

Experiment Number	Initial concentration [O_2]	Initial concentration [NO]	Initial Rate mol/(L x s)
1	0.040	0.020	0.028
2	0.040	0.040	0.057
3	0.040	0.080	0.114
4	0.080	0.040	0.227
5	0.020	0.040	0.014

a. Write the rate law that is consistent with the data.
   - In experiments 1 & 2 [O_2] is held constant and [NO] is doubled, the rate also doubles in response ∴ the reaction is first order with respect to NO.
   - In experiments 2 & 4 [NO] is held constant and [O_2] is doubled, the rate increases by a factor of four (0.227 ÷ 0.057 = 3.98) or is quadrupled in response ∴ the reaction is second order with respect to O_2.
   - The rate law is now written as Rate = $k[NO][O_2]^2$.

b. Calculate the value of the specific rate constant, k, and specify units.
   - Using experiment 1,

     rate = $k[NO][O_2]^2$

     $$\frac{\text{rate}}{[NO][O_2]^2} = k = \frac{0.028}{[0.020][0.40]^2} = \frac{4.5 \times 10^{-7}}{M^2 \bullet s} \text{ or } 4.5 \times 10^{-7} \, M^{-1}s^{-1}$$

   - Note that the right side of the rate law equation is equal to kM^3, therefore M^2 must be in the denominator of k along with a time unit.

c. What is the overall order for the reaction?
   - Third order overall.

3. Dinitrogen pentoxide decomposes at 45°C to yield nitrogen dioxide and oxygen gas according to the equation

$$2\ N_2O_5(g) \rightarrow 4NO_2(g) + O_2(g)$$

Data are collected in the table.

Experiment Number	Initial concentration [N₂O₅]	Initial Rate mol/(Lx s)
1	1.75×10^{-3}	2.06×10^{-9}
2	3.50×10^{-3}	8.24×10^{-9}
3	7.00×10^{-3}	3.30×10^{-8}

a. Write the rate law that is consistent with the data.
   - In experiments 1 & 2 [N₂O₅] is doubled, the rate increases by a factor of four (since both are multiplied by 10-9 you can simply compare $8.24 \div 2.06 = 4$) or is quadrupled in response $\therefore$ the reaction is second order with respect to N₂O₅.
   - The rate law is now written as Rate = $k[N_2O_5]^2$.

b. Calculate the value of the specific rate constant, k, and specify units.
   - Using experiment 1,
   $$\text{rate} = k[N_2O_5]^2$$
   $$\frac{\text{rate}}{[N_2O_5]^2} = k = \frac{2.06 \times 10^{-9}}{[1.75 \times 10^{-3}]^2} = \frac{6.73 \times 10^{-4}}{M \bullet s} \text{ or } 6.73 \times 10^{-4}\ M^{-1}s^{-1}$$

c. What is the overall order for the reaction?
   - Second order overall.

4. Nitrogen monoxide reacts with chlorine according to the equation

$$2 NO(g) + Cl_2(g) \rightarrow 2 NOCl(g)$$

The following initial rates of reaction have been observed for certain reactant concentrations:

Experiment Number	Initial concentration [NO]	Initial concentration [Cl₂]	Initial Rate mol/(Lx h)
1	0.50	0.50	1.14
2	1.00	0.50	4.56
3	1.00	1.50	41.0

a. Write the rate law that is consistent with the data.
   - In experiments 1 & 2 [Cl₂] is held constant and [NO] is doubled, the rate quadruples in response ∴ the reaction is second order with respect to NO.
   - In experiments 2 & 3 [NO] is held constant and [Cl₂] is tripled, the rate increases by a factor of nine (41.0 ÷ 4.56 = 8.99) in response ∴ the reaction is second order with respect to Cl₂.
   - The rate law is now written as Rate = $k[NO]^2[Cl_2]^2$.

b. Calculate the value of the specific rate constant, k, and specify units.
   - Using experiment 1,
     rate = $k[NO]^2[Cl_2]^2$
   - $$\frac{rate}{[NO]^2[Cl_2]^2} = k = \frac{1.14}{[0.50]^2[0.50]^2} = \frac{18.2}{M^3 \bullet h} \text{ or } M^{-3} h^{-1}$$
   - Note that the right side of the rate law equation is equal to kM^4, therefore M^3 must be in the denominator of k along with a time unit.

c. What is the overall order for the reaction?
   - Fourth order overall.

5. Hydrogen reacts with nitrogen monoxide to form dinitrogen monoxide, laughing gas, according to the equation

$H_2 + 2NO(g) \rightarrow N_2O(g) + H_2O(g)$

The following data was collected:

[NO], M	0.40	0.80	0.80
[H$_2$], M	0.35	0.35	0.70
Initial Rate mol/(L x s)	5.040×10^{-5}	2.016×10^{-4}	4.032×10^{-4}

a. Write the rate law that is consistent with the data.
   - Ah, the table was literally turned! Students may want to rearrange the data into a format they are comfortable with. At the very least, they should label the columns Experiments 1, 2 & 3.
   - Also, note that the exponents on the rates are not the same. It is a good problem-solving strategy to instruct students to re-write the values so that they all have the same exponent so that comparison of the rates is easier.
   - In experiments 1 & 2 [H$_2$] is held constant and [NO] is doubled, the rate quadruples in response $\therefore$ the reaction is second order with respect to NO.
   - In experiments 2 & 3 [NO] is held constant and [H$_2$] is doubled, the rate doubles in response $\therefore$ the reaction is first order with respect to Cl$_2$.
   - The rate law is now written as Rate $= k[H_2][NO]^2$.

b. Calculate the value of the specific rate constant, k, and specify units.
   - Using experiment 1,
     rate $= k[H_2][NO]^2$
   - $\dfrac{\text{rate}}{[H_2][NO]^2} = k = \dfrac{5.040 \times 10^{-5}}{[0.35][0.40]^2} = \dfrac{9.0 \times 10^{-4}}{M^2 \bullet s}$ or $M^{-2}\,s^{-1}$
   - Note that the right side of the rate law equation is equal to kM^3, therefore M^2 must be in the denominator of k along with a time unit.

c. What is the overall order for the reaction?
   - Third order overall.

Chemical Reaction Rates I
Solving Kinetics Problems Involving Differential Rate Law

Chemical kinetics is the study of the speed or rate of a chemical reaction under various conditions. Collisions must occur in order for chemical reactions to take place. These collisions must be of sufficient energy to make and break bonds. The collisions must also be "effective" which means they not only have sufficient energy during the collision but that the molecules also collide in the proper orientation. The speed of a reaction is expressed in terms of its Ârate" — some measurable quantity is changing with time.

$$\text{Rate} = \frac{\text{Change in concentration of a reactant or a product}}{\text{Change in time}}$$

Most commonly, we discuss four different types of rate when discussing chemical reactions: relative rate, instantaneous rate, differential rate, and integrated rate. Although these sound complicated, they are actually quite simple. This activity will focus primarily on the last two types, differential rate and integrated rate.

First, we need to discuss factors that affect the rate of a chemical reaction:
1. **Nature of the reactants**—Some reactant molecules react quickly, others react very slowly.
2. **Concentration of reactants**—more molecules in a given volume means more collisions which means more bonds are broken and more bonds are formed.
3. **Temperature**—"heat 'em up & speed 'em up"; the faster the molecules move, the more likely they are to collide and the more energetic those collisions will be.
   - An increase in temperature produces more successful collisions because more collisions will meet the required activation energy. There is a general increase in reaction rate with increasing temperature.
   - In fact, a general rule of thumb is that increasing the temperature of a reaction by 10°C doubles the reaction rate.
4. **Catalysts**—accelerate chemical reactions by allowing for more effective collisions, thus lowering the activation energy. The forward and reverse reactions are both accelerated to the same degree. Catalysts are not themselves transformed in the reaction, so they may be used over and over again.
5. **Surface area of reactants**—exposed surfaces affect speed.
   - Except for substances in the gaseous state or in solution, reactions occur at the boundary, or interface, between two phases.
   - The greater the surface area exposed, the greater the chance of collisions between particles, hence, the reaction should proceed at a much faster rate. Example: coal dust is very explosive as opposed to a piece of charcoal. Solutions provide the ultimate surface area exposure!

DIFFERENTIAL RATE

This is sometimes referred to as the "differential rate law" or simply "rate law". In most reactions once molecules begin colliding, product molecules are formed and are now available to initiate the reverse reaction. To simplify matters, we will deal only with the initial rates of the forward reaction throughout our discussions.

Differential rate laws or "rate expressions" express how a measured **rate depends on reactant concentrations**. *To find the exact relation between rate and concentration, we must do some experiments and collect information.*

Consider this generalized reaction:

$$aA + bB \rightarrow xX$$

The rate expression will *always* have the form:
Initial reaction rate = $k[A]_o{}^m[B]_o{}^n$
k = rate constant
$[A]$ = concentration of reactant A in mol/L
$[B]$ = concentration of reactant B in mol/L
m = order of reaction for reactant A
n = order of reaction for reactant B
the little subscript "o" means "original" or initial concentration at time "zero".

Exponents m and n can be zero, whole numbers or fractions and *must be determined by experimentation!*

The rate constant, k, is temperature dependent and must be evaluated by experiment. The order with respect to a certain reactant is the *exponent* on its concentration term in the rate expression. The overall order of the reaction is the sum of all the exponents on all the concentration terms in the rate expression. The significance of the most common orders is explained below:

1. **Zero order**: The change in concentration of reactant has no effect on the rate. These are not very common. General form of rate equation: **Rate = k**

2. **First order**: Rate is directly proportional to the reactants concentration; doubling [reactant], doubles the rate. These are very common. Nuclear decay reactions usually fit into this category. General form of rate equation: **Rate = k [A]**

3. **Second order**: Rate is quadrupled when reactant concentration (symbolized by [reactant]) is doubled and increases by a factor of 9 when [reactant] is tripled etc. These are common, particularly in gas-phase reactions. General form of rate equation: **Rate = k [A]^2**

Now that we have mastered the format and the lingo, let's apply what we have learned!

Experiments were performed involving reactants A and B and the following data was collected:

Experiment Number	Initial Rate mol/(Lx hr)	Initial concentration $[A]_o$	Initial concentration $[B]_o$
1	0.50×10^{-2}	0.50	0.20
2	0.50×10^{-2}	0.75	0.20
3	0.50×10^{-2}	1.00	0.20
4	1.00×10^{-2}	0.50	0.40
5	1.50×10^{-2}	0.50	0.60

When analyzing a set of data, you should be aware of the following:

In any well designed experiment, there should be a control. In experiments 1, 2 & 3, the concentration of reactant B was held constant at 0.20 M while the concentration of reactant A was varied. Controls are good starting places!

A good experimenter then reverses the variables and varies B while holding A constant. Experiments 1, 4 & 5 held the concentration of reactant A constant and varied the concentration of reactant B. When this has been done properly, deducing the orders of the reactions is simplified.

Start by attacking the controls! Focus only on those experiments and how the varied reactant concentration affected the rate. Ignore the concentration of the controlled reactant since it is not changing and thus not contributing to any change in the rate.

Look for a doubling, the number 2 is easy to work with, you can easily multiply by, divide by, square or cube the number 2.

Question: "Where do I start?" **Answer**: "With the controls!" Experiments 1, 2 & 3 hold [B] constant, start there. . . The rate stays the same regardless of how the concentration of [A] changes, therefore, it is zero order with respect to A. Also, scan the exponents on the rates. If any of the exponents are different, make an adjustment to the numbers so that all of the exponents are the same.

Experiment Number	Initial Rate mol/(Lx hr)	Initial concentration $[A]_o$	Initial concentration $[B]_o$
1	0.50×10^{-2}	0.50	0.20
2	0.50×10^{-2}	0.75	0.20
3	0.50×10^{-2}	1.00	0.20

Question: "Now what?" **Answer**: "Move on to Experiments 1, 4 & 5 where the concentration of reactant A is held constant." Compare experiments 1 & 4. The [A] is held constant and the rate doubles with a doubling of [B]. Think: rate $= k[B]^?$ which becomes $2(\text{rate}) = k[2B]^?$ once each are doubled. The only possible value for the exponent we seek is 1, since it's the only way $2 = [2]^1$. Since the exponent is one, the reaction is first order with respect to reactant B.

For further proof, compare Experiments 1 & 5 and you see that the rate triples with a tripling of [B]. This indicates a **direct relationship** and therefore, the rate is first order with respect to [B].

Experiment Number	Initial Rate mol/(Lx hr)	Initial concentration $[A]_o$	Initial concentration $[B]_o$
1	0.50×10^{-2}	0.50	0.20
4	1.00×10^{-2}	0.50	0.40
5	1.50×10^{-2}	0.50	0.60

Summary: Initial reaction rate $= k[A]_o{}^0[B]_o{}^1 = k[B]_o{}^1$; remember any number raised to the zero power is one!

The overall reaction order is 1 because the sum of all the exponents is 1.

Next, we need to solve for the numerical value of the rate constant, k.

Use a set of the data from the table above to calculate k:

FROM EXPERIMENT 1:

$$\text{rate} = k\ [B]^1$$

$$0.0050\ \frac{\text{mol}}{L \bullet hr} = k\ [0.20\ \frac{\text{mol}}{L}]^1$$

$$\text{therefore,}\quad k = \frac{2.5 \times 10^{-2}}{hr}$$

You should get the same value with any set of data!

What if all that logic gives you a headache? You can always rely on good old algebra skills. This method is especially useful if the experiment was designed poorly resulting in one reactant's concentration *never* being held constant. You still have to compare two experiments and controls are good starting points.

Here's the algebraic setup:

$$\frac{\text{rate 1}}{\text{rate 2}} = \frac{k[\text{reactant}]^m [\text{reactant}]^n}{k[\text{reactant}]^m [\text{reactant}]^n}$$

Select a trial where one reactant concentration is held constant SO THAT IT CANCELS; the k's will also cancel

Using trials 1 & 4:

$$\frac{0.50 \times \cancel{10^{-2}}}{\cancel{1.00} \times 10^{-2}} = \frac{k [0.50]^m [0.20]^n}{\cancel{k} \cancel{[0.50]^m} [0.40]^n} \quad \text{so....} \quad \tfrac{1}{2} = [\,\tfrac{1}{2}\,]n \text{ and } \therefore \text{ n must be ONE to make that true!}$$

A word about the units for k...

The units on k are very important. They are also simple to deduce if you respect the fact that [A] translates to the "concentration of reactant A" in $M = \dfrac{\text{mol}}{L} = \text{mol L}^{-1}$. Rate on the other hand, is ALWAYS communicated as a change in molarity (or atmospheres) per unit time. Think of it as $\dfrac{M}{\text{time}} =$ M time^{-1} = mol L^{-1} time^{-1}. The rate constant, k MUST have the units assigned to it that make BOTH sides of the equation end up as M/time!

Rate Law Expression showing the correct units on k where k is a numerical value determined from experimental data analysis.			Reasoning for the units on k : Initial Rate = k [reactant(s)]$^{order(s)}$ $M/t = M/t$ The units on k MUST make this true!
$\dfrac{M}{time} =$	$\dfrac{k}{time}$	$[A]$	$\dfrac{M}{time} = k\,M$; the right side is currently just M, so k must have the reciprocal time unit attached to it so that the right side becomes M/t. k is reported as a number/time unit.
$\dfrac{M}{time} =$	$\dfrac{k}{M \cdot time}$	$[A]^2$	$\dfrac{M}{time} = k\,M^2$; the right side is currently M^2, so k must have the reciprocal time unit attached to it AND a *reciprocal molarity* so that the right side becomes M/t. k is reported as a number/M • time unit.
$\dfrac{M}{time} =$	$\dfrac{k}{M \cdot time}$	$[A][B]$	$\dfrac{M}{time} = k\,M^2$; the right side is also currently M^2, so k must have the reciprocal time unit attached to it AND a *reciprocal molarity* so that the right side becomes M/t. k is reported as a number/M • time unit.
$\dfrac{M}{time} =$	$\dfrac{k}{M^2 \cdot time}$	$[A][B]^2$	$\dfrac{M}{time} = k\,M^3$; the right side is currently M^3, so k must have the reciprocal time unit attached to it AND a *reciprocal molarity squared* so that the right side becomes M/t. k is reported as a number/M^2 • time unit.

PURPOSE

In this activity you will determine the order of a reactant from concentration-rate data using the method of differential rate law.

MATERIALS

calculator   paper and pencil

PROCEDURE

Solve the problems found on your student answer page. Be sure to show all work paying attention to the proper use of significant digits and units.

Name _____

Period _____

Chemical Reaction Rates I
Solving Kinetics Problems Involving Differential Rate Law

CONCLUSION QUESTIONS

1. The rate of the reaction between CO and NO_2

    $$CO(g) + NO_2(g) \rightarrow CO_2(g) + NO(g)$$

    was studied at 650 K starting with various concentrations of CO and NO_2, and the data in the table were collected.

Experiment Number	Initial Rate mol/(Lx h)	Initial concentration [CO]	Initial concentration [NO₂]
1	3.4×10^{-4}	0.510	0.0350
2	6.8×10^{-4}	0.510	0.0700
3	1.7×10^{-4}	0.510	0.0175
4	6.8×10^{-4}	0.102	0.0350
5	10.2×10^{-4}	0.153	0.0350

a. Write the rate law that is consistent with the data.

b. Calculate the value of the specific rate constant, k, and specify units.

c. What is the overall order for the reaction?

2. The initial rate of the reaction of nitrogen monoxide and oxygen was measured at 35°C for various initial concentrations of NO and O_2.

2 NO(g) + O_2(g) → 2 NO_2(g)

Experiment Number	Initial concentration [O_2]	Initial concentration [NO]	Initial Rate mol/(Lx s)
1	0.040	0.020	0.028
2	0.040	0.040	0.057
3	0.040	0.080	0.114
4	0.080	0.040	0.227
5	0.020	0.040	0.014

a. Write the rate law that is consistent with the data.

b. Calculate the value of the specific rate constant, k, and specify units.

c. What is the overall order for the reaction?

3. Dinitrogen pentoxide decomposes at 45°C to yield nitrogen dioxide and oxygen gas according to the equation

$$2 N_2O_5(g) \rightarrow 4NO_2(g) + O_2(g)$$

Data are collected in the table.

Experiment Number	Initial concentration $[N_2O_5]$	Initial Rate mol/(L x s)
1	1.75×10^{-3}	2.06×10^{-9}
2	3.50×10^{-3}	8.24×10^{-9}
3	7.00×10^{-3}	3.30×10^{-8}

a. Write the rate law that is consistent with the data.

b. Calculate the value of the specific rate constant, k, and specify units.

c. What is the overall order for the reaction?

4. Nitrogen monoxide reacts with chlorine according to the equation

$$2 NO(g) + Cl_2(g) \rightarrow 2 NOCl(g)$$

The following initial rates of reaction have been observed for certain reactant concentrations:

Experiment Number	Initial concentration [NO]	Initial concentration [Cl₂]	Initial Rate mol/(Lx h)
1	0.50	0.50	1.14
2	1.00	0.50	4.56
3	1.00	1.50	41.0

a. Write the rate law that is consistent with the data.

b. Calculate the value of the specific rate constant, *k*, and specify units.

c. What is the overall order for the reaction?

5. Hydrogen reacts with nitrogen monoxide to form dinitrogen oxide, laughing gas, according to the equation

$H_2 + 2NO(g) \rightarrow N_2O(g) + H_2O(g)$

The following data was collected:

[NO], M	0.40	0.80	0.80
[H₂], M	0.35	0.35	0.70
Initial Rate mol/(L x s)	5.040×10^{-5}	2.016×10^{-4}	4.032×10^{-4}

a. Write the rate law that is consistent with the data.

b. Calculate the value of the specific rate constant, k, and specify units.

c. What is the overall order for the reaction?

Chemical Reaction Rates II
Solving Kinetics Problems Involving Integrated Rate Law

OBJECTIVE

Students will use a graphing calculator and concentration-time data to determine the order of a reactant using the graphical methods of integrated rate law.

LEVEL

Chemistry

NATIONAL STANDARDS

UCP.1, UCP.2, UCP.3, B.3, B.6

TEKS

15(B)

CONNECTIONS TO AP

AP Chemistry:
    III. Reaction types D. Kinetics 1. Concept of rate of reaction 2. Use of experimental data and graphical analysis to determine reactant order, rate constants, and reaction rate laws 3. Effect of temperature changes on rates 4. Energy of activation; the role of catalysts

TIME FRAME

two 45 minute class periods

MATERIALS

    graphing calculator   pencil

TEACHER NOTES

This activity is designed to provide an introduction to the types of problems students will face on the AP Chemistry exam. This lesson should follow the lesson on differential rate law found in this book. The mathematical skills and graphing calculator skills are those students should have mastered in their mathematics class since they are required of Algebra I students on the TAKS test. Do not be intimidated by the use of the graphing calculator, your students use it regularly in their math classes. Students should complete the conclusion questions for homework. The kinetics topics involving mechanisms and the Arrhenius equation for calculating the activation energy are not addressed in this lesson.

Prior to this lesson, students should have an understanding of basic collision theory, especially reaction coordinate diagrams (a.k.a. potential energy diagrams). From these diagrams they should be able to compare the energy of the reactants, the energy of the products, and the activation energy. Emphasize that collisions between molecules must occur for chemical reactions to take place. Also emphasize that a catalyst lowers the activation energy by providing an alternate collision pathway, one that requires less energy. Remind students that a catalyst is neither a reactant nor a product, so it can participate in chemical reactions over and over again. Be sure students also have an understanding of a reaction intermediate and that it is a highly unstable species that is formed at the peak of the potential energy diagram.

ANSWERS TO THE CONCLUSION QUESTIONS

1. Data for the decomposition of N_2O_5 in a solution at 55°C are as follows:

[N_2O_5] (mol/L)	Time (min)
2.08	3.07
1.67	8.77
1.36	14.45
0.72	31.28

```
LinReg
y=ax+b
a=.0330040062
b=.325928499
r²=.9824939052
r=.9912083057
```

Use graphical methods of the integrated rate law to determine the following:

a. What is the order for the reaction?
   - The statistics displayed next to the data above are from the linear regression of L1,L4 which translates to $\frac{1}{[N_2O_5]}$ vs. time. This regression had the highest r value ∴ the reaction is second order with respect to N_2O_5.

b. What is the rate constant for the reaction?
   - The slope = k, so $k = \dfrac{0.0330}{M \bullet min}$

c. Write the rate law that is consistent with the data?
   - Rate = $k \, [N_2O_5]^2$

d. What is the concentration of N_2O_5 at 2.00 minutes?
   - Press [2nd][TRACE][ENTER] and enter 2 as the x value.
   - When $x = 2.00$ min, $y = 0.392$ which is the reciprocal of the correct concentration.
   - $\dfrac{1}{[N_2O_5]} = \dfrac{1}{[0.392]} = 2.55 \times 10^{-3}$

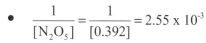 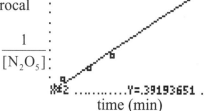

time (min)

e. At what time is the concentration of N_2O_5 equal to 0.55M?
   - First, write an equation in Y2 (Press [Y=][▼] until you arrive at Y2=) that is the reciprocal of 0.55, then press [GRAPH]. If you do not see the horizontal line representing 0.55^{-1}, then press [WINDOW] and change the Ymax. Press [GRAPH] to view the graph and see if your horizontal line appears. The reciprocal of 0.55 is less than 2, so we chose 2. Students do this in math class all the time. This is not a new skill for them.

- Now the horizontal line appears, but it still does not intersect our regression equation. Press WINDOW again and adjust the Xmax. This may take some trial and error, we settled on 50 as an Xmax. Press GRAPH. The intersection should now be on your screen.
- Next calculate the x and y values at the intersection. Press 2nd TRACE 5 ENTER ENTER ENTER. The x value is the time in minutes which is equal to 1.82 minutes.

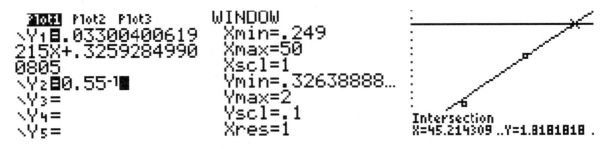

2. Sucrose, $C_{12}H_{22}O_{11}$, decomposes in dilute acid solution to form the two isomers glucose and fructose that both have the chemical formula of $C_6H_{12}O_6$. The rate of this reaction has been studied in acid solution, and the data in the table were obtained.

Time (min)	$[C_{12}H_{22}O_{11}]$ (mol/L)
0	0.316
25	0.289
50	0.264
150	0.183
200	0.153

```
LinReg
  y=ax+b
  a=-.0036360999
  b=-1.151078371
  r²=.999983142
  r=-.999991571
```

Use graphical methods of the integrated rate law to determine the following:

a. What is the order for the reaction?
   - The statistics displayed next to the data above are from the linear regression of L1,L3 which translates to $\ln[C_{12}H_{22}O_{11}]$ vs. time. This regression had the highest r value $\therefore$ the reaction is first order with respect to $C_{12}H_{22}O_{11}$.

b. What is the rate constant for the reaction?
   - The | slope | = k, so $k = \dfrac{3.64 \times 10^{-3}}{\text{min}}$

c. Write the rate law that is consistent with the data?
   - Rate = $k\,[C_{12}H_{22}O_{11}]$

d. What is the concentration of $[C_{12}H_{22}O_{11}]$ at 75.0 minutes?
   - When x = 75.0 min, y = -1.42 which is the natural log of the correct concentration.
   - $e^{-1.42} = 0.242M$

e. At what time is the concentration of $[C_{12}H_{22}O_{11}]$ equal to 0.115M?

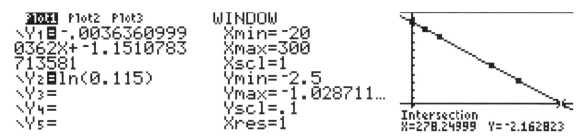

- First, write an equation in Y2 that is the natural log of 0.115M, then press GRAPH. Adjust your window until the horizontal line appears.
- Next calculate the x and y values at the intersection.
- The x value is the time in minutes which is equal to 278 minutes.

3. Ammonia decomposes when heated according to the equation

$$NH_3(g) \rightarrow NH_2(g) + H(g)$$

The data in the table for this reaction were collected at 2500K.

[NH3] (mol/L)	Time (h)
8.00 x 10-7	0
7.20 x 10-7	15
6.55 x 10-7	30
5.54 x 10-7	60

LinReg
y=ax+b
a=9248.644296
b=1249888.237
r²=.9999967698
r=.9999983849

Use graphical methods of the integrated rate law to determine the following:

a. What is the order for the reaction?
- The statistics displayed next to the data above are from the linear regression of L1,L4 which translates to $\dfrac{1}{[NH_3]}$ vs. time. This regression had the highest r value $\therefore$ the reaction is second order with respect to NH_3.

b. What is the rate constant for the reaction?
- The | slope | = k, so $k = \dfrac{9250}{M \bullet h}$

c. Write the rate law that is consistent with the data?
- Rate $= k\,[NH_3]^2$

d. What is the concentration of NH_3 at 90.0 hours?
- Adjust your window as necessary.
- When $x = 90.0$ min, $y = 2,082,266$ which is the reciprocal of the correct concentration.
- $\dfrac{1}{[NH_3]} = \dfrac{1}{[2,082,266]} = 4.80 \times 10^{-7}$

e. At what time is the concentration of NH_3 equal to 7.00×10^{-7} M?

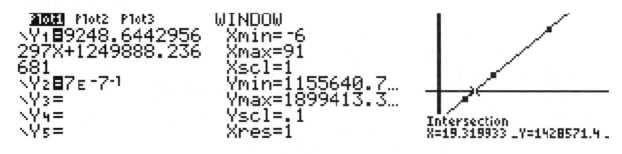

- First, write an equation in Y2 that is the reciprocal of 7.00×10^{-7}, then press GRAPH. Adjust your window until you see the intersection.
- Next calculate the x and y values at the intersection. Press [2nd][TRACE][5][ENTER][ENTER][ENTER]. The x value is the time in minutes which is equal to 19.3 minutes.

4. The reaction $2HOF(g) \rightarrow 2HF(g) + O_2(g)$ occurs at 45°C. The following data was collected:

Time (min)	[HOF] (mol/L)
0	0.850
5	0.754
10	0.666
15	0.587
20	0.526

```
LinReg
y=ax+b
a=-.0242047564
b=-.1632588305
r²=.9995185618
r=-.9997592519
```

a. What is the order for the reaction?
- The statistics displayed next to the data above are from the linear regression of L1,L3 which translates to ln[HOF] vs. time. This regression had the highest r value ∴ the reaction is first order with respect to HOF.

b. What is the rate constant for the reaction?
- The | slope | $= k$, so $k = \dfrac{0.163}{min}$

c. Write the rate law that is consistent with the data?
- Rate $= k$ [HOF]

d. What is the concentration of [HOF] at 60.0 minutes?
  - Adjust your window as necessary.
  - When $x = 60.0$ min, $y = -1.62$ which is the natural log of the correct concentration.
  - $e^{-1.62} = 0.198$M

e. At what time is the concentration of [HOF] equal to 0.250M?

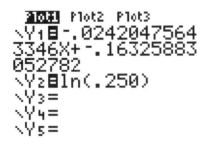 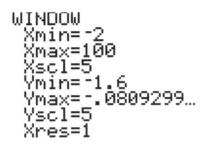 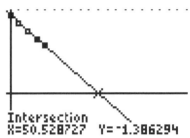

  - First, write an equation in Y2 that is the natural log of 0.250M, then press GRAPH. Adjust your window until the horizontal line appears. This one was tricky since it was the Xmin that needed adjusting since the natural logs are all negative!
  - Next calculate the x and y values at the intersection.
  - The x value is the time in minutes which is equal to 50.5 minutes.

5. Data for the decomposition of dinitrogen oxide into its elements at 900°C is as follows:

Time (min)	[N₂O] (mol/L)
15	0.0835
30	0.0680
80	0.0350
120	0.0022

LinReg
y=ax+b
a= -.0127547971
b= -2.30383746
r²=.9988809817
r= -.9994403342

a. What is the order for the reaction?
  - The statistics displayed next to the data above are from the linear regression of L1,L3 which translates to $\ln[N_2O]$ vs. time. This regression had the highest r value $\therefore$ the reaction is first order with respect to N_2O.

b. What is the rate constant for the reaction?
  - The | slope | = k, so $k = \dfrac{0.0128}{\text{min}}$

c. Write the rate law that is consistent with the data?
  - Rate = $k\,[N_2O]$

d. What is the concentration of [N₂O] at 240.0 minutes?
    - Adjust your window as necessary.
    - When x = 240.0 min, y = -5.36 which is the natural log of the correct concentration.
    - $e^{-5.36} = 0.0047M$

e. At what time is the concentration of [N₂O] equal to 0.0500M?

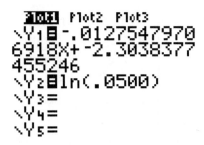

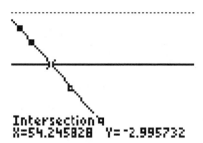

    - First, write an equation in Y2 that is the natural log of 0.0500M, then press GRAPH. You should not have to adjust your window.
    - Next calculate the x and y values at the intersection.
    - The x value is the time in minutes which is equal to 54.2 minutes.

Chemical Reaction Rates II
Solving Kinetics Problems Involving Integrated Rate

Law Chemical kinetics is the study of the speed or rate of a chemical reaction under various conditions. Collisions must occur in order for chemical reactions to take place. These collisions must be of sufficient energy to make and break bonds. The collisions must also be "effective" which means they not only have sufficient energy during the collision but that the molecules also collide in the proper orientation. The speed of a reaction is expressed in terms of its "rate" — some measurable quantity is changing with time.

$$\text{Rate} = \frac{\text{Change in concentration of a reactant or a product}}{\text{Change in time}}$$

Most commonly, we discuss four different types of rate when discussing chemical reactions: relative rate, instantaneous rate, differential rate, and integrated rate. Although these sound complicated, they are actually quite simple. This activity will focus primarily on the last two types, differential rate and integrated rate.

First, we need to discuss factors that affect the rate of a chemical reaction:

1. **Nature of the reactants**—Some reactant molecules react quickly, others react very slowly.

2. **Concentration of reactants**—more molecules in a given volume means more collisions which means more bonds are broken and more bonds are formed.

3. **Temperature**—"heat 'em up & speed 'em up"; the faster the molecules move, the more likely they are to collide and the more energetic those collisions will be.
   - An increase in temperature produces more successful collisions because more collisions will meet the required activation energy. There is a general increase in reaction rate with increasing temperature.
   - In fact, a general rule of thumb is that increasing the temperature of a reaction by 10°C doubles the reaction rate.

4. **Catalysts**—accelerate chemical reactions by allowing for more effective collisions, thus lowering the activation energy. The forward and reverse reactions are both accelerated to the same degree. Catalysts are not themselves transformed in the reaction, so they may be used over and over again.

5. **Surface area of reactants**—exposed surfaces affect speed.
   - Except for substances in the gaseous state or in solution, reactions occur at the boundary, or interface, between two phases.

- The greater the surface area exposed, the greater the chance of collisions between particles, hence, the reaction should proceed at a much faster rate. Example: coal dust is very explosive as opposed to a piece of charcoal. Solutions provide the ultimate surface area exposure!

INTEGRATED RATE

Differential rate law expresses how the **rate depends on concentration**. This is the most common and what we have been discussing. Integrated rate law expresses how the **concentrations depend on time**. If the data you have been given contains "Rate" and "Concentration" data, you use the differential rate law methods. If the data you have been given contains "Concentration" and "Time" data, you use the integrated rate law methods. Why on Earth do we need two methods? It is a result of experimental convenience.

When we wish to know how long a reaction must proceed to reach a predetermined concentration of some reagent, we can construct curves or derive an equation that relates concentration and time. This sounds scary, but is quite simple as long as you appreciate some elegant patterns. First the graphs…

When graphing the concentration of a reactant vs. time, one of the following two shapes is observed:

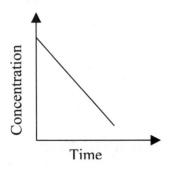

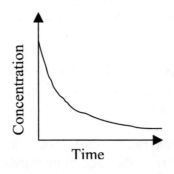

Zero order, always a straight line. No further analysis is needed—just read the graph.

If the graph is a curve, it could be 1st or 2nd order! Further analysis is needed to determine which one. This involves creating two additional graphs in search of a linear relationship.

Laying the Foundation in Chemistry

GRAPHICAL METHODS FOR DISTINGUISHING FIRST AND SECOND ORDER REACTIONS

Why are we in search of a linear relationship? $y = mx + b$ is the friendly little equation format for a straight line that you already know and love. It allows you to quickly solve for anything you need. You need to know HOW to determine zero, first, and second order relationships from concentration vs. time graphs.

Question: How do I get a linear relationship?

Answer: Set up your axes so that time is always on the x-axis. Next, sketch your two new graphs so that you can determine whether the reactant is first or second order. Plot the natural log (ln [A], NOT log) of the concentration on the y-axis of the first graph and the reciprocal concentration on the y-axis of the second graph. You are in search of linear data! Here comes the elegant part… If you do the set of graphs in this order with the y-axes being "**c**oncentration", "**n**atural log of concentration" and "**r**eciprocal concentration", the alphabetical order of the y-axis variables leads to 0, 1, 2 orders respectively for that reactant.

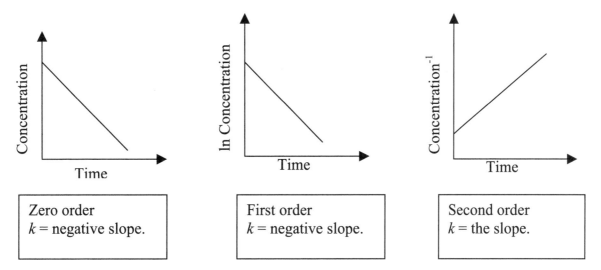

Zero order
k = negative slope.

First order
k = negative slope.

Second order
k = the slope.

You can now easily solve for either time or concentration once you know the order of the reactant. Just remember $y = mx + b$. Pick up the variables that gave you the straight line and insert them in place of x and y in the equation. "A" is reactant A and A_o is the initial concentration of reactant A at time zero [the *y*-intercept].

	Y	=	mx	+	b
zero order	[A]	=	-kt	+	$[A_o]$
first order	ln[A]	=	-kt	+	ln $[A_o]$
second order	1/[A]	=	kt	+	$1/[A_o]$

It is also useful to remember that | slope | $= k$, since the rate constant is NEVER negative. If you are asked to write the rate expression [or rate law] it is simply Rate = k[A]$^{\text{order you determined from analyzing the graphs}}$

PURPOSE

In this activity you will determine the order of a reactant from the method of integrated rate law. You will also use a graphing calculator and concentration-time data to determine the order of a reactant using the graphical methods of differential rate law.

MATERIALS

calculator   paper and pencil

PROCEDURE

Solve the problems found on your student answer page. Be sure to show all work paying attention to the proper use of significant digits and units.

Name _____

Period _____

Chemical Reaction Rates II
Solving Kinetics Problems Involving Integrated Rate

ANALYSIS

The following TI-83 or TI-83+ calculator steps may prove useful when applying the graphical methods of the integrated rate law.

To determine the order of a reactant:

- Press [2nd][+][4][ENTER][ENTER] to clear all lists.
- Press [2nd][0] and scroll down to DIAGNOSTICS ON and press [ENTER][ENTER] so that your calculator will display the linear correlation coefficient, r, for each linear regression you perform.
- Press [MODE] and use your arrow keys to select FLOAT, press [ENTER][2nd][MODE] to quit.
- Press [CLEAR] to clear your screen.
- To enter data, press [STAT] and [ENTER] to select EDIT. Use the summary below to prepare your analysis of the data:

  L1: time (x variable throughout!)
  L2: concentration   [A]   straight line = zero order
  L3: ln concentration   ln [A] straight line = first order
  L4: reciprocal concentration   1/[A]   straight line = second order

- Perform 3 linear regressions. Begin by pressing [STAT][▶][4], LinReg(ax + b) is now displayed.
- Press [2nd][1][,][2nd][2] to select your x-values from L1 (time) and your y-values from L2. Make note of the r value. You seek the best linear fit for your data. The best fit will have an r value closest to ± 1.
- Press [2nd][ENTER] to re-display the LinReg command, press [◀][2nd][3] to replace L2 with L3, press [ENTER] to execute the command and make note of the r value.
- Press [2nd][ENTER] to re-display the LinReg command, press [◀][2nd][4] to replace L3 with L4, press [ENTER] to execute the command and make note of the r value.
- Examine your r values and decide which set of data gives the best linear fit.
- Paste the regression equation into [Y=] by pressing [2nd][ENTER] until you return to the LinReg command that yielded the best fit.
- Once the L1, L$_{whichever you chose}$ combination is properly displayed, press [,] [VARS][▶] to Y-VARS then [1][1]. If you were successful, you'll see LinReg(ax +b) L$_1$, L$_{whichever you chose}$, Y$_1$ displayed on your screen. Press [ENTER] to execute the command and paste the regression equation into Y1.

More helpful hints:
Recall that the order of the reaction is 0; 1; 2 respectively for each L1, L2; L1,L3; L1,L4 combination respectively and | slope | = k and Rate = k[reactant]^{order}

Next, since you have created a line, *never* forget: y = mx + b   (TI uses an "a" instead of an "m")

If L1,L3 was your best "r" then, the reaction is first order and

$$y = mx + b \qquad \text{becomes}$$

$$\ln[\text{conc.}] = -kt + \ln[\text{conc.}_o]$$

*Notice that you must respect the sign of k when substituting into the $y = mx + b$ format.

Do the same substitutions into $y = mx + b$ for the other formats!

To determine the concentration at a given time:
- Set up your STAT PLOT by pressing [2nd][Y=]. Make sure only one plot is on and that you choose ⌐·· and select the list combination that generated the regression line you pasted into Y1.
- Press [ZOOM][9] to display the graph.
- Press [2nd][TRACE][ENTER] to select CALCULATE and VALUE. An X= is displayed in the lower left-hand corner of the screen. Enter the time you were given in the problem and press [ENTER]. Record the *y*-value displayed. Recall what you placed on your *y*-axis. It is most likely *not* L2 or the concentration. It is more likely either the natural log value of the concentration or the reciprocal of the concentration. *Solve for the actual concentration before recording your answer.*
- If you get an ERR:INVALID message, press [WINDOW] and re-set your Xmax value. The calculator can only calculate what it displays. Choose a value for Xmax that includes your desired time value.

To determine the time at a given concentration:
- Press [Y=] then [▼] to Y2. Recall the command that generated your regression equation in Y1. Write an equation that is either the concentration, ln concentration or reciprocal concentration based on whether your *y*-value was L2, L3, or L4 respectively.
- Press [2nd][TRACE][5] to select CALCULATE and INTERSECT. Press [ENTER][ENTER][ENTER]. Record the *x*-value displayed since it is the time value.
- If you get an ERR:INVALID message, press [WINDOW] and re-set your Ymax or Ymin values.

CONCLUSION QUESTIONS

1. Data for the decomposition of N_2O_5 in a solution at 55°C are as follows:

[N_2O_5] (mol/L)	Time (min)
2.08	3.07
1.67	8.77
1.36	14.45
0.72	31.28

Use graphical methods of the integrated rate law to determine the following:

a. What is the order for the reaction?

b. What is the rate constant for the reaction?

c. Write the rate law that is consistent with the data?

d. What is the concentration of N_2O_5 at 2.00 minutes?

e. At what time is the concentration of N_2O_5 equal to 0.55M?

2. Sucrose, $C_{12}H_{22}O_{11}$, decomposes in dilute acid solution to form the two isomers glucose and fructose that have the chemical formula of $C_6H_{12}O_6$. The rate of this reaction has been studied in acid solution, and the data in the table were obtained.

Time (min)	$[C_{12}H_{22}O_{11}]$ (mol/L)
0	0.316
25	0.289
50	0.264
150	0.183
200	0.153

Use graphical methods of the integrated rate law to determine the following:

a. What is the order for the reaction?

b. What is the rate constant for the reaction?

c. Write the rate law that is consistent with the data?

d. What is the concentration of $[C_{12}H_{22}O_{11}]$ at 75.0 minutes?

e. At what time is the concentration of $[C_{12}H_{22}O_{11}]$ equal to 0.115M?

3. Ammonia decomposes when heated according to the equation

$$NH_3(g) \rightarrow NH_2(g) + H(g)$$

$[NH_3]$ (mol/L)	Time (h)
8.00×10^{-7}	0
7.20×10^{-7}	15
6.55×10^{-7}	30
5.54×10^{-7}	60

The data in the table for this reaction were collected at 2500K.

Use graphical methods of the integrated rate law to determine the following:

a. What is the order for the reaction?

b. What is the rate constant for the reaction?

c. Write the rate law that is consistent with the data?

d. What is the concentration of NH_3 at 90.0 hours?

e. At what time is the concentration of NH_3 equal to 7.00×10^{-7} M?

4. The reaction $2HOF(g) \rightarrow 2HF(g) + O_2(g)$ occurs at 45°C.

The following data was collected:

Time (min)	[HOF] (mol/L)
0	0.850
5	0.754
10	0.666
15	0.587
20	0.526

a. What is the order for the reaction?

b. What is the rate constant for the reaction?

c. Write the rate law that is consistent with the data?

d. What is the concentration of [HOF] at 60.0 minutes?

e. At what time is the concentration of [HOF] equal to 0.250M?

5. Data for the decomposition of dinitrogen oxide into its elements at 900°C is as follows:

Time (min)	$[N_2O]$ (mol/L)
15	0.0835
30	0.0680
80	0.0350
120	0.0022

a. What is the order for the reaction?

b. What is the rate constant for the reaction?

c. Write the rate law that is consistent with the data?

d. What is the concentration of $[N_2O]$ at 240.0 minutes?

e. At what time is the concentration of $[N_2O]$ equal to 0.0500M?

The Iodine Clock Reaction
Investigating Chemical Kinetics: A Microscale Approach

OBJECTIVE
Students will time a classic iodine clock reaction and subsequently graph their data to determine the order of a reactant involved in the reaction.

LEVEL
Chemistry

NATIONAL STANDARDS
UCP.1, UCP.2, UCP.3, B.3, B.6, E.1, E.2, G.2

TEKS
1(A), 2(A), 2(B), 2(C), 2(D), 2(E), 15(B)

CONNECTIONS TO AP
AP Chemistry:
  III. Reaction types D. Kinetics 1. Concept of rate of reaction 2. Use of experimental data and graphical analysis to determine reactant order, rate constants, and reaction rate laws 3. Effect of temperature changes on rates 4. Energy of activation; the role of catalysts

TIME FRAME
45 minutes

MATERIALS
(For a class of 28 working in pairs)

28 1 x 12 well strips	14 timing devices
distilled water	14 microtip pipets for distilled water
14 sets of filled microtip pipets—one each of	cotton swabs for drying the well strips
solutions A, B & C	between trials

TEACHER NOTES
This activity is designed to follow an introductory classroom lesson on chemical kinetics. A teacher demonstration is necessary to show the students the proper shake down method for mixing the solutions. During the demonstration, impress upon them the necessity of holding the strips gently but firmly. The wells are mixed all at once with a snapping motion; there is **no shaking up and down**. (This will cause spillage.) Also inform students that they will have to shake the reaction strip in order to remove the chemicals at the end of each trial.

Since the wells of the well strips are small, it is essential that microtip pipets are used. These can either be purchased or made from a thin stem pipet. The stem of thin stem pipets can be "pulled" and trimmed so that they deliver very small drops. See Figure 1.

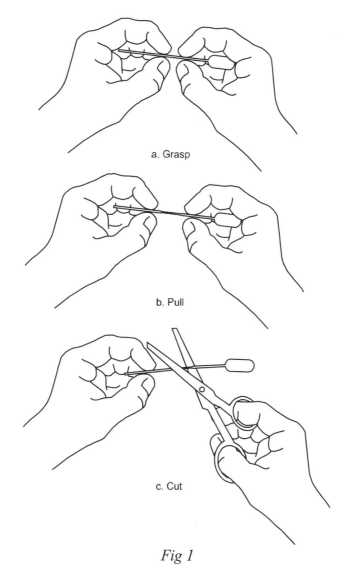

a. Grasp

b. Pull

c. Cut

Fig 1

It is best that you familiarize yourself with the information presented to students on the student pages. If students follow the directions properly, the solutions will turn blue in sequence from left to right AND in groups of three since the contents of wells 1-3, 4-6, etc. are the same. The entire strip should change color in less than three minutes.

PREPARATION OF SOLUTIONS
Set up the following solutions in polyethylene pipets. One pipet holds enough solution for each lab group to run four trials.

Solution A:
2.0 mL of 30% hydrogen peroxide + 93 mL of distilled water. (15 mL of
3% peroxide and 80 mL of water will also work but reaction times will be longer)

Solution B:
Starch solution with 0.005 M sodium thiosulfate ($Na_2S_2O_3$): Boil 100 mL distilled water and spray in laundry starch until a faint bluish translucence is noticeable. Cool. Add 0. 124 g $Na_2S_2O_3 \cdot 5$ H_2O in 100 mL.

Solution C:
Add 1.74 g KI, 1.4g $NaC_2H_3O_2$, and 3 mL 6 M acetic acid to total 197 mL of water.

Additional information for the reactions of this lab:

The reaction rate is measured by determining the time required for the reaction to consume a small amount of thiosulfate that is initially present in solution. Thiosulfate is relatively inert toward hydrogen peroxide, but is very rapidly oxidized by triiodide:

$$3\ I^- + H_2O_2 + 2\ H^+ \rightarrow I_3^- + 2\ H_2O$$

$$I_3^- + 2\ S_2O_3^{2-} \rightarrow 3\ I^- + S_4O_6^{2-}$$

No appreciable amount of triiodide forms until the thiosulfate has been completely consumed. At this point, the starch reacts with the triiodide, suddenly turning dark blue (forming a starch-iodine complex). During this reaction, only a negligible amount of the peroxide reacts so its initial value is essentially unchanged. The H^+ concentration is buffered and remains virtually unchanged. The I^- oxidized by the peroxide is immediately regenerated by the reaction of triiodide with thiosulfate so the I^- concentration is unchanged during the measured time interval.

Students should be able to determine that the reaction is first order with respect to H_2O_2.

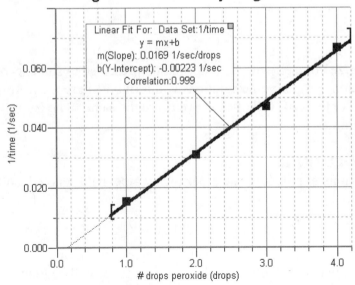

POSSIBLE ANSWERS TO THE CONCLUSION QUESTIONS AND SAMPLE DATA

Data Table											
Well # 1	2	3	4	5	6	7	8	9	10	11	12

Time required for the contents of each well to turn blue (seconds):

	1	2	3	4	5	6	7	8	9	10	11	12
Trial 1												
Trial 2												
Trial 3												

For each trial conducted, three wells contained the same number of H_2O_2 drops. This means that with each complete trial, you actually ran three individual trials simultaneously for a total of 9 total trials for each number of drops of H_2O_2. Average the nine times recorded and use the average time when graphing your data.

AVERAGE:	15.0		21.2		32.1		64.4

1. What is the order of $[H_2O_2]$ in this reaction? Use the results from your graph to justify your answer.
   - The graph indicates that this reaction is first order with respect to H_2O_2. The graph of 1/time vs. number of drops of H_2O_2 gives a straight line with a correlation of 0.999 and the regression line passes very nearly through the origin.
   - If the reaction was zero order with respect to H_2O_2, the slope of this graph would also be zero.
   - If the reaction was second order with respect to H_2O_2, the graph would be a parabola.

2. Briefly describe how you would use variations of this protocol to determine the order for the other reactants.
   - Repeat the experiment twice. During the first repeat, hold the concentrations of H_2O_2 constant as well as that of H^+ and vary the concentration of I^-. For the second repeat, hold the concentrations of H_2O_2 and I^- constant and vary the concentration of H^+.
   - Graph the data from each trial to determine the order of the varied reactant.

3. Having determined the order for the three reactants, how would you determine the value of k?
   - Select a trial from one of the sets of data and solve for k
     $$\text{RATE} = k[H_2O_2]^x[I^-]^y[H^+]^z$$
   - $$k = \frac{\text{RATE}}{[H_2O_2]^x[I^-]^y[H^+]^z}$$

4. A student fails to use a cotton swab to dry the wells between trials 2 and 3. What effect does this have on the times recorded for the third trial? Explain. Begin your explanation by stating the time will "increase, decrease or remain the same".

   - The time required would increase since the water adhering to the inside of the well would dilute the solutions.

5. Use your regression equation to determine the time it would take for 6 drops of H_2O_2 to react.

- $y = mx + b$

$$\frac{1}{time} = 0.017 \frac{1/sec}{drop}(\#drops)+0$$

$$\frac{1}{time} = 0.017 \frac{1/sec}{drop}(6.0 \text{ drops})+0$$

$$\frac{1}{time} = \frac{0.10}{sec} \quad \text{take the reciprocal of both sides;}$$

time = 10. seconds (2 SF)

The Iodine Clock Reaction
Investigating Chemical Kinetics: A Microscale Approach

In this laboratory activity, you will study the rate of a chemical reaction. The reaction is called a "clock" reaction because the formation of one of the products is accompanied by an abrupt color change which makes timing the reaction easy and accurate.

The rate of a reaction is governed by the following relationship:

$$\text{RATE} = \frac{\Delta \, [\text{reactant}]}{\Delta \, \text{time}} = k[A]^x[B]^y[C]^z.....$$

The quantities in brackets are the concentrations of the reactants expressed in moles per liter. Each reactant's concentration is raised to an appropriate power, which we refer to as orders. The orders show how the concentration of each substance affects the rate of the reaction. Together with the rate constant, k, they give the overall rate of the reaction. The numerical values of x, y & z **must be determined by experimentation**. These numbers determine the order of the reaction. Added together, they give the overall order of the reaction.

The reaction to be studied in this experiment is the acid-buffered oxidation of iodide ion to the triiodide ion by hydrogen peroxide:

$$3 \, I^- + H_2O_2 + 2 \, H^+ \rightarrow I_3^- + 2 \, H_2O$$

The activity is designed to hold the concentration of all of the other reactants constant so that any time variations you observe are due to changing the concentration (drop count) of H_2O_2.

PURPOSE
In this activity you will collect concentration (the number of drops of H_2O_2) and time data for a chemical clock reaction. You will graph your results and determine the order of H_2O_2 in the overall rate law.

MATERIALS
two 1 x 12 well strips   timing device
distilled water   1 microtip pipet for distilled water
1 filled microtip pipet of each solution A, B & C   cotton swabs

---

Safety Alert
1. Wear goggles and aprons throughout this laboratory exercise.
2. Should any reagents come in contact with your skin, wash the affected area immediately.

PROCEDURE

1. Arrange two 1 x 12-well strips such that the numbers on the wells are read from left to right.

2. Holding the dropper of distilled water perfectly vertical; place 2 drops of distilled water in each well of the first strip and 3 drops of distilled water in each well of the second strip. To practice the shakedown method, hold the strips firmly together, shake them once vigorously in a downward motion. This is done by dropping your hands as fast as you can and stopping abruptly. There is *no upward motion* so this is a shake*down* method.

3. Continue to practice this technique until you are able to successfully transfer the entire contents of one strip into the other without spilling. Dry your wells before proceeding to step 4.

4. Place drops of solutions into the two 1 x 12 well-strips according to the tables below:

   Solution A contains hydrogen peroxide. Solution B contains sodium thiosulfate. Solution C contains potassium iodide, sodium acetate and acetic acid.

STRIP 1: SOLUTION "A" and WATER

Well #	1	2	3	4	5	6	7	8	9	10	11	12
Drops Sol'n A	4	4	4	3	3	3	2	2	2	1	1	1
Drops Water	0	0	0	1	1	1	2	2	2	3	3	3

STRIP 2: SOLUTIONS "B & C"

Well #	1	2	3	4	5	6	7	8	9	10	11	12
Drops Sol'n B	1	1	1	1	1	1	1	1	1	1	1	1
Drops Sol'n C	4	4	4	4	4	4	4	4	4	4	4	4

Gently invert Strip 2 onto Strip 1 so that the wells line up. Capillary action will keep the solutions in the wells even though you are inverting the strips.

5. Holding the strips firmly together, shake them once vigorously in a downward motion. This is done by dropping your hands as fast as you can and stopping abruptly. There is *no upward motion* so this is a shake*down* method. Your partner should start timing immediately!

6. Gently rock Strip 1, the reaction strip, to further mix the reagents. Record the time that each well takes to turn blue in the data table provided on your student answer page.

7. Once finished, dispose of the contents from the strips into the sink and rinse the strips with distilled water. Dry each well with cotton swabs to prevent dilution for the next trial.

8. Repeat steps 4-7 *twice*.

9. Proceed to the Analysis section found on your student answer page to determine the order of $[H_2O_2]$ in this reaction.

Name _____

Period _____

The Iodine Clock Reaction
Investigating Chemical Kinetics: A Microscale Approach

DATA AND OBSERVATIONS

Data Table												
Well #	1	2	3	4	5	6	7	8	9	10	11	12
Time required for the contents of each well to turn blue (seconds):												
Trial 1												
Trial 2												
Trial 3												
For each trial conducted, three wells contained the same number of H_2O_2 drops. This means that with each complete trial, you actually ran three individual trials simultaneously for a total of 9 total trials for each number of drops of H_2O_2. Average the nine times recorded and use the average time when graphing your data.												
AVERAGE:												

ANALYSIS

The rate of reaction can be represented by the following equation:

$$\text{RATE} = \frac{\Delta\,[\text{reactant}]}{\Delta\,\text{time}} = k[H_2O_2]^x[I^-]^y[H^+]^z$$

The concentrations of I^- and H^+ are held constant in this experiment—all the wells in Strip 2 were filled with the same number of drops of solutions B and C. Thus, we may write the rate law:

$$\text{RATE} = \frac{\Delta\,[H_2O_2]}{\Delta\,\text{time}} = k\,[H_2O_2]^x$$

where $[I^-]^y[H^+]^z$ is multiplied by k. Multiplying three constants together creates a new constant we will call the pseudo-rate constant, k'.

Recall that the expression for the rate is:

$$\text{RATE} = \frac{\Delta\,[H_2O_2]}{\Delta\,\text{time}} = \frac{1}{\Delta\,\text{time}}[H_2O_2] = k\,[H_2O_2]^x$$

Plotting $\dfrac{1}{\Delta t}$ is the same as plotting a constant multiplied by the reaction rate.

But the rate is equal to $k'[H_2O_2]^x$. Therefore, a plot of $\dfrac{1}{\Delta t}$ versus $[H_2O_2]$ gives an indication of the exponent, x.

If the slope equals zero, x = 0. If there is a straight line through the origin, x = 1. If there is a parabola, x = 2.

1. Use a graphing calculator or computer graphing software to plot a graph of your results. Plot 1/time on the *y*-axis and concentration (the number of drops of H_2O_2 solution used) on the *x*-axis. Use the information above to determine the order (exponent) of $[H_2O_2]$.

2. Either print this graph or plot your results on a sheet of graph paper and attach it to the back of your student pages.

CONCLUSION QUESTIONS

1. What is the order of $[H_2O_2]$ in this reaction? Use the results from your graph to justify your answer.

2. Briefly describe how you would use variations of this protocol to determine the order for the other reactants.

3. Having determined the order for the three reactants, how would you determine the value of *k*?

4. A student fails to use a cotton swab to dry the wells between trials 2 and 3. What effect does this have on the times recorded for the third trial? Explain. Begin your explanation by stating the time will "increase, decrease or remain the same".

5. Use your regression equation to determine the time it would take for 6.0 drops of H_2O_2 to react. Show all of your work paying attention to significant digits.

General Chemical Equilibrium
Solving Equilibrium Problems
Using the RICE-Table Method

OBJECTIVE
Students will be introduced to the concept of general equilibrium. Students will solve gas phase and concentration equilibrium problems using the RICE table problem-solving method. Students will also learn to equate K_c to K_p.

LEVEL
Chemistry

NATIONAL STANDARDS
UCP.1, UCP.2, UCP.3, UCP.4, B.3, B.6

TEKS
12(A), 15(A)

CONNECTIONS TO AP
AP Chemistry:
  III. Reaction types A. Reaction types C. Equilibrium 1. Concept of dynamic equilibrium, physical and chemical; equilibrium constants 2. Quantitative treatment b. Equilibrium constants for reactions in solution

TIME FRAME
90 minutes plus homework

MATERIALS
  calculator   pencil and paper

TEACHER NOTES
This activity is designed as a classroom lesson and will require a teacher lecture. Students should be guided through the general concepts and definitions of chemical equilibrium as well as the RICE-table method of problem solving. This activity should follow a unit on kinetics. Solving stoichiometry problems using the table method presented in an earlier lesson in this book will provide a smooth transition for the RICE-table method of solving equilibrium problems.

There are four topics that are considered difficult in the AP Chemistry curriculum, mostly due to the lack of exposure during the first-year chemistry course. These topics include: kinetics, equilibrium, thermodynamics and electrochemistry. This lesson is designed to introduce the basics of gas phase and concentration equilibrium. Much of what is presented here is probably absent from your high school chemistry textbook. However, the first question in the free-response section of the AP Chemistry exam is always an equilibrium problem. Your students will benefit from an introduction to chemical equilibrium in your first-year course.

Although the AP exam has not yet required the use of the quadratic formula in their equilibrium problems, inclusion of this component will reinforce student's algebra skills and give students that ever important connection between math and science. If students have graphing calculators available, it is advisable to install a quadratic formula solving program for their use. Students are allowed to use programmed calculators on the AP Chemistry exam.

Many of these programs are available for free download at http://www.ticalc.org/pub/83plus/basic/math/. This will require the use of a graph link device and the free TI Connect software available at http://education.ti.com/us/product/accessory/connectivity/down/download.html at the time of this printing. Students are quite adept at downloading and installing programs from the computer as evidenced by the game repertoire many of their calculators possess! If you are uncomfortable with this process, make an announcement in class, and ask students to bring their calculator already programmed so they can share with other students through a calculator to calculator link.

Four examples are provided within the student pages. Answers to Example 1 are given below, all of the other examples have solutions worked out on the student pages.

(a) $2NH_{3(g)} \rightleftharpoons N_{2(g)} + 3H_{2(g)}$; $K = \dfrac{[H_2]^3[N_2]}{[NH_3]^2}$

(b) $4NH_{3(g)} + 7O_{2(g)} \rightleftharpoons 4NO_{2(g)} + 6H_2O_{(g)}$; $K = \dfrac{[NO_2]^4[H_2O]^6}{[NH_3]^4[O_2]^7}$

(c) $2NO_{(g)} + Cl2_{(g)} \rightleftharpoons 2NOCl_{(g)}$; $K = \dfrac{[NOCl]^2}{[NO]^2[Cl_2]}$

ANSWERS TO THE CONCLUSION QUESTIONS

1. Write an equilibrium constant expression for each of the following unbalanced reactions:

First, balance each equation, and then write the equilibrium expression leaving out pure solids and pure liquids.

a. $PCl_{5(g)} \rightleftharpoons PCl_{3(g)} + Cl_{2(g)}$; $K = \dfrac{[PCl_3][Cl_2]}{[PCl_5]}$

b. $NH_4HS_{(g)} \rightleftharpoons H_2S_{(g)} + NH_{3(g)}$; $K = \dfrac{[NH_3][H_2S]}{[NH_4HS]}$

c. $NH_4NO_{3(s)} \rightleftharpoons NH_4^+{}_{(aq)} + NO_3^-{}_{(aq)}$; $K = [NH_4^+][NO_3^-]$

d. $HCOOH_{(aq)} + H_2O \rightleftharpoons H_3O^+{}_{(aq)} + HCOO^-{}_{(aq)}$; $K = \dfrac{[H_3O^+][HCOO^-]}{[HCOOH]}$

e. $Bi_2S_{3(s)} \rightleftharpoons 2\,Bi^{3+}{}_{(aq)} + 3\,S^{2-}{}_{(aq)}$; $K = [Bi^{3+}]^2[S^{2-}]^3$

2. The equilibrium constant K_p for the following reaction is 11.5 at 300°C when the amounts of reactant and products are given in atmospheres. Suppose a tank initially contains PCl_5 with a pressure of 3.00 atm at 300°C. (a) What is the partial pressure of chlorine gas once equilibrium has been established? (b) What is the value of K_c for this reaction at 300°C?

$$PCl_{5(g)} \rightleftharpoons PCl_{3(g)} + Cl_{2(g)}$$

(a)

R:	$PCl_5(g)$	$\rightleftharpoons$	$PCl_3(g)$	+	$Cl_2(g)$
I:	3.00 atm		0		0
C:	$-x$		$+x$		$+x$
E:	$3.00-x$		x		x

$$K_p = \frac{(PCl_3)(Cl_2)}{(PCl_5)} = 11.5 = \frac{(x)(x)}{(3.00-x)} = \frac{(x^2)}{(3.00-x)}$$

$$11.5(3.00-x) = x^2$$

$34.5 - 11.5x = x^2$ set the equation equal to zero with the value for x^2 being positive

$0 = x^2 + 11.5x - 34.5$ now you must use the quadratic formula to solve for x.

Allow students to use a calculator program!

$x = 2.47$ atm $= P_{Cl_2}$ (and 2.47 atm $= P_{PCl_3}$ while $P_{PCl_5} = 3 - x = 3 - 2.47$ atm $= 0.53$ atm)

(b)

$$K_p = K_c(RT)^{\Delta n}; \Delta n = +1$$

$$K_c = \frac{K_p}{(RT)^{\Delta n}} = \frac{11.5}{(0.0821 \times 573)^{+1}} = 0.244$$

3. An aqueous solution of ethanol and acetic acid, each with a concentration 0.810 M, is heated to 125°C. At equilibrium, the acetic acid concentration is 0.645 M.

(a) Calculate K at 125°C for the reaction.

$$C_2H_5OH_{(aq)} + CH_3COOH_{(aq)} \rightleftharpoons H_2O_{(\ell)} + CH_3CO_2C_2H_{5(aq)}$$
ethanol   acetic acid   ethyl acetate

(b) What is K for the following reaction that occurs at 125°C?

$$2CH_3CO_2C_2H_{5(aq)} + 2H_2O_{(\ell)} \rightleftharpoons 2CH_3COOH_{(aq)} + 2C_2H_5OH_{(aq)}$$

(a)

R: $C_2H_5OH_{(aq)} + CH_3COOH_{(aq)} \rightleftharpoons H_2O_{(\ell)} + CH_3CO_2C_2H_{5\,(aq)}$

I:	0.810	0.810	0
C:	$-x$	$-x$	$+x$
E:	$0.810-x$	$0.810-x$	x
	$0.810-0.645$	$0.810-0.645$	*0.645

*The problem stated that the equilibrium concentration of $CH_3CO_2C_2H_5$ was 0.645M, so that IS the value of x.

$$K = \frac{[CH_3CO_2C_2H_5]}{[C_2H_5OH][CH_3COOH]} = \frac{[0.645]}{[0.810-0.645]^2} = 23.7$$

(b) The second reaction is simply the original reaction, but reversed and doubled.

$$K_{new} = \frac{1}{K^2} = \frac{1}{23.7^2} = 0.00178 = 1.78 \times 10^{-3}$$

4. The equilibrium constant, K_c, for the following reaction

$$H_2(g) + I_2(g) \rightleftharpoons 2HI(g)$$

is determined to be 57.85 at 450°C.

(a) If 1.50 mol of each reactant is placed in a 2.00-L flask at 450°C what are the concentrations of H_2, I_2, and HI when equilibrium has been achieved?

(b) What is K_p for this reaction at 450°C?

(a)

R: $H_2(g)$ + $I_2(g) \rightleftharpoons$ $2\ HI\ (g)$

I: $\dfrac{1.5mol}{2.00\text{-}L} = 0.75M$ 0.75 0

C: $-x$ $-x$ $+2x$

E: $0.75-x$ $0.75-x$ $2x$

$$K = \frac{[HI]^2}{[H_2][I_2]} = \frac{[2x]^2}{[0.75-x]^2} = 57.85$$

$$\sqrt{57.85} = 7.60 = \frac{[2x]}{[0.75-x]}$$

$$7.60(0.75-x) = 2x;\ x = 0.594$$

$$[H_2] = [I_2] = 0.75 - 0.594 = 0.156\frac{mol}{L}$$

$$[HI] = 2x = 2(0.594) = 1.12\frac{mol}{L}$$

(b)

$$K_p = K_c(RT)^{\Delta n};\ \Delta n = 0 \therefore K_p = K_c = 57.85$$

5. Ammonium hydrogen sulfide decomposes on heating.

$$NH_4HS_{(s)} \rightleftharpoons H_2S_{(g)} + NH_{3(g)}$$

(a) If K_p is 0.11 at 25° C when the partial pressures are expressed in atmospheres, what is the total pressure in the flask at equilibrium?

(b) What is K_c for this reaction at 25°C?

(a)

R: $NH_4HS(s) \rightleftharpoons H_2S(g) + NH_3(g)$

I: 0   0

C: $+x$ $+x$

E: x x

$$K_p = (H_2S)(NH_3) = 0.11 = x^2$$

$$x = \sqrt{0.11} = 0.33\,atm = P_{H_2S} = P_{NH_3}$$

(b) $K_p = K_c(RT)^{\Delta n}$; $\Delta n = +2$

$$K_c = \frac{K_p}{(RT)^{\Delta n}} = \frac{0.11}{(0.0821 \times 298)^{+2}} = 1.59 \times 10^{-4}$$

6. The following equilibrium constants are given at 500 K:

$$H_2(g) + Br_2(g) \rightleftharpoons 2\,HBr(g) \qquad\qquad K_p = 7.9 \times 10^{11}$$

$$H_2(g) \rightleftharpoons 2H(g) \qquad\qquad K_p = 4.8 \times 10^{-41}$$

$$Br_2(g) \rightleftharpoons 2Br(g) \qquad\qquad K_p = 2.2 \times 10^{-15}$$

(a) Calculate K_p for the reaction of H atoms and Br atoms to give HBr at 500 K.

To begin, rearrange the equations until you can add all three equations together to give $2H(g) + 2Br(g) \rightleftharpoons 2HBr(g)$ as the sum.

$\cancel{H_2(g)} + \cancel{Br_2(g)} \rightleftharpoons 2\,HBr(g)$ $K_p = 7.9 \times 10^{11}$ (nothing done here)

$2H(g) \rightleftharpoons \cancel{H_2(g)}$ $K_p = 4.8 \times 10^{-41}$ (reversed; take reciprocal of K)

$2Br(g) \rightleftharpoons \cancel{Br_2(g)}$ $K_p = 2.2 \times 10^{-15}$ (reversed; take reciprocal of K)

$2H(g) + 2Br(g) \rightleftharpoons 2\,HBr(g)$ $K_{new} = 7.9 \times 10^{11} \times \dfrac{1}{4.8 \times 10^{-41}} \times \dfrac{1}{2.2 \times 10^{-15}}$

Next, the coefficients must be halved so the summary K is raised to the ½ power which is the square root of the value to give $K_{pnew} = \sqrt{7.9 \times 10^{11} \times \dfrac{1}{4.8 \times 10^{-41}} \times \dfrac{1}{2.2 \times 10^{-15}}} = 2.74 \times 10^{33}$

(b) Calculate K_c for this reaction at 500K.

$$K_p = K_c (RT)^{\Delta n}; \; \Delta n = -1$$

$$K_c = \frac{K_p}{(RT)^{\Delta n}} = \frac{2.74 \times 10^{33}}{(0.0821 \times 500)^{-1}} = 1.12 \times 10^{35}$$

7. The equilibrium constant for the dissociation of iodine molecules to iodine atoms is 3.76×10^{-3} at 1000 K.

$$I_2(g) \; \rightleftharpoons \; 2 \, I(g)$$

(a) Suppose 0.150 mole of I_2 is placed in a 15.5-L flask at 1000K. What are the concentrations of I_2 and I when the system comes to equilibrium?

(b) What is K_p for this reaction at 1000 K?

(a)

R: $\quad I_2(g) \qquad\qquad \rightleftharpoons \qquad\qquad 2 \, I(g)$

I: $\quad \dfrac{1.5 \text{mol}}{15.5\text{-L}} = 0.00968 \text{M} \qquad\qquad 0$

C: $\quad -x \qquad\qquad\qquad\qquad\qquad +2x$

E: $\quad 0.00968 - x \qquad\qquad\qquad\quad 2x$

$$K = \frac{[I]^2}{[I_2]} = \frac{[2x]^2}{[0.00968 - x]} = 3.76 \times 10^{-3}$$

$$3.76 \times 10^{-3} (0.00968 - x) = 4x^2$$

$$\frac{3.64 \times 10^{-5} - 3.76 \times 10^{-3} x}{4} = x^2 \quad \text{set equation equal to zero and use a quadratic solving program}$$

$$0 = x^2 + 9.40 \times 10^{-4} x - 9.10 \times 10^{-6}$$

$$x = [I_2] = 2.58 \times 10^{-3} \frac{\text{mol}}{\text{L}}$$

$$[I] = 2x = 2(2.58 \times 10^{-3}) = 5.16 \times 10^{-3} \frac{\text{mol}}{\text{L}}$$

(b) $K_p = K_c (RT)^{\Delta n}; \; \Delta n = +1$

$$K_p = 3.76 \times 10^{-3} (0.0821 \times 1,000)^{+1} = 0.309$$

8. K_c for the decomposition of ammonium hydrogen sulfide is 1.8×10^{-4} at 15°C.

$$NH_4HS(s) \rightleftharpoons NH_3(g) + H_2S(g)$$

(a) When 5.00 grams of the pure salt decomposes in a sealed 3.0-L flask at 15°C, what are the equilibrium concentrations of NH_3 and H_2S?

(b) What are the equilibrium concentrations at 15°C if 10.0 grams of the pure salt decomposes in the sealed flask?

(a)

R:	$NH_4HS_{(s)} \rightleftharpoons$	$NH_{3\,(g)}$	+	$H_2S_{(g)}$
I:		0		0
C:		$+x$		$+x$
E:		x		x

$K = [NH_3][H_2S] = x^2 = 1.8 \times 10^{-4}$

$[NH_3] = [H_2S] = \sqrt{1.8 \times 10^{-4}} = 0.0134 \dfrac{mol}{L}$

(b) There was no change in temperature, so K_c remains constant at 1.8×10^{-4}.

The amount of solid has no effect, it is not a part of the equilibrium expression.

9. Hemoglobin (Hb) can form a complex with both O_2 and CO. For the reaction

$$HbO_2(aq) + CO(g) \rightleftharpoons HbCO\ (aq) + O_2(g)$$

at body temperature, K is about 2.0×10^2. If the ratio $\dfrac{[HbCO]}{[HbO_2]}$ comes close to 1.0, death is probable.

(a) What partial pressure of CO in the air is likely to be fatal? Assume the partial pressure of O_2 is 0.20 atm.

(b) What is K for the reverse reaction?

(a) We will let $\dfrac{[HbCO]}{[HbO_2]} = 1$ in the equilibrium expression.

$$\therefore K = 1 \frac{[O_2]}{[CO]} = \frac{[O_2]}{[CO]}$$

$$K = 200 = \frac{[0.20]}{[CO]}$$

$$[CO] = \frac{[0.20]}{200} = 0.001 = 1.0 \times 10^{-3}$$

(b)

$$K_{reverse} = \frac{1}{K} = \frac{1}{200} = 0.005 = 5.0 \times 10^{-3}$$

10. Lexan is a plastic used to make compact discs, eyeglass lenses, and bullet-proof glass. One of the compounds used to make Lexan is phosgene ($COCl_2$), an extremely poisonous gas. Phosgene decomposes by the reaction

$$COCl_2(g) \rightleftharpoons CO(g) + Cl_2(g)$$

for which $K_p = 7.2 \times 10^{-11}$ at 80°C.

(a) If pure phosgene at an initial pressure of 1.0 atm decomposes, calculate the equilibrium pressures of all species.

(b) What is K_c for this reaction at 80°C?

(a)

R: $COCl_2(g) \rightleftharpoons CO(g) + Cl_2(g)$

I: 1.0 atm   0   0

C: $-x$ $+x$ $+x$

E: $1.0-x$ x x

$$K_p = \frac{(COCl_2)(Cl_2)}{(PCl_5)} = 7.2 \times 10^{-11} = \frac{(x)(x)}{(1.0-x)} = \frac{(x^2)}{(1.0-x)}$$

$7.2 \times 10^{-11}(1.0-x) = x^2$

$7.2 \times 10^{-11} - 7.2 \times 10^{-11}x = x^2$ set the equation equal to zero with the value for x^2 being positive

$0 = x^2 + 7.2 \times 10^{-11}x - 7.2 \times 10^{-11}$

$x = 8.49 \times 10^{-6}$atm $= P_{Cl_2} = P_{CO}$ while $P_{COCl_2} = 1.0 - x = 1.0 - 8.49 \times 10^{-6} = 1.0$atm

(b) $K_p = K_c(RT)^{\Delta n}; \Delta n = +1$

$$K_c = \frac{K_p}{(RT)^{\Delta n}} = \frac{7.2 \times 10^{-11}}{(0.0821 \times 353)^{+1}} = 2.48 \times 10^{-12}$$

General Chemical Equilibrium
Solving Equilibrium Problems
Using the RICE-Table Method

THE NATURE OF THE EQUILIBRIUM STATE

Equilibrium is defined as the condition when the rate of the forward reaction equals the rate of the reverse reaction. Equilibrium reactions are reversible which is indicated by the presence of double

arrows, $\rightleftharpoons$, between the reactants and the products. The double arrows indicate that the reaction is proceeding in both the forward and reverse direction and once equilibrium is established, the rate of each direction is equal (Figure 1). We also say that once equilibrium is reached, it is **dynamic.** This means reactants are forming products and products are forming reactants simultaneously such that the net concentration of each remains constant. Notice, this does NOT say the concentrations of the products and reactants are equal! That is a very common misconception.

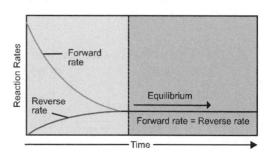

Fig. 1

Examine Figure 2 which represents the following generalized chemical reaction:

$$A + B \rightleftharpoons C + D$$

Notice that equilibrium is reached once the concentrations of the reactants and products remain constant, but again, not necessarily equal.

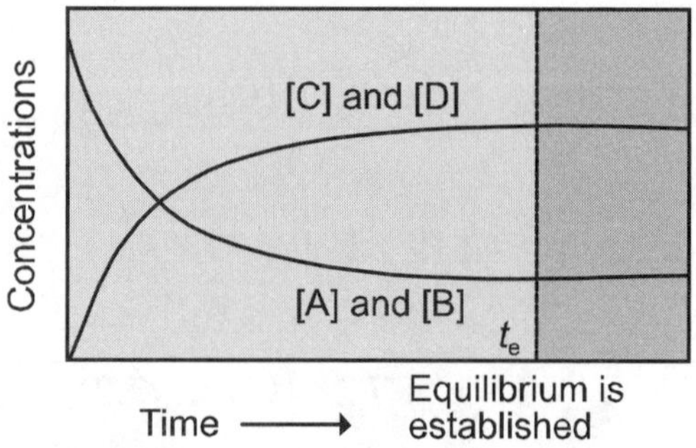

What you should be able to interpret from the graphs in Figures 1 & 2:

Fig. 2

1. Equilibrium is established as soon as the curves become "flat" on either graph. This indicates a constant value has been reached for the y-axis variable.
   - For Figure 1, the rates for the forward and reverse reactions converge and the rates are equal.
   - For Figure 2, the concentrations decrease for the reactants and increase for the products until a constant value is achieved, not necessarily an equal value. The time at which equilibrium is established is denoted as $t_{equilibrium}$ or t_e.

2. Equilibrium is not necessarily established at some midpoint on the graph.
   - For Figure 1, notice that the two rates did not converge at the midpoint between the two original values. The forward reaction has a much larger rate constant than the reverse reaction so it "pushes" the equilibrium position down on the graph.
   - For Figure 2, the time at which equilibrium is established can occur anywhere along the x-axis.

THE EQUILIBRIUM POSITION

Whether the reaction lies far to the right [products favored] or to the left [reactants favored] depends on three main factors:

- The initial concentrations of the reactants, the products, or both. The more molecules present, the more collisions that occur and the faster the reaction.
- The relative energies of reactants and products. Nature goes to minimum energy, always.
- The degree of organization of reactants and products. Nature goes to maximum disorder, always.

THE EQUILIBRIUM EXPRESSION

The first step in solving an equilibrium problem is to write the equilibrium constant expression. For the general reaction

$$aA + bB \rightleftharpoons cC + dD$$

The equilibrium constant expression, K, (a.k.a. Law of Mass Action expression) is written as follows:

$$K = \frac{[C]^c [D]^d}{[A]^a [B]^b}$$

- [ ] indicates concentration in molarity, M, (moles/liter)
- K_c indicates that all of the quantities in the problem are given in molarity. This applies to reactions in aqueous solution and sometimes gases.
- K_p indicates that all of the products and reactants are gases and the "p" is for partial pressures expressed in a pressure unit such as atmospheres.
- "K" values are always written without units.
- **Pure solids** do not appear in the expression
- **Pure liquids** do not appear in the expression
- **Water** does not appear in the expression since it is a pure liquid.

Pure solids and liquids, including water, do not appear in the equilibrium expression since their concentrations do not appreciably change. In other words, they are so very concentrated compared to the reactants and products in the dilute aqueous phase.

Never, ever forget that K is both a number and a relationship. You have to use both in order to be successful when working these problems.

What is the significance of K? Always remember, that generally $K = \dfrac{[\text{products}]}{[\text{reactants}]}$.

K > 1 means that the reaction **favors the products** at equilibrium. This means the concentration of the products is greater than the concentration of the reactants when the system reaches equilibrium.

K < 1 means that the reaction **favors the reactants** at equilibrium. This means that concentration of the reactants is greater than the concentration of the products when the system reaches equilibrium.

Example 1

Write the equilibrium expressions for the following chemical reactions:

1. $2NH_3(g) \rightleftarrows N_2(g) + 3H_2(g)$

2. $4NH_3(g) + 7O_2(g) \rightleftarrows 4NO_2(g) + 6H_2O(g)$

3. $2NO(g) + Cl_2(g) \rightleftarrows 2NOCl(g)$

FACTORS AFFECTING THE VALUE OF K

1. **Temperature:** As long as the temperature is constant, the numerical value of K is constant. Change the temperature, and you will have to calculate a new K.

2. **Changing the stoichiometric coefficients:** When the stoichiometric coefficients of a balanced equation are multiplied by some factor, the K is raised to the power of the multiplication factor, n, **(K^n)**. Doubling the coefficients means that K is squared; tripling the coefficients means that K is cubed; cutting the coefficients in half means that you raise K to the ½ power or take the square root of the original K, etc.

3. **Reversing Equations:** When the equations are reversed, simply take the reciprocal

    (1/K or K^{-1}) of K since the general $K = \dfrac{[products]}{[reactants]}$ has just been "flipped".

4. **Adding Chemical Reactions Together:** When equations are added together to create a summary equation, multiply the respective Ks of each equation together.
    ($K_{summary\ reaction} = K_1 \times K_2 \times K_3 \ldots$)

THE RELATIONSHIP BETWEEN K_P AND K_C

Since gas quantities can be expressed in either pressure units (atm) or concentration units (mol/L), we need a way to compare K_p and K_c. Think of this as the "politically correct" equation. This helps us remember which K is listed first in the relationship so we don't get it backwards! The equation, one worth memorizing, is:

$$K_p = K_c(RT)^{\Delta n}$$

Where:

Δn = the change in the number of moles of gas as the chemical reaction is read from left to right. Δn = (total moles gas produced) – (total moles gas reacting). Increasing the number moles of gas from left to right makes Δn a positive term, decreasing the number of moles of gas from left to right makes Δn a negative term.

R = universal gas law constant. Use the $0.0821 \dfrac{L \cdot atm}{mol \cdot K}$ version of R.

T = temperature in Kelvin.

From the equation you see that $K_p = K_c$ **if** Δn is equal to zero since the entire (RT) term taken to the zero power becomes ONE. That means that any time the number of moles of gaseous product = number of moles of gaseous reactant, $K_p = K_c$.

A FOOL PROOF METHOD FOR SOLVING EQUILIBRIUM PROBLEMS:
THE RICE-TABLE METHOD

This method of problem-solving allows you to see relationships quickly and formulate a plan of attack when faced with an equilibrium problem.

THE STEPS:
Set up your "RICE" table so that the letters **RICE** are on separate lines of your paper at the left margin:

R — stands for the chemical **Reaction**. Be sure it is balanced. It can be net ionic as well as molecular.

I — stands for **Initial** concentration [mol/L] or initial pressures in atmospheres if solving for Kp. You may have to convert information given into molarities or use a gas law to calculate a pressure in atmospheres.

C — stands for **Change** in concentration [mol/L] or other amounts so that this data matches your initial data. We let "x" represent unknown quantities.

E — stands for **Equilibrium** concentrations or other amounts. This line represents the sum of lines **I** and **C**.

From the balanced equation, write your equilibrium expression below your RICE table. Remember to omit any pure solids or liquids that may appear in the equation.

Place the initial concentrations or amounts under the appropriate reactant or product in the balanced equation, forming columns for that reactant or product. It is not uncommon for some of the "players" in the balanced reaction to start with initial concentrations or amounts of zero.

Calculate the change in concentration or amount using the coefficients from the balanced equation as the coefficients on "x". If you lose 2 CO_2, then you would write -2x as its change in concentration or amount. If you gain 2 CO_2, then you would write +2x as its change in concentration or amount.

Add the "**I**" and the "**C**" rows together under each reactant or product to get an expression for the "**E**" row. Simplify the expression whenever possible.

Substitute these simplified expressions into the K expression and calculate "x".

Return to the table and plug the value for "x" into the "**E**" line to calculate an exact concentration or amount.

Example 2

Consider the following reaction:

$$H_{2\,(g)} + CO_{2\,(g)} \rightleftharpoons H_2O_{\,(g)} + CO_{\,(g)}$$

When $H_2(g)$ is mixed with $CO_2(g)$ at 1,000 K, equilibrium is achieved according to the equation above. In one experiment, the following equilibrium concentrations were measured.

[H_2] = [CO_2] = 0.40 mol/L
[H_2O] = 0.20 mol/L
[CO] = 0.30 mol/L

(a) Using the equilibrium concentrations given above, calculate the value of Kc, the equilibrium constant for the reaction.

Solution: This one is an easy start. The equilibrium concentrations were given in the problem so, proceed directly to the equilibrium constant expression.

$$K = \frac{[H_2O][CO]}{[H_2][CO_2]} = \frac{[0.20][0.30]}{[0.40][0.40]} = \frac{[0.06]}{[0.16]} = 0.38$$

(b) Determine Kp, in terms of Kc for this system.

Solution: Time for the politically correct equation!

$$\mathbf{K_p = K_c(RT)^{\Delta n}}$$

$$K_p = 0.38 \left[\left(0.0821 \frac{L \cdot atm}{mol \cdot K} \right) (1,000 + 273K) \right]^0 = 0.38 \times 1.0 = 0.38$$

Remember, K_p only equals K_c when Δn is equal to zero since there is no change in the number of moles of gas between the products and reactants. It's quite alright to simply state $K_p = K_c$ as long as you <u>also</u> state that Δn is equal to ZERO.

Example 3

In a second experiment, the system in Example 2 is heated from 1,000 K to a higher temperature and the value of the equilibrium constant K_c, changes. At this new temperature, $K_c = 1.5$. Calculate the equilibrium concentrations for all of the reactants and products involved in the reaction if the initial concentrations of hydrogen and carbon dioxide are both equal to 0.50M.

Solution: Construct a RICE table and insert the initial concentrations.

R:	$H_{2(g)}$	+	$CO_{2(g)}$	$\rightleftharpoons$	$H_2O_{(g)}$	+	$CO_{(g)}$
I:	0.50 M		0.50 M		0		0
C:	$-x$		$-x$		$+x$		$+x$
E:	$0.50-x$		$0.50-x$		$0+x$		$0+x$

Note:
"I" + "C" = "E" for each reactant and product in the RICE table.

Plug the new equilibrium values into the K expression and set it equal to the value of K given at this new temperature:

$$K = \frac{[H_2O][CO]}{[H_2][CO_2]} = \frac{[x][x]}{[0.50-x][0.50-x]} = \frac{[x]^2}{[0.50-x]^2} = 1.5$$

$$\therefore \sqrt{\frac{[x]^2}{[0.50-x]^2}} = \sqrt{1.5}$$

$$\frac{x}{0.50-x} = 1.2$$

$$x = 1.2(0.50 - x)$$

$$2.2x = 0.60$$

$$\therefore x = [0.27] = [H_2O] = [CO]$$

$$and \ [H_2] = [CO_2] = 0.50 - x = 0.50 - 0.27 = [0.23]$$

Example 4

K_c for the decomposition of nitrogen monoxide is 2.5×10^{-5} at 1,700 K.

$$2\,NO(g) \rightleftharpoons N_2(g) + O_2\,(g)$$

A 0.015 mol sample of nitrogen monoxide is placed in a 12.0-L flask and heated to 1,700 K. Calculate the equilibrium concentrations of all reactants and products once equilibrium has been reached.

(a)

R: $2\,NO(g) \qquad \rightleftharpoons \qquad N_2(g) \quad + \quad O_2(g)$

I: $\dfrac{0.015\ mol}{12.0\ L} = 0.00125 \qquad\qquad 0 \qquad\qquad 0$

C: $-2x \qquad\qquad\qquad\qquad +x \qquad +x$

E: $0.00125 - 2x \qquad\qquad\qquad x \qquad\quad x$

$$K_p = \frac{[N_2][O_2]}{[NO]} = 2.5 \times 10^{-5} = \frac{[x][x]}{[0.00125 - 2x]} = \frac{\left[x^2\right]}{[0.00125 - 2x]}$$

$2.5 \times 10^{-5}\,(0.00125 - 2x) = x^2$

$3.125 \times 10^{-8} - 5.0 \times 10^{-5}x = x^2$ set the equation equal to zero with the value for x^2 being positive

$0 = x^2 + 5.0 \times 10^{-5}x - 3.125 \times 10^{-8}$ now you must use the quadratic formula to solve for x.

It is suggested that you use a graphing calculator with a program that solves the quadratic formula. You will get two roots, usually one will be negative, which is ridiculous since x represents a concentration. Regardless, only one of the roots will "make sense" when inserted back into your RICE table.

$x = [N_2] = [O_2] = 1.53 \times 10^{-4}\ \dfrac{mol}{L}$

$[NO] = 0.00125 - 2x = 0.00122 = 1.2 \times 10^{-3}\ \dfrac{mol}{L}$

PURPOSE
In this activity you will learn to write equilibrium constant expressions and solve equilibrium problems using the RICE-table method.

MATERIALS
calculator   paper and pencil

PROCEDURE
1. Solve the problems found on your student answer page. Be sure to show all work paying attention to the proper use of significant digits and units.

Name _____

Period _____

General Chemical Equilibrium
Solving Equilibrium Problems
Using the RICE-Table Method

CONCLUSION QUESTIONS

Show all work and pay special attention to significant digits and units where appropriate.

1. Write an equilibrium constant expression for each of the following unbalanced reactions:
   First, balance each equation, and then write the equilibrium expression leaving out pure solids and pure liquids.

   a. $PCl_{5(g)} \rightleftharpoons PCl_{3(g)} + Cl_{2(g)}$

   b. $NH_4HS_{(s)} \rightleftharpoons H_2S_{(g)} + NH_{3(g)}$

   c. $NH_4NO_{3(s)} \rightleftharpoons NH_4^+{}_{(aq)} + NO_3^-{}_{(aq)}$

   d. $HCOOH_{(aq)} + H_2O_{(\ell)} \rightleftharpoons H_3O^+{}_{(aq)} + HCOO^-{}_{(aq)}$

   e. $Bi_2S_{3(s)} \rightleftharpoons 2\,Bi^{3+}{}_{(aq)} + 3\,S^{2-}{}_{(aq)}$

2. The equilibrium constant K_p for the following reaction is 11.5 at 300°C when the amounts of reactant and products are given in atmospheres. Suppose a tank initially contains PCl_5 with a pressure of 3.00 atm at 300°C.

   (a) What is the partial pressure of chlorine gas once equilibrium has been established?

   (b) What is the value of K_c for this reaction at 300°C?

   $$PCl_{5(g)} \rightleftharpoons PCl_{3(g)} + Cl_{2(g)}$$

3. An aqueous solution of ethanol and acetic acid, each with a concentration 0.810 M, is heated to 125°C. At equilibrium, the acetic acid concentration is 0.645 M.

   (a) Calculate K at 125°C for the reaction.

   $$\underset{\text{ethanol}}{C_2H_5OH_{(aq)}} + \underset{\text{acetic acid}}{CH_3COOH_{(aq)}} \rightleftharpoons H_2O_{(\ell)} + \underset{\text{ethyl acetate}}{CH_3CO_2C_2H_{5(aq)}}$$

(b)   What is K for the following reaction that occurs at 125°C?

$$2CH_3CO_2C_2H_{5(aq)} + 2H_2O_{(\ell)} \rightleftharpoons 2CH_3COOH_{(aq)} + 2C_2H_5OH_{(aq)}$$

4. The equilibrium constant for the following reaction

$$H_2(g) + I_2(g) \rightleftharpoons 2HI(g)$$

is determined to be 57.85 at 450°C.

(a)   If 1.50 mol of each reactant is placed in a 2.00-L flask at 450°C what are the concentrations of H_2, I_2, and HI when equilibrium has been achieved?

(b)   What is K_p for this reaction at 450°C?

5. Ammonium hydrogen sulfide decomposes on heating.

$$NH_4HS_{(g)} \rightleftharpoons H_2S_{(g)} + NH_{3(g)}$$

(a)   If K_p is 0.11 at 25° C when the partial pressures are expressed in atmospheres, what is the total pressure in the flask at equilibrium?

(b)   What is K_c for this reaction at 25°C?

6. The following equilibrium constants are given at 500 K:

$H_2(g) + Br_2(g) \rightleftharpoons 2\ HBr(g)$ $K_p = 7.9 \times 10^{11}$

$H_2(g) \rightleftharpoons 2H(g)$ $K_p = 4.8 \times 10^{-41}$

$Br_2(g) \rightleftharpoons 2Br(g)$ $K_p = 2.2 \times 10^{-15}$

(a)   Calculate K_p for the reaction of H atoms and Br atoms to give HBr at 500 K.

(b)   Calculate K_c for this reaction at 500K.

7. The equilibrium constant for the dissociation of iodine molecules to iodine atoms is 3.76×10^{-3} at 1000 K.

$$I_2(g) \rightleftharpoons 2\,I(g)$$

(a) Suppose 0.150 mole of I_2 is placed in a 15.5-L flask at 1000K. What are the concentrations of I_2 and I when the system comes to equilibrium?

(b) What is K_p for this reaction at 1000 K?

8. K_c for the decomposition of ammonium hydrogen sulfide is 1.8×10^{-4} at 15°C.

$$NH_4HS(s) \rightleftharpoons NH_3(g) + H_2S(g)$$

(a) When 5.00 grams of the pure salt decomposes in a sealed 3.0-L flask at 15°C, what are the equilibrium concentrations of NH_3 and H_2S?

(b) What are the equilibrium concentrations at 15°C if 10.0 grams of the pure salt decomposes in the sealed flask?

9. Hemoglobin (Hb) can form a complex with both O_2 and CO. For the reaction

$$HbO_2(aq) + CO(g) \rightleftharpoons HbCO(aq) + O_2(g)$$

at body temperature, K is about 2.0×10^2. If the ratio $\dfrac{[HbCO]}{[HbO_2]}$ comes close to 1.0, death is probable.

(a) What partial pressure of CO in the air is likely to be fatal? Assume the partial pressure of O_2 is 0.20 atm.

(b) What is K for the reverse reaction?

10. Lexan is a plastic used to make compact disks, eyeglass lenses, and bullet-proof glass. One of the compounds used to make Lexan is phosgene ($COCl_2$), an extremely poisonous gas. Phosgene decomposes by the reaction

$$COCl_2(g) \rightleftharpoons CO(g) + Cl_2(g)$$

for which $K_p = 7.2 \times 10^{-11}$ at 80°C.

(a) If pure phosgene at an initial pressure of 1.0 atm decomposes, calculate the equilibrium pressures of all species.

(b) What is Kc for this reaction at 80°C?

Disturbing Equilibrium
Observing Le Chatelier's Principle

TEACHER PAGES

OBJECTIVE
Students will observe the effect of a change to conditions on a system at equilibrium and then explain the effects observed by application of Le Chatelier's Principle.

LEVEL
Chemistry

NATIONAL STANDARDS
UCP.1, UCP.2, UCP.4, A.1, A.2, B.3

TEKS
1(A), 2(B), 2(D), 2(E), 5(A), 11(B), 12(A), 13(A)

CONNECTIONS TO AP
AP Chemistry:
   III. Reactions C. Equilibrium 1. Concept of dynamic equilibrium, physical and chemical; Le Chatelier's principle; equilibrium constants

TIME FRAME
45 minutes

MATERIALS
(For a class of 28 working in pairs)

saturated potassium nitrate solution	potassium nitrate crystals
0.10 M potassium thiocyanate	0.10 M iron III chloride
potassium chloride crystals	deionized water
ice	14 goggles
84 test tubes	14 droppers
14 test tube racks	14 100 mL beakers
14 50 mL graduated cylinders	14 250 mL beakers
14 scoopulas	14 white sheets of paper

TEACHER NOTES

Preparation of Solutions:

0.1 M KNO_3 – place 5.06 g of KNO_3 into a 500 mL volumetric flask and add distilled water to the 500 mL mark.

0.1 M KSCN – place 4.86 g of KSCN into a 500 mL volumetric flask; add distilled water to the 500 mL mark.

0.1 M $FeCl_3$ – place 8.11 g of $FeCl_3$ into a 500 mL volumetric flask; add distilled water to the 500 mL mark.

Saturated solution of KNO_3 – place a 500 mL beaker on a magnetic stir plate and continue adding potassium nitrate until crystals no longer dissolve. Test the solution for saturation by running part I of the experiment.

Students should read the lab and complete the pre-lab questions prior to this class period. As students enter the classroom, check for completion of the pre-lab and then briefly discuss any questions students may have before the lab begins.

This lab is set up to be a partial inquiry lab involving the concept of Le Chatelier's Principle. The students will initially perform only Part I of the lab including the corresponding conclusion questions. The lab instructs them to stop and wait for your discussion. During the discussion time make sure that students understand how and why the reaction shifted when various stresses were applied to the system. The dialogue will naturally lead into a complete discussion of Le Chatelier's Principle. Students will be responsible for taking notes in the space provided on their student answer page during this discussion.

Student Notes following Part I:

Le Chatelier's Principle states that "a system at equilibrium, or changing toward equilibrium, responds in the way that tends to relieve or "undo" any stress placed upon it". In other words, the reaction will shift its equilibrium position in the direction that minimizes the change.

We can use Le Chatelier's Principle to make predictions as to how a system will react in response to changes in external conditions. Three common methods for changing the chemical equilibrium are: 1) adding or removing reactant or product, 2) changing the pressure, and 3) changing the temperature.

Changing Concentrations of Reactants or Products

Consider the following example for changing reactant or product concentrations, $N_{2(g)} + H_{2(g)} \leftrightarrow 2NH_{3(g)}$. If we add more hydrogen gas to the system, equilibrium will shift so that the hydrogen concentration would be reduced and move toward its original value. In other words, the equilibrium would shift toward producing more ammonia (product) to re-establish equilibrium. Likewise, if we add more ammonia to the system at equilibrium, the system will shift left toward making more reactants. The ammonia will decompose to produce more nitrogen and hydrogen gas. If we were to remove NH_3, the system will adjust by producing more ammonia.

Effects of Volume and Pressure Changes

If a system is at equilibrium and the pressure is increased by reducing the volume, Le Chatelier's Principle indicates that the system will shift its equilibrium position to reduce the pressure. Therefore, at constant temperature, equilibrium will shift in the direction of the least number of moles of gas. Less moles of gas translate into fewer collisions of gas particles which in turn would mean less pressure.

Effect of Temperature Changes

When the temperature is decreased, equilibrium will shift in the direction that produces heat. Therefore if we had an exothermic reaction: Reactants ↔ products + heat, the equilibrium would shift in the direction that produces heat. Since the heat is on the product side, the equilibrium would shift to the right or in favor of the products. When the temperature is increased, equilibrium will shift in the direction that absorbs the heat, in this case to the left or toward the reactants.

Post Lab

After completing the lab exercise, the teacher should follow up with students by asking questions about what they experienced in the previous lab. They should be able to describe the effects of temperature and concentration changes. The notes during this post-lab discussion should include the effects of changing pressure, adding a catalyst and addition of an inert gas on the equilibrium position. A one-page handout modeled from a retired AP exam question would be a great summation to these activities.

REFERENCES:

Brown, LeMay, and Bursten, 8th edition. *Chemistry the Central Science*. Prentice Hall, New Jersey, 1997.

Whitten, Davis, and Peck, 5th edition. *General Chemistry*. Saunders College Publishing, New York, 1996.

POSSIBLE ANSWERS TO THE CONCLUSION QUESTIONS AND SAMPLE DATA

Systems	Observations
Saturated KNO_3 cooled	More crystals have formed at the bottom of the tube
Saturated KNO_3 warmed	The crystals disappeared
Fe^{3+} / SCN^- mixture	Turned a yellowish-orange color
Fe^{3+} / SCN^- mixture + additional Fe^{3+}	Turned a darker reddish brown
Fe^{3+} / SCN^- mixture + additional SCN^-	Turned a darker reddish brown
Fe^{3+} / SCN^- mixture + KCl	Little lighter yellow

PRE-LAB QUESTIONS

1. Define reversible reaction:
   - The reaction may take place in the forward or the reverse direction.

2. Define irreversible reaction:
   - The reaction only occurs in the forward direction.

3. When studying a written equation, what will be the indication that the reaction is reversible?
   - Arrows pointed in both directions.

4. Define dynamic equilibrium.
   - Changes are occurring within the system yet there is no net visible change observed.

5. Explain what will happen to the equilibrium reaction shown in equation 1 if a stress is placed on the system such as a temperature change or a change in concentration?
   - The system will adjust until it reaches equilibrium again.

CONCLUSION QUESTIONS

** A balanced equation for the equilibrium that existed before the saturated potassium nitrate was cooled is:

$$KNO_{3(s)} \leftrightarrow KNO_{3(aq)}$$

1. Describe what happened to the potassium nitrate when placed in the cold water?
   - It formed crystals.

2. Refer to the equation above. Based on your observations, did the equilibrium shift toward the reactants or toward the products when the temperature was cooled? ** hint, look at the phases of reactants and products in the above equation. Rewrite the equation including the energy term.
   - Toward the reactants because more solid crystals were formed and solid potassium nitrate is located on the reactant side of the equation
   - $KNO_{3(s)}$ + heat $\leftrightarrow$ $KNO_{3(aq)}$

3. What visible evidence was present to prove that the system had re-established equilibrium just before removing it from the ice water?
   - The formation of crystals had stopped.

4. In which direction did the equilibrium shift when you again increased the temperature? Explain your answer.
   - It shifted to the products. The crystals returned to aqueous solution.

5. Was there a stress applied to the system? If so, what was the stress?
   - There was a temperature change.

6. Using Le Chatelier's Principle, explain what happened in the potassium nitrate system when the solution was placed in ice and then returned to room temperature.
   - The decrease in temperature caused the equilibrium to shift toward the reactants and produce more reactant. The increase in temperature caused the equilibrium to shift toward the products and produce more reactants

** A balanced equation for the equilibrium that existed after the iron (III) and thiocyanate ions were combined in the beaker is: $Fe^{+3}_{(aq)}$ + $SCN^-_{(aq)}$ $\leftrightarrow$ $[FeSCN]^{2+}_{(aq)}$

The actual complex formed is: $[Fe(H_2O)_5(NCS)]^{2+}$ with an octahedral shape—however, for simplicity we will use the version written for the students.

7. What explanation can be given for writing the iron and thiocyanate reaction as it is written?
   - The equation shows only those ions involved in the reaction. The ones not shown are spectator ions.

8. Cite evidence proving that there was a shift in equilibrium when iron (III) chloride was added to the test tube? In which direction did it shift? Justify your answer.
   - The solution got darker and more concentrated.

- It shifted towards the products producing more $[FeSCN]^{2+}$
- Increasing the concentration of a reactant will cause a shift to the products.

9. Cite evidence proving that there was a shift in equilibrium when potassium thiocyanate was added to the test tube? In which direction did it shift? Justify your answer.
    - The solution got darker and more concentrated.
    - It shifted towards the products producing more $[FeSCN]^{2+}$
    - Increasing the concentration of a reactant will cause a shift to the products.

10. Explain the effect that the addition of potassium chloride had on the system.
    - None. K^+ and Cl^- were spectator ions. The solution got lighter due to the dilution.

11. Explain the changes observed in the thiocyanato iron (III) ion system in terms of Le Chatelier's Principle.
    - The increase in concentration of reactant ions caused a shift in equilibrium to produce more products.

Disturbing Equilibrium
Observing Le Chatelier's Principle

Chemical reactions can be described as reversible or irreversible. A reversible reaction is one in which the products can react to give the reactants, just as the reactants can react to give the products. Another way of saying this is that both the forward and reverse reactions can occur. A common example of a reversible reaction is the reaction between iron (III) ion and thiocyanate ion to produce a complex ion known as isothiocyanatoiron (III) ion. Refer to equation 1.

Equation 1: $Fe^{+3} + SCN^- \leftrightarrow [FeSCN]^{2+}$

An irreversible reaction is a reaction in which the products do not react to give the reactants once again. A familiar example is the burning of wood. The products are carbon ash, carbon dioxide, water and a number of other gaseous and liquid substances. These products do not recombine to form wood. The reaction proceeds only in the forward direction.

Consider the reversible reaction shown in Equation 1 above. The rate of the forward reaction depends on the temperature and the concentrations of the reactants, Fe^{+3} and SCN^-. Similarly, the rate of the reverse reaction depends on the temperature and the concentration of $[FeSCN]^{+2}$. What will happen if the system reaches equilibrium? Both reactions will continue to occur, but the net effect will produce no change in the concentrations of the reactants or products. For every pair of Fe^{+3} and SCN^- ions that combine to form $[FeSCN]^{+2}$, one will break apart to give Fe^{+3} and SCN^-. Consequently, although changes are occurring constantly within the system, no net change will be observed. This condition is known as a state of dynamic equilibrium.

PURPOSE
In this activity you will observe the effect of changing conditions on two different equilibrium systems and explain the observed effects by application of Le Chatelier's Principle.

MATERIALS

saturated potassium nitrate solution

0.10 M potassium thiocyanate

potassium chloride crystals

ice

6 test tubes

test tube rack

50 mL graduated cylinder

scoopula or spoon

potassium nitrate crystals

0.10 M iron III chloride

deionized water

goggles

droppers

100 mL beaker

250 mL beaker

white sheet of paper

Safety Alert

1. Iron (III) chloride and potassium iodide are irritants. Iron (III) chloride will also stain clothes so make sure to wear an apron
2. Potassium thiocyanate is slightly toxic by ingestion and an irritant.

PROCEDURE
PART 1: EFFECT OF TEMPERATURE ON PHYSICAL EQUILIBRIUM
NOTE: This is a qualitative study. The amounts given are just estimates. Remember, a quick estimate of a 1mL in a test tube is about one pinky-finger width (20 drops). A "small amount" of a solid is about the size of a green pea.

1. Answer the pre-lab questions on your student answer page.

2. Add 2 mL of saturated potassium nitrate solution to a clean test tube. Using a scoopula, add a small amount of potassium nitrate crystals to act as seed crystals. (These crystals should not dissolve; if they do, inform the teacher that the solution is not saturated.)

3. Cool the test tube in a beaker of ice water for 10 minutes. Watch the test tube closely for the first five minutes. Then observe once every minute for the remaining five minutes. Record whether the number of crystals increased, decreased or remained the same in the data table.

4. Remove the tube from the ice water bath and place it in the test tube rack. Record what happens as the solution warms to room temperature. Record whether the number of crystals increased, decreased or remained the same in the data table.

5. Answer conclusion questions 1-5 on your student answer page.

STOP FOR CLASS NOTES ONCE THIS SECTION IS COMPLETE!

PART II: CONCENTRATION OF IONS – EFFECT ON CHEMICAL EQUILIBRIUM

1. Using a graduated cylinder, measure 50.0 mL of distilled water and pour this volume into a 100 mL beaker. Add 1 mL of 0.1 M iron (III) chloride, and 1 mL of 0.1 M potassium thiocyanate to the beaker with the water. Mix the solution by swirling. The reaction that you see is described by Equation 1 in the introduction to this experiment. The color that appears is due to the presence of the isothiocyanatoiron (III) ions, $[FeSCN]^{+2}$. Record your observations in data table 1.

2. Pour about 5 mL of the mixture from step 1 above into each of four identical clean test tubes numbered 1-4. Hold the tubes in front of a white sheet of paper and observe the color. All four test tubes should appear the same color.

3. Tube 1 will serve as the control in this experiment. To tube 2, add 20 drops of 0.1 M iron (III) chloride. To tube 3, add 20 drops of 0.1 M potassium thiocyanate. Tap each tube to mix the solutions. To tube 4, add a small amount of solid potassium chloride and tap the tube to dissolve the crystals.

4. Answer the remaining conclusion questions on the student answer page.

Name _____

Period _____

Disturbing Equilibrium
Observing Le Chatelier's Principle

DATA AND OBSERVATIONS

Systems	Observations
Saturated KNO_3 cooled	
Saturated KNO_3 warmed	
Fe^{3+} / SCN^- mixture	
Fe^{3+} / SCN^- mixture + additional Fe^{3+}	
Fe^{3+} / SCN^- mixture + additional SCN^-	
Fe^{3+} / SCN^- mixture + KCl	

PRE-LAB QUESTIONS

1. Define reversible reaction:

2. Define irreversible reaction:

3. When studying a written equation, what will be the indication that the reaction is reversible?

4. Define dynamic equilibrium.

5. Explain what will happen to the equilibrium reaction shown in equation 1 if a stress is placed on the system such as a temperature change or a change in concentration?

CLASS NOTES

CONCLUSION QUESTIONS

** A balanced equation for the equilibrium that existed before the saturated potassium nitrate was cooled is:

$$KNO_{3(s)} \leftrightarrow KNO_{3(aq)}$$

1. Describe what happened to the potassium nitrate when placed in the cold water?

2. Refer to the equation above. Based on your observations, did the equilibrium shift toward the reactants or toward the products when the temperature was cooled? ** hint, look at the phases of reactants and products in the above equation. Rewrite the equation including the energy term.

3. What visible evidence was present to prove that the system had re-established equilibrium just before removing it from the ice water?

4. In which direction did the equilibrium shift when you again increased the temperature? Explain your answer.

5. Was there a stress applied to the system? If so, what was the stress?

6. Using Le Chatelier's Principle, explain what happened in the potassium nitrate system.

** A balanced equation for the equilibrium that existed after the iron (III) and thiocyanate ions were combined in the beaker is: $Fe^{+3}_{(aq)} + SCN^-_{(aq)} \leftrightarrow [FeSCN]^{2+}_{(aq)}$

7. What explanation can be given for writing the iron and thiocyanate reaction as it is written?

8. Cite evidence proving that there was a shift in equilibrium when iron (III) chloride was added to the test tube? In which direction did it shift? Justify your answer.

9. Cite evidence proving that there was a shift in equilibrium when potassium thiocyanate was added to the test tube? In which direction did it shift? Justify your answer.

10. Explain the effect that the addition of potassium chloride had on the system.

11. Explain the changes observed in the isothiocyanatoiron (III) ion system in terms of Le Chatelier's Principle.

Acid-Base Equilibrium
Solving pH Problems for Weak Acids and Bases

OBJECTIVE
Students will be introduced to the concept of weak acid-base equilibrium. Students will solve acid-base equilibrium problems using the RICE table problem-solving method.

LEVEL
Chemistry

NATIONAL STANDARDS
UCP.2, UCP.3, UCP.4, A.1, A.2, B.2

TEKS
14(A)

CONNECTIONS TO AP
AP Chemistry:

   III. Reaction types A. Reaction types 1. Acid-base reactions; concepts of Arrhenius, Bronsted-Lowry C. Equilibrium 1. Concept of dynamic equilibrium, physical and chemical; equilibrium constants 2. Quantitative treatment b. Equilibrium constants for reactions in solution (1) Constants for acids, bases; pK; pH

TIME FRAME
90 minutes plus homework

MATERIALS
   calculator   paper and pencil

TEACHER NOTES
This activity is designed as a classroom lesson and will require a teacher lecture. Students should be guided through the concepts and theories of general acid-base chemistry prior to this activity. This activity should follow the lesson on general equilibrium expressions and the RICE-table method of problem solving. You should guide the students through the example problems found on the student answer pages.

There are many types of equilibrium, including: concentration, gas, acid-base and solubility. This lesson is designed as an introduction to the basic skills and concepts involving weak acid-base equilibrium. Much of what is presented here is probably absent from your high school chemistry textbook. However, the first question in the free-response section of the AP Chemistry exam is always an equilibrium problem. Your students will benefit from an introduction to chemical equilibrium in your first-year course.

ANSWERS TO THE CONCLUSION QUESTIONS
It is suggested that answers only be rounded at the end of the problem.

1. A dilute solution of household ammonia contains 0.3124 mol of NH_3 per liter of solution. The K_b of ammonia is 1.8×10^{-5} at 25°C.

    (a) Calculate the hydroxide ion concentration of the solution at 25°C.

    (b) Calculate the pH of the solution at 25°C.

    (a)

    R: $NH_3(aq) + H_2O(\ell) \rightleftharpoons NH_4^+(aq) + OH^-(aq)$

    I: $\dfrac{0.3214 \text{ mol}}{L} = 0.3124M$ $\qquad$ 0 $\qquad$ 0

    C: $-x$ $\qquad\qquad\qquad$ $+x$ $\qquad$ $+x$

    E: $0.3124 - \cancel{x}$ $\qquad\qquad$ x $\qquad$ x

    neglect x since $100K_b = 0.0018$ which is very small compared to 0.3214

    $$K_b = 1.8 \times 10^{-5} = \frac{\left[NH_4^+\right]\left[OH^-\right]}{\left[NH_3\right]} = \frac{\left[x^2\right]}{\left[0.3124\right]} \therefore x = \left[OH^-\right] = 2.371 \times 10^{-3} M$$

    (b) There are several approaches possible, but each should yield the correct answer.

    $pOH = -\log\left[OH^-\right] = -\log\left[2.371 \times 10^{-3}\right] = 2.6250 \therefore pH = 14 - 2.6250 = 11.3750$

    NOTE: Four significant digits were reported on the initial number of moles, thus 4 decimal places should be reported on pH or pOH.

2. (a) Calculate the ionization constant for HF if a 0.10M solution of HF is 8.1% ionized at 25°C.

(b) Calculate the pH of the solution.

(a)

R: $HF(aq) \rightleftharpoons H^+(aq) + F^-(aq)$

I:	0.10M	0	0
C:	$-x$	$+x$	$+x$
E:	$0.10 - 0.0081$	$0.0081*$	0.0081
	$= 0.092$		

It is given that HF ionizes 8.1%, so solve for $\left[H^+\right]$ at equilibrium:

$$8.1\% = \frac{\left[H^+\right]}{\left[HF\right]} \times 100 \therefore 0.081 = \frac{\left[H^+\right]}{\left[HF\right]} = 0.081 = \frac{\left[H^+\right]}{\left[0.10\right]}$$

$\left[H^+\right] = 0.0081 M*$ plug into the RICE table above as the equilibrium concentration, x.

$$K_a = \frac{\left[H^+\right]\left[F^-\right]}{\left[HF\right]} = \frac{\left[0.0081\right]^2}{\left[0.092\right]} = 7.1 \times 10^{-4}$$

(b) $pH = -\log\left[H^+\right] = -\log\left[0.0081\right] = 2.09$

NOTE: Two significant digits were reported on the initial molarity, thus two decimal places should be reported on the pH.

3. Formic acid is the irritant that causes the body's reaction to an ant's sting. It has a Ka value of 1.8×10^{-4} at 25°C.

   (a) Calculate the concentration of hydrogen ion in a 0.575M solution of formic acid at 25°C.

   (b) Calculate the percent ionization of a 0.575M solution at 25°C.

(a) It is acceptable to write HA for any weak acid and A⁻ as the anion or conjugate base of that weak acid.

Let HA = formic acid

R: HA(aq) $\rightleftharpoons$ H⁺(aq) + A⁻(aq)

I: 0.575 M   0   0

C: $-x$ $+x$ $+x$

E: $0.575 - \cancel{x}$ x x

   Neglect x since 0.575 is much larger than $100K_a$.

$$K_a = \frac{\left[H^+\right]\left[A^-\right]}{\left[HA\right]} = \frac{\left[x\right]^2}{\left[0.575\right]} = 1.8 \times 10^{-4} \therefore x = \left[H^+\right] = \left[A^-\right] = 1.04 \times 10^{-4}$$

(b) Now that $\left[H^+\right]$ is known, calculate the percent of ionization for acetic acid:

$$\% \text{ionization} = \frac{\left[H^+\right]}{\left[HA\right]} \times 100 \therefore \% \text{ ionization} = \frac{\left[1.04 \times 10^{-4}\right]}{\left[0.575\right]} \times 100 = 0.018\%$$

Laying the Foundation in Chemistry

4. Acetic acid has a K_a value of 1.8×10^{-5} at 25°C.

(a) Calculate the percent ionization of a 0.265M acetic acid solution at 25°C.

(b) Calculate the [OH⁻] of the solution at 25°C.

(a) It is acceptable to write HA for any weak acid and A⁻ as the anion or conjugate base of that weak acid.

Let HA = acetic acid.

R: HA(aq) $\rightleftharpoons$ H⁺(aq) + A⁻(aq)

I: 0.265M   0   0

C: $-x$ $+x$ $+x$

E: $0.265 - x$ x x

Neglect x since 0.265 is much larger than $100K_a$.

$$K_a = \frac{\left[H^+\right]\left[A^-\right]}{\left[HA\right]} = \frac{\left[x\right]^2}{\left[0.265\right]} = 1.8 \times 10^{-5} \therefore x = \left[H^+\right] = \left[A^-\right] = 2.18 \times 10^{-3}$$

Now that $\left[H^+\right]$ is known, calculate the percent of ionization for acetic acid:

$$\% \text{ ionization} = \frac{\left[H^+\right]}{\left[HA\right]} \times 100 \therefore \% \text{ ionization} = \frac{\left[2.18 \times 10^{-3}\right]}{\left[0.265\right]} \times 100 = 0.824\%$$

(b) $K_w = \left[H^+\right]\left[OH^-\right] = 1 \times 10^{-14} \therefore \left[OH^-\right] = \frac{1 \times 10^{-14}}{\left[2.18 \times 10^{-3}\right]} = 4.59 \times 10^{-12}$

TEACHER PAGES

5. Monochloroacetic acid, $HC_2H_2ClO_2$, is a skin irritant that is used in "chemical peels" intended to remove the top layer of dead skin from the face and ultimately improve the complexion. The value of the K_a for monochloroacetic acid is 1.35×10^{-3}.

(a) Calculate the pH of a 0.25M solution of monochloroacetic acid at 25°C.

(b) Calculate the percent ionization of a 0.25M solution of monochloroacetic acetic acid at 25°C.

(a) HA = monochloroacetic acid.

R: $HA(aq) \rightleftharpoons H^+(aq) + A^-(aq)$

I: 0.25M   0   0

C: $-x$ $+x$ $+x$

E: $0.25 - x$ x x

You can neglect x since 0.25 is larger than $100K_a$. It is a close call, but using the quadratic formula yields the exact same value for x when the sig. fig. rules are applied.

$$K_a = \frac{[H^+][A^-]}{[HA]} = \frac{[x]^2}{[0.25-x]} = 1.35 \times 10^{-3} \therefore x = [H^+] = [A^-] = 0.018$$

$$pH = -\log[H^+] = -\log[0.018] = 1.74$$

(b) Now that $[H^+]$ is known, calculate the percent of ionization for monochloroacetic acid:

$$\% \text{ ionization} = \frac{[H^+]}{[HA]} \times 100 \therefore \% \text{ ionization} = \frac{[0.018]}{[0.25]} \times 100 = 7.3\%$$

6. Codeine ($C_{18}H_{21}NO_3$) is a derivative of morphine that is used as an analgesic, narcotic or antitussive. It was once commonly used in cough syrups but is now available only by prescription because of its addictive properties.

(a) If the pH of a 6.3×10^{-3} M solution of codeine is 9.83 at 30°C, calculate the [H^+] of the solution.

(b) Calculate the K_b at 30°C

(a) If pH = 9.83, then $\left[H^+\right] = 10^{-9.83} = 1.5 \times 10^{-10}$ M

(b) Now that $\left[H^+\right]$ is known, calculate $\left[OH^-\right]$ using the K_w relationship:

$$K_w = \left[H^+\right]\left[OH^-\right] = 1 \times 10^{-14} \therefore \left[OH^-\right] = \frac{1 \times 10^{-14}}{1.5 \times 10^{-10}} = 6.7 \times 10^{-5}$$

Since the formula for codeine is not given, and you are asked for a K_b, use the generic

equation for the dissociation of a base: $BOH \rightleftharpoons B^+ + OH^-$

$$\therefore K_b = \frac{\left[B^+\right]\left[OH^-\right]}{[BOH]} = \frac{\left[6.7 \times 10^{-5}\right]^2}{\left[6.3 \times 10^{-3}\right]} = 7.1 \times 10^{-7}$$

7. The Ka of acetic acid at 25°C is $1.8 \times 10\text{-}5$.

(a) What are the equilibrium concentrations of H_3O^+, $C_2H_3O_2^-$ and $HC_2H_3O_2$ in a 0.350 M aqueous solution of acetic acid?

(b) What is the pH of the solution at 25°C?

(a) It is acceptable to write HA for any weak acid and A^- as the anion or conjugate base of that weak acid. Let HA = acetic acid.

R: $HA(aq) \rightleftharpoons H^+(aq) + A^-(aq)$

I: 0.350 M   0   0

C: $-x$ $+x$ $+x$

E: $0.350 - \cancel{x}$ x x

Neglect x since 0.350 is much larger than $100K_a$.

$$K_a = \frac{\left[H^+\right]\left[A^-\right]}{\left[HA\right]} = \frac{\left[x\right]^2}{\left[0.350\right]} = 1.8 \times 10^{-5} \therefore x = \left[H^+\right] = \left[A^-\right] = 2.50 \times 10^{-3}\,M \text{ and } \left[HA\right] = 0.350\,M$$

(b) Now that $\left[H^+\right]$ is known, calculate the pH of the solution:

$$pH = \text{-}\log\left[H^+\right] = -\log\left[2.50 \times 10^{-3}\right] = 2.600$$

8. The weak base methylamine, CH_3NH_2, has a K_b of 4.2×10^{-4} at 25°C. It reacts with water according to the equation

$$CH_3NH_2(aq) + H_2O(\ell) \rightleftharpoons CH_3NH_3^+(aq) + OH^-(aq)$$

(a) Calculate the equilibrium concentration of OH^- in a 0.15M solution of the base.

(b) What are the pH and pOH of the solution?

(a)

R: $CH_3NH_2(aq) + H_2O(\ell) \rightleftharpoons CH_3NH_3^+(aq) + OH^-(aq)$

I: 0.15 M   0   0

C: $-x$ $+x$ $+x$

E: $0.15 - \cancel{x}$ x x

neglect x since $100K_b = 0.0018$ which is very small compared to 0.3214

$$K_b = 4.2 \times 10^{-4} = \frac{\left[NH_4^+\right]\left[OH^-\right]}{\left[NH_3\right]} = \frac{\left[x^2\right]}{\left[0.15\right]} \therefore x = \left[OH^-\right] = 7.9 \times 10^{-3}\,M$$

(b) There are several approaches possible, but each should yield the correct answer.

$$pOH = -\log\left[OH^-\right] = -\log\left[7.9 \times 10^{-3}\right] = 2.10 \therefore pH = 14 - 2.10 = 11.90$$

9. A 1.00 M solution of a weak acid has a pH of 3.849 at 25°C.

(a) Calculate the $[H^+]$ in the 1.00 M solution.

(b) Calculate the K_a for the weak acid at 25°C

(a)

R: $\quad HA(aq) \rightleftharpoons H^+(aq) + A^-(aq)$

I: $\quad 1.00\ M \qquad\qquad 0 \qquad\quad 0$

C: $\quad -x \qquad\qquad\quad +x \qquad\quad +x$

E: $\quad 1.00-10^{-3.849} \qquad 10^{-3.849} \quad \therefore \quad$ also $10^{-3.849}$

$\therefore \left[H^+\right] = \left[A^-\right] = 1.42\times10^{-4}$

(b) $\ K_a = \dfrac{\left[H^+\right]\left[A^-\right]}{[HA]} = \dfrac{\left[10^{-3.849}\right]^2}{\left[1.00-10^{-3.849}\right]} = 2.00\times10^{-8}$

10. Phosphoric acid has a K_{a1} of 7.5×10^{-3} at $25°$

   (a) Calculate the $[H_2PO_4^{2-}]$ present at equilibrium in a 0.35 M solution.

   (b) Calculate the pH of the solution at 25°C.

   (a) It is acceptable to write HA for any weak acid and A^- as the anion or conjugate base of that weak acid. Let HA = phosphoric acid and A^- = dihydrogenphosphate ion.

R:	HA(aq)	$\rightleftharpoons$	H^+(aq)	+	A^-(aq)
I:	0.35 M		0		0
C:	$-x$		$+x$		$+x$
E:	$0.35 - x$		x		x

You cannot neglect x since 0.35 is less than $100K_a$ or 0.75.

$$K_a = \frac{\left[H^+\right]\left[A^-\right]}{[HA]} = \frac{[x]^2}{[0.35-x]} = 7.5 \times 10^{-3}$$

$$7.5 \times 10^{-3}(0.35-x) = x^2$$

$$0 = x^2 + 7.5 \times 10^{-3}x - 2.6 \times 10^{-3}$$

$$\therefore x = \left[H^+\right] = \left[A^-\right] = 0.047\,M$$

   (b) Now that $\left[H^+\right]$ is known, calculate the pH of the solution:

$$pH = -\log\left[H^+\right] = -\log[0.047] = 1.33$$

Acid-Base Equilibrium
Solving pH Problems for Weak Acids and Bases

How do we classify a compound as an acid or base? Recall that the word **acid** is from the Latin word *acidus* which means sour. The word **alkali** is an Arabic word for the ashes that come from burning certain plants and water solutions of these materials feel slippery and taste bitter. (Of course, you should *never* taste chemicals in the laboratory.) In this activity, we explore several different kinds of equilibrium constants, K. We will discuss K_a, K_b, and K_w. But, first it is important that we do a quick review of general acid-base theory. There are three major acid-base theories and you should know each by name:

ARRHENIUS THEORY
acid—a substance that produces a proton (H^+) in water forming the hydronium ion, H_3O^+ in aqueous solutions
base—a substance that produces an hydroxide ion in water (OH^-); note that ammonia is a MAJOR exception since it is a base, but has no OH^- to produce!

BRÖNSTED-LOWRY THEORY
acid—a substance that produces a proton (H^+) in water forming the hydronium ion, H_3O^+ in aqueous solutions
base—a substance that accepts a proton in water; this better explains the basic behavior of ammonia since $NH_3 + H^+ \rightleftharpoons NH_4^+$, the ammonium ion in aqueous solutions

LEWIS THEORY
acid—a substance that accepts an electron pair into an empty orbital (think "acids accept"); BF_3 is the classic example of a Lewis acid.
base—a substance that donates an electron pair.

Lewis acid   Lewis base

Additional acid-base terms to know:

monoprotic—describes acids donating one H^+ such as HCl, $HC_2H_3O_2$, or HNO_3
diprotic—describes acids donating two H^+ ions such as H_2SO_4
polyprotic—describes acids donating more than one H^+ ions such as H_3PO_4
amphiprotic—describes a substance that can act as either an acid or a base. This means it can either lose a proton or gain one. Water can do this to form hydroxide or hydronium ions. Other examples include HCO_3^-, HSO_4^-, HPO_4^{2-}
amphoteric—describes an oxide that shows both acidic and basic properties. Soluble aluminum compounds are famous for this behavior.

ACID-BASE DISSOCIATION CONSTANTS

IF an acid or base is strong, it completely dissociates and we assume the reaction goes to completion. IF an acid or base is weak, it does not dissociate completely and equilibrium is established. *The vast majority of acids and bases are weak.* Their equilibrium constants, K, are referred to as acid or base dissociation constants, K_a and K_b respectively. You will have to use the K_a or K_b to determine the pH in an equilibrium problem, or vice versa.

Many common weak acids are oxyacids, like phosphoric acid and nitrous acid. Other common weak acids are organic acids—those that contain a carboxyl group, –COOH, like acetic acid and benzoic acid. Acetic acid is written $HC_2H_3O_2$ or CH_3COOH or H_3CCOOH if you really want to emphasize the bonding. It simply means a methyl [–CH₃] group is attached to the carboxylic acid group shown below.

$$-C\diagdown\!\!\!\!\!\underset{O-H}{\overset{O}{\diagup}}$$

The generalized chemical reaction showing the dissociation of a weak acid is:

$$HA + H_2O \rightleftharpoons H_3O^+ + A^- \quad \text{or more simply} \quad HA \rightleftharpoons H^+ + A^-$$

"A⁻" is the anion of the acid, which is whatever remains once the proton is removed. It is important to remember that K is a relationship (the equilibrium expression) and a number. For weak acids, the number is much less than one since the equilibrium position is FAR to the left.

$$K_a = \frac{[H^+][A^-]}{[HA]} \quad <<< 1$$

There is no need to memorize any K_a values.

Weak bases that are not Arrhenius bases (without OH⁻) react with water to produce a hydroxide ion. Common examples of weak non-Arrhenius bases are ammonia (NH_3), methylamine (CH_3NH_2), and ethylamine ($C_2H_5NH_2$). The lone pair of electrons on N forms a bond with H^+. Most weak bases involve N. No need to memorize these, just have an awareness.

The generalized chemical reaction showing the dissociation of a weak base is:

$$B + H_2O \rightleftharpoons HB^+ + OH^-$$

You may want to write the water as HOH, so that it is easier to see that the base is accepting a proton and leaving the hydroxide ion behind.

$$B + HOH \rightleftharpoons HB^+ + OH^-$$

The equilibrium expression for the dissociation of a weak base is known as the K_b.

$$K_b = \frac{[HB^+][OH^-]}{[B]} \quad <<< 1$$

THE SIGNIFICANCE OF K_W

Around 1900 a man named Fredrich Kohlrausch found that no matter how pure a sample of water, it still conducts a minute amount of electric current. In order to do this water must be able to self-ionize. The water molecule is **amphiprotic**, meaning it can behave as either an acid or a base. It may dissociate with itself to a slight extent. *Only about 2 out of a billion water molecules are ionized at any instant!*

$$H_2O_{(l)} \quad + \quad H_2O_{(l)} \quad \rightleftharpoons \quad H_3O^+_{(aq)} \quad + \quad OH^-_{(aq)}$$

The equilibrium expression for the above reaction is referred to as K_w, **the ionization constant for water**. K_w is also the relationship between the acid and base dissociation constants of a conjugate acid-base pair.

$$K_w = [H_3O^+][OH^-] = K_a \times K_b = 1.0 \times 10^{-14} \quad @ \; 25°C$$

THE PH SCALE

K_w is the entire basis for the pH scale. The scale is really intended to communicate the strengths of weak acids and bases. The pH of a solution is calculated as the negative base 10 logarithm of the hydronium ion concentration or simply the $[H^+]$:

$$pH = -\log [H^+] \quad \text{which also means that } [H_3O^+] \text{ or } [H^+] = 10^{-pH}$$

Since $K_w = [H^+][OH^-] = 1 \times 10^{-14}$ and a pure sample of water has $[H^+] = [OH^-]$ and each is equal to 1×10^{-7}, we can now account for the fact that the pH of a neutral solution is 7 and that the range of the entire scale is 0-14.

$$\text{most acidic} \quad \textit{neutral} \quad \text{most basic}$$
$$0 \quad \leftarrow \quad 7.0 \quad \rightarrow \quad 14$$

If more hydrogen ion is present, $[H^+] > [OH^-]$, and the pH is acidic (less than 7). If more hydroxide ion is present, $[H^+] < [OH^-]$, and the pH is basic (greater than 7).

The pOH of a solution is calculated as the negative base 10 logarithm of the hydroxide ion concentration:

$$pOH = -\log [OH^-]$$

and best yet, pH + pOH = 14

The pH scale is a logarithmic scale which means that a difference of two units (an acid with a pH of 3 versus a pH of 5) is a difference of 10^2 units, or 100, in terms of the strength of the acid or base. So, an acid with a pH 3 is 100 times more acidic than an acid with a pH of 5. Logarithmic numbers have a characteristic and a mantissa. The characteristic is the part to the left of the decimal while the mantissa is the part to the right of the decimal. The characteristic is a place holder while the mantissa is communicating the accuracy of the measurement. For this reason, the decimal places communicate the significant figures. If you are given a pH of 3.58 in a problem and are asked to solve for the concentration of the weak acid, you should report its concentration to two significant figures. Conversely, if you are asked to find the pH of a weak base and given its concentration as 1.25M, then you should report the pH with three decimal places EVEN if it comes out to be something like 10.565.

CALCULATING THE PH OF ACIDS AND BASES

First, determine if the acid or base in the question is strong or weak. If it is strong, take the negative log of the acid or base concentration to calculate either the pH or pOH. If it is weak, you enter the land of equilibrium problems and a RICE table is in order. Always start by writing the balanced equation, setting up the acid equilibrium expression (K_a), or the base equilibrium expression (K_b), and defining initial concentrations, changes, and final concentrations in terms of "x". Once you have done this substitute values and variables into the K_a or K_b expression and solve for "x".

Example 1
Calculate the pH of a 0.015 M HCl solution.

Solution: Since this is a strong acid go directly to pH = -log [H^+] = -log [0.015] = 1.82 (two significant figures given for the concentration requires two decimal places on the pH)

Example 2
The K_a of acetic acid is 1.8×10^{-5}. (a) Calculate the pH of a 1.5×10^{-3} M solution of acetic acid. (b) What is the equilibrium concentration of acetic acid?

Solution:
(a) Since this is a weak acid, a RICE table is in order.

Reaction	$HC_2H_3O_2$	$\rightleftharpoons$	H^+	+ $C_2H_3O_2^-$
Initial	1.5×10^{-3}		0	0
Change	$-x$		$+x$	$+x$
Equilibrium	$(1.5 \times 10^{-3} - x)$		x	x

$$K_a = \frac{[H^+][C_2H_3O_2^-]}{[HC_2H_3O_2]} = \frac{[x^2]}{[1.5\times10^{-3}-x]} = 1.8 \times 10^{-5}$$

$$1.8 \times 10^{-5}\left(1.5\times10^{-3}-x\right) = x^2$$

Set the equation equal to zero so that x^2 is positive.

$$0 = x^2 + 1.8\times10^{-5}x - 2.7\times10^{-8}$$

Solve using the quadratic formula. It is recommended that you use a graphing calculator with a quadratic formula solving program since you are allowed to do that on the AP Chemistry exam.

$$x = [C_2H_3O_2^-] = [H^+] = 1.6\times10^{-4}$$

$$\therefore pH = -\log[H^+] = -\log\left[1.6\times10^{-4}\right] = 3.80$$

(2 decimal places since 2 significant digits in the original concentration)

(b) $[HC_2H_3O_2]$ at equilibrium $= 1.5\times10^{-3} - x = 1.5\times10^{-3} - 1.6\times10^{-4} = 1.3\times10^{-3}\dfrac{mol}{L}$

Notice that we had to use the quadratic formula. Many equilibrium problems allow you to "neglect x" which greatly simplifies the calculation and avoids the quadratic formula all together. *You may always neglect subtracting x when the original concentration of the weak acid or base is greater than $100K_a$ or $100K_b$.*

Examine Example 1 again. If our original concentration had been something like 0.15M, then we could have neglected subtracting x since $100K_a = 1.8 \times 10^{-3} = 0.0018$ and $0.15 > 0.0018$.

The determination of the pH of a **weak base** is very similar to the determination of the pH of a weak acid. Follow the same steps. Remember, however, that x is the **[OH⁻]** and taking the negative log of x will give you the **pOH** and not the pH! Then use the relationship that pH + pOH + 14.

Example 3
Calculate the pH of a 1.0 M solution of methylamine, CH_3NH_2 ($K_b = 4.38 \times 10^{-4}$).

Solution: Since this is a weak base, again a RICE table is in order. Remember to determine if you can "neglect x", just be sure and indicate in your work that you at least thought about "x" by crossing x out and writing "neglect x". Also, with bases that do NOT contain an OH⁻ ion, it is easier to write a correct equation if you include the water and even write it as HOH.

Reaction	CH_3NH_2	+	HOH $_{(\ell)}$	$\rightleftharpoons$	OH⁻	+	$CH_3NH_3^+$
Initial	1.0 M				0		0
Change	-x				+x		+x
Equilibrium	(1.0 − x)				x		x

$$K_b = \frac{[OH^-][CH_3NH_3^+]}{[CH_3NH_2]} = \frac{[x^2]}{[1.00-x]} = \frac{[x^2]}{[1.00]} = 4.38 \times 10^{-4} \qquad \text{Then solve for "}x\text{"}$$

"neglect x"

$$x = \sqrt{(4.38 \times 10^{-4})(1.00)} = 0.0209$$

$x = [OH^-]$ from our table above so, pOH = $-$ log [0.0209] = 1.68

pH = 14 $-$ pOH = 14 $-$ 1.68 = 12.32

CALCULATING PERCENT DISSOCIATION

Ah, this is the easy part! It's simply the $\left[\dfrac{\text{amount dissociated}}{\text{original concentration}}\right] \times 100\%$.

To find the percent dissociated from Example 1, recall that x = $[H^+]$ which indicates the "amount dissociated" so…

$$\% \text{ dissociation} = \left(\frac{\text{amount dissociated}}{\text{original concentration}}\right) \times 100\% = \left(\frac{1.6 \times 10^{-4}}{1.5 \times 10^{-3}}\right) \times 100\% = 10.7\%$$

We can also calculate K_a from percent dissociation.

Example 4
In a 0.100 *M* aqueous solution, lactic acid is 3.7% dissociated. Calculate the value of K_a for this acid.

Solution: If an acid is 3.7 % dissociated, then

$$3.7\% = \left(\frac{H^+ \text{ dissociated}}{\text{original concentration}}\right) \times 100\%; \text{ first divide both sides by 100}$$

$$0.037 = \left(\frac{H^+ \text{ dissociated}}{0.100 \text{ M}}\right) \text{ so, } (0.037)(0.100M) = 0.0037M = [H^+]$$

$$K_a = \frac{[H^+][A^-]}{[HA]} = \frac{[x^2]}{[0.100\text{-}x]} = \frac{[0.0037^2]}{[0.100]} = 1.4 \times 10^{-4}$$

PURPOSE

You will be introduced to the concept of weak acid-base equilibrium. You will learn to solve acid-base equilibrium problems using the RICE table problem-solving method and subsequently calculate the pH of the resulting solution.

MATERIALS

calculator   paper and pencil

PROCEDURE

Solve the problems found on your student answer page. Be sure to show all work paying attention to the proper use of significant digits and units.

Name _____

Period _____

Acid-Base Equilibrium
Solving pH Problems for Weak Acids and Bases

CONCLUSION QUESTIONS

Use your own paper and show all work and pay special attention to significant digits and units where appropriate.

1. A dilute solution of household ammonia contains 0.3124 mol of NH_3 per liter of solution. The K_b of ammonia is 1.8×10^{-5} at 25°C.

   (a) Calculate the hydroxide ion concentration of the solution at 25°C.

   (b) Calculate the pH of the solution at 25°C.

2. (a) Calculate the ionization constant for HF if a 0.10M solution of HF is 8.1% ionized at 25°C.

   (b) Calculate the pH of the solution.

3. Formic acid is the irritant that causes the body's reaction to an ant's sting. It has a Ka value of 1.8×10^{-4} at 25°C.

   (a) Calculate the concentration of hydrogen ion in a 0.575M solution at 25°C.

   (b) Calculate the percent ionization of a 0.575M solution at 25°C.

4. Acetic acid has a K_a value of 1.8×10^{-5} at 25°C.

   (a) Calculate the percent ionization of a 0.265M acetic acid solution at 25°C.

   (b) Calculate the [OH⁻] of the solution at 25°C.

5. Monochloroacetic acid, $HC_2H_2ClO_2$, is a skin irritant that is used in "chemical peels" intended to remove the top layer of dead skin from the face and ultimately improve the complexion. The value of the K_a for monochloroacetic acid is 1.35×10^{-3}.

   (a) Calculate the pH of a 0.25M solution of monochloroacetic acid at 25°C.

   (b) Calculate the percent ionization of a 0.25M solution of monochloroacetic acetic acid at 25°C.

6. Codeine ($C_{18}H_{21}NO_3$) is a derivative of morphine that is used as an analgesic, narcotic or antitussive. It was once commonly used in cough syrups but is now available only by prescription because of its addictive properties.

    (a) If the pH of a 6.3×10^{-3} M solution of codeine is 9.83 at 30°C, calculate the $[H^+]$ of the solution.

    (b) Calculate the K_b at 30°C.

7. The K_a of acetic acid at 25°C is 1.8×10^{-5}.

    (a) What are the equilibrium concentrations of H_3O^+, $C_2H_3O_2^-$ and $HC_2H_3O_2$ in a 0.350 M aqueous solution of acetic acid?

    (b) What is the pH of the solution at 25°C?

8. The weak base methylamine, CH_3NH_2, has a K_b of 4.2×10^{-4} at 25°C. It reacts with water according to the equation

    $$CH_3NH_2(aq) + H_2O(\ell) \rightleftharpoons CH_3NH_3^+(aq) + OH^-(aq)$$

    (a) Calculate the equilibrium concentration of OH^- in a 0.15M solution of the base.

    (b) What are the pH and pOH of the solution?

9. A 1.00 M solution of a weak acid has a pH of 3.849 at 25°C.

    (a) Calculate the $[H^+]$ in the 1.00 M solution.

    (b) Calculate the K_a for the weak acid at 25°C.

10. Phosphoric acid has a K_{a1} of 7.5×10^{-3} at 25°C.

    (a) Calculate the $[H_2PO_4^{2-}]$ present at equilibrium in a 0.35 M solution.

    (b) Calculate the pH of the solution at 25°C.

How Weak is Your Acid?
Determining Ka

OBJECTIVE
Students will experimentally determine the Ka of several weak acids, gather data from the class and rank the acids in order of strength.

LEVEL
Chemistry

NATIONAL STANDARDS
UCP.2, UCP.3, UCP.4, A.1, A.2, B.2, E.1, E.2

TEKS
1(A), 2(A), 2(B), 2(C), 2(D), 2(E), 14(A)

CONNECTIONS TO AP
AP Chemistry:

   III. Reaction types A. Reaction types 1. Acid-base reactions; concepts of Arrhenius, Bronsted-Lowry C. Equilibrium 1. Concept of dynamic equilibrium, physical and chemical; equilibrium constants 2. Quantitative treatment b. Equilibrium constants for reactions in solution (1) Constants for acids, bases; pK; pH

TIME FRAME
45 minutes

MATERIALS
(For a class of 28 working in pairs)

14 pH meters (or CBL with pH probe)	several dropping bottles of 1% phenolphthalein
14 125 mL Erlenmeyer flasks	several samples of weak acids
14 100 mL beakers	2 liters of distilled water
14 50 mL graduated cylinders	14 dropper bottles of 0.10 M NaOH
14 stirring rods	

TEACHER NOTES

This lab will require that students are familiar with the concept of strong and weak acids and have calculated pH of weak acids using RICE tables. This experiment is designed to follow the lesson on Ka also found in this book. By performing this experiment students will be able to calculate the Ka for several weak acids using logical mathematical relationships that are presented in the introduction. You will need to discuss the pre-lab with the students and model the problem solving to ensure their understanding of this exercise. The basic premise for this lab is that the pH = pKa when you are half-way to the equivalence point. In the second year chemistry class, the Henderson-Hasselbach equation easily shows this relationship: $pH = pKa + \log\left[\dfrac{base}{acid}\right]$ since at the half-equivalence point, the ratio of base to acid is 1 and the log of 1 is equal to zero, therefore pH = pKa. However, first year students should have no trouble finding the Ka value from their knowledge of equilibrium expressions.

In order for students to visualize the magnitude of the different Ka values, class data will be recorded and the acids will be organized from strongest to weakest. As an extension to this lab, you could have the students calculate % error given the theoretical Ka value and explain any discrepancies in their values. You will need to look up the Ka values for the students to use if you choose this option.

Preparation for this experiment is minimal and is easily set up in approximately half an hour. If pH meters are not available, pH paper (wide range) may be used to test the pH of each acid. You may wish for students to determine the pH using both the meter and the paper for comparison.

Because 14 different weak solid acids and or acidic salts may be difficult to obtain, assign several groups the same solid, have them collaborate and then report one Ka value to the class.

Suggested solid samples are listed in the sample data table that follows.

Preparation of solutions:

1% phenolphthalein: Measure 0.50 grams of solid phenolphthalein, place in a 50. mL volumetric flask and add about 25 mL of 95% ethanol. Swirl to dissolve the indicator. Continue adding ethanol until a total volume of 50. mL is reached. Mix well and place in small labeled dropper bottles.

0.10 M NaOH: Measure 4.00 grams of sodium hydroxide pellets and place into a large container (greater than 1 liter) add 500 mL of distilled water and mix well. Continue adding distilled water until a total volume of 1000 mL is reached. Mix well and place in small labeled dropper bottles. An alternative distribution method is to have small 50 mL beakers labeled with disposable pipets and allow students to obtain the NaOH from the stock container as they need it.

POSSIBLE ANSWERS TO THE CONCLUSION QUESTIONS AND SAMPLE DATA

The Ka values listed are theoretical values—student answers will vary. Many salts may be used but you should avoid using salts that are diprotic in nature for this experiment. The Ka values listed are often the Ka for the second ionization of the acids and will not match to the Ka values given in tables of standard values.

Sample	pH	pKa	Calculated Ka
$KHSO_4$ potassium bisulfate	1.92	1.92	1.2×10^{-2}
$KHSO_3$ potassium bisulfite	7.19	7.19	6.4×10^{-8}
$KHC_8H_4O_4$ potassium hydrogen phthalate	5.41	5.41	3.9×10^{-6}
KH_2PO_4 potassium dihydrogen phosphate	7.21	7.21	6.2×10^{-8}
$KHC_4H_4O_6$ potassium hydrogen tartrate	4.34	4.34	4.6×10^{-5}
$NaHSO_4$ sodium bisulfate	1.92	1.92	1.2×10^{-2}
$NaHSO_3$ sodium bisulfite	7.19	7.19	6.4×10^{-8}
NaH_2PO_4 sodium dihydrogen phosphate	7.21	7.21	6.2×10^{-8}
$NaHC_4H_4O_6$ sodium hydrogen tartrate	4.34	4.34	4.6×10^{-5}

1. Arrange the acids listed in the data table in order of increasing strength.
   - $H_2PO_4^-$; HSO_3^-; $HC_8H_4O_4^-$; $HC_4H_4O_6^-$; HSO_4^-
   - These are listed as the acids in solution–students should recognize that the anion is the acting acid not the entire salt.

2. Using the formula given to you by your teacher, write the chemical equation showing the ionization of your salt in water.
   - Using $NaHC_4H_4O_6$ as the sample salt—student answers will vary with sample
   - $NaHC_4H_4O_{6(s)} \rightarrow Na^+_{(aq)} + HC_4H_4O_6^-_{(aq)}$

3. Show the equation for the dissociation of the anion with water.
   - $HC_4H_4O_6^-_{(aq)} + H_2O_{(l)} \rightleftharpoons H_3O^+_{(aq)} + C_4H_4O_6^{2-}_{(aq)}$

4. Write the equilibrium expression for the reaction shown in number 3.
   - $Ka = \dfrac{[H_3O^+][C_4H_4O_6^{2-}]}{[HC_4H_4O_6^-]}$

5. Explain, using your Ka expression, how the Ka value can be determined from the procedure followed in the laboratory.
   - In the lab we neutralized exactly half of the acid. At this point, the concentration of $HC_4H_4O_6^-$ and the concentration of $C_4H_4O_6^{2-}$ are equal to each other, thus the ratio of acid to base is 1 and their values "cancel out". When they are equal, the Ka = the hydronium ion concentration. By measuring the pH at this point, we are able to calculate the Ka value.

6. Why was it not necessary to know the exact amount of the acid used in this experiment?
   - You only needed to know that ½ of it was neutralized.

7. Why was it not necessary to know the exact amount of NaOH used during this experiment?
   - We only needed enough to neutralize ½ of the acid—the amount does not matter.

8. Why was it necessary to know the exact amount of distilled water used in this experiment?
   - All of the assumptions in the lab were dependent on knowing exactly where the halfway point is, so this volume is critical.

9. While performing this experiment, a student accidentally spilled a few milliliters of distilled water during the initial step of dissolving the acid. Describe how this mistake will affect the calculated Ka value, if at all.
   - If the student realized this mistake and took only half of the volume left then there would be no effect on the calculated Ka value.
   - If the student continued without making adjustments and poured exactly 25.00 mL into the flask and neutralized this amount with the base, the student effectively neutralized more than half of the solution. The measured pH will be greater than expected resulting in a smaller value for the Ka value.

10. Could a similar experiment be performed to find the Kb of a weak base? Explain any modifications needed.
    - A similar experiment could be performed to determine the Kb value. Modifications to this experiment may include: using an acid similar to 0.10 M HCl to neutralize the base; use the relationship that the pOH = pKb so when the pH is measured you would need to subtract from 14 before continuing with the calculation to find the Kb value; if phenolphthalein is still used as an indicator the solution will begin with a bright pink color and you would titrate until a very faint pink color exists.

11. The student washed and rinsed the Erlenmeyer flask before beginning the experiment. What effect if any, would not drying the flask have on the experimental Ka value?
    - If the student does not dry the flask, the amount of water will be greater making the acid a bit more dilute than it should be. The pH will be lower than expected resulting in a larger Ka value.

REFERENCES:

Laboratory Experiments for Advanced Placement Chemistry, Determination of the Dissociation Constant for Weak Acids, Vonderbrink, Flinn Scientific, Inc., 1999, pg 95-99.

How Weak is Your Acid?
Determining Ka

The equilibrium constant for a weak acid, Ka, indicates relative strength of a weak acid. Strength is determined by the amount of ionization of the acid. Strong acids, like HCl, are said to ionize 100%. Weak acids only partially ionize. The larger the Ka value, the more products are present at equilibrium and thus, the greater the amount of dissociation.

In this experiment you will determine the Ka value for a weak acid using your knowledge of salt hydrolysis and acid-base reactions. The basic premise is that when the conjugate acid and the conjugate base in solution are of equal concentration, the Ka value will be equal to the hydrogen ion concentration. This is called the half-equivalence point.

For example, if your teacher asked you to find the Ka value for a sample of potassium bisulfate, $KHSO_4$, the first step in solving this problem is to write the ionization equation for the salt in water.

$$KHSO_{4(s)} \rightarrow K^+_{(aq)} + HSO_4^-{}_{(aq)}$$

Determine which component will react with water. In solution, the bisulfate ion is able to donate a proton like an acid to produce hydronium ions in solution.

$$HSO_4^-{}_{(aq)} + H_2O_{(l)} \rightleftharpoons H_3O^+{}_{(aq)} + SO_4^{2-}{}_{(aq)}$$

The salts in this lab will behave in a similar fashion. Set up the Ka expression for this reaction.

$$Ka = \frac{\left[H_3O^+\right]\left[\cancel{S}O_4^{2-}\right]}{\left[H\cancel{S}O_4^-\right]}$$

If the sulfate ion and the bisulfate ion are equal in concentration, the Ka will equal the hydronium ion concentration. The two ions are equal in concentration when exactly half of the acid is neutralized. In this experiment you will separate an acid solution into two equal volumes, neutralize one of the solutions, recombine, and measure the pH of the total solution. This will represent the pH when exactly half of the acid has been neutralized. Remember that $pH = -\log[H_3O^+]$; therefore, we know that the pH = pKa. To find the Ka value of the acid simply take the antilog of the pH value or perhaps more recognizable, 10-pH!

If the pH of your solution is 2.55, then the pKa value is also 2.55. pKa = -log [Ka]

By taking the antilog of -2.55, we get 2.82×10^{-3} as the Ka value for our acid.

PURPOSE

In this lab the Ka value for several weak acids will be determined. Class data will be collected and the weak acids will be organized according to strength.

MATERIALS

pH meter (or CBL with pH probe)
100 mL beaker
50 mL graduated cylinder
stirring rod
125 mL Erlenmeyer flask

1% phenolphthalein
sample of weak acid
distilled water
0.10 M NaOH

Safety Alert
1. Wear goggles and aprons.
2. Weak acids and bases are often irritating to the skin. If skin contact is made, flush the area with water.
3. Phenolphthalein is prepared with alcohol. Keep it away from flames. Wash hands thoroughly after handling.

PROCEDURE

1. Obtain an acid sample from your teacher and record the correct chemical formula in the space provided on the student answer page.

2. Mass about 0.20 grams of the sample and place it into a 100 mL beaker.

3. Using the graduated cylinder obtain exactly 50.0 mL of distilled water. Add all of the water to the beaker with the acid. Stir to dissolve the sample.

4. Pour exactly 25.0 mL of the acid solution into a clean, dry Erlenmeyer flask. Add a few drops of phenolphthalein indicator and swirl to mix.

5. Add NaOH dropwise with continued swirling until a faint pink color persists for about 5 seconds. (At this point, all 25.00 mL will be neutralized representing exactly half of the original 50.00 mL of solution.)

6. Pour the entire contents from the flask back into the beaker with the remaining acid solution.

7. Measure the pH of this solution.

8. Dispose of the contents of the beaker by flushing down the drain with water. Clean and rinse all glassware.

9. Calculate the Ka of the acid in the space provided on your student answer page. Be sure to show all of your calculations. Record your results with the class data as instructed by your teacher.

10. Complete the ANALYSIS and CONCLUSION questions.

Name _____

Period _____

How Weak is Your Acid?
Determining Ka

DATA AND OBSERVATIONS

Sample formula: _____

ANALYSIS

Show work for the calculation of you Ka here:

Record class data in the table below:

Sample	pH	pKa	Calculated Ka

CONCLUSION QUESTIONS

1. Arrange the acids listed in the data table in order of increasing strength.

2. Using the formula given to you by your teacher, write the chemical equation showing the ionization of your salt in water.

3. Show the equation for the dissociation of the anion with water.

4. Write the equilibrium expression for the reaction shown in number 3.

5. Explain, using your Ka expression, how the Ka value can be determined from the procedure followed in the laboratory.

6. Why was it not necessary to know the exact amount of the acid used in this experiment?

7. Why was it not necessary to know the exact amount of NaOH used during this experiment?

8. Why was it necessary to know the exact amount of distilled water used in this experiment?

9. While performing this experiment, a student accidentally spilled a few milliliters of distilled water during the initial step of dissolving the acid. Describe how this mistake will affect the calculated Ka value, if at all.

10. Could a similar experiment be performed to find the Kb of a weak base? Explain any modifications needed.

11. The student washed and rinsed the Erlenmeyer flask before beginning the experiment. What effect if any, would not drying the flask have on the experimental Ka value?

What Do You Mean, "It Is Soluble After All?"
Exploring Solubility Equilibrium

OBJECTIVE

Students will learn to apply general equilibrium problem-solving techniques to solubility equilibrium.

LEVEL

Chemistry

NATIONAL STANDARDS

UCP.1, UCP.2, UCP.3, UCP.4, B.3, B.6

TEKS

12(A), 15(A)

CONNECTIONS TO AP

AP Chemistry:
   III. Reaction types A. Reaction types C. Equilibrium 1. Concept of dynamic equilibrium, physical and chemical; equilibrium constants 2. Quantitative treatment b. Equilibrium constants for reactions in solution

TIME FRAME

90 minutes

MATERIALS

   calculator   paper and pencil

TEACHER NOTES

This activity is designed as a classroom lesson and will require a teacher lecture. Students should be guided through the general concepts and definitions of chemical equilibrium as it relates to solubility. This activity should follow the lesson on general equilibrium expressions and the RICE-table method of problem solving. You should guide the students through the example problems found on the student answer pages. It is also assumed that students know their solubility rules by this time in the course.

There are many types of equilibrium including: concentration, gas, acid-base and solubility. This lesson is designed as an introduction to the basic skills and concepts involving solubility equilibrium. Much of what is presented here is probably absent from your high school chemistry textbook. However, the first question in the free-response section of the AP Chemistry exam is always an equilibrium problem. Your students will benefit from an introduction to solubility equilibrium in your first-year course.

ANSWERS TO THE CONCLUSION QUESTIONS

Emphasize to students that the value of a constant does *not* figure into their decision regarding how many significant digits to report in their answers.

1. In a saturated solution of $Ag_2C_2O_4$, the concentration of silver ion is 2.1×10^{-4} mol/L. From this information, calculate the solubility product constant, K_{sp} of $Ag_2C_2O_4$.

$$Ag_2C_2O_4(s) \rightleftharpoons 2Ag^+ + C_2O_4^{2-}$$
$$ 2x \quad\quad x$$

$$K_{sp} = \left[Ag^+\right]^2\left[C_2O_4^{2-}\right] = [2x]^2[x]$$

$$\left[Ag^+\right] = 2.1\times10^{-4}\ \frac{mol}{L} = 2x \therefore x = \left[C_2O_4^{2-}\right] = 1.05\times10^{-4}\ \frac{mol}{L}$$

$$K_{sp} = \left[2.1\times10^{-4}\right]^2\left[1.05\times10^{-4}\right] = 4.6\times10^{-12}$$

2. What amount of chloride ion must be exceeded before silver chloride will precipitate out of a solution in which the concentration of silver ion is 2.7×10^{-5} mol/L? K_{sp} for AgCl is 1.8×10^{-10} at 25°C.

Solve for $\left[Cl^-\right]$

$$AgCl(s) \rightleftharpoons Ag^+ + Cl^-$$

$$K_{sp} = 1.8\times10^{-10} = \left[Ag^+\right]\left[Cl^-\right]$$

$$\therefore \left[Cl^-\right] = \frac{K_{sp}}{\left[Ag^+\right]} = \frac{1.8\times10^{-10}}{\left[2.7\times10^{-5}\right]} = 6.7\times10^{-6}\ \frac{mol}{L}$$

TEACHER PAGES

3. K_{sp} for HgS is 1.6×10^{-54} at 25°C. If mercury (II) nitrate has a concentration of 2.0×10^{-25} mol/L, what is the maximum amount of sulfide ion that can exist in this solution at 25°C?

Solve for $\left[S^{2-} \right]$

$$HgS(s) \rightleftharpoons Hg^{2+} + S^{2-}$$

$$K_{sp} = 1.8 \times 10^{-10} = \left[Hg^{2+} \right]\left[S^{2-} \right]$$

$$[Hg(NO_3)_2] = 2.0 \times 10^{-25}\ \frac{mol}{L} \therefore \left[Hg^{2+} \right] = 2.0 \times 10^{-25}\ \frac{mol}{L} \left(and\ \left[NO_3^- \right] = 4.0 \times 10^{-25}\ \frac{mol}{L} \right)$$

$$\left[S^{2-} \right] = \frac{K_{sp}}{\left[Hg^{2+} \right]} = \frac{1.6 \times 10^{-54}}{\left[2.0 \times 10^{-25} \right]} = 8.0 \times 10^{-30}\ \frac{mol}{L}$$

4. Consider the reaction $SrCO_3(s) \rightleftharpoons Sr^{2+} + CO_3^{2-}$. Given that the molar concentrations of the two ions in this reaction are both equal to 2.6×10^{-5} mol/L at 25°C, what is the value of K_{sp} for SrCO₃ at 25°C?

$$SrCO_3(s) \rightleftharpoons Sr^{2+} + CO_3^{2-}$$

$$K_{sp} = \left[Sr^{2+} \right]\left[CO_3^{2-} \right]$$

$$\left[Sr^{2+} \right] = \left[CO_3^{2-} \right] = 2.6 \times 10^{-5}\ \frac{mol}{L}$$

$$\therefore K_{sp} = \left[2.6 \times 10^{-5} \right]\left[2.6 \times 10^{-5} \right] = \left[2.6 \times 10^{-5} \right]^2 = 6.8 \times 10^{-10}$$

5. If the solubility product constant for AgBr is 5.0×10^{-13} at 25°C, and the molar concentration of Ag^+ and Br^- are each 7.5×10^{-5} mol/L in a solution of AgBr at 25°C, will there be a precipitate of AgBr formed?

Yes, since $Q_{sp} > K_{sp}$ ∴ a precipitate forms.

$$AgBr(s) \rightleftharpoons Ag^+ + Br^-$$

$$\left[Ag^+\right] = \left[Br^-\right] = 7.5 \times 10^{-5} \frac{mol}{L}$$

$$Q_{sp} = \left[Ag^+\right]\left[Br^-\right] = \left[7.5 \times 10^{-5}\right]^2 = 5.6 \times 10^{-9}$$

Compare Q_{sp} to K_{sp} of 5.0×10^{-13}

6. Given that K_{sp} for $Ca_3(PO_4)_2$ is 1.3×10^{-32} at 25°C, what are the molar concentrations of Ca^{2+} and PO_4^{3-} in a saturated solution at 25°C?

$$Ca_3\left(PO_4\right)_2(s) \rightleftharpoons 3\,Ca^{2+} + 2\,PO_4^{3-}$$

$$\qquad\qquad\qquad\qquad 3x \qquad\quad 2x$$

$$K_{sp} = \left[Ca^{2+}\right]^3\left[PO_4^{3-}\right]^2 = [3x]^3[2x]^2 = \left(27x^3\right)\left(4x^2\right) = 108x^5$$

$$K_{sp} = 1.3 \times 10^{-32} = 108x^5$$

$$x = \sqrt[5]{\frac{1.3 \times 10^{-32}}{108}} = 1.6 \times 10^{-7}$$

$$\left[Ca^{2+}\right] = 3x = 4.9 \times 10^{-7}$$

$$\left[PO_4^{3-}\right] = 2x = 3.2 \times 10^{-7}$$

T E A C H E R   P A G E S

7. The solubility product of $PbCl_2$ at 25°C is 1.6×10^{-5}. If $[Cl^-] = 4.7 \times 10^{-3}$, what is the concentration of lead ion in equilibrium with the chloride ion at 25°C?

$$PbCl_2(s) \rightleftharpoons Pb^{2+} + 2Cl^-$$

$$K_{sp} = 1.6 \times 10^{-5} = \left[Pb^{2+} \right] \left[Cl^- \right]^2$$

$$\left[Pb^{2+} \right] = \frac{K_{sp}}{\left[Cl^- \right]^2} = \frac{1.6 \times 10^{-5}}{\left[4.7 \times 10^{-3} \right]^2} = 0.72 \frac{mol}{L}$$

8. Calculate the molar solubility at 25°C of $Ba(OH)_2$ if it has a K_{sp} of 5.0×10^{-3} at 25°C?

$$Ba(OH)_2(s) \rightleftharpoons Ba^{2+} + 2\,OH^-$$
$$ x \qquad 2x$$

$$K_{sp} = \left[Ba^{2+} \right]\left[OH^- \right]^2 = [x][2x]^2 = 4x^3 = 5.0 \times 10^{-3}$$

$$x = \text{molar solubility} = \sqrt[3]{\frac{5.0 \times 10^{-3}}{4}} = 0.11 \frac{mol}{L}$$

9. Will a precipitate form when 75.0 mL of 0.020M $BaCl_2$ and 125mL of 0.040M Na_2SO_4 are mixed together at 25°C? Ksp of $BaSO_4 = 1.5 \times 10^{-9}$ at 25°C.

$$0.0750\cancel{L} \times \frac{0.020 \text{ mol}}{\cancel{L}} BaCl_2 = 0.0015 \text{ mol } BaCl_2 = 0.0015 \text{ mol } Ba^{2+}$$

$$0.125\cancel{L} \times \frac{0.040 \text{ mol}}{\cancel{L}} Na_2SO_4 = 0.0050 \text{ mol } Na_2SO_4 = 0.0015 \text{ mol } SO_4^{2-}$$

Total volume of solution = 200mL = 0.200L

$$BaSO_4(s) \rightleftharpoons Ba^{2+} + SO_4^{2-}$$

$$Q_{sp} = \left[Ba^{2+} \right]\left[SO_4^{2-} \right]$$

$$Q_{sp} = \left[\frac{0.0015}{0.200} \right]\left[\frac{0.0050}{0.200} \right] = 1.9 \times 10^{-4}$$

$Q_{sp} > K_{sp}$ ∴ precipitate forms.

10. Will a precipitate form when 100.0 mL of 0.020M $Pb(NO_3)_2$ is added to 100.0 mL of 0.020M NaCl at 25°C? K_{sp} of $PbCl_2 = 1.6 \times 10^{-5}$ at 25°C.

$$0.1000\cancel{L} \times \frac{0.020 \text{ mol}}{\cancel{L}} Pb(NO_3)_2 = 0.0020 \text{ mol } Pb(NO_3)_2 = 0.0020 \text{ mol } Pb^{2+}$$

$$0.1000\cancel{L} \times \frac{0.020 \text{ mol}}{\cancel{L}} NaCl = 0.0020 \text{ mol } NaCl = 0.0020 \text{ mol } Cl^-$$

Total volume of solution = 200mL = 0.200L

$$PbCl_2(s) \rightleftharpoons Pb^{2+} + 2 Cl^-$$

$$Q_{sp} = \left[Pb^{2+} \right]\left[Cl^- \right]^2$$

$$Q_{sp} = \left[\frac{0.0020}{0.200} \right]\left[\frac{0.0020}{0.200} \right]^2 = 1.0 \times 10^{-6}$$

$Q_{sp} < K_{sp}$ ∴ NO precipitate forms.

What Do You Mean, "It Is Soluble After All?"
Exploring Solubility Equilibrium

In this activity, we explore a new kind of K, K_{sp}, where the "sp" stands for "solubility product". This is where we confess our sins as chemistry teachers [gasp!]. Remember those solubility rules we keep harping about? Well, some of those compounds deemed insoluble actually dissolve a bit. We haven't *exactly* been lying to you, since the future attorneys among you probably read the fine print where "soluble" has a qualifier of 3.0 grams dissolving in 100. mL of water. This qualifier means that if only 2.9 grams dissolves in 100. mL of water the compound or salt is deemed insoluble. What does this mean? It means that nothing is completely insoluble, even silver chloride!

DETERMINIG THE SOLUBILITY PRODUCT CONSTANT, K_{SP}

Saturated solutions of salts represent another type of chemical equilibria. Slightly soluble salts establish a dynamic equilibrium with the hydrated cations and anions in solution. No ions are initially present when the solid is first added to water. As dissolution proceeds, the concentration of ions increases until equilibrium is established. This occurs when the solution is saturated. All of these equilibria follow this general reaction pattern:

$$\text{solid} + \text{H}_2\text{O}_{(\ell)} \rightleftharpoons \text{cation}^+ + \text{anion}^-$$

Since water is a pure liquid, it does NOT appear in the equilibrium constant expression. And neither does the solid. That means that the K_{sp} expression is simply the product of the ions in solution since there is no denominator.

Consider a saturated solution of AgCl, the simplified equation would be (notice the water is left out of the simplified equation):

$$\text{AgCl}_{(s)} \rightleftharpoons \text{Ag}^+_{(aq)} + \text{Cl}^-_{(aq)}$$

The solubility product expression would be:

$$K_{sp} = [\text{Ag}^+][\text{Cl}^-] = 1.6 \times 10^{-10}$$

Remember, K of any variety represents both a number *and* a relationship! K_{sp} values are published in many different sources including your textbook.

DETERMINING K_{SP} FROM EXPERIMENTAL MEASUREMENTS

In the laboratory a K_{sp} is determined by careful laboratory measurements using various spectroscopic methods.

Example 1

Lead (II) chloride dissolves to a slight extent in water according to the equation:

$$\text{PbCl}_{2(s)} \rightleftharpoons \text{Pb}^{+2}_{(aq)} + 2\text{Cl}^-_{(aq)}$$

Calculate the Ksp if the lead ion concentration has been found to be 1.62×10^{-2}M.

Solution:
If lead=s concentration is equal to "x", then chloride=s concentration is equal to "$2x$". You must respect the stoichiometry of the reaction when assigning x values.

$$K_{sp} = [Pb^{+2}][Cl^-]^2 = (1.62 \times 10^{-2})(3.24 \times 10^{-2})^2 = 1.70 \times 10^{-5}$$

ESTIMATING SALT SOLUBILITY FROM K$_{SP}$

Example 2
The K_{sp} for $CaCO_3$ is 3.8×10^{-9} at 25°C. Calculate the solubility of calcium carbonate in pure water in
a) moles per liter

b) grams per liter

Solution:
a) First, write the balanced chemical equation for the dissolution:

$$CaCO_{3\,(s)} \rightleftharpoons Ca^{2+}_{(aq)} + CO_3^{2-}_{(aq)}$$

Next, write the K_{sp} expression and set it equal to the value you were given in the problem. Substitute "x" for each ion, this one is easy since the ion:ion ratio is 1:1.

$$K_{sp} = [Ca^{2+}][CO_3^{2-}] = 3.8 \times 10^{-9} \text{ therefore,}$$
$$K_{sp} = [x][x] = x^2 = 3.8 \times 10^{-9} \text{ finally, solve for "x"}$$
$$x = \sqrt{3.8 \times 10^{-9}} = 6.2 \times 10^{-5}\text{M}$$

Why is this the solubility of $CaCO_{3\,(s)}$ in moles per liter? Think back to the RICE tables, in this chemical equation, all of the substances have coefficients of one. So, when one $CaCO_3$ formula unit dissolves it forms one each of the ions. That means that the $[CaCO_3]$ = [either ion]. Any time you've solved for "x", you have the solubility of the solid in moles/liter.

b) Just use dimensional analysis:

$$\left(\frac{6.2 \times 10^{-5}\text{moles}}{\text{liter}} \right) \times \left(\frac{100.\ \text{g}}{\text{mole}} \right) = 6.2 \times 10^{-2}\ \frac{\text{g}}{\text{liter}}$$

Don't forget solubility changes with temperature! Some substances become less soluble at low temperatures while some become more soluble. This graph is a classic.

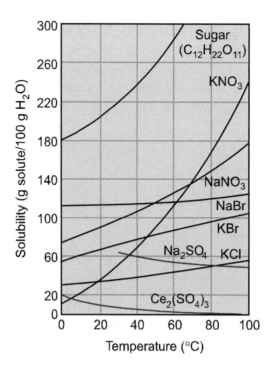

TO PRECIPITATE OR NOT TO PRECIPITATE?

First, we must be introduced to the concept of the reaction quotient, Q. We can use the value of Q to decide whether or not a precipitate will form. We can also determine the concentration of the ion in question, when precipitation begins. Q, the reaction quotient, is calculated exactly as you calculate K. What does the value of Q mean when compared to the value of K?

1. $Q < K_{sp}$, the system is not at equilibrium (*un*saturated)
2. $Q = K_{sp}$, the system is at equililibrium (saturated)
3. $Q > K_{sp}$, the system is not at equililibrium (*super*saturated)

Precipitates form when the solution is supersaturated!

Example 3
At 25°C, the value of K_{sp} for $PbF_{2\ (s)}$ is 4.0×10^{-8}. If 50.0 mL of 0.020 M $KF_{(aq)}$ is added to 50.0 mL of 0.060 M $Pb(NO_3)_{(aq)}$, will a precipitate form? Assume that the volumes are additive.

Solution:
"Assume the volumes are additive" is your clue that the molarities of the ions need to be recalculated. The total volume of the solution is 50.0 mL + 50.0 mL = 100.0 mL = 0.100L

$$\text{\# moles of } F^- = M \times L = (0.020 \text{ mol/L}) (0.050 \text{ L}) = 0.0010 \text{ moles } F^-$$

$$[F^-] = 0.0010 \text{ moles}/ 0.100 \text{ L} = 0.010 \text{ M}$$

$$\text{\# moles of } Pb^{2+} = M \times L = (0.060 \text{ mol/L} (0.050 \text{ L}) = 0.0030 \text{ moles } Pb^{2+}$$

$$[Pb^{2+}] = 0.0030 \text{ moles}/0.100 \text{ L} = 0.030 \text{ M}$$

Calculate Q_{sp} and compare it to K_{sp} to determine if a precipitate occurs...

$$PbF_{2\,(s)} \rightleftharpoons Pb^{2+}_{\,(aq)} + 2\,F^-_{\,(aq)}$$

$$Q_{sp} = [Pb^{2+}][\,F^-]^2 = [0.030][0.010]^2 = 3.0 \times 10^{-6}$$

Compare Q_{sp} to K_{sp}. Since Q_{sp} is greater than K_{sp}, a precipitate forms.

SOLUBILITY AND THE COMMON ION EFFECT

Experiments show that the solubility of any salt is always less in the presence of a "common ion". Why? LeChatelier=s Principle, that=s why! Be reasonable and use approximations when you can. The pH can also affect solubility. Evaluate the balanced chemical equation to see which reagent would "react" with the addition of acid or base.

Example 4
Would magnesium hydroxide be more soluble in an acid or a base? Why?

$$Mg(OH)_{2(s)} \rightleftharpoons Mg^{2+}_{\,(aq)} + 2\,OH^-_{\,(aq)}$$

Addition of a base would shift the equilibrium to the left, decreasing the solubility since more solid magnesium hydroxide forms as a result of the increase in OH⁻ ions. Addition of an acid will remove OH⁻ ions as water is formed. This will shift the equilibrium to the right, decreasing the amount of solid magnesium hydroxide making it *more* soluble.

PURPOSE

In this activity you will learn to write equilibrium constant expressions and solve equilibrium problems using the RICE-table method.

MATERIALS

calculator   paper and pencil

PROCEDURE

1. Solve the problems found on your student answer page. Be sure to show all work paying attention to the proper use of significant digits and units.

Name _____

Period _____

What Do You Mean, "It Is Soluble After All?"
Exploring Solubility Equilibrium

CONCLUSION QUESTIONS

1. In a saturated solution of $Ag_2C_2O_4$, the concentration of silver ion is 2.1×10^{-4} mol/L. From this information, calculate the solubility product constant, K_{sp} of $Ag_2C_2O_4$.

2. What amount of chloride ion must be exceeded before silver chloride will precipitate out of a solution in which the concentration of silver ion is 2.7×10^{-5} mol/L? K_{sp} for AgCl is 1.8×10^{-10} at 25°C.

3. K_{sp} for HgS is 1.6×10^{-54} at 25°C. If mercury (II) nitrate has a concentration of 2.0×10^{-25} mol/L, what is the maximum amount of sulfide ion that can exist in this solution at 25°C?

4. Consider the reaction $SrCO_3(s) \rightleftharpoons Sr^{2+} + CO_3^{2-}$. Given that the molar concentrations of the two ions in this reaction are both equal to 2.6×10^{-5} mol/L at 25°C, what is the value of K_{sp} for $SrCO_3$ at 25°C?

5. If the solubility product constant for AgBr is 5.0×10^{-13} at 25°C, and the molar concentration of Ag^+ and Br^- are each 7.5×10^{-5} mol/L in a solution of AgBr at 25°C, will there be a precipitate of AgBr formed?

6. Given that K_{sp} for $Ca_3(PO_4)_2$ is 1.3×10^{-32} at 25°C., what are the molar concentrations of Ca^{2+} and PO_4^{3-} in a saturated solution at 25°C?

7. The solubility product of $PbCl_2$ at 25°C is 1.6×10^{-5}. If $[Cl^-] = 4.7 \times 10^{-3}$, what is the concentration of lead ion in equilibrium with the chloride ion at 25°C?

8. Calculate the molar solubility at 25°C of $Ba(OH)_2$ if it has a K_{sp} of 5.0×10^{-3} at 25°C?

9. Will a precipitate form when 75.0 mL of 0.020M $BaCl_2$ and 125mL of 0.040M Na_2SO_4 are mixed together at 25°C? Ksp of $BaSO_4 = 1.5 \times 10^{-9}$ at 25°C.

10. Will a precipitate form when 100.0 mL of 0.020M $Pb(NO_3)_2$ is added to 100.0 mL of 0.020M NaCl at 25°C? K_{sp} of $PbCl_2 = 1.6 \times 10^{-5}$ at 25°C.

Neutral or Not?
Exploring Salt Hydrolysis

OBJECTIVE

Students will prepare a salt solution of assigned molarity and determine the pH of the solution using a pH meter. They will then be required to explain via equations and calculations why some salts are neutral and others are not.

LEVEL

Chemistry

NATIONAL STANDARDS

UCP.2, UCP.3, A.1, A.2, B.2, B.3, E.1, E.2, F.4

TEKS

1(A), 1(B), 2(A), 2(B), 2(D), 2(E), 4(C), 8(A), 11(B), 12(C), 14(A), 14(C)

CONNECTIONS TO AP

AP Chemistry:
    III. Reactions C. Equilibrium 2. Quantitative treatment b. Equilibrium constants for reactions in solution 3. Common ion effect; buffers; hydrolysis

TIME FRAME

45 minutes class time; 45 minutes homework

MATERIALS

(For a class of 28 working in pairs)

14 pH meters* (students could share 4 per meter)

variety of solid salts in small labeled containers

14 100 mL beakers

several balances

2 liters of distilled water (placed in labeled squirt bottles)

14 50mL graduated cylinders

14 stirring rods

weighing paper

*If pH meters are not readily available, wide-range indicator paper (0-14) may also be used. In fact, it is often interesting to have students use both pH meters and pH paper. Students can then compare the two results and even calculate the difference between them.

TEACHER NOTES

This experiment should follow a study of acids and bases so that students will already be familiar with the proper set up and use of the pH meter.

The first part of this exercise requires that students prepare a 0.25 M solution of an assigned salt. You will need to provide them with the chemical formula (including waters of hydration) so that the appropriate molar mass is found. For example, calcium acetate is commonly sold as $Ca(C_2H_3O_2)_2 \cdot H_2O$. If the water of hydration is not considered, the molarity will not be calculated correctly.

Before beginning, review with students the proper technique for preparing the solution. The sample should be massed and then placed into a 50 mL volumetric flask or 50 mL graduated cylinder. Distilled water is added to the line on the volumetric or to the 50 mL mark on the graduated cylinder. If students use a weighing paper, the paper is washed into the volumetric flask with distilled water. It is always a good idea to add about half of the water, mix, and allow the salt to completely dissolve before filling to final volume. Remind students to mix the final solution well before measuring the pH. Students should pour their solution into the 100 mL beaker before trying to measure the pH.

The students are asked to gather information from four lab groups with salts different than their own. Therefore, you must have at least five different salts available or modify the number of samples needed. A list of possible suggestions is found below but many other salts could be used. If using other salts, make sure that you check the pH of the salts before distributing. Hopefully, this lab will remove the misconception that all salts are neutral and will pique students' curiosity as to why this is so.

The second part of the lesson requires that you guide the students through the chemical logic that explains their observations. The term for this phenomenon is salt hydrolysis. Teacher notes are provided for you to use in your discussion, however you may wish to share them with the students at the conclusion of the laboratory exercise. In order for students to be successful with this portion of the activity, they must have experience with K_a, K_b, and pH calculations prior to this lesson. By introducing salt hydrolysis and their pH calculation, students will have a good background when attempting buffer calculations in AP Chemistry.

Laboratory comments:
A variety of solid salts may be used for this activity. Some suggestions are:

Neutral:	**Basic:**	**Acidic:**
sodium chloride	sodium bicarbonate	ammonium chloride
ammonium acetate	potassium bicarbonate	ammonium sulfate
potassium sulfate	calcium acetate	

It is always a good idea to run the experiment ahead of time to ensure that the results are as you expect. The pH of the distilled water must be checked in advance. The pH meters must be calibrated for accuracy.

TEACHER NOTES: AFTER COLLECTION OF DATA

The products of any acid-base neutralization reaction are a salt and water. The hydronium ion from the acid reacts with the hydroxide ion from the base to give water. The anion from the acid reacts with the cation from the base to give a salt. This salt can be neutral, acidic or basic. Determining the qualitative pH of the aqueous salt solution is easy as long as you remember the strong acids and the strong bases.

Consider $KHCO_3$. First, determine which acid reacted with which base to form this salt. This salt could have been formed from potassium hydroxide and carbonic acid. Next, determine the relative strength of the acid and the base. KOH is a strong base and H_2CO_3 is a weak acid. "Strong wins" and determines the pH of the resulting salt. K_2CO_3 is therefore a basic or alkaline salt. When this salt reacts with water, a hydrolysis reaction occurs that produces a basic solution.

By considering the strength of acid and base "parents" of your salt, the following combinations are possible:

> strong base + strong acid = neutral salt
> strong base + weak acid = basic salt
> weak base + strong acid = acidic salt
> weak base + weak acid = ??? depends on the K_a and K_b values—whichever is greater wins!

This makes the process of determining relative pH fairly straightforward. But how can we determine the true pH of a salt solution quantitatively? We can calculate it!

Salts are electrolytes and therefore ionize in solution. For the purpose of this lab exercise, all of the salts used will be soluble, so we will assume that they all ionize 100%. If we know the concentration of the original salt then we know the concentration of the components of that salt.

Example: 0.50 M NaCl

$$NaCl_{(s)} \rightarrow Na^+_{(aq)} + Cl^-_{(aq)}$$

The sodium ion has a concentration of 0.50M and so does the chloride ion since the salt ionizes 100% and all species are in a one to one mole ratio.

The tough step comes when determining the relative strengths of the ions in solution. If the cation was produced from a strong base then it is a weak conjugate acid and will not react with water significantly. The same is true for the anion—if it was formed from a strong acid, it will be a very weak conjugate base and will not react with water. This situation yields a neutral salt as we see with sodium chloride. Sodium ion comes from the strong base, sodium hydroxide. Chloride ion originates from the strong acid, hydrochloric acid. Neither ion reacts with water so the solution is neutral.

Consider the example of potassium bicarbonate, $KHCO_3$, from earlier. The salt will ionize 100% as follows:

$$KHCO_{3(s)} \rightarrow K^+_{(aq)} + HCO_3^-_{(aq)}$$

The cation comes from a strong base, KOH, and is therefore a very weak conjugate acid and does not react with water. Since "strong wins" the salt solution will be basic, so we know there must be an OH⁻ as the product. It may be helpful for students to write water as HOH until they get the hang of this. The ion that was contributed by the strong is also the source of the spectator ion. That means the ion from the weak, is the participant in the hydrolysis reaction, or HCO_3^- in this case. This is because bicarbonate ion comes from carbonic acid, a weak acid, and is therefore a fairly strong conjugate base. The anion will react with water or undergo hydrolysis as follows:

$$HCO_3^- + HOH \rightleftharpoons H_2CO_3 + OH^- \text{ becomes } HCO_3^- + H_2O \rightleftharpoons H_2CO_3 + OH^-$$

Let's try another example: Will ammonium chloride be acidic, basic or neutral? Justify your answer.

$$NH_4Cl_{(s)} \rightarrow NH_4^+{}_{(aq)} + Cl^-{}_{(aq)}$$

The cation comes from ammonia, a weak base, therefore the ammonium ion is a fairly strong conjugate acid and will hydrolyze with water. The chloride ion comes from hydrochloric acid, a strong acid, and is therefore a very weak conjugate base and will not react with water. Remember, "strong wins" therefore it is an acidic salt and H^+ (or H_3O^+ if you prefer) is a product and strong is the source of the spectator ion, so the Cl- ion does not participate in the hydrolysis reaction.

$$NH_4^+ + HOH \rightleftharpoons NH_3 + H_3O^+ \text{ becomes } NH_4^+ + H_2O \rightleftharpoons NH_3 + H_3O^+$$

Now, how could you identify the exact pH of a salt?
- You must be given the concentration of the salt solution.
- You must ionize the salt and determine if either ion will react with water.
    - If no ion reacts with water, the solution is neutral—pH = 7.00
    - Remember to work backwards and ask yourself which acid reacted with which base, and that "strong wins" and determines the pH of the salt. Also remember strong is the source of the spectator ion and thus does not participate in the hydrolysis reaction.
    - Thus the ion from the weak acid or base reacts with water so, write the hydrolysis reaction that matches your prediction, remembering if the solution is acidic, H_3O^+ is a product and if the solution is basic OH^- is a product. (It may be helpful to write water as HOH so students can better visualize the reaction.)
- You must find the appropriate K_a or K_b value from a list of ionization constants (a sample list is found at the conclusion of the notes).
- You must calculate the needed ionization constant using: $K_w = K_a \times K_b$
- You must set up the equilibrium expression, substitute values, and solve for the concentration of hydronium or the hydroxide ion, whichever is appropriate.
- Calculate the pH using the pH formula. Remember, if you solve for the hydroxide ion you must first calculate the pOH and then subtract from 14.00 to obtain the pH.

Let's try an example: Calculate the pH of a 0.50 M solution of $KHCO_3$

The salt will ionize 100% as follows: $KHCO_{3(s)} \rightarrow K^+{}_{(aq)} + HCO_3^-{}_{(aq)}$

The cation comes from a strong base, KOH, "strong wins" and the resulting solution will be basic (so OH^- will be a product unless the anion is from a strong acid—in this case it is not). Strong is the source of the spectator ion since K^+ is a very weak conjugate acid and does not react with water. That means the bicarbonate ion participates in the hydrolysis reaction since the bicarbonate ion from carbonic acid, a weak acid, and is a fairly strong conjugate base.

The bicarbonate anion will react with water or undergo hydrolysis as follows:

$$HCO_3^- + HOH \rightleftharpoons H_2CO_3 + OH^- \quad \text{becomes} \quad HCO_3^- + H_2O \rightleftharpoons H_2CO_3 + OH^-$$

Since we see OH^- as a product, we must find the K_b value for bicarbonate ion. The K_a for carbonic acid is 4.2×10^{-7}. Use the relationship for Kw as follows: $K_b = \dfrac{K_w}{K_a} = \dfrac{1 \times 10^{-14}}{4.2 \times 10^{-7}} = 2.38 \times 10^{-8}$

The next step is to set up the K_b expression and solve:

$$K_b = \frac{[H_2CO_3][OH^-]}{[HCO_3^-]} = 2.38 \times 10^{-8} = \frac{x^2}{0.50}$$

$$x^2 = 1.19 \times 10^{-8}$$

$$x = [OH^-] = 1.09 \times 10^{-4} \text{ M}$$

$$pOH = -\log(1.09 \times 10^{-4}) = 3.96$$

$$pH = 14.00 - 3.96 = 10.04 \text{ basic}$$

Instruct students to complete their ANALYSIS and CONCLUSION questions for the next class session. You may want to copy this table and hand it to students or refer them to a resource for finding these values.

Substance	Ionization Constant
H_2CO_3	$K_{a1} = 4.2 \times 10^{-7}$
HCO_3-	$K_{a2} = 4.8 \times 10^{-11}$
$HC_2H_3O_2$	$K_a = 1.8 \times 10^{-5}$
H_2SO_4	$K_{a1} = $ very large!
HSO_4-	$K_{a2} = 1.2 \times 10^{-2}$
H_3PO_4	$K_{a1} = 7.5 \times 10^{-3}$
H_2PO_4-	$K_{a2} = 6.2 \times 10^{-8}$
$HPO_4^{2}-$	$K_{a3} = 3.6 \times 10^{-13}$
NH_3	$K_b = 1.8 \times 10^{-5}$

POSSIBLE ANSWERS TO THE CONCLUSION QUESTIONS AND SAMPLE DATA

HYPOTHESIS

Student answers will vary. One possible hypothesis might be worded as follows:

- I expect that my salt, potassium bicarbonate, will be basic since it was formed from a strong base and a weak acid.

DATA AND OBSERVATIONS

Chemical Formula for Salt	pH Observed
$KHCO_3$	> 7.00
NH_4Cl	< 7.00
$NH_4C_2H_3O_2$	7.00
$NaHCO_3$	> 7.00
$NaCl$	7.00

ANALYSIS

1. Show the calculations for the preparation of your salt solution:

$$50\,mL \times \frac{0.25\,mol}{L} \times \frac{1\,L}{1000\,mL} \times \frac{100.12\,g\,KHCO_3}{1\,mol\,KHCO_3} = 1.25\,g\,KHCO_3$$

- Measure out the calculated mass of the solid and place it into the 50 mL volumetric flask or the 50 mL graduated cylinder. Be sure to rinse the weighing paper into the flask, add distilled water until about ½ filled. Mix well until the salt dissolves and continue adding water until the 50 mL mark is reached. Mix again.

2. Calculate the pH for each of the salts listed in your data table.
   a. Formula for salt: $\underline{KHCO_3}$

   - Calculation of pH:

   $KHCO_3 \rightarrow K^+ + HCO_3^-$ (100% dissociation)

   $HCO_3^- + H_2O \xrightleftharpoons{\hspace{1cm}} H_2CO_3 + OH^-$

   $$K_b = \frac{K_w}{K_a} = \frac{1 \times 10^{-14}}{4.2 \times 10^{-7}} = 2.38 \times 10^{-8}$$

$$K_b = \frac{[H_2CO_3][OH^-]}{[HCO_3^-]} = 2.38 \times 10^{-8} = \frac{x^2}{0.25}$$

$$x^2 = 5.95 \times 10^{-9}$$

$$x = [OH^-] = 7.72 \times 10^{-5} \text{ M}$$

$$pOH = -\log(7.72 \times 10^{-5}) = 4.11$$

$$pH = 14.00 - 4.11 = 9.89 \text{ basic}$$

b. Formula for salt: <u>NH$_4$Cl</u>

- Calculation of pH:

$$NH_4Cl \rightarrow NH_4^+ + Cl^-$$

$$NH_4^+ + H_2O \rightleftharpoons NH_3 + H_3O^+$$

$$K_a = \frac{K_w}{K_b} = \frac{1 \times 10^{-14}}{1.8 \times 10^{-5}} = 5.56 \times 10^{-10}$$

$$K_a = \frac{[NH_3][H_3O^+]}{[NH_4^+]} = 5.56 \times 10^{-10} = \frac{x^2}{0.25}$$

$$x^2 = 1.39 \times 10^{-10}$$

$$x = [H_3O^+] = 1.18 \times 10^{-5} \text{ M}$$

$$pH = -\log(1.18 \times 10^{-5}) = 4.93 \text{ acidic}$$

c. Formula for salt: <u>NH$_4$C$_2$H$_3$O$_2$</u>

- Calculation of pH:

$$NH_4C_2H_3O_2 \rightarrow NH_4^+ + C_2H_3O_2^-$$
$$K_a \text{ for acetic acid} = 1.8 \times 10^{-5}$$

$$K_b \text{ for ammonia} = 1.8 \times 10^{-5}$$

Both ions present have the same affinity to react with water and the salt is therefore neutral.

d. Formula for salt: <u>NaHCO$_3$</u>

- Calculation of pH:

$$NaHCO_3 \rightarrow Na^+ + HCO_3^-$$

$$HCO_3^- + H_2O \rightleftharpoons H_2CO_3 + OH^-$$

$$K_b = \frac{K_w}{K_a} = \frac{1 \times 10^{-14}}{4.2 \times 10^{-7}} = 2.38 \times 10^{-8}$$

$$K_b = \frac{[H_2CO_3][OH^-]}{[HCO_3^-]} = 2.38 \times 10^{-8} = \frac{x^2}{0.25}$$

$$x^2 = 5.95 \times 10^{-9}$$

$$x = [OH^-] = 7.72 \times 10^{-5} \text{ M}$$

$$pOH = -\log(7.72 \times 10^{-5}) = 4.11$$

$$pH = 14.00 - 4.11 = 9.89 \text{ basic}$$

e. Formula for salt: <u>NaCl</u>

- Calculation of pH:

$$NaCl \rightarrow Na^+ + Cl^-$$

Neither ion in solution has a strong attraction to hydrolyze with the water and therefore the solution is neutral.

CONCLUSION QUESTIONS

1. Which of the salts listed in your data table are neutral? Explain why this must be true.
   - $NH_4C_2H_3O_2$ and NaCl are both neutral. Ammonium acetate is formed from a weak acid and a weak base that have the same numerical value for their respective ionization constants and are therefore neutral. Sodium chloride produces ions in solution that do not have an affinity to react with water and therefore they remain in solution yielding a neutral pH.

2. Which of the salts listed in your data table are basic? Explain.
   - $KHCO_3$ and $NaHCO_3$ are both basic. The bicarbonate ion is a fairly strong conjugate base and reacts with water to produce the hydroxide ion. The production of hydroxide yields a basic solution.

3. Which of the salts listed in your data table are acidic? Explain.
   - NH_4Cl was acidic. The ammonium ion is a fairly strong conjugate acid and reacts with water to produce the hydronium ion. The production of the hydronium ion yields an acidic solution.

4. Compare the pH obtained experimentally to the calculated pH for each of the salts listed in your data table. Explain any discrepancies.
   - Student answers will vary. Students should include some of the following as explanations:
     - If the solutions were not accurately prepared, the molarities of the original salt solution may differ from 0.25 M and therefore have yielded an incorrect calculated pH.
     - The pH of the distilled water may have been too low to yield accurate results.
     - The pH meters may not have been properly calibrated to yield accurate readings.

5. A student performed this same experiment but accidentally spilled some of the measured acidic salt while transferring it to the volumetric flask. The student did not think that a small quantity of spilled salt would matter much, so he continued with the rest of the experiment. To his surprise, the measured pH and the calculated pH were not the same. Using your knowledge of salt hydrolysis, explain whether the experimental pH was too low or too high as a result of his error.
   - The experimental pH was too high. The true molarity would be smaller yielding a smaller value for the hydronium ion concentration. When the (–) log of the hydronium ion is calculated, the result is a larger pH.

Neutral or Not?
Exploring Salt Hydrolysis

Salts are produced in the laboratory by way of a neutralization reaction between an acid and a base. The name for this type of chemical reaction seems a misnomer in some respects since the salts that are formed from these reactions are not always neutral. Salts often hydrolyze with water to yield acidic or basic solutions.

PURPOSE
In this lab you will prepare a salt solution of assigned molarity, measure the pH of the solution and then compare with others in your class. You will then use calculations to verify the accuracy of your pH.

MATERIALS

pH meter   distilled water bottle
variety of solid salts   graduated cylinder
100 mL beakers   stirring rod
balance   weighing paper

Safety Alert
1. Wear safety goggles at all times in the laboratory.
2. Some salts may be skin irritants. Flush with copious amounts of water if a sample comes in contact with your skin.

PROCEDURE

1. Obtain a salt sample from your teacher. Note the correct chemical formula. Make a hypothesis as to whether your salt will be acidic, basic or neutral and record this on your student answer page.

2. You need to prepare 50 mL of a 0.25M solution of the salt that you were assigned. Show all of the calculations necessary to properly prepare this solution in the space provided on your student answer page. Pay particular attention to significant figures and units.

3. Have your teacher check the accuracy of your calculation before preparing your solution.

4. Prepare your sample using a clean, dry 50 mL volumetric flask or a clean, dry 50 mL graduated cylinder. Mix your solution thoroughly following proper techniques. Transfer the solution into a clean 100 mL beaker.

5. Measure the pH of your salt solution and record the value in the data table.

6. Clean the lab area and dispose of your salt sample as instructed by your teacher. Be sure to turn the pH meter off after rinsing with distilled water.

7. Gather the data for four different salts from other lab groups. Be sure to record both their formulas and their pHs in your data table.

8. Listen and take good notes as your teacher explains why salts are not always neutral. Follow along as the pH for a salt is calculated.

9. Complete the ANALYSIS and CONCLUSION questions found on your student answer page.

Name _____

Period _____

Neutral or Not?
Exploring Salt Hydrolysis

HYPOTHESIS

DATA AND OBSERVATIONS

Chemical Formula for Salt	pH Observed

ANALYSIS

Substance	Ionization Constant
H_2CO_3	$K_{a1} = 4.2 \times 10^{-7}$
HCO_3-	$K_{a2} = 4.8 \times 10^{-11}$
$HC_2H_3O_2$	$K_a = 1.8 \times 10^{-5}$
H_2SO_4	$K_{a1} =$ very large!
HSO_4-	$K_{a2} = 1.2 \times 10^{-2}$
H_3PO_4	$K_{a1} = 7.5 \times 10^{-3}$
H_2PO_4-	$K_{a2} = 6.2 \times 10^{-8}$
$HPO_4^{2}-$	$K_{a3} = 3.6 \times 10^{-13}$
NH_3	$K_b = 1.8 \times 10^{-5}$

1. Show the calculations for the preparation of your salt solution:

2. Calculate the pH for each of the salts listed in your data table.
   a. Formula for salt:

   - Calculation of pH:

   b. Formula for salt:

   - Calculation of pH:

   c. Formula for salt:

   - Calculation of pH:

   d. Formula for salt:

   - Calculation of pH:

   e. Formula for salt:

   - Calculation of pH:

CONCLUSION QUESTIONS

1. Which of the salts listed in your data table are neutral? Explain why this must be true.

2. Which of the salts listed in your data table are basic? Explain.

3. Which of the salts listed in your data table are acidic? Explain.

4. Compare the pH obtained experimentally to the calculated pH for each of the salts listed in your data table. Explain any discrepancies.

5. A student performed this same experiment but accidentally spilled some of the measured acidic salt while transferring it to the volumetric flask. The student did not think that a small quantity of spilled salt would matter much, so he continued with the rest of the experiment. To his surprise, the measured pH and the calculated pH were not the same. Using your knowledge of salt hydrolysis, explain whether the experimental pH was too low or too high as a result of his error.

Titrations—Titrations
Determining the Percent of Acetic Acid in Vinegar

OBJECTIVE
Students will perform a titration using a buret and indicator to determine the percent of acetic acid in vinegar.

LEVEL
Chemistry

NATIONAL STANDARDS
UCP.3, A.1, A.2, B.3

TEKS
1(B), 1(C), 11(A), 11(B), 14(A), 14(C)

CONNECTIONS TO AP
AP Chemistry:
  III. Reactions A. Reaction types 1. Acid-base reactions B. Stoichiometry

TIME FRAME
60 minutes (This lab can be done in a 45 minute period if it is set up before the class and students do not have to clean the burets.)

MATERIALS
(For a class of 28 working in pairs)

14 ring stands
28 burets (25.0 mL or 50.0 mL)
14 dropping bottles containing 1% methanolic
  solution of phenolphthalein*
14 pieces of white paper
potassium biphthalate (KHP) to standardize NaOH
distilled water

2.0 L white vinegar—any store brand
14 double buret clamps
14 250 mL Erlenmeyer flasks
42 beakers for waste and solution transfer
2.0 L NaOH solution (approximately 0.5 M)
14 buret caps
large container for sodium hydroxide solution
labels

TEACHER NOTES
If potassium biphthalate (also called potassium hydrogen phthalate) solid is available it is advisable that you standardize the NaOH solution before distributing to students. If an analytical balance is available, use it to mass the KHP and be sure to report the molarity of the NaOH to four significant digits. By using the KHP to standardize the solution, problems such as CO_2 dissolved in the water and the hygroscopic nature of the NaOH can be ignored while mixing the solution.

To Prepare the ~ 0.5M NaOH Solution:

Mass 20. grams of sodium hydroxide pellets into a 1.0 L volumetric flask. Add enough distilled water to dissolve the NaOH. After the NaOH is dissolved, fill the flask to the mark with distilled water and transfer to a large container. Plastic gallon milk bottles provide adequate temporary storage, but do not store NaOH long term in a plastic bottle. Repeat the procedure and combine the two NaOH solutions. Mix thoroughly. *Note:* if you need to prepare more than one class's worth of NaOH you will either need to combine all the batches into one large container before standardizing or standardize and distribute each batch individually.

To Prepare the Phenolphthalein Solution:

You may purchase prepared solutions of phenolphthalein or prepare by mixing it from the pure powdered phenolphthalein. Mass 1.0 gram of powdered phenolphthalein in a beaker. Add methanol until the total mass is 100 grams. Stir and place in dropping bottles. Be sure to wash your hands well after mixing. Phenolphthalein has been used as a laxative.

To Standarize the NaOH with KHP:

Using an analytical balance, tare a waxed weighing paper or other container. Mass approximately 2.0 g of KHP on the weighing paper and transfer it to a 250 mL Erlenmeyer flask. Record the actual mass of KHP to as many decimal places as are available on your balance. Use a squirt bottle to wash the residue from the weighing paper into the flask. Add approximately 50 mL distilled water and 3 drops of phenolphthalein solution. Allow the KHP to dissolve.

Coat a buret with the NaOH solution by pouring approximately 5 mL of the solution into the buret and tipping and rolling the buret until the solution comes in contact with all the inside surfaces. Dispense this NaOH rinse through the tip of the buret into a waste beaker. Repeat this procedure two more times. Fill the coated buret with the NaOH solution and place in the right side of the double buret clamp. Place a white piece of paper under the KHP flask and titrate the KHP solution to a faint pink endpoint. Record the volume of NaOH required to reach the endpoint. Repeat the entire procedure two more times. Calculate the molarity of the NaOH solution for each trial and average the molarities. Write the average molarity of the NaOH on the bottle.

Sample Calculation:

At the endpoint of the titration: $\text{mol KHP} = \text{mol NaOH}$

$$\frac{2.156 \text{ g KHP}}{1} \times \frac{\text{mol KHP}}{202.2 \text{ g KHP}} = 0.01066 \text{ mol KHP and } \therefore 0.01066 \text{ mol NaOH}$$

$$\frac{0.01066 \text{ mol NaOH}}{1} \times \frac{1}{22.01 \text{ mL NaOH}} \times \frac{1000 \text{ mL}}{\text{L}} = 0.4843 \frac{\text{mol}}{\text{L}}$$

If this is the first time students have actually performed a titration, you should demonstrate the correct technique of a two buret titration. Be sure to show them how to manipulate the stopcock with the left hand and swirl the flask with the right hand. It is a good idea to "over shoot" the endpoint and back titrate a few times to show the students how this is done. Read the volume on the buret by placing a piece of white paper behind it to intensify the meniscus. Record the volume of both solutions to the nearest 0.01 mL. Since you have already standardized your NaOH, you can use the demonstration titrations to determine the actual percentage of acetic acid in the vinegar. It will be your choice whether to give this value to the students and have them calculate a percent error, or whether you use this value as part of your grading rubric.

Time can be saved on lab day by having the burets coated and set up beforehand. If there are several sections of chemistry classes performing the lab, time may be saved by clearly labeling the burets and capping them between classes. Allow the last class to clean the burets. Be sure to demonstrate the coating technique and emphasize the importance of this step for all classes.

POSSIBLE ANSWERS TO THE CONCLUSION QUESTIONS AND SAMPLE DATA

Molarity of NaOH: _____ 0.4843 M _____

Data Table				
	Titration 1	Titration 2	Titration 3	Titration 4
Initial volume of Vinegar	0.45 mL	11.35 mL	22.50 mL	33.65 mL
Final volume of vinegar	11.35 mL	22.50 mL	33.65 mL	45.00 mL
Volume of vinegar used	10.90 mL	11.15 mL	11.15 mL	11.35 mL
Initial volume of NaOH	0.30 mL	21.00 mL	0.50 mL	20.00 mL
Final volume of NaOH	21.00 mL	42.50 mL	20.00 mL	39.20 mL
Volume of NaOH used	20.70 mL	21.50 mL	19.50 mL	19.20 mL

ANALYSIS

1. Calculate the molarity of the vinegar for each titration.

   - Calculation of molarity of vinegar: $M_aV_a=M_bV_b$
     Example:

     $(M_{acid})(10.90\,mL) = (0.4843\,M)(20.70\,mL) = 0.9197\,M$

     a. Titration 1: 0.9197 M

     b. Titration 2: 0.0339 M

     c. Titration 3: 0.8470 M

     d. Titration 4: 0.8193 M

     e. Average molarity of vinegar: 0.8800 M

2. Calculate the percent deviations for your titration trials. The formula is given below. Show your work.

   $|M_{titration_1} - M_{ave.}|=\text{Deviation}_1$

   $|M_{titration_2} - M_{ave.}|=\text{Deviation}_2$

   $|M_{titration_3} - M_{ave.}|=\text{Deviation}_3$

   $|M_{titration_4} - M_{ave.}|=\text{Deviation}_4$

   $$\frac{\text{Deviation}_1 + \text{Deviation}_2 + \text{Deviation}_3 + \text{Deviation}_4}{4}=\text{Deviation}_{ave.}$$

   $$\frac{\text{Deviation}_{ave}}{M_{ave}} \times 100\%=\%\text{Deviation}$$

   $$\text{Deviation}_{ave}=\frac{|0.9197\text{-}0.8800|+|0.9339\text{-}0.8800|+|0.8470\text{-}0.8800|+|0.8193\text{-}0.8800|}{4}=0.04683$$

   $$\%\text{Deviation}=\frac{0.04683}{0.8800} \times 100\%=5.321\%$$

3. Convert the molarity (mol/L) of the vinegar to grams vinegar/L. Show your work.

   - $$\frac{0.8800\ \cancel{mol}}{L} \times \frac{60.06\ g\ acetic\ acid}{\cancel{mol}} = \frac{52.85\ g\ acetic\ acid}{L}$$

4. Since vinegar is a very dilute solution, you may assume that the density of the solution is 1.0 g/mL. Therefore, the mass of one liter of solution would be __1000__ grams.

5. Calculate the mass percentage of acetic acid in the vinegar.

- $$\frac{\text{gram}_{\text{acetic acid}}}{\text{gram}_{\text{solution}}} \times 100\%$$

  $$\frac{52.85\,\text{g}_{\text{acetic acid}}}{1000\,\text{g}_{\text{solution}}} \times 100\% = 5.285\%$$

CONCLUSION QUESTIONS

1. Write the chemical formula for acetic acid.

- $HC_2H_3O_2$ or CH_3COOH

2. The structural formula for acetic acid is shown below. Circle the ionizable hydrogen.

-

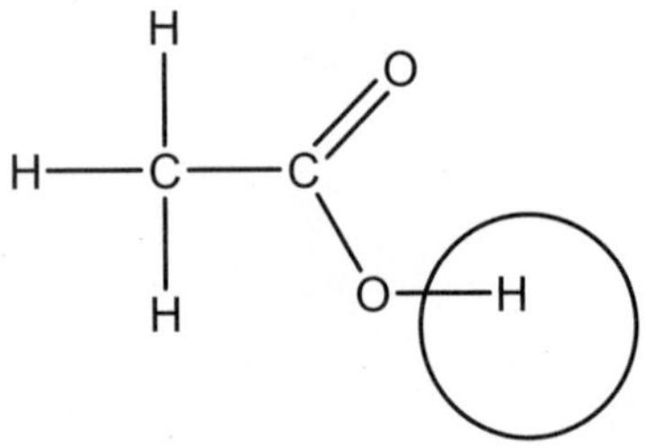

3. Using the structural formula, explain why acetic acid has four hydrogens, but is a monoprotic acid.

- The hydrogens that are on the carbon are not ionizable.

4. Acetic acid is a weak acid. What is the difference between a weak acid and a strong acid.

- In strong acids, the ionizable hydrogens ionize 100%. In weak acids, only a small percentage of hydrogens will ionize.

5. What would happen to the strength of the acid if the three hydrogens that were on the adjoining carbon were replaced with fluorines? Explain your answer.

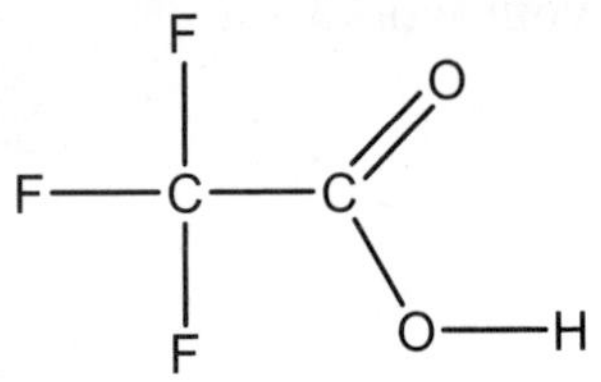

-

- The fluorines are much more electronegative than the hydrogens. They pull the electron density away from the ionizable hydrogen, thus weakening the bond between the oxygen and hydrogen. This causes more of the hydrogens to ionize and makes the acid stronger.

6. How would the following laboratory mistakes affect your calculated percent of acetic acid in vinegar? You should state clearly whether the error will cause the calculated percent error to increase, decrease, or remain the same. Your answer must be mathematically justified.

a. The buret was wet with water and you failed to coat it with NaOH solution.
   - The NaOH solution would be more dilute than recorded. Therefore it would take more mL of NaOH to titrate to an endpoint and the percentage of acetic acid calculated would be lower than the actual value.

   - $$(M_{NaOH})(V_{NaOH}) = (M_{HOAc})(V_{HOAc})$$
     If V_{NaOH} is greater with all other values the same, then M_{HOAc} will be less.

   - Teacher note: The accepted abbreviation of acetic acid is HOAc. Students need to be exposed to this.

b. You added 60 mL distilled water to the Erlenmeyer flask before you started the titration rather than 50 mL.
   - The amount of distilled water added to the flask does not affect the number of moles of either acetic acid or sodium hydroxide. Therefore there would be no effect. There is no change in the number of moles.

   - $$(M_{NaOH})(V_{NaOH}) = (M_{HOAc})(V_{HOAc})$$
     There is no change in the number of moles.

Titrations—Titrations
Determining the Percent of Acetic Acid in Vinegar

Vinegar is a solution of acetic acid. Different brands and types of vinegar range from approximately 3.0 % to 10% vinegar by mass. In this lab you will analyze a store brand of vinegar using a technique called titration.

PURPOSE
In this activity you will titrate a sample of vinegar using a standardized solution of NaOH and calculate the percent of acetic acid in vinegar.

MATERIALS
ring stand

2 burets (25.0 mL or 50.0 mL)

dropping bottle containing 1% methanolic solution of phenolphthalein

piece of white paper

2.0 L white vinegar—any store brand

distilled water

double buret clamp

2 buret caps

1 Erlenmeyer flasks (250 mL)

3 beakers (one for waste and two for transferring NaOH and vinegar solutions (size does not matter)

NaOH solution (approximately (0.5 M)

labels

Safety Alert
1. Wear your goggles.
2. Wash your hands before leaving the lab.

PROCEDURE
1. You will work with a partner in this lab. Each of you will perform two titrations. You will combine your data so that you have the data for four titrations. Alternate doing the titrations.

2. Obtain approximately 60 mL of vinegar in a beaker and label the beaker. Obtain approximately 100 mL of standardized NaOH solution in another beaker and label it. You will use these solutions to coat and fill your burets.

3. Each of you will need to prepare one buret for the titration. Burets need to be coated with the solution that will go in them. One of you will coat and fill a buret with vinegar. The other person should coat and fill a buret with the NaOH solution. Coat a buret with solution by pouring approximately 5 mL of the solution into the buret and tipping and rolling the buret until the solution comes in contact with all the inside surfaces. Dispense through the tip into a waste beaker. Repeat two more times. Fill the coated buret with the solution and place in the double buret clamp. Open the stopcock and run the solution through the tip to dispense several drops into the waste beaker. Put the vinegar (acid) in the left side of the double buret clamp and the NaOH (base) in the right side clamp. Label each buret. If you always put the acid on the left and the base on the right, they will be alphabetical and there is less chance of getting them mixed up. However the burets should always be labeled as well.

4. Your teacher has standardized the NaOH. Record the molarity of the NaOH on your student answer page.

5. Once both burets are filled, record the initial volumes of each in your data table. It is not necessary to have the volumes at the zero point. It is only necessary to know the initial volumes.

6. Drain approximately 10.0 mL of vinegar into a 250-mL Erlenmeyer flask. Add approximately 50 mL distilled water and 3 drops phenolphthalein indicator.

7. Titrate the vinegar with the NaOH solution to a faint pink endpoint. Be sure to use your left hand to manipulate the stopcock and your right hand to swirl the flask. This technique gives you better control.

8. If you over shoot the endpoint, back titrate with the vinegar solution until the pink color in the flask disappears and then titrate with the NaOH until a good endpoint is reached. A good endpoint will have a pale pink color which is sustained for 30 seconds.

9. Record final volumes from each buret and determine the volume of each solution used.

10. Change partners and repeat the procedure. It may be necessary to refill the burets between titrations. You will need to do a total of four titrations.

11. Follow your teacher's instructions for cleaning your lab station. If your teacher instructs, clean the burets by draining the solution into the waste beakers and then filling them with distilled water and allowing the water to drain out through the tip into a beaker. Repeat three times. Your teacher may or may not want you to clean the burets between classes.

Name _____

Period _____

Titrations—Titrations
Determining the Percent of Acetic Acid in Vinegar

DATA AND OBSERVATIONS

Molarity of NaOH_____

Data Table				
	Titration 1	Titration 2	Titration 3	Titration 4
Initial volume of Vinegar				
Final volume of vinegar				
Volume of vinegar used				
Initial volume of NaOH				
Final volume of NaOH				
Volume of NaOH used				

ANALYSIS

1. Calculate the molarity of the vinegar for each titration.
   a. Titration 1:

   b. Titration 2:

   c. Titration 3:

   d. Titration 4:

   e. Average molarity of vinegar

2. Calculate the percent deviations for your titration trials. The formula is given below. Show your work.

$$|M_{titration_1} - M_{ave.}| = Deviation_1$$

$$|M_{titration_2} - M_{ave.}| = Deviation_2$$

$$|M_{titration_3} - M_{ave.}| = Deviation_3$$

$$|M_{titration_4} - M_{ave.}| = Deviation_4$$

$$\frac{Deviation_1 + Deviation_2 + Deviation_3 + Deviation_4}{4} = Deviation_{ave.}$$

$$\frac{Deviation_{ave}}{M_{ave}} \times 100\% = \%Deviation$$

3. Convert the molarity (mol/L) of the vinegar to grams vinegar/L. Show your work.

4. Since vinegar is a very dilute solution, you may assume that the density of the solution is 1.0 g/mL. Therefore, the mass of one liter of solution would be _____ grams.

5. Calculate the mass percentage of acetic acid in the vinegar.

$$\frac{gram_{acetic\,acid}}{gram_{solution}} \times 100\%$$

CONCLUSION QUESTIONS

1. Write the chemical formula for acetic acid.

2. The structural formula for acetic acid is shown below. Circle the ionizable hydrogen.

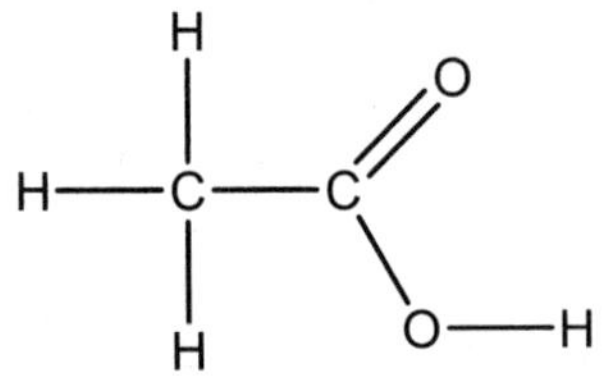

3. Using the structural formula, explain why acetic acid has four hydrogens, but is a monoprotic acid.

4. Acetic acid is a weak acid. What is the difference between a weak acid and a strong acid.

5. What would happen to the strength of the acid if the three hydrogens that were on the adjoining carbon were replaced with fluorines? Explain your answer.

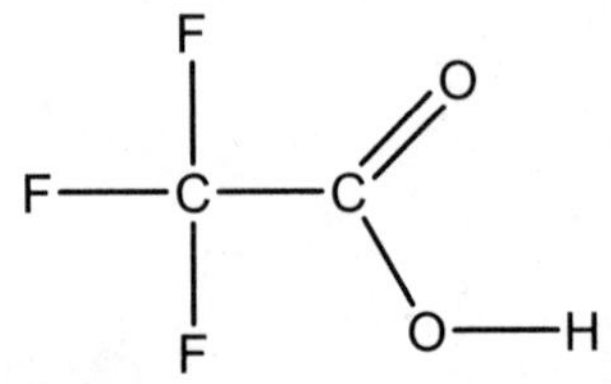

6. How would the following laboratory mistakes affect your calculated percent of acetic acid in vinegar? You should state clearly whether the error will cause the calculated percent error to increase, decrease, or remain the same. Your answer must be mathematically justified.

    a. The buret was wet with water and you failed to coat it with NaOH solution.

    b. You added 60 mL distilled water to the Erlenmeyer flask before you started the titration rather than 50 mL.

Thermodynamics
Differentiating Thermodynamic Data
ΔH, ΔS, ΔG

OBJECTIVE

Students will be able to calculate enthalpy, entropy and free energy from different types of data and explain the differences between these three terms.

LEVEL

Chemistry

NATIONAL STANDARDS

UCP.1, UCP.2, B.5, B.6, G.2

TEKS

5(A), 5(B), 15(A)

CONNECTIONS TO AP

AP Chemistry:

III. Reactions E. Thermodynamics 1. State functions 2. First law: change in enthalpy; heat of formation; heat of reaction; Hess's law; heats of vaporization and fusion; calorimetry 3. Second law: entropy; free energy of formation; free energy of reaction; dependence of change in free energy on enthalpy and entropy changes

TIME FRAME

2 to 3 45-minute lecture sessions
45 minutes of homework

MATERIALS

student note sheet   calculator

TEACHER NOTES

The student notes provided can be used as the basis for your lecture on this topic. Many first year chemistry texts introduce students to the concept of enthalpy but fail to complete the thermodynamic story. This lesson provides the students with the opportunity to calculate enthalpy, entropy and free energy and explore the relationship between all of these through the use of the Gibb's free energy equation. It is assumed that students will have prior knowledge of endothermic and exothermic reaction diagrams and have previously performed calculations using specific heat.

This lesson will serve as an introduction to thermodynamics and should be completed before performing the lab on *Hess's Law* also in this book.

For more information, refer to a freshman college chemistry text of your choice. If students need more practice you should assign questions from the college text.

Example problems from the student notes:
Point out to students that enthalpy cannot be measured directly—it must be calculated. There are several ways to calculate enthalpy values depending on what information is available.

Example 1: Stoichiometry (the energy term is used just like any other substance when making comparisons)

The combustion of propane can be represented by the following equation:

$$C_3H_{8(g)} + 5\ O_{2(g)} \rightarrow 3\ CO_{2(g)} + 4\ H_2O_{(l)} + 2220\ kJ$$

Is the reaction endothermic or exothermic? Hint: is energy released or absorbed?
- Exothermic—the energy term is a product so energy is released

What is the value of ΔH with proper sign?
- $\Delta H = -2220$ kJ (the sign is negative for an exothermic reaction)

If 85.00 grams of propane were burned, how much energy would be released?
- $85.00\ \cancel{g}\ C_3H_8 \times \dfrac{1\ \cancel{mol}\ C_3H_8}{44.0\ \cancel{g}\ C_3H_8} \times \dfrac{-2220\ kJ}{1\ \cancel{mol}\ C_3H_8} = -4290\ kJ$

Example 2:
In a coffee cup calorimeter, 50.0 mL of 1.0 M NaOH and 50.0 mL of 1.0 M HCl are mixed. Both solutions were originally at 23.5°C. After the reaction, the final temperature is 29.9°C. Assuming that all solutions have a density of 1.0 g/cm^3, and a specific heat capacity of 4.184 J/g°C, calculate the enthalpy change for the neutralization of HCl by NaOH. Assume that no heat is lost to the surroundings or the calorimeter.
- Enthalpy $= q = m \bullet Cp \bullet \Delta T$
  m $= 100.0$ grams (since the density of the solution is 1.00 g/mL, 100.0 mL $= 100.0$ g)
  $C_p = 4.184$ J/ g°C
  $\Delta T = 29.9°C - 23.5°C = 6.4\ °C$
  $q = (100.0\ g)\ (4.184\ J/\ g°C)\ (6.4\ °C)$
  $= -2700$ J (negative sign because energy was released since the temperature increased)

Example 3:
Using the standard heats of formation given, calculate the $\Delta H°_{rxn}$ for:

$$CH_{4(g)} + 2\ O_{2(g)} \rightarrow CO_{2(g)} + 2\ H_2O_{(g)}$$

Substance	$\Delta H°_f$ (kJ/mol)
Methane, $CH_{4(g)}$	-74.81
Carbon dioxide, $CO_{2(g)}$	-393.5
Water vapor, $H_2O_{(g)}$	-241.8

In solving this problem, have students get used to writing the equation first and then substituting in numbers.

TEACHER PAGES

- $\Delta H^{\circ}_{rxn} = \Sigma\, \Delta H^{\circ}_{products} - \Sigma\, \Delta H^{\circ}_{reactants}$

- $\Delta H^{\circ}_{rxn} = [1(-393.5) + 2(-241.8)] - [1(-74.81) + 2(0)] = -802.3$ kJ   (exothermic)

- You may have to remind students that oxygen is a diatomic molecule and therefore has an enthalpy of zero. The most common mistake that students make is to subtract in the wrong order (reactants minus products) thereby ending up with the incorrect sign. Students also neglect to distribute the negative sign across *each* term within the $\Sigma\, \Delta H^{\circ}_{reactants}$ term.

Example 4:

Given the following equations and the enthalpies:

$$C_2H_{4(g)} + 3\,O_{2(g)} \rightarrow 2\,CO_{2(g)} + 2\,H_2O_{(l)} \qquad \Delta H^{\circ}_1 = -1411 \text{ kJ/mol reaction}$$

$$C_2H_5OH_{(l)} + 3\,O_{2(g)} \rightarrow 2\,CO_{2(g)} + 3\,H_2O_{(l)} \qquad \Delta H^{\circ}_2 = -1367 \text{ kJ/mol reaction}$$

Calculate the heat of reaction for the following equation:

$$C_2H_{4(g)} + H_2O_{(l)} \rightarrow C_2H_5OH_{(l)} \qquad \Delta H^{\circ}_{rxn} = \text{??}$$

- The second equation must be reversed to yield ethanol. Change the sign of ΔH°_2.

$$2\,CO_{2(g)} + 3\,H_2O_{(l)} \rightarrow C_2H_5OH_{(l)} + 3\,O_{2(g)} \qquad \Delta H^{\circ}_2 = +1367 \text{ kJ/mol reaction}$$

$$C_2H_{4(g)} + 3\,O_{2(g)} \rightarrow 2\,CO_{2(g)} + 2\,H_2O_{(l)} \qquad \Delta H^{\circ}_1 = -1411 \text{ kJ/mol reaction}$$

- Cancel common terms, add up the enthalpies to yield: $\Delta H^{\circ}_{rxn} = -44$ kJ/mol reaction

$$C_2H_{4(g)} + H_2O_{(l)} \rightarrow C_2H_5OH_{(l)} \qquad \Delta H^{\circ}_{rxn} = \text{??}$$

Example 5:

Use the bond energies given to estimate the enthalpy of reaction for the same reaction that we calculated above:

$$C_2H_{4(g)} + H_2O_{(l)} \rightarrow C_2H_5OH_{(l)} \qquad \Delta H^{\circ}_{rxn} = \text{??}$$

Bond	Bond Energy (kJ/mol of bonds)
C-C	346
C=C	602
C-H	413
O-H	463
C-O	358

In solving this problem, write the equation:
$$\Delta H^{\circ}_{rxn} = \Sigma \text{ Bond energy}_{reactants} - \Sigma \text{ Bond energy }_{products}$$

Or maybe easier to use the term:
$$\Delta H^{\circ}_{rxn} = \Sigma \text{ Bonds broken} - \Sigma \text{ Bond formed}$$

Students need to draw the structures for each substance and tally the number and type of bonds:

Broken: (reactants)	Formed: (products)
(1) C=C	(1) C-C
(4) C-H	(5) C-H
(2) O-H	(1) C-O
	(1) O-H

Simplify the list by canceling common terms:

Broken: (reactants)	Formed: (products)
(1) C=C	(1) C-C
	(1) C-H
(1) O-H	(1) C-O

Calculate:
$$\Delta H^{\circ}_{rxn} = [1(602) + 1(463)] - [1(346) + 1(413) + 1(358)]$$
$$= 1065 - 1117$$
$$= -52 \text{ kJ}$$

(Note: this is very close to the value that was calculated above—bond energies are only average values and will not always yield the exact results)

1. It is often useful to draw structural formulas to work bond energy problems. Then students can visualize the bonds that need to be broken and the bonds that are forming. Remember that it is an endothermic process to break bonds and an exothermic process to make bonds, it is quite easy to determine whether the reaction is exothermic or endothermic.

C_2H_4 H_2O C_2H_5OH

In looking at this structural reaction, you can see that you need to break 4 C-H bonds, 1 C=C, and 2 O-H bonds on the left side of the reaction. That is (4 x 413 kJ) + (602 kJ) + (2 x 463 kJ) for a total of 3180 kJ of energy put into the system to break bonds. On the right hand side of the reaction, you are forming 5 C-H bonds, 1 C-C bond, 1 C-O bond, and one O-H bond. That is (5 x 413 kJ) + (346 kJ) + (358 kJ) + (463 kJ) for a total of 3232 kJ of energy being given off. There is more energy given off than put into the reaction, so the net energy is -52 kJ.

Since tables of bond energies are an average of bond energies in many compounds, energy calculations based on bond energies are approximate.

Example 6:
Predict whether the entropy, ΔS, value increases or decreases for the following situations.

1. melting of ice at room temperature
    - entropy increases—a phase change from organized solid to more disorganized liquid

2. condensation of water vapor on a mirror
    - entropy decreases—a phase change from a disorganized gas to a more organized liquid

3. an iron nail rusting outside
    - entropy decreases—4 moles of pure iron in the solid state react with 3 moles of pure oxygen gas to produce 2 moles of a much more ordered solid compound
    - $4Fe_{(s)} + 3\ O_{2(g)} \rightarrow 2\ Fe_2O_{3(s)}$
    - There are 7 moles of reactants to produce 2 moles of product.

4. $N_{2(g)} + 3\ H_{2(g)} \rightarrow 2\ NH_{3(g)}$
    - entropy decreases—4 moles of gas produce 2 moles of gas

5. $NaHCO_{3(s)} + HC_2H_3O_{2(l)} \rightarrow NaC_2H_3O_{2(aq)} + H_2O_{(l)} + CO_{2(g)}$
    - Entropy increases—a solid and a liquid yield an aqueous solution, a liquid and a gas—or 2 moles of reactant yield 3 moles of product

Example 7:
Calculate the standard molar entropy for nitrogen gas given the following information:

$$N_2H_{4(l)} + 2\ H_2O_{2(l)} \rightarrow N_{2(g)} + 4\ H_2O_{(g)} \qquad\qquad \Delta S^{\circ}_{rxn} = +\ 605.9\ J/K$$

Substance	Entropy S$^{\circ}$, J/mol·K
$N_2H_{4(l)}$	121.2
$H_2O_{2(l)}$	109.6
$N_{2(g)}$	??
$H_2O_{(g)}$	188.7

- The first step in solving this problem is to write the entropy formula. Point out to students that solving for entropy is just as easy as solving for enthalpy—just watch for different units. If the students question why the Δ sign disappeared in the equation remind them that the values are absolute entropies since a perfect crystal is 0.

- $\Delta S^{\circ}_{rxn} = \Sigma\ S^{\circ}_{products} - \Sigma\ S^{\circ}_{reactants}$

- Notice that this question is not asking for the entropy of the reaction, but rather for one of the products in the reaction. Have students pay close attention to what answer is being sought. The answer is given here along with the equation.

    $605.9 = [1(N_2) + 4(188.7)] - [1(121.2) + 2(109.6)]$

    $605.9 = (x + 754.8) - (340.4)$

    $946.3 = x + 754.8$

    $x = N_2 = 191.5$ J/mol • K

Example 8:

Given the following information, determine whether the reaction is spontaneous at room temperature, 25°C.

$$NH_{3(g)} + HCl_{(g)} \rightarrow NH_4Cl_{(s)} \qquad \Delta H° = -176 \text{ kJ} \qquad \Delta S° = -285 \text{ J/ K}$$

- In solving this problem with students, first have them predict the spontaneity. Enthalpy is negative so the energy change is favored. Entropy, however is very negative, indicating a more ordered state and is not favored. In a reaction like this, the spontaneity is temperature dependent. Using the free energy equation, solve. Be careful to convert temperature to kelvins and Joules for entropy to kJ!

- $\Delta G° = \Delta H° - T \Delta S°$

- $\Delta G° = (-176 \text{ kJ}) - [(298K)(-.285 \text{ kJ/K})]$

- $\Delta G° = -91.1$ kJ   The negative sign indicates that the reaction is spontaneous at room temperature.

- Ask students what would happen if the temperature were increased to 500°C? Students should realize that as the temperature increases, the value of free energy will decrease and there will be a point at which the reaction will no longer be spontaneous. This is the true essence and beauty of Gibb's equation! Using Gibb's equation, we can predict spontaneity for all types of variations of enthalpy and entropy.

POSSIBLE ANSWERS TO THE ANALYSIS QUESTIONS

1. Calculate the enthalpy change for the following reaction if 25.00 grams of calcium carbonate were used. Is this reaction endothermic or exothermic?

$$CaCO_{3(s)} + 176 \text{ kJ} \rightarrow CaO_{(s)} + CO_{(g)}$$

- $25.00 \text{ g CaCO}_3 \times \dfrac{1 \text{ mol CaCO}_3}{100. \text{ g CaCO}_3} \times \dfrac{176 \text{ kJ}}{1 \text{ mol CaCO}_3} = 44.0 \text{ kJ}$

- The reaction is endothermic since the energy term is written as a reactant.

2. Calculate the $\Delta H°_f$ for ethane, $C_2H_{6(l)}$, given the following information:

$$2C_2H_{6(l)} + 7O_{2(g)} \rightarrow 4CO_{2(g)} + 6H_2O_{(l)} + 3386 \text{ kJ}$$

Substance	Enthalpy ΔH (kJ/mol)
$C_2H_{6(l)}$	??
$CO_{2(g)}$	-393.5
$H_2O_{(l)}$	-285.8

- $\Delta H^{\circ}_{rxn} = \Sigma \Delta H^{\circ}_{products} - \Sigma \Delta H^{\circ}_{reactants}$

- The ΔH°_{rxn}, is given in the equation. Be careful of the sign—this is an exothermic reaction so the value is negative.

- $-3386 \text{ kJ} = [4(-393.5) + 6(-285.8)] - [2(C_2H_6) + 7(0)]$

- $-3386 = (-1574 + -1714.8) - (2x)$
  $-97.2 = 2x$
  $x = \Delta H^{\circ}_f$ for ethane $= -48.6$ kJ/mol

3. Assume that no heat is lost to the surroundings and calculate the enthalpy change involved when 200. mL of 0.10M potassium hydroxide and 200. mL of 0.10M hydrochloric acid are mixed. Both solutions began at room temperature, 24.5 °C, and have densities very close to 1.00 g/mL. The highest temperature reached when the two solutions were mixed was 31.0 °C. Assume that the specific heat of the solution is equal to that of water, 4.184 J/g • K.

- Enthalpy $= q = m \cdot C_p \cdot \Delta T$
  m $= 400.0$ grams (since the density of the solution is 1.00 g/mL, 400.0 mL = 400.0 g)
  C_p $= 4.184$ J/ g°C
  ΔT $= 31.0°C - 24.5°C = 6.4$ °C
  q $= (400.0 \text{ g}) (4.184 \text{ J/ g°C}) (6.5 °C)$
  $= -10,900$ J (negative sign because energy was released since the temperature increased)

4. Find the enthalpy of reaction for:
   $2 SO_{3(g)} \rightarrow 2 SO_{2(g)} + O_{2(g)}$ $\Delta H = ??$

Given:

$S_{(s)} + O_{2(g)} \rightarrow SO_{2(g)}$ $\Delta H = -296.8$ kJ/mol

$S_{(s)} + 3/2 O_{2(g)} \rightarrow SO_{3(g)}$ $\Delta H = -395.6$ kJ/mol

Is the reaction endothermic or exothermic?

- Use Hess's Law of summation:

- Reverse equation 2: $SO_{3(g)} \rightarrow S_{(s)} + 3/2 O_{2(g)}$ $\Delta H = + 395.6$ kJ/mol

- Multiply both equations by 2:

  $2S_{(s)} + 2O_{2(g)} \rightarrow 2SO_{2(g)}$ $\Delta H = 2(-296.8 \text{ kJ/mol}) = -593.6$

  $2SO_{3(g)} \rightarrow 2S_{(s)} + 3O_{2(g)}$ $\Delta H = 2 (+ 395.6 \text{ kJ/mol}) = 791.2$

- Cancel terms and sum the enthalpies:

$$2SO_{3(g)} \rightarrow 2SO_{2(s)} + O_{2(g)} \qquad \Delta H = 791.2 + (-593.6) = 197.6 \text{ kJ/mol}$$

- The reaction is endothermic since the enthalpy has a positive value.

5. Using the bond energies, calculate the ΔH_{rxn} for:

$$CH_3Cl + Cl_2 \rightarrow CHCl_3 + H_2$$

What does this value tell you about the reaction?

Bond	Bond Energy (kJ/mol)
C-H	413
C-Cl	339
Cl-Cl	242
H-H	436

- $\Delta H^{\circ}_{rxn} = \Sigma$ Bond energy$_{reactants}$ $- \Sigma$ Bond energy$_{products}$

- Students need to draw the structures for each substance and tally the number and type of bonds:

- Broken: (reactants)   Formed: (products)
  (3) C-H   (1) C-H
  (1) C-Cl   (3) C-Cl
  (1) Cl-Cl   (1) H-H

- Simplify the list by canceling common terms:
  Broken: (reactants)   Formed: (products)
  (2) C-H
                            (2) C-Cl
  (1) Cl-Cl   (1) H-H

- Calculate:
  $\Delta H^{\circ}_{rxn} = [2(413) + 1(242)] - [2(339) + 1(436)]$
  $= 1068 - 1114$
  $= -46 \text{ kJ}$

- The reaction is exothermic when enthalpy is negative.

6. Predict whether the entropy of the system increases or decreases for the following situations and explain your reasoning:
   a. Boiling a pot of water
      - Entropy increases—moving from a liquid to a gas the particles become more disorganized.

   b. $2 NH_4NO_{3(s)} \rightarrow 2 N_{2(g)} + 4 H_2O_{(g)} + O_{2(g)}$
      - Entropy increases—2 moles of solid produce 7 moles of gas

7. Calculate the entropy for the following reaction and state the significance of your answer.

$$4 \text{ Al}_{(s)} + 3 \text{ O}_{2(g)} \rightarrow 2 \text{ Al}_2\text{O}_{3(s)}$$

Substance	Entropy S° (J/mol · K)
Al	28.3
O₂	205.0
Al₂O₃	50.92

- $\Delta S^{\circ}_{rxn} = \Sigma S^{\circ}_{products} - \Sigma S^{\circ}_{reactants}$

- $\Delta S^{\circ}_{rxn} = [2(50.94)] - [4(28.3) + 3(205.0)]$
  $= 101.88 - 728.2$
  $\Delta S^{\circ}_{rxn} = -672 \text{ J/K}$

- The entropy of this system is decreasing since the value for entropy is negative. This also makes logical sense looking at the number of moles and the states of matter, 4 moles of solid and 3 moles of gas form 2 moles of solid.

8. Will the following reaction occur spontaneously at 25°C? Justify your answer.

$$3 \text{ O}_{2(g)} \rightarrow 2 \text{ O}_{3(g)} \qquad \Delta H_{rxn} = 285 \text{ kJ} \qquad \Delta S_{rxn} = -137 \text{ J/K}$$

- The enthalpy value is positive—this is not favored; the entropy value is negative—this is not favored. This reaction will not occur spontaneously at any temperature since both driving forces are not favored.

9. The following reaction is nonspontaneous at 25° C since its entropy value is 16.1 J/K. At what temperature will the following reaction become spontaneous?

$$\text{CH}_{4(g)} + \text{N}_{2(g)} + 163.8 \text{ kJ} \rightarrow \text{HCN}_{(g)} + \text{NH}_{3(g)}$$

When looking for the point at which the temperature changes, the free energy value is equal to zero—this change would occur at just beyond equilibrium.

- $\Delta G^{\circ} = \Delta H^{\circ} - T \Delta S^{\circ}$
  $0 = (+163.8 \text{ kJ}) - [(? \text{ K}) (+0.0161 \text{ kJ/K})]$

  $K = \dfrac{163.8 \text{ kJ}}{0.0161 \text{ kJ/K}} = 10,174 \text{ K}$

10. Explain the significance of positive and negative values for each thermodynamic quantity—ΔH, ΔS and ΔG.

	ΔH	ΔS	ΔG
+ (positive)	Endothermic	More disorder	Not spontaneous
– (negative)	Exothermic	More order	Spontaneous

Thermodynamics
Differentiating Thermodynamic Data
ΔH, ΔS, ΔG

Do you often wonder about the energy changes that take place around you? Why do some things react and others do not? Why do some reactions get hot and others get cold? Energy is a key factor in everything that we do. From the food that we consume to the launching of a rocket—energy conversions are taking place constantly.

PURPOSE
In this activity you will calculate and explain the relevance of enthalpy, entropy and free energy.

MATERIALS
  student note sheet   calculator

CLASS NOTES

BASIC VOCABULARY

Energy (E) — The ability to do work or produce heat. The sum of all of the potential and kinetic energy in a system is known as the internal energy of the system. Energy is measured in Joules.

> **Potential energy** — Stored energy—in chemistry, this is usually referring to the energy stored in bonds.

> **Kinetic energy** — The energy of motion. Usually, the energy of particles in a sample is proportional to the Kelvin temperature. Kinetic energy depends on the mass and the square of the velocity of the object. $KE = \frac{1}{2}mv^2$

Law of Conservation of Energy — Energy is never created or destroyed.
Energy of the universe is constant.
First Law of Thermodynamics

Heat (q) — The transfer of energy in a physical or chemical process is known as heat. Heat always flows from a warmer object to a cooler one. Heat transfers because of a temperature difference. Remember that temperature is not a measure of energy—it just reflects the motion of particles. Heat (q) is measured in Joules while temperature is measured in degrees.

Specific heat capacity (Cp) — The energy required to raise the temperature of 1 gram of a substance by one degree Celsius.

$$\text{specific heat (Cp)} = \frac{\text{quantity of heat transferred (Joule or cal)}}{(g \text{ of material}) (\text{degrees of temperature change})}$$

Molar heat capacity — same as above but specific to one mole of substance
        (J/mol • K   or   J/mol • °C)

Units of Energy:

calorie — The amount of heat needed to raise the temperature of 1.00 gram of water by 1.00 • °C.

kilocalorie — This is the food calorie represented by a capital C.

Joule — This is the SI unit for energy. 1 calorie = 4.184 Joules

$$KE = \frac{1}{2}mv^2 = \frac{kg \cdot m^2}{s^2} = 1.0\,J$$

Enthalpy (H) — A measure of heat content at constant pressure.

Enthalpy of reaction (ΔH_{rxn}) — heat absorbed or released by a chemical reaction

Enthalpy of combustion (ΔH_{comb}) — heat released when one mole of a substance is burned in the presence of oxgen

Enthalpy of formation (ΔH_f) — heat absorbed or released when one mole of a compound is formed from elements in their standard states

Enthalpy of fusion (ΔH_{fus}) — heat absorbed to melt 1 mole of solid to liquid at the normal melting point

Enthalpy of vaporization (ΔH_{vap}) — heat absorbed to change 1 mole of liquid to gas at the normal boiling point

Enthalpy of neutralization (ΔH_{neut}) — heat absorbed or released when a neutralization reaction occurs

Endothermic — The net absorption of energy (heat) by the system. Energy is a reactant; **ΔH is +;** when baking soda and vinegar are mixed the reaction is very cold to the touch. (Think! Endo –energy absorbed *inside* the system)

Exothermic — The net release of energy (heat) by the system. Energy is a product; **ΔH is -;** when you are burning methane gas in the lab burner it produces heat; light sticks give off light which is also energy. (Think! Exo—energy is released *outside* the system)

State Function — A property that is independent of past or future behavior. It does not matter which road brought you to school today—you started at your house and ended here! There are probably lots of ways for that to happen just as there will be many ways to solve some of these problems. State function symbols are written in all capitals.

Entropy (S) — A measure of disorder in the system or a measure of chaos.
The Second Law of Thermodynamics—the universe tends toward a state of greater disorder in spontaneous reactions.

Gibbs Free Energy (G) — The amount of free energy available to do work.

Thermodynamics — The study of energy and its interconversions in physical and chemical processes.

Now, let's take a look at each thermodynamic property in detail.

ENTHALPY:

ΔH, enthalpy, is the difference between the potential energies of the products and the potential energies of the reactants. Enthalpy is one of the driving forces in a chemical reaction. In chemistry we calculate the change in enthalpy to find out if a system is endothermic or exothermic. Exothermic reactions are represented by a negative sign for enthalpy and endothermic reactions are represented by a positive sign for enthalpy. Exothermic reactions are favored in nature because systems like to have minimum energy states for stability. ΔH is a state function and is equal to q at constant pressure.

$\Delta H = -$ (exothermic) this is favored   **$\Delta H = +$ (endothermic)**

Enthalpy can be calculated from several sources including:

- Stoichiometry
- Calorimetry
- Tables of standard values
- Hess's Law
- Bond energies

STOICHIOMETRY:

Using a balanced chemical equation which includes the energy term, there are several pieces of information that you can determine.

Example 1:

The combustion of propane can be represented by the following equation:

$$C_3H_{8(g)} + 5\ O_{2(g)} \rightarrow 3\ CO_{2(g)} + 4\ H_2O_{(l)} + 2220\ kJ$$

1. Is the reaction endothermic or exothermic? Hint: is energy released or absorbed?
2. What is the value of ΔH with proper sign?
3. If 85.00 grams of propane were burned, how much energy would be released?

CALORIMETRY:

Calorimetry is the process of measuring heat based on observing a temperature change throughout a reaction. In a lab setting we use a simple polystyrene coffee-cup calorimeter to find the energy of a particular system. Using a polystyrene cup and reactants that begin at the same initial temperature, a reaction is performed and the highest temperature of the system is recorded. The only **data** needed to calculate the energy released or absorbed during a reaction includes: mass or volume of the reactants; initial and final temperatures.

Using the specific heat equation

$$q = m \cdot Cp \cdot \Delta T$$

you can calculate the energy released or absorbed. (**Note**: $q = \Delta H$ at these conditions since the pressure remains constant) One assumption that is made in the chemistry laboratory is that all of the heat lost is equal to the heat gained in a reaction. We ignore the fact that some heat may be lost to the polystyrene cup or to the environment.

q = quantity of heat (Joules or calories)
m = mass **in grams**
$\Delta T = T_f - T_i$ (final – initial)
Cp = specific heat capacity (J/g • °C)

Example 2:
In a coffee cup calorimeter, 50.0 mL of 1.0 M NaOH and 50.0 mL of 1.0 M HCl are mixed. Both solutions were originally at 23.5°C. After the reaction, the final temperature is 29.9°C. Assuming that all solutions have a density of 1.0 g/cm³, and a specific heat capacity of 4.184 J/g°C, calculate the enthalpy change for the neutralization of HCl by NaOH. Assume that no heat is lost to the surroundings or the calorimeter.

Tables of Standard Values:
A table of standard values shows enthalpies of formation, ΔH_f^o, for various compounds. The superscript 0 represents standard thermodynamic conditions. Each time you see this superscript on any thermodynamic quantity it means the value was determined at 25°C and 1 atmosphere of pressure and 1.0 M solutions. Elements and diatomic molecules in their most stable states have a standard heat of formation value of zero. Using the table of standard values you can calculate the reaction enthalpy with the following equation:

$$\Delta H^o_{rxn} = \Sigma \, \Delta H^o_{products} \, - \, \Sigma \, \Delta H^o_{reactants}$$

This equation translates into: the summation of the enthalpies of the products minus the summation of the enthalpies of the reactants. The equation must be balanced and you must multiply by the number of moles of each product and reactant to obtain the summation. The values listed in the table are given for only 1 mole of each compound.

Example 3:
Using the standard heats of formation given, calculate the ΔH^o_{rxn} for:

$$CH_{4(g)} + 2\,O_{2(g)} \rightarrow CO_{2(g)} + 2\,H_2O_{(g)}$$

Substance	ΔH^o_f (kJ/mol)
Methane, $CH_{4(g)}$	-74.81
Carbon dioxide, $CO_{2(g)}$	-393.5
Water vapor, $H_2O_{(g)}$	-241.8

Hess's Law:
Since enthalpy is a state function and does not depend on the pathway, we can sum up the enthalpy for a series of equations to find the enthalpy for an overall reaction. The equation for Hess's Law:

$$\Delta H^o_{rxn} = \Delta H^o_1 + \Delta H^o_2 + \ldots.$$

Helpful Hints:

- First decide how to rearrange equations so that reactants and products are on the appropriate sides of the arrows.
- If the equations are reversed, the sign of ΔH must be reversed.
- If the equations are multiplied to get a correct coefficient, multiply the ΔH by this coefficient since ΔH's are in kJ/mol (division applies similarly).
- Check to ensure that everything cancels out to give you the exact equation you want.
- Hint** It is often helpful to begin by working backwards from the answer that you want. In other words—write the final equation first!

Example 4:
Given the following equations and the enthalpies:

$C_2H_{4(g)} + 3\ O_{2(g)} \rightarrow 2\ CO_{2(g)} + 2\ H_2O_{(l)}$ $\qquad$ $\Delta H^{\circ}_1 = -1411$ kJ/mol reaction

$C_2H_5OH_{(l)} + 3\ O_{2(g)} \rightarrow 2\ CO_{2(g)} + 3\ H_2O_{(l)}$ $\qquad$ $\Delta H^{\circ}_2 = -1367$ kJ/mol reaction

Calculate the heat of reaction for the following equation:

$C_2H_{4(g)} + H_2O_{(l)} \rightarrow C_2H_5OH_{(l)}$ $\qquad$ $\Delta H^{\circ}_{rxn} = ??$

BOND ENERGIES:
Bond energy is defined as the energy needed to break one mole of bonds in the gaseous substance to form gaseous products. It always requires energy to break a chemical bond. Bond energy is essentially a measure of bond strength. The stronger the bond, the greater the amount of energy necessary to break the bond. Triple bonds require more energy to break than do double or single bonds. The equation used to calculate the enthalpy of reaction from bond energies is:

$$\Delta H^{\circ}_{rxn} = \Sigma \text{ Bond energy}_{reactants} - \Sigma \text{ Bond energy}_{products}$$

Essentially, the total energy of the **bonds broken** minus the total energy of the **bonds formed**.

Example 5:
Use the bond energies given to estimate the enthalpy of reaction for the same reaction that we calculated above:

$C_2H_{4(g)} + H_2O_{(l)} \rightarrow C_2H_5OH_{(l)}$ $\qquad$ $\Delta H^{\circ}_{rxn} = ??$

Bond	Bond Energy (kJ/mol of bonds)
C-C	346
C=C	602
C-H	413
O-H	463
C-O	358

ENTROPY

It was mentioned above that nature tends to favor exothermic reactions. Yet there are some endothermic reactions in nature that do spontaneously occur. When mixing baking soda and vinegar the reaction is quite cold to the touch. What is the driving force behind a reaction like this? Entropy, ΔS, is the second driving force for chemical reactions. The ΔS value represents a measure of disorder or chaos. Nature prefers chaos or disorganization. In the baking soda and vinegar reaction a solid and a liquid are mixed and a gas is produced the reaction gets more disorganized. This increase in entropy is enough to offset the unfavorable endothermic enthalpy and make the reaction spontaneous.

$\Delta S = +$ (more disorder) this is favored   **$\Delta S = -$ (less disorder)**

You can often predict the entropy of a system based on physical evidence. The list below outlines some of the situations in which entropy increases.

1. The entropy of a substance always increases as it changes state from solid to liquid to gas. Solids are the most organized state while gasses are the most disorganized state.
2. When a pure solid or liquid dissolves in a solvent, the entropy of the substance increases.
3. Entropy generally increases with increasing molecular complexity. For example, consider the crystal structure of KCl vs. $CaCl_2$. $CaCl_2$ would have greater entropy since there are more moving electrons causing disorder in this molecule.
4. Reactions that increase the number of moles of particles from the reactants to the products often increase the entropy of the system.

Example 6:
Predict whether the entropy, **ΔS,** value increases or decreases for the following situations.

1. melting of ice at room temperature
2. condensation of water vapor on a mirror
3. an iron nail rusting outside
4. $N_{2(g)} + 3 H_{2(g)} \rightarrow 2 NH_{3(g)}$
5. $NaHCO_{3(s)} + HC_2H_3O_{2(l)} \rightarrow NaC_2H_3O_{2(aq)} + H_2O_{(l)} + CO_{2(g)}$

Entropy values can also be calculated from data much like enthalpy values. When using a table of standard values for entropy, the units are usually written in J/mol · K and even elements and diatomic molecules will have an entropy value. In fact, the Third Law of Thermodynamics states that the entropy of a perfect crystal is zero. This provides the reference for our absolute entropy values.

Table of Standard Values: $\Delta S^{\circ}_{rxn} = \Sigma S^{\circ}_{products} - \Sigma S^{\circ}_{reactants}$ (does this look familiar?)

Example 7:
Calculate the standard molar enthalpy for nitrogen gas given the following information:

$N_2H_{4(l)} + 2 H_2O_{2(l)} \rightarrow N_{2(g)} + 4 H_2O_{(g)}$ $\Delta S^{\circ}_{rxn} = + 605.9 \text{ J/K}$

Substance	Entropy S° J/mol · K
$N_2H_{4(l)}$	121.2
$H_2O_{2(l)}$	109.6
$N_{2(g)}$	??
$H_2O_{(g)}$	188.7

FREE ENERGY:

We know that the two driving forces behind any chemical reaction are enthalpy and entropy. We know that minimum enthalpy (exothermic reactions) is favored and that maximum entropy (greater disorder) is favored. How can you determine if a chemical reaction will really occur spontaneously? Josiah Willard Gibbs may have asked the same question before formulating the Gibbs free energy equation. Gibbs identified a relationship between enthalpy and entropy in terms of another state function, G. The value obtained from solving for G is significant in that it represents the amount of free energy available to do useful work in a system. Elements in their free state have ΔG°_f equal to zero just as in enthalpy. A negative value for free energy represents a spontaneous reaction.

$\Delta G = -$ spontaneous $\Delta G = +$ not spontaneous $\Delta G = 0$ equilibrium

Free energy can also be calculated from the same type of standard values as enthalpy and entropy. The most useful equation, however, is the Gibbs equation where the relationship between the two driving forces and temperature is represented.

$$\Delta G^{\circ} = \Delta H^{\circ} - T \Delta S^{\circ}$$

There are two tricks to this equation:

1. The temperature must be in kelvins.
2. Watch carefully for the units! ΔH is usually given in kJ/mole and ΔS is usually given in J/mol · K.

Example 8:

Given the following information, determine whether the reaction is spontaneous at room temperature, 25°C.

$NH_{3(g)} + HCl_{(g)} \rightarrow NH_4Cl_{(s)}$ $\Delta H^{\circ} = -176$ kJ/mol $\Delta S^{\circ} = -285$ J/mol · K

SUMMARY OF THERMODYNAMIC DATA AND THEIR RELATIONSHIPS:

ΔH	ΔS	Result (ΔG)
negative	positive	spontaneous at all temperatures
positive	positive	spontaneous at high temperatures
negative	negative	spontaneous at low temperatures
positive	negative	not spontaneous, ever

Thermodynamics
Differentiating Thermodynamic Data
ΔH, ΔS, ΔG

ANALYSIS

Use your class notes and solve the following. Be sure to show all of your work for full credit.

1. Calculate the enthalpy change for the following reaction if 25.00 grams of calcium carbonate were used. Is this reaction endothermic or exothermic?

$$CaCO_{3(s)} + 176\ kJ \rightarrow CaO_{(s)} + CO_{(g)}$$

2. Calculate the $\Delta H°_f$ for ethane, $C_2H_{6(l)}$, given the following information:

$$2C_2H_{6(l)} + 7O_{2(g)} \rightarrow 4CO_{2(g)} + 6H_2O_{(l)} + 3386\ kJ$$

Substance	Enthalpy ΔH (kJ/mol)
$C_2H_6(l)$	??
$CO_2(g)$	-393.5
$H_2O(l)$	-285.8

3. Assume that no heat is lost to the surroundings and calculate the enthalpy change involved when 200. mL of 0.10M potassium hydroxide and 200. mL of 0.10M hydrochloric acid are mixed. Both solutions began at room temperature, 24.5 °C, and have densities very close to 1.00 g/mL. The highest temperature reached when the two solutions were mixed was 31.0 °C. Assume that the specific heat of the solution is equal to that of water, 4.184 J/g • K.

4. Find the enthalpy of reaction for:

    $$2 \, SO_{3(g)} \; \rightarrow \; 2 \, SO_{2(g)} \; + O_{2(g)} \qquad\qquad \Delta H = \; ??$$

    Given:

    $$S_{(s)} + O_{2(g)} \rightarrow SO_{2(g)} \qquad\qquad \Delta H = -296.8 \text{ kJ/mol}$$

    $$S_{(s)} + 3/2 \, O_{2(g)} \rightarrow SO_{3(g)} \qquad\qquad \Delta H = -395.6 \text{ kJ/mol}$$

    Is the reaction endothermic or exothermic?

5. Using the bond energies, calculate the ΔH_{rxn} for:

    $$CH_3Cl + Cl_2 \; \rightarrow \; CHCl_3 \; + H_2$$

    What does this value tell you about the reaction?

Bond	Bond Energy (kJ/mol)
C-H	413
C-Cl	339
Cl-Cl	242
H-H	436

6. Predict whether the entropy of the system increases or decreases for the following situations and explain your reasoning:

    a. Boiling a pot of water

    b. $2 \, NH_4NO_{3(s)} \rightarrow 2 \, N_{2(g)} + 4 \, H_2O_{(g)} + O_{2(g)}$

7. Calculate the entropy for the following reaction and state the significance of your answer.

$$4\ Al_{(s)} + 3\ O_{2(g)} \rightarrow 2\ Al_2O_{3(s)}$$

Substance	Entropy S^o (J/mol • K)
Al	28.3
O_2	205.0
Al_2O_3	50.92

8. Will the following reaction occur spontaneously at 25°C? Justify your answer.

$$3\ O_{2(g)} \rightarrow 2\ O_{3(g)} \qquad \Delta H_{rxn} = 285\ kJ \qquad \Delta S_{rxn} = -137\ J/K$$

9. The following reaction is nonspontaneous at 25° C since its entropy value is 16.1 J/K. At what temperature will the following reaction become spontaneous?

$$CH_{4(g)} + N_{2(g)} + 163.8\ kJ \rightarrow HCN_{(g)} + NH_{3(g)}$$

10. Explain the significance of positive and negative values for each thermodynamic quantity—ΔH, ΔS and ΔG.

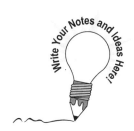

How Hot is a Candle?
Determining the Heat of Combustion of Paraffin

OBJECTIVE
Students will use a simple calorimeter to determine the molar heat of combustion of paraffin.

LEVEL
Chemistry

NATIONAL STANDARDS
UCP.2, UCP.3, A.1, A.2, B.5, E.1, E.2

TEKS
1(B), 5(A), 5(B), 5(C)

CONNECTIONS TO AP
AP Chemistry:
    III. Reactions E. Thermodynamics 1. State functions 2. First Law

TIME FRAME
45 minutes

MATERIALS
(For a class of 28 working in pairs)

14 ring stands	14 rings
14 empty aluminum soda cans, washed with the rings attached	14 glass rods, wooden dowels, or pencils containers for ice water
balances	14 tea candles
ice	14 graphing calculators
matches	14 stainless steel temperature probes
14 CBL2s or LabPros	

TEACHER NOTES
Basic calorimetry concepts are extremely important for students to understand. The concepts of the conservation of energy, heat exchange in a calorimeter, and definitions of molar heat of combustion, standard conditions, standard molar enthalpy of formation, and enthalpy of reactions are all essential ideas if students are to be successful in answering AP level thermodynamic questions.

ANSWERS TO PRE-LAB QUESTIONS

1. How much heat will it take to raise 750. g of water from 25.0°C to 75.0°C? The specific heat of water is 4.18 J/g °C.

$$q = mc\Delta T$$

- $q = \dfrac{750.\ g}{} \times \dfrac{4.18\ J}{g\ C^\circ} \times \dfrac{50.0\ C^\circ}{} \times \dfrac{kJ}{1000\ J} = 157\ kJ$

2. Sally Student combines 50.0 g of water at 20.0°C with 100. g of water at 70.0°C. What is the final temperature of the mixture?

- The essential concept is that the energy is conserved. Heat lost will equal heat gained.

$$(50.0\,g)(c)(T_f\text{-}20.0\,^\circ C) = (100.g)(c)(70.0\,^\circ C\text{-}T_f)$$

Since c is on both sides of the equation, it will cancel.

$$50.0\,gT_f\text{-}1000g^\circ C = 7000g^\circ C\text{-}100gT_f$$

- $150.0gT_f = 8000g^\circ C$

$$T_f = \dfrac{8000\,g^\circ C}{150.0\,g}$$

$$T_f = 53.3\,^\circ C$$

3. Given the following equation:
$$C_3H_8 + 5\,O_2 \rightarrow 3\,CO_2 + 4\,H_2O \quad \Delta H = \text{-}2200kJ/mol$$
How many kilojoules of heat are released when 3.64 g of propane, C_3H_8, are burned?

- This is essentially a stoichiometric problem. The enthalpy of reaction is expressed in kJ/mol or reaction (regardless of how the reaction is written). Many textbooks express the ΔH of reaction simply in kJ. For teachers who are teaching AP Chemistry, you should go to the discussion of this issue on wwww.apcentral.collegeboard.com.

- $\dfrac{3.64\ g\ C_3H_8}{} \times \dfrac{mol\ C_3H_8}{44.1\ g\ C_3H_8} \times \dfrac{2200\ kJ}{mol\ reaction} = 182\ kJ$

4. If the heat from problem 3 were absorbed by 2500 g of water at 18°C, what would be the final temperature of the water?

$$q = mc\Delta T$$

- $182kJ = \dfrac{2.50kg}{} \times \dfrac{4.18J}{g\,^\circ C} \times \dfrac{(T_f - 18^\circ C)}{} = 35.4^\circ C$

5. What is the molar mass of paraffin, $C_{25}H_{52}$?
   - 352 g/mol

POSSIBLE ANSWERS TO THE CONCLUSION QUESTIONS AND SAMPLE DATA

Mass of aluminum can	14.71 g
Mass of aluminum can and water	131.71 g
Mass of water	117.00 g
Mass of candle and glass plate before burning	15.22 g
Mass of candle and glass plate after burning	15.04 g
Mass of candle burned	0.18 g
Initial temperature of water	17.1 $^\circ$C
Final temperature of water	27.7 $^\circ$C
ΔT of water	10.6 $^\circ$C
Heat gained by water	$q = mc\Delta T$ $q = (\dfrac{117.00\cancel{g}}{})(\dfrac{4.18J}{\cancel{g}^\circ\cancel{C}})(\dfrac{10.6^\circ\cancel{C}}{}) = 5184J$
Heat lost from candle	Heat gained by water = Heat lost from candle 5184 J
Heat of combustion of the candle (kJ/g)	$(\dfrac{5184\cancel{J}}{})(\dfrac{kJ}{1000\cancel{J}})(\dfrac{1}{0.18g}) = \dfrac{28.8kJ}{g}$
Molar heat of combustion of paraffin (kJ/mol)	$(\dfrac{28.8kJ}{\cancel{g}})(\dfrac{352\cancel{g}}{mol}) = \dfrac{10137kJ}{mol}$
Theoretical molar heat of paraffin	14,800 kJ/mol
% Error $\dfrac{\lvert O-A \rvert}{A} \times 100\%$	$\dfrac{\left\lvert 10137\dfrac{kJ}{mol} - 14800\dfrac{kJ}{mol} \right\rvert}{14000\dfrac{kJ}{mol}} \times 100\% = 31.5\%$

1. Write a balanced equation for the combustion of paraffin.
   - $C_{25}H_{52} + 38\,O_2 \rightarrow 25\,CO_2 + 26\,H_2O$

2. How does your experimental heat of combustion of paraffin (kJ/g) compare with the heat of combustion of propane (kJ/g)? (See pre-lab questions.)
   - The molar heat of combustion of propane is -2200 kJ/mol; the theoretical molar heat of combustion of paraffin is -14800 kJ/mol. In order to answer the question, both of these quantities must be converted to kJ/g by dividing the heat by the molar mass of each substance. In units of kJ/g, propane has a heat of combustion of 50 kJ/g and theoretically, paraffin has a heat of combustion of 42 kJ/g. However, the question asks the students to compare their experimental data to the actual data of propane, so their answers may vary. The important issue is whether they converted the quantities to kJ/g to do the comparison.

3. How does your theoretical heat of combustion of paraffin (kJ/g) compare with the heat of combustion of propane (kJ/g)? (See pre-lab questions.)
   - This is answered in question 2.

4. How does the theoretical molar heat of combustion of paraffin (kJ/mol) compare with the molar heat of combustion of propane (kJ/mol)?
   - Measured in kJ/mol, the molar heat of combustion of paraffin is much higher than the molar heat of combustion of propane.

5. As a consumer of energy, which is the more useful quantity to you, heat of combustion (kJ/g) or molar heat of combustion (kJ/mol)? Explain your answer.
   - Consumers would generally buy their fuel by mass or by volume. In either case, it would be more practical to use kJ/g.

6. This lab has several sources of error. List as many as possible. Suggest possible corrections for these errors
   - The most significant error in the lab is the loss of heat to the surroundings. That causes the experimental value for the molar heat of combustion to be lower than it should be. Students should not be penalized for this. Other errors include the splashing of melted paraffin when the candle is blown out, errors in measurement, and, unfortunately, errors in calculations. These certainly deserve penalties in grade.

How Hot is a Candle?
Determining the Heat of Combustion of Paraffin

The quantity of heat released when a given amount of a substance burns is called the heat of combustion. The amount of heat released when one mole of that substance is burned is called the molar heat of combustion. These measurements are usually taken at thermodynamic standard conditions (298 K and 1 atm), and it is symbolized by ΔH°_{comb}. In this lab you will investigate the molar heat of combustion of paraffin, $C_{25}H_{52}$. The basic formula in use is $q = mc\Delta T$. Be sure you understand this concept.

PURPOSE
In this activity you will collect the heat produced from the burning of a small candle in a simple calorimeter and use the data to compute the molar heat of combustion of paraffin.

MATERIALS
ring stand

empty aluminum soda cans, washed with the ring pull tab still attached

balance

ice

matches

CBL2 or LabPro

ring

glass rod, wooden dowel, or pencil

container for ice water

tea candle

TI-83+ graphing calculator

stainless steel temperature probe

Safety Alert
1. Wear eye goggles.
2. You will be using matches and a candle. Be cautious of an open flame.
3. Dispose of used matches in waste baskets only after dipping them in water.

PROCEDURE
1. Answer the Pre-Lab questions on your student answer page.

2. Obtain one of the aluminum cans your teacher has provided. Mass the can and record the value in the data table on the student answer page. Place a glass rod, wooden dowel, or pencil through the pop top and support on a ring and stand as shown in the diagram.

3. Obtain ice water from the container as your teacher instructs. Pour approximately 100 mL of cold water into the can. Be sure that no ice is transferred into the can with the water. If you spill any water during the transfer, wipe it off. Mass the can with the water and record the mass in the data table.

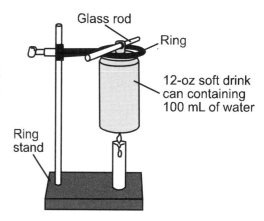

4. Mass your candle and record the mass in the data table.

5. Set up your CBL or LabPro and calculator with a stainless steel temperature probe.
   * Go to APPS and select CHEMBIO
   * Select SET UP PROBES from the MAIN MENU
   * Enter "1" as the number of probes
   * Select TEMPERATURE from the SELECT PROBE menu
   * Enter "1" as the channel number
   * Select USE STORED from the CALIBRATION menu.

6. Set up the calculator and CBL for data collection
   * Select COLLECT DATA from the MAIN MENU
   * Select TIME GRAPH from the DATA COLLECTION menu
   * Time between samples: 2
   * Number of samples: 150
   * USE TIME SETUP unless you made a mistake.
   * $Y_{min} = 0$
   * $Y_{max} = 50$
   * $Y_{scl} = 1$
   * Do not start collecting data until you are instructed to do so in step 6.

7. Place the temperature probe in the aluminum can. Start the program by pressing [ENTER]. Allow 20 seconds to pass while the temperature equilibrates. Light your candle and place it under the can as close to the can as possible without putting the candle out. Stir the water gently with the probe until the CBL screen indicates that the sampling is finished by displaying "DONE".

8. Quickly and carefully blow out the candle, being very careful not to splash paraffin. Mass the candle again and record. Calculate the mass of candle burned.

9. On your calculator, use the **DRAW** function to insert a horizontal line. Press [2nd][PRGM] to get to DRAW. Arrow down to 3: Horizontal to put in a horizontal line. This line will help you determine the change in temperature more accurately. Record the change in temperature in your data table.

Name _____

Period _____

How Hot is a Candle?
Determining the Heat of Combustion of Paraffin

PRE-LAB QUESTIONS

1. How much heat will it take to raise 750. g of water from $25.0^\circ C$ to $75.0^\circ C$? The specific heat of water is 4.18 J/g $^\circ C$.

2. Sally Student combines 50.0 g of water at $20.0^\circ C$ with 100. g of water at $70.0^\circ C$. What is the final temperature of the mixture?

3. Given the following equation:

$$C_3H_8 + 5 O_2 \rightarrow 3 CO_2 + 4 H_2O \quad \Delta H = -2200kJ$$

How many kilojoules of heat are released when 3.64 g of propane, C_3H_8, are burned?

4. If the heat from problem 3 were absorbed by 2500 g of water at $18^\circ C$, what would be the final temperature of the water?

5. What is the molar mass of paraffin, $C_{25}H_{52}$?

DATA AND OBSERVATIONS

Mass of aluminum can	
Mass of aluminum can and water	
Mass of water	
Mass of candle and glass plate before burning	
Mass of candle and glass plate after burning	
Mass of candle burned	
Initial temperature of water	
Final temperature of water	
ΔT of water	
Heat gained by water	
Heat lost from candle	
Heat of combustion of the candle (kJ/g)	
Molar heat of combustion of paraffin (kJ/mol)	
Theoretical molar heat of paraffin	14,800 kJ/mol
% Error $\dfrac{\lvert O-A \rvert}{A} \times 100\%$	

CALCULATION HINTS:

The data table clearly leads you through most of the calculations. Refer back to your pre-lab problems. The heat gained by the water theoretically is the same as the heat lost from the candle. Pay attention to the units needed for the last two calculations.

CONCLUSION QUESTIONS AND ERROR ANALYSIS:

1. Write a balanced equation for the combustion of paraffin.

2. How does your <u>experimental</u> heat of combustion of paraffin (kJ/g) compare with the heat of combustion of propane (kJ/g)? (See pre-lab questions.)

3. How does your <u>theoretical</u> heat of combustion of paraffin (kJ/g) compare with the heat of combustion of propane (kJ/g)? (See pre-lab questions.)

4. How does the theoretical molar heat of combustion of paraffin (kJ/mol) compare with the molar heat of combustion of propane (kJ/mol)?

5. As a consumer of energy, which is the more useful quantity to you, heat of combustion (kJ/g) or molar heat of combustion (kJ/mol)? Explain your answer.

6. This lab has several sources of error. List as many as possible. Suggest possible corrections for these errors.

Hess's Law
Determining the Heat of Formation of Magnesium Oxide

OBJECTIVE
Students will perform two mini experiments and use the data calculated from each to determine the enthalpy of formation for magnesium oxide using Hess's Law.

LEVEL
Chemistry

NATIONAL STANDARDS
UCP.1, UCP.2, UCP.3, B.3, B.5, B.6, E.1, E.2, G.2

TEKS
1(A), 2(A), 2(B), 2(C), 2(D), 2(E), 3(A), 5(A), 5(B)

CONNECTIONS TO AP
AP Chemistry:
 III. Reactions E. Thermodynamics 1. State functions 2. First law: change in enthalpy; heat of formation; heat of reaction; Hess's law; heats of vaporization and fusion; calorimetry

TIME FRAME
45 minutes

MATERIALS
(For a class of 28 working in pairs)

10.0 grams of Mg metal shavings	several balances
15.0 grams of MgO powder	14 glass stirring rods
2 liters of 1.0 M HCl	14 CBL2 units w/ temperature probes or 14
28 foam cups to be used as calorimeter	thermometers
weighing papers or weighing boats	14 100.mL graduated cylinders

TEACHER NOTES
This lab is written for use with a CBL2 but can easily be performed with laboratory thermometers since the students only need the initial temperature and the highest temperature for each reaction. Digital thermometers are a nice alternative.

This lab exercise should be performed when studying thermodynamics. The lab could be used to introduce or reinforce Hess's law. Student notes covering this topic are found in the lesson titled *Thermodynamics* in this book.

If time permits, you may have the students calculate the calorimeter constant. It was intentionally omitted from this lesson for first year students but it always included for AP students.

POSSIBLE ANSWERS TO THE CONCLUSION QUESTIONS AND SAMPLE DATA

SAMPLE DATA TABLE

	Mg + HCl	**MgO + HCl**
Volume of 1.0 M HCl (mL)	60.0 mL	60.0 mL
Mass of solid (grams)	0.65 g	1.05 g
Initial temperature of HCl (°C)	22.5 °C	22.5 °C
Highest temperature of mixture (°C)	54.5 °C	27.0 °C

ANALYSIS

For each of the following calculations be sure to show all work for full credit.

PART I:

1. Calculate the mass of the solution using 1.04 g/mL as the density of 1.0 M HCl.

   - $60.0 \, mL \times \dfrac{1.04 \, g}{mL} = 62.4 \, g \, HCl$ Total mass of solution: $62.4 \, g + 0.65 \, g = 63.05 \, g$ solution

2. Calculate the ΔT for the reaction.
   - $\Delta T = $ final – initial $= 54.5 \, °C - 22.5 \, °C = 32.0 \, °C$

3. Calculate the amount of heat given off by the reaction between magnesium and hydrochloric acid.
   - Enthalpy $= q = m \cdot Cp \cdot \Delta T$
     $= (63.05 \, g) \, (4.184 \, J/g \cdot °C) \, (32.0 \, °C)$
     $= 8442 \, J$

4. Calculate the number of moles of magnesium used in the reaction.

   - $0.65 \, g \, Mg \times \dfrac{1 \, mol \, Mg}{24.3 \, g \, Mg} = 0.0267 \, mol \, Mg$

5. Find the molar heat (kJ/mol) for the reaction between magnesium and hydrochloric acid.
   - $8442 \, J = 8.442 \, kJ$
   - $\dfrac{8.442 \, kJ}{0.0267 \, mol} = 316 \, kJ/mol \; = \Delta H = -316 \, kJ/mol$

6. Write a balanced chemical equation including the molar heat for this reaction.
   - $Mg_{(s)} + 2HCl_{(aq)} \rightarrow MgCl_{2(aq)} + H_{2(g)} + 316 \, kJ$

PART II:

1. Calculate the mass of the solution using 1.04 g/mL as the density of 1.0 M HCl.

   - $60.0\,\cancel{mL} \times \dfrac{1.04\,g}{\cancel{mL}} = 62.4\,g\,HCl$ Total mass of solution: $62.4\,g + 1.05\,g = 63.45\,g$ solution

2. Calculate the ΔT for the reaction.
   - $\Delta T = final - initial = 27.0\,°C - 22.5\,°C = 4.5\,°C$

3. Calculate the amount of heat given off by the reaction between magnesium oxide and hydrochloric acid.
   - Enthalpy $= q = m \cdot Cp \cdot \Delta T$
     $= (63.45\,g)\,(4.184\,J/g \cdot °C)\,(4.5\,°C)$
     $= 1194\,J$

4. Calculate the number of moles of magnesium oxide used in the reaction.
   - $1.05\,\cancel{g}\,MgO \times \dfrac{1\,mol\,MgO}{40.3\,\cancel{g}\,MgO} = 0.0261\,mol\,MgO$

5. Find the molar heat (kJ/mol) for the reaction between magnesium oxide and hydrochloric acid.
   - $1194\,J = 1.194\,kJ$ $\dfrac{1.194\,kJ}{0.0261\,mol} = 45.7\,kJ/mol$ $\Delta H = -45.7\,kJ/mol$

6. Write a balanced chemical equation including the molar heat for this reaction.
   - $MgO_{(s)} + 2HCl_{(aq)} \rightarrow MgCl_{2(aq)} + H_2O_{(l)} + 45.7\,kJ$

7. Using Hess's law calculate the heat of formation for MgO from the combustion of magnesium with oxygen as shown in the introduction. Combine the equations you wrote for Part I and Part II of this lab with the formation of water equation shown below. Use your calculated molar heat values as the ΔH for your reactions.
   - Desired equation: $Mg_{(s)} + \frac{1}{2} O_{2(g)} \rightarrow MgO_{(s)}$ $\Delta H = ?$

$H_{2(g)} + \frac{1}{2} O_{2(g)} \rightarrow H_2O_{(l)}$	$\Delta H = -286\,kJ/mol$
$Mg_{(s)} + 2HCl_{(aq)} \rightarrow MgCl_{2(aq)} + H_{2(g)}$	$\Delta H = -316.0\,kJ$
$MgO_{(s)} + 2HCl_{(aq)} \rightarrow MgCl_{2(aq)} + H_2O_{(l)}$	$\Delta H = -45.7\,kJ$

   - Reverse the 3rd equation: (change the sign of enthalpy)

$MgCl_{2(aq)} + H_2O_{(l)} \rightarrow MgO_{(s)} + 2HCl_{(aq)}$	$\Delta H = +45.7\,kJ$
$H_{2(g)} + \frac{1}{2} O_{2(g)} \rightarrow H_2O_{(l)}$	$\Delta H = -286\,kJ/mol$
$Mg_{(s)} + 2HCl_{(aq)} \rightarrow MgCl_{2(aq)} + H_{2(g)}$	$\Delta H = -316.0\,kJ$

   - Cancel common terms and add the enthalpies together:

$Mg_{(s)} + \frac{1}{2} O_{2(g)} \rightarrow MgO_{(s)}$	$\Delta H = -556.3\,kJ = -556\,kJ$

8. Calculate your percent error. Give plausible explanations for any discrepancies. (Use a table of thermodynamic values or a CRC to find the theoretical heat of formation for MgO.)

$$\% \, error = \frac{|theoretical - experimental|}{theoretical} \times 100 \qquad \% \, error = \frac{|602 - 556|}{602} \times 100 = 7.6\% \, error$$

- Answers will vary but usually the value that is obtained in the lab is lower than expected.
- This can be explained by the *assumption* that no heat was lost to the calorimeter or to the surroundings. There will be some heat lost to both of these.

TEACHER PAGES

Hess's Law
Determining the Heat of Formation of Magnesium Oxide

The reaction of magnesium ribbon as it burns in the presence of oxygen produces a lot of energy in the form of light and heat! In fact, the temperature has been found to reach as high as 2400°C. How do we find the heat of formation of such an exothermic reaction in the laboratory? Germain Hess discovered that if you could find the enthalpy of reaction for several "simple" reaction then you could add those enthalpies to obtain the enthalpy of reaction for a more complicated reaction. Hess's law may be shown by the equation:

$$\Delta H^{\circ}_{rxn} = \Delta H^{\circ}_1 + \Delta H^{\circ}_2 + ...$$

By performing two simple reactions in the laboratory and using a table of standard formation values for a third reaction, we will be able to calculate the heat of formation of magnesium oxide.

$$Mg_{(s)} + \tfrac{1}{2} O_{2(g)} \rightarrow MgO_{(s)} \qquad \Delta H^{\circ}_{rxn} = ?$$

Remember that enthalpy is a state function. It does not matter how you get from reactants to products the answer will be the same. When using Hess's law to find enthalpy there are a few things to note:

1. The equations given must be rearranged in order to obtain the desired reaction.

2. When an equation is reversed, the sign of ΔH must also be reversed.

3. When an equation is multiplied (or divided) by any quantity the ΔH value must also be multiplied (or divided) by the same amount.

PURPOSE
In this activity you will perform two simple reactions and use the data from these reactions along with standard values for a third reaction to find the heat of formation for magnesium oxide using Hess's law.

MATERIALS
Mg metal shavings
MgO powder
1.0 M HCl
coffee cup calorimeter
weighing papers or weighing boats

balance
glass stirring rod
CBL2 unit w/ temperature probes or
    thermometer
100. mL graduated cylinder

Safety Alert

1. Hydrochloric acid is a strong acid. Neutralize spills with baking soda and then wipe the area down with a clean cloth.
2. If HCl splashes onto skin or in eyes flush with cold, running water for 10 to 15 minutes. Notify the teacher.
3. When measuring out the magnesium oxide powder be careful not to inhale the fine powder.
4. Keep magnesium metal away from open flames
5. Goggles and aprons should always be worn when handling chemicals in the laboratory

PROCEDURE

PART I: MAGNESIUM SHAVINGS

1. Measure 60.0 mL of 1.0 M HCl and carefully pour it into the calorimeter. Avoid splashing.

2. Mass approximately 0.60 grams of magnesium shavings. Record the exact mass of the shavings in the data table on the student answer page.

3. Prepare the CBL2 unit, graphing calculator and temperature probe for data collection. Be sure that all cables are firmly in place and that the temperature probe is placed into channel 1.

4. Turn on the CBL2 and graphing calculator. Press APPS and select DataMate.

5. Set up the calculator and CBL2 for data collection:
    a. Select 1: SETUP.
    b. Using the arrow keys, arrow down to MODE and press ENTER.
    c. Select 2: TIME GRAPH.
    d. Select 2: CHANGE TIME SETTINGS.
    e. Enter "3" as the time between samples in seconds. Press ENTER.
    f. Enter "120" as the number of samples. This will allow the CBL2 to collect data for 6 minutes.
    g. Press ENTER then select 1: OK.

6. Place the temperature probe in the calorimeter. Press 2: START to begin data collection. Allow the probe to collect 3 or 4 readings until the temperature is constant.

7. Once the temperature is constant, carefully add all of the magnesium pieces and begin stirring gently with the glass stirring rod. Continue stirring until the six minutes have expired and the CBL2 has stopped collecting data.

8. When the CBL2 has completed data collection, remove the temperature probe and rinse and dry it. Dispose of the calorimeter contents by pouring them down the drain and rinsing with water.

9. A graph of the time and temperature data will appear on the calculator screen. Use the left and right arrows to trace the graph and find the initial and the highest temperature for the reaction. Record these two temperatures in the data table on the student answer page.

PART II: MAGNESIUM OXIDE

1. Rinse the calorimeter thoroughly and dry.

2. Measure 60.0 mL of 1.0 M HCl and carefully pour it into the calorimeter. Avoid splashing.

3. Mass approximately 1.00 gram of magnesium oxide powder. Record the exact mass of the powder in the data table on the student answer page.

4. Set up the calculator and CBL2 for data collection.
   a. Select 1: SETUP.
   b. Using the arrow keys, arrow down to MODE and press ENTER.
   c. Select 2: TIME GRAPH.
   d. Select 2: CHANGE TIME SETTINGS.
   e. Enter "3" as the time between samples in seconds. Press ENTER.
   f. Enter "80" as the number of samples. This will allow the CBL2 to collect data for 6 minutes.
   g. Press ENTER then select 1: OK.

5. Place the temperature probe in the calorimeter. Press 2: START to begin data collection. Allow the probe to collect 3 or 4 readings until the temperature is constant.

6. Once the temperature is constant, carefully add all of the magnesium oxide powder and begin stirring gently with the glass stirring rod. Continue stirring until the six minutes have expired and the CBL2 has stopped collecting data.

7. When the CBL2 has completed data collection, remove the temperature probe and rinse and dry it. Dispose of the calorimeter contents by pouring them down the drain and rinsing with water.

8. A graph of the time and temperature data will appear on the calculator screen. Use the left and right arrows to trace the graph and find the initial and the highest temperature for the reaction. Record these two temperatures in the data table on the student answer page.

9. Clean your lab area and be sure to wash your hands with soap and water.

Name _____

Period _____

Hess's Law
Determining the Heat of Formation of Magnesium Oxide

DATA AND OBSERVATIONS

DATA TABLE

	Mg + HCl	MgO + HCl
Volume of 1.0 M HCl (mL)		
Mass of solid (grams)		
Initial temperature of HCl (°C)		
Highest temperature of mixture (°C)		

ANALYSIS

For each of the following calculations be sure to show all work for full credit.

PART I:

1. Calculate the mass of the solution using 1.04 g/mL as the density of 1.0 M HCl.

2. Calculate the ΔT for the reaction.

3. Calculate the amount of heat given off by the reaction between magnesium and hydrochloric acid.

4. Calculate the number of moles of magnesium used in the reaction.

5. Find the molar heat (kJ/mol) for the reaction between magnesium and hydrochloric acid.

6. Write a balanced chemical equation including the molar heat for this reaction.

PART II:

1. Calculate the mass of the solution using 1.04 g/mL as the density of 1.0 M HCl.

2. Calculate the ΔT for the reaction.

3. Calculate the amount of heat given off by the reaction between magnesium oxide and hydrochloric acid.

4. Calculate the number of moles of magnesium oxide used in the reaction.

5. Find the molar heat (kJ/mol) for the reaction between magnesium oxide and hydrochloric acid.

6. Write a balanced chemical equation including the molar heat for this reaction.

7. Using Hess's law calculate the heat of formation for MgO from the combustion of magnesium with oxygen as shown in the introduction. Combine the equations you wrote for Part I and Part II of this lab with the formation of water equation shown below. Use your calculated molar heat values as the ΔH for your reactions.

$$H_{2(g)} + \frac{1}{2} O_{2(g)} \rightarrow H_2O_{(l)} \quad \Delta H = -286 \text{ kJ/mol}$$

8. Calculate your percent error. Give plausible explanations for any discrepancies. (Use a table of thermodynamic values or a CRC to find the theoretical heat of formation for MgO.)

OIL RIG
Balancing Redox Equations

OBJECTIVE
Students will use the half-reaction method to balance oxidation reduction equations

LEVEL
Chemistry

NATIONAL STANDARDS
UCP.1, UCP.2, B.3

TEKS
5(A), 10(A)

CONNECTIONS TO AP
AP Chemistry:
   III. Reactions A. Reaction types 3. Oxidation Reduction reactions a. oxidation number b. the role of the electron in oxidation-reduction.

TIME FRAME
30 minutes of homework before the lesson; 45 minutes of classroom instruction; 45 minutes of homework after the lesson; 30 minutes of class time to go over the homework.

MATERIALS
   pencil and paper   student answer pages for each student.
   white boards   dry erase markers
   paper towels for erasing

TEACHER NOTES
For a description of how to make and use the white boards, see the lesson on *Dimensional Analysis* found in this book. You will want to make sure that students are doing the procedure correctly. Besides allowing for kinesthetic learning, the white boards allow the teacher to easily check student work.

If you have not used the half-reaction method to balance redox equations, you will want to practice this technique, being very sure that you can successfully complete each of the problems. The half-reaction method is superior to the ion-charge method in that it provides additional practice in writing net ionic equations. In addition, it always works. It forces students to think about how a certain chemical species actually exists in a given reaction.

Oxidation reduction reactions are ones that transfer electrons from one chemical species to another. Many reactions are actually redox reactions, but are also classified in other ways. For example, synthesis reactions involving elements, some decomposition reactions, combustion reactions, and single replacement reactions are also redox. The first part of the student exercise will ask the students to indicate which reactions are oxidation reduction reactions. They will also have to indicate which chemical species is being oxidized and which is being reduced. This is accomplished by determining the oxidation states of each atom and determining which states are changed during the course of the reaction. If no oxidation states are changed, then it is not a redox reaction. Students should already be familiar with determining oxidation states. It is suggested that Exercise 1 be assigned as homework prior to the lesson on balancing redox equations via the half-reaction method.

ANSWERS TO EXERCISE 1:

1. Redox: $\overset{0}{2Na} + \overset{0}{Cl_2} \rightarrow 2\overset{+}{Na}\overset{-}{Cl}$

   Oxidized $\quad \overset{0}{Na}$

   Reduced $\quad \overset{0}{Cl_2}$

2. Redox: $2\overset{+}{K}\overset{5+}{Cl}\overset{2-}{O_3} \rightarrow 2\overset{+}{K}\overset{-}{Cl} + 3\overset{0}{O_2}$

   Oxidized $\quad \overset{2-}{O}$

   Reduced $\quad \overset{5+}{Cl}$

3. Redox: $\overset{8/3-}{C_3}\overset{+}{H_8} + 5\overset{0}{O_2} \rightarrow 3\overset{4+}{C}\overset{2-}{O_2} + 4\overset{+}{H_2}\overset{2-}{O}$

   Oxidized $\quad \overset{\frac{3}{8}-}{C}$

   Reduced $\quad \overset{0}{O_2}$

4. No: $\overset{+}{Ag}\overset{5+}{N}\overset{2-}{O_3} + \overset{+}{K}\overset{-}{Cl} \rightarrow \overset{+}{Ag}\overset{-}{Cl} + \overset{+}{K}\overset{5+}{N}\overset{2-}{O_3}$

5. Redox: $\overset{3+}{Fe_2}\overset{2-}{O_3} + 3\overset{0}{H_2} \rightarrow 2\overset{0}{Fe} + 3\overset{+}{H_2}\overset{2-}{O}$

   Oxidized $\quad \overset{0}{H_2}$

   Reduced $\quad \overset{3+}{Fe}$

6. No: $\overset{+}{Na}\overset{2-}{O}\overset{+}{H} + \overset{+}{H}\overset{-}{Cl} \rightarrow \overset{+}{Na}\overset{-}{Cl} + \overset{+}{H_2}\overset{2-}{O}$

7. Redox: $$2\overset{0}{Mg}+\overset{0}{O_2}\rightarrow 2\overset{2+}{Mg}\overset{2-}{O}$$

Oxidized $\overset{0}{Mg}$

Reduced $\overset{0}{O_2}$

8. No: $$\overset{2+}{Ca}(\overset{2-}{O}\overset{+}{H})_2\rightarrow \overset{2+}{Ca}\overset{2-}{O}+\overset{+}{H_2}\overset{2-}{O}$$

9. Redox: $$\overset{0}{Cu}+2\overset{+}{Ag}\overset{5+}{N}\overset{2-}{O_3}\rightarrow \overset{2+}{Cu}(\overset{5+}{N}\overset{2-}{O_3})_2+2\overset{0}{Ag}$$

Oxidized $\overset{0}{Cu}$

Reduced $\overset{+}{Ag}$

10. Redox: $$2\overset{+}{K}\overset{7+}{Mn}\overset{2-}{O_4}+10\overset{2+}{Fe}\overset{6+}{S}\overset{2-}{O_4}+8\overset{+}{H_2}\overset{6+}{S}\overset{2-}{O_4}\rightarrow 2\overset{2+}{Mn}\overset{6+}{S}\overset{2-}{O_4}+\overset{3+}{Fe_2}(\overset{6+}{S}\overset{2-}{O_4})_3+\overset{+}{K_2}\overset{6+}{S}\overset{2-}{O_4}+4\overset{+}{H_2}\overset{2-}{O}$$

Oxidized $\overset{2+}{Fe}$

Reduced $\overset{7+}{Mn}$

Suggested teaching procedures for balancing reactions with the half-reaction method

Reactions in acid or neutral medium:
Start with a very simple redox equation. It needs to be one that students could easily balance by inspection. For example, if you run an electric current through molten sodium chloride, it will decompose into sodium metal and chlorine gas. This is the reverse of the first reaction that the students did in Exercise 1. Write the skeleton equation on the board and write in oxidation states. Ask the students what is being oxidized and what is being reduced.

$\overset{+}{Na}\overset{-}{Cl}\rightarrow \overset{0}{Na}+\overset{0}{Cl_2}$ The sodium is being reduced; the chlorine is being oxidized.

Start with the reduction reaction and write a balanced half-reaction. Use electrons to balance the charge. Follow that with the oxidation half-reaction.

$\overset{+}{Na}+1e^-\rightarrow \overset{0}{Na}$ reduction

$2\overset{-}{Cl}\rightarrow \overset{0}{Cl_2}+2e^-$ oxidation

The total number of electrons lost must equal the total number of electrons gained. Since the reduction reaction needs only one electron, and the oxidation loses two electrons, the reduction reaction needs to be multiplied by 2. Then the two reactions are added together so that the electrons are cancelled.

$$2(\overset{+}{Na} + 1e^- \rightarrow \overset{0}{Na})$$

$$2\overset{-}{Cl} \rightarrow \overset{0}{Cl_2} + 2e^-$$

$$2\overset{+}{Na} + 2e^- \rightarrow 2\overset{0}{Na})$$

$$2\overset{-}{Cl} \rightarrow \overset{0}{Cl_2} + 2e^-$$

Adding the two equations together, you obtain a final balanced equation.

$$2NaCl \rightarrow 2\overset{0}{Na} + \overset{0}{Cl_2}$$

This equation is very easy to balance by inspection. However, equation #10 is much more difficult to balance by inspection. For the more difficult equations, the half-reaction method will allow you to balance the equation correctly and relatively easily. Follow the step to see how easy it is!

1. Write the oxidation numbers on each atom to determine which atoms are reduced and which atoms are oxidized.

$$\overset{+}{K}\overset{7+}{Mn}\overset{2-}{O_4} + \overset{2+}{Fe}\overset{6+}{S}\overset{2-}{O_4} + \overset{+}{H_2}\overset{6+}{S}\overset{2-}{O_4} \rightarrow \overset{2+}{Mn}\overset{6+}{S}\overset{2-}{O_4} + \overset{3+}{Fe_2}(\overset{6+}{S}\overset{2-}{O_4})_3 + \overset{+}{K_2}\overset{6+}{S}\overset{2-}{O_4} + \overset{+}{H_2}\overset{2-}{O}$$

From this step, it is clear that the manganese(VII) is reduced and the iron(II) is oxidized. Deal with the reduction reaction first. Keep chemical species intact in the form that they are in. This reaction is most likely in solution, so the potassium permanganate is dissolved and the potassium ion is a spectator ion.

$$MnO_4^- \rightarrow Mn^{2+}$$

2. The Mn is balanced, but the oxygen is not. Use H_2O to balance the oxygens and then use H^+ to balance the hydrogens.

$$8H^+ + MnO_4^- \rightarrow Mn^{2+} + 4H_2O$$

3. Finally, use electrons to balance the charge on both sides of the equation. Now you have a totally balanced half-reaction.

$$8H^+ + MnO_4^- + 5e^- \rightarrow Mn^{2+} + 4H_2O$$

4. Repeat the procedure with the oxidation reaction.

$$2Fe^{2+} \rightarrow Fe_2^{3+} + 2e^-$$

5. Both equations need to transfer the same number of electrons, so the reduction reaction will need to be multiplied by 2 and the oxidation reaction will need to be multiplied by 5.

$$2(8H^+ + MnO_4^- + 5e^- \rightarrow Mn^{2+} + 4H_2O) = 16H^+ + 2MnO_4^- + 10e^- \rightarrow 2Mn^{2+} + 8H_2O$$

$$5(2Fe^{2+} \rightarrow Fe_2^{3+} + 2e^-) = 10Fe^{2+} \rightarrow 5Fe_2^{3+} + 10e^-$$

6. Add the two half-reactions together to get the balanced net ionic equation.

$$16H^+ + 2MnO_4^- + 10e^- \rightarrow 2Mn^{2+} + 8H_2O$$

$$10Fe^{2+} \rightarrow 5Fe_2^{3+} + 10e^-$$

$$16H^+ + 2MnO_4^- + 10Fe^{2+} \rightarrow 2Mn^{2+} + 5Fe_2^{3+} + 8H_2O$$

7. Substitute the coefficients from the net ionic equation back into the original equation. Balance all spectator ions. Double check each atom, saving hydrogen for next to last and oxygens for last. If the oxygens are balanced after checking all other atoms, the equation is balanced.

$$2\overset{+}{K}\overset{7+}{Mn}\overset{2-}{O}_4 + 10\overset{2+}{Fe}\overset{6+}{S}\overset{2-}{O}_4 + 8\overset{+}{H}_2\overset{6+}{S}\overset{2-}{O}_4 \rightarrow 2\overset{2+}{Mn}\overset{6+}{S}\overset{2-}{O}_4 + 5\overset{3+}{Fe}_2(\overset{6+}{S}\overset{2-}{O}_4)_3 + \overset{+}{K}_2\overset{6+}{S}\overset{2-}{O}_4 + 8\overset{+}{H}_2\overset{2-}{O}$$

Reactions in basic (alkaline) medium:

$$KMnO_4 + H_2O_2 \rightarrow MnO_2 + O_2 + KOH + H_2O$$

Reactions in base are a little more difficult to balance, but the basic steps start out the same.

1. Write the oxidation numbers on every atom.

$$\overset{+}{K}\overset{7+}{Mn}\overset{2-}{O}_4 + \overset{+}{H}_2\overset{-}{O}_2 \rightarrow \overset{4+}{Mn}\overset{2-}{O}_2 + \overset{0}{O}_2 + \overset{+}{K}\overset{2-}{O}\overset{+}{H} + \overset{+}{H}_2\overset{2-}{O}$$

2. Write a balanced half-reaction for both the reduction reaction and the oxidation reaction, just as you did before. Use water to balance oxygens and H^+ to balance hydrogens.

$$4H^+ + MnO_4^- + 3e^- \rightarrow MnO_2 + 2H_2O$$

$$H_2O_2 \rightarrow 2H_2O + 2H^+ + 2e^-$$

3. Find the least common multiple (LCM) of the electrons and multiple each half-reaction so that the electrons cancel. Add the two half-reactions.

$$2(4H^+ + MnO_4^- + 3e^- \rightarrow MnO_2 + 2H_2O)$$

$$3(H_2O_2 \rightarrow O_2 + 2H^+ + 2e^-)$$

After cross multiplying and collecting like terms, the resulting net ionic equation is as shown below.

$$8H^+ + 2MnO_4^- + 6e^- \rightarrow 2MnO_2 + 4H_2O$$

$$\underline{3H_2O_2 \rightarrow 3O_2 + 6H^+ + 6e^-}$$

$$2H^+ + 2MnO_4^- + 3H_2O_2 \rightarrow 2MnO_2 + 4H_2O + 3O_2$$

4. However, this reaction is in a basic, not an acidic medium. To get rid of the H^+, add 2 OH^- to each side of the equation. This will react with the H^+ and turn it into water. Then adjust the water so that it is on only one side of the equation.

$$2H^+ + 2MnO_4^- + 3H_2O_2 \rightarrow 2MnO_2 + 4H_2O + 3O_2$$

$$\underline{+2OH^- \qquad\qquad\qquad \rightarrow \quad +2OH^-}$$

$$2H_2O + 2MnO_4^- + 3H_2O_2 \rightarrow 2MnO_2 + 4H_2O + 3O_2 + 2OH^-$$

$$2MnO_4^- + 3H_2O_2 \rightarrow 2MnO_2 + 2H_2O + 3O_2 + 2OH^-$$

5. Substitute the coefficients into the basic equation and balance any remaining spectator ions. Check your work.

$$2KMnO_4 + 3H_2O_2 \rightarrow 2MnO_2 + 3O_2 + 2KOH + 2H_2O$$

ANSWERS TO EXERCISE 2:

Be sure that you can balance each of these equations by the half-reaction method before class.

1. $MnO_2 + HCl \rightarrow MnCl_2 + Cl_2 + H_2O$

- $\overset{4+}{Mn}\overset{2-}{O_2} + \overset{+}{H}\overset{-}{Cl} \rightarrow \overset{2+}{Mn}\overset{-}{Cl_2} + \overset{0}{Cl_2} + \overset{+}{H_2}\overset{-}{O}$

  $4H^+ + \overset{4+}{Mn}\overset{2-}{O_2} + 2e^- \rightarrow \overset{2+}{Mn} + 2H_2O$

- $\underline{2\overset{-}{Cl} \rightarrow \overset{0}{Cl_2} + 2e^-}$

  $4\overset{+}{H} + MnO_2 + 2\overset{-}{Cl} \rightarrow \overset{2+}{Mn} + 2H_2O + \overset{0}{Cl_2}$

- $MnO_2 + 4HCl \rightarrow MnCl_2 + Cl_2 + 2H_2O$ (Note: the discrepancy between the coefficients on the H^+ and the Cl^- is because two of the chloride ions are not oxidized, but are spectator ions. When this happens, always use the larger of the two coefficients.)

2. $K_2Cr_2O_7 + HI \rightarrow KI + CrI_3 + I_2 + H_2O$

- $\overset{+}{K_2}\overset{6+}{Cr_2}\overset{2-}{O_7} + \overset{+}{H}\overset{-}{I} \rightarrow \overset{+}{K}\overset{-}{I} + \overset{3+}{Cr}\overset{-}{I_3} + \overset{0}{I_2} + \overset{+}{H_2}\overset{2-}{O}$

- $14H^+ + Cr_2O_7^{2-} + 6e^- \rightarrow 2\overset{3+}{Cr} + 7H_2O$

$$2I^- \rightarrow \overset{0}{I_2} + 2e^-$$

- $14H^+ + Cr_2O_7^{2-} + 6e^- \rightarrow 2\overset{3+}{Cr} + 7H_2O$

$$\underline{3(2I^- \rightarrow \overset{0}{I_2} + 2e^-)}$$

$14H^+ + Cr_2O_7^{2-} + 6I^- \rightarrow 3\overset{0}{I_2} + 2\overset{3+}{Cr} + 7H_2O$

- $K_2Cr_2O_7 + 14HI \rightarrow 2KI + 2CrI_3 + 3I_2 + 7H_2O$ (Note that again you have some iodide not being oxidized. You also have potassium as a spectator ion that needs adjustment at the end.)

3. $KClO_3 \rightarrow KClO_4 + KCl$

- $\overset{+}{K}\overset{5+}{Cl}\overset{2-}{O_3} \rightarrow \overset{+}{K}\overset{7+}{Cl}\overset{2-}{O_4} + \overset{+}{K}\overset{-}{Cl}$

$3(H_2O + ClO_3^- \rightarrow ClO_4^- + 2H^+ + 2e^-)$

- $\underline{6H^+ + ClO_3^- + 6e^- \rightarrow Cl^- + 3H_2O}$

$4ClO_3^- \rightarrow 3ClO_4^- + Cl^-$

- $4KClO_3 \rightarrow 3KClO_4 + KCl$

4. $I_2 + HNO_3 \rightarrow HIO_3 + NO + H_2O$

- $\overset{0}{I_2} + \overset{+}{H}\overset{5+}{N}\overset{2-}{O_3} \rightarrow \overset{+}{H}\overset{5+}{I}\overset{2-}{O_3} + \overset{2+}{N}\overset{2-}{O} + \overset{+}{H_2}\overset{2-}{O}$

$10(4H^+ + NO_3^- + 3e^- \rightarrow NO + 2H_2O)$

$$\underline{3(6H_2O + \overset{0}{I_2} \rightarrow 2HIO_3 + 10H^+ + 10e^-)}$$

- $40H^+ + 10NO_3^- + 18H_2O + 3I_2 \rightarrow 6HIO_3 + 30H^+ + 10NO + 20H_2O$

$10H^+ + 10NO_3^- + 3I_2 \rightarrow 6HIO_3 + 10NO + 2H_2O$

- $3I_2 + 10HNO_3 \rightarrow 6HIO_3 + 10NO + 2H_2O$

5. $NO_2 + H_2O \rightarrow HNO_3 + NO$

- $\overset{4+}{N}\overset{2-}{O_2} + \overset{+}{H_2}\overset{2-}{O} \rightarrow \overset{+}{H}\overset{5+}{N}\overset{2-}{O_3} + \overset{2+}{N}\overset{2-}{O}$

  $2H^+ + NO_2 + 2e^- \rightarrow NO + H_2O$

- $\underline{2(H_2O + NO_2 \rightarrow NO_3^- + 2H^+ + e^-)}$

  $H_2O + 3NO_2 \rightarrow 2NO_3^- + NO$

- $3NO_2 + H_2O \rightarrow 2HNO_3 + NO$

6. $PH_3 + N_2O \rightarrow H_3PO_4 + N_2$

- $\overset{3-}{P}\overset{+}{H_3} + \overset{+}{N_2}\overset{2-}{O} \rightarrow \overset{+}{H_3}\overset{5+}{P}\overset{2-}{O_4} + \overset{0}{N_2}$

  $4(2H^+ + N_2O + 2e^- \rightarrow N_2 + H_2O)$

- $\underline{4H_2O + PH_3 \rightarrow H_3PO_4 + 8H^+ + 8e^-}$

  $4N_2O + PH_3 \rightarrow H_3PO_4 + 4N_2$

- $PH_3 + 4N_2O \rightarrow H_3PO_4 + 4N_2$

7. $NH_3 + O_2 \rightarrow NO + H_2O$

- $\overset{3-}{N}\overset{+}{H_3} + \overset{0}{O_2} \rightarrow \overset{2+}{N}\overset{2-}{O} + \overset{+}{H_2}\overset{2-}{O}$

  $5(4H^+ + O_2 + 4e^- \rightarrow 2H_2O)$

- $\underline{4(H_2O + NH_3 \rightarrow NO + 5H^+ + 5e^-)}$

  $5O_2 + 4NH_3 \rightarrow 4NO + 6H_2O$

- $4NH_3 + 5O_2 \rightarrow 4NO + 6H_2O$

8. $HPO_3 + C \rightarrow H_2O + CO + P_4$

- $\overset{+}{H}\overset{5+}{P}\overset{2-}{O_3} + \overset{0}{C} \rightarrow \overset{+}{H_2}\overset{2-}{O} + \overset{2+}{C}\overset{2-}{O} + \overset{0}{P_4}$

  $20H^+ + 4HPO_3 + 20e^- \rightarrow P_4 + 12H_2O$

- $\underline{10(H_2O + C \rightarrow CO + 2H^+ + 2e^-)}$

  $4HPO_3 + 10C \rightarrow 10CO + P_4 + 2H_2O$

- $4HPO_3 + 10C \rightarrow 2H_2O + 10CO + P_4$

9. $C_4H_6 + O_2 \rightarrow CO_2 + H_2O$

- $\overset{3/2-}{C}_4 \overset{+}{H}_6 + \overset{0}{O_2} \rightarrow \overset{4+}{C}\overset{2-}{O_2} + \overset{+}{H_2}\overset{2-}{O}$

    $11\left(4H^+ + O_2 + 4e^- \rightarrow 2H_2O\right)$

- $\underline{\quad 2\left(8H_2O + C_4H_6 \rightarrow 4CO_2 + 22H^+ + 22e^-\right) \quad}$

    $44H^+ + 11O_2 + 16H_2O + 2C_4H_6 \rightarrow 8CO_2 + 44H^+ + 22H_2O$

    $11O_2 + 2C_4H_6 \rightarrow 8CO_2 + 6H_2O$

- $2C_4H_6 + 11O_2 \rightarrow 8\,CO_2 + 6H_2O$

10. $KOH + Cl_2 \rightarrow KClO_3 + KCl + H_2O$

- $\overset{+}{K}\overset{2-}{O}\overset{+}{H} + \overset{0}{Cl_2} \rightarrow \overset{+}{K}\overset{5+}{C}\overset{2-}{lO_3} + \overset{+}{K}\overset{-}{Cl} + \overset{+}{H_2}\overset{2-}{O}$

    $5(\frac{1}{2}Cl_2 + 1e^- \rightarrow Cl^-)$

- $\underline{3H_2O + \frac{1}{2}Cl_2 \rightarrow ClO_3^- + 6H^+ + 5e^-}$

    $3H_2O + 3Cl_2 \rightarrow ClO_3^- + 6H^+ + 5Cl^-$

    $\underline{\quad + \quad 6OH^- \rightarrow 6OH^- \quad}$

    $6OH^- + 3Cl_2 \rightarrow ClO_3^- + 3H_2O + 5Cl^-$

- $6KOH + 3\,Cl_2 \rightarrow KClO_3 + 5KCl + 3\,H_2O$

(Note: In this case the Cl_2 disproportionates. If the student uses Cl_2 instead of ½ Cl_2 in each half-reaction, the coefficients will be doubled and can be divided at the end. Also, this is a basic reaction, so OH^- must be added at the end to neutralize the H^+.)

OIL RIG
Balancing Redox Equations

Early in the year you learned to classify chemical reactions in order to help you predict the products that would occur. You learned about composition (synthesis), decomposition, single replacement, double displacement, and combustion reactions. There is another classification method that involves determining whether electrons have been transferred from one chemical to another. These reactions are called oxidation reduction reactions, or REDOX if you are talking "chemist's slang". Many of the reactions that we have studied, and classified in another way, are also redox reactions.

In this lesson you will learn to classify reactions as "redox", and learn a new method of balancing some of the more difficult reactions.

We identify which is substance is oxidized and which is reduced by using the mnemonic "OIL RIG". We are talking about electrons, so just remember:

Oxidation
Is
Losing electrons

Reduction
Is
Gaining electrons

Of course it might not seem logical to you that when you reduce something you gain electrons. However, remember that the electrons are negative, so when you gain a negative particle, the oxidation state goes down, or is reduced in number.

PURPOSE
In this activity you will learn to recognize redox reactions and will learn the half-reaction method of balancing these equations.

MATERIALS
paper	pencil
white board	dry erase marker
paper towel	

PROCEDURE
1. The first exercise on your student page asks you to mark the oxidation states on each atom, determine whether any of those states change from reactants to products, and indicate which chemical is oxidized and which is reduced. Complete this exercise now. The rules for determining oxidation states are below, just in case you have forgotten them.

RULES FOR DETERMINING OXIDATION STATES

Definition: An oxidation state is the charge that an atom would have IF all of its bonds were ionic. That means that sometimes oxidation states are real, and sometimes they are just an accounting method that we have for keeping track of electrons in a chemical reaction.

1. The oxidation state of a free element is zero.

2. The oxidation state for a monatomic ion is equal to its charge.

3. The algebraic sum of the oxidation states of all of the atoms in a compound must be zero.

4. The oxidation state for alkali metals in compounds is +1; the oxidation state for alkaline earth metals in compounds is +2.

5. In compounds, the more electronegative element is always negative.

6. In compounds, hydrogen is usually +1 unless it is more electronegative than the element that it is bonded with. In that case, it is a hydride and has a charge of -1.

7. In compounds, oxygen is usually -2. However, if it is a peroxide, it is -1, and, if it is a superoxide, it is - ½. If it is bonded to fluorine, oxygen will be +2. These compounds are rare.

8. Oxidation states do not have to be the ones found on the periodic table. They do not even have to be whole numbers. **Rule three may not be violated!** Remember, oxidation states are the charge that an atom would have IF all of its bonds were ionic. Many compounds are covalent

Name _____

Period _____

OIL RIG
Balancing Redox Equations

EXERCISE 1: IS THIS A REDOX REACTION OR NOT?

In this exercise, you need to mark the oxidation states of each atom in every compound in the equation. If any of these oxidation states change from the products to the reactants, then the reaction is a redox and you should write the word "redox" in the blank that is found at the left of the reaction. If it is not a redox reaction, write "no" beside the reaction. If the reaction is redox, determine which atoms are reduced and which are oxidized, and write those atoms in the proper blanks. If the reaction is not redox, leave the blanks for "oxidized" and "reduced" blank.

1. _____ $2\,Na + Cl_2 \rightarrow 2\,NaCl$

   Oxidized _____

   Reduced _____

2. _____ $2\,KClO_3 \rightarrow 2\,KCl + 3\,O_2$

   Oxidized _____

   Reduced _____

3. _____ $C_3H_8 + 5O_2 \rightarrow 3CO_2 + 4\,H_2O$

   Oxidized _____

   Reduced _____

4. _____ $AgNO_3 + KCl \rightarrow AgCl + KNO_3$

   Oxidized _____

   Reduced _____

5. _____ $Fe_2O_3 + H_2 \rightarrow Fe + H_2O$ (This equation is not balanced. Balance it!)

   Oxidized _____

   Reduced _____

6. _____ $NaOH + HCl \rightarrow NaCl + H_2O$

   Oxidized _____

   Reduced _____

7. _____ $2\,Mg + O_2 \rightarrow 2\,MgO$

   Oxidized _____

   Reduced _____

8. _____ $Ca(OH)_2 \rightarrow CaO + H_2O$

   Oxidized _____

   Reduced _____

9. _____ $Cu + 2\,AgNO_3 \rightarrow Cu(NO_3)_2 + 2\,Ag$

   Oxidized _____

   Reduced _____

10. _____ $KMnO_4 + FeSO_4 + H_2SO_4 \rightarrow MnSO_4 + Fe_2(SO_4)_3 + K_2SO_4 + H_2O$
    (This equation is not balanced. You may want to try it, but your teacher will show you how to do it in class.)

    Oxidized _____

    Reduced _____

EXERCISE 2: BALANCING REDOX REACTIONS—YOUR TURN

Use your white board to balance the first four reactions. Let your teacher see your work, step by step. Do the last six equations on separate paper—one equation per page—to turn in.

Step 1: Write oxidation numbers on each atom in the equation.

Step 2: Determine the atoms that are reduced and the atoms that are oxidized.

Step 3: Write balanced half-reactions as instructed by your teacher. Balance the atom that is oxidized or reduced first. Use water to balance oxygens. Use H^+ to balance hydrogens.

Step 4: Multiply each equation in such a way that the number of electrons lost are the same as the number of electrons gained.

Step 5: Add the two half-reactions to obtain the net ionic equation. Place the coefficients from the net ionic equation into the original equation.

Step 6: Adjust spectator ions if necessary. Check all atoms for conservation. Check hydrogens next to last and oxygens last.

1. $MnO_2 + HCl \rightarrow MnCl_2 + Cl_2 + H_2O$

2. $K_2Cr_2O_7 + HI \rightarrow KI + CrI_3 + I_2 + H_2O$

3. $KClO_3 \rightarrow KClO_4 + KCl$

4. $I_2 + HNO_3 \rightarrow HIO_3 + NO + H_2O$

5. $NO_2 + H_2O \rightarrow HNO_3 + NO$

6. $PH_3 + N_2O \rightarrow H_3PO_4 + N_2$

7. $NH_3 + O_2 \rightarrow NO + H_2O$

8. $HPO_3 + C \rightarrow H_2O + CO + P_4$

9. $C_4H_6 + O_2 \rightarrow CO_2 + H_2O$

10. $KOH + Cl_2 \rightarrow KClO_3 + KCl + H_2O$

It's Electrifying!
Comparing Galvanic and Electrolytic Cells

OBJECTIVE
Students will explore properties of galvanic and electrolytic cells and calculate reduction potentials for galvanic cells.

LEVEL
Chemistry

NATIONAL STANDARDS
UCP.2, UCP.3, B.3, B.6

TEKS
5(A), 5(B), 10(A), 10(B), 11(B), 11(C)

CONNECTIONS TO AP
AP Chemistry:
   III. Reactions A. Reaction types 3. Oxidation-reduction reactions a. Oxidation number b. The role of the electron in oxidation-reduction c. Electrochemistry: electrolytic and galvanic cells; Faraday's laws; standard half-cell potentials

TIME FRAME
45 minutes

MATERIALS
   student notes page   table of reduction potentials
   calculator

TEACHER NOTES
This lesson on electrochemistry should follow the lesson OIL RIG, also found in this book. After completing OIL RIG students should be familiar with oxidation-reduction reactions, half reactions, and the terminology used throughout this lesson. This lesson will address applications of electrochemistry, calculation of standard cell potentials, and the use of Faraday's constant in solving electrochemical cell problems. Though much of this information may not be in your standard high school chemistry text, the calculations in this exercise will lay the foundation for success in solving electrochemistry problems in the AP course.

When teaching the students to read the table of standard reduction potentials be sure to point out that this table has many formats. The students should focus on the reduction potential values and note that the more positive values are those elements that are most easily reduced.

A good introductory demonstration is to have a potato clock (or something similar) running on your desk as students arrive. Clocks such as this are usually composed of two different metal electrodes and the "juices" of the potato contain ions which allow for electric current to flow. Novelty stores often stock such items for a nominal fee. You might also place several different types of batteries out for students to observe.

A simple and fun lab exercise is to have a "Lemon-Battery Contest" between the students. Have the students bring any types of fruits, vegetables, and wires that they would like from home on the day of the lab. You will place several different metal samples out for the students use as electrodes and allow them to connect all types of circuits. You will have to furnish alligator clips and multimeters (either CBL or digital ones). Check with your physics teacher; they probably have what you need. The clips and multimeters could also be purchased at Radio Shack. By connecting different metal samples, students begin to discover that some metal combinations produce more voltage than others. Ultimately, they begin to develop the table of reduction potentials. Many may place their own jewelry into the fruits and vegetables in an attempt to obtain the highest voltage. You might award a small prize to the group with the highest voltage read on the teacher multimeter. (The cells that are created in this lab are not standard galvanic cells — the solutions are not composed of the metal ions, just the ions of the fruit/vegetable, therefore, standard voltages should not be expected.)

Construct an electrolytic cell and allow students to observe the process during your discussion. One simple demonstration is to place a Petri dish on the overhead projector about ½ full of 0.1 M KI solution. Add a few drops of phenolphthalein and a few drops of starch and stir well. Using a 9-volt battery, two alligator clips and two graphite pencil leads, observe the electrolysis of KI solution. The reactions that occur are:

Anode: $2\ I^- \rightarrow I_2 + 2\ e^-$ (the formation of I_2 yields a blue-black color in solution with starch)

$I^- + I_2 \rightarrow I_3^-$ (the triiodide ion yields a brown color in solution)

Cathode: $2e^- + 2\ H_2O \rightarrow H_2 + 2\ OH^-$ (the OH^- will yield a pink color in solution with phenolphthalein)

There are two good lessons derived from this demonstration. First it is a great way to have students observe an electrolytic cell and second it leads nicely into a discussion about which substances react in an aqueous solution. Often water is more easily oxidized and/or reduced than the other ions that might be in solution.

Additional comments:

The terms galvanic and voltaic cells are used interchangeably. They both indicate a spontaneous chemical reaction that produces electricity. Why two terms? History. In 1780, Luigi Galvani discovered that when two different metals (copper and zinc for example) were connected together and then touched to different parts of a frog's leg nerve, they made the leg contract. He called this "animal electricity."

This discovery paved the way for all electrical batteries.

The Voltaic Pile was invented by Alessandro Volta in 1800. He demonstrated that when certain metals and chemicals come into contact with each other they can produce an electrical current. He placed together several pairs of copper and zinc discs separated by paper soaked in salt water, and an electrical current was produced. This was the first chemical battery.

For his contributions to the study of electricity, the SI unit called the volt was named after Volta.

POSSIBLE ANSWERS TO THE ANALYSIS AND CONCLUSION QUESTIONS

1. List the two main types of electrochemical cells and briefly state the purpose of each.
   - Galvanic—use chemical energy to produce electrical energy
   - Electrolytic—use electrical energy to produce pure forms of elements

2. Refer to the Table of Standard Reduction Potentials to answer the following questions.
   a. State the general location of metals and nonmetals on the table.
      - Metals are toward the bottom of the table while nonmetals are closer to the top of the table. This indicates that nonmetals easily undergo reduction (gain of electrons) while metals easily undergo oxidation (loss of electrons).

   b. Arrange the following metals in order of increasing activity. Explain your reasoning.
      - Fe, Mg, Na, Au, Ag
      - Na, Mg, Fe, Ag, Au
      - The reduction potentials (measured in voltages) in order are: -2.714, -2.37, -0.44, +0.799, +1.50
      - The smallest values are the ones that are most easily oxidized and are thus, the most reactive metals.

   c. Write the reduction half-reaction for water with its voltage.
      - $2 H_2O_{(l)} + 2e- \rightarrow 2 OH^-_{(aq)} + H_{2(g)}$ $E° = -0.828$ volts

   d. Write the oxidation half-reaction for water with its voltage.
      - $2 H_2O_{(l)} \rightarrow 4 H^+_{(aq)} + O_{2(g)} + 4e-$ $E° = -1.299$ volts

   e. What would you expect to see if copper wire were placed into a solution of silver nitrate? Support your answer with appropriate equations.
      - Copper wire is Cu(s) reacting with silver ions, Ag^+ (aq)
      - If the reaction were to occur, the balanced equation would be:
      - $Cu_{(s)} + Ag^+_{(aq)} \rightarrow Cu^{2+}_{(aq)} + Ag_{(s)}$
      - In this reaction copper is oxidized and silver is reduced
      - $Ag^+ + e- \rightarrow Ag$ (+ 0.80 volts) $Cu \rightarrow Cu^{2+} + 2 e-$ (-0.34 volts) (remember to change the sign when you reverse the half-reaction) Since the addition of these two voltages is a positive value we would expect to see solid silver metal form around the outside of the copper wire. As copper (II) ions move into solution, the solution should change from a colorless solution to a light blue solution.

   f. Combining which of the following would yield the greatest cell potential? Support your answer. Ni, Fe, Cu, Li, Ag, Al, Au
      - The reduction potentials (measured in voltages) in order are: -0.25; -0.44; +0.34; -3.05; +0.80; -1.66; +1.50

- Combining lithium and gold would produce the greatest potential difference. The two metals are at opposite ends of the reduction potential table. If combined, the standard cell potential should be: 3.05 + 1.50 = 4.55 volts.

3. Sketch and label a diagram of a galvanic cell composed of tin and zinc. Both electrodes are in 1.0 M solutions of an appropriate metal salt and the experiment is carried out at 25°C. Be sure to include and label the following in your apparatus: anode, cathode, electron flow, salt bridge, ion flow, and voltmeter.

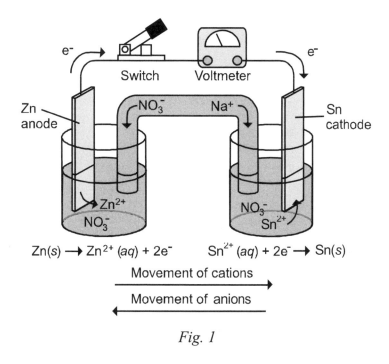

Fig. 1

a. Write the half-cell reactions occurring at each electrode.
   - Cathode: $Sn^{2+} + 2e- \rightarrow Sn$ (-0.14 V)
   - Anode: $Zn \rightarrow Zn^{2+} + 2e-$ (+0.76 V)

b. Write the balanced overall cell reaction.
   - $Zn_{(s)} + Sn^{2+} \rightarrow Zn^{2+} + Sn_{(s)}$

c. Calculate the standard reduction potential for this reaction.
   - $E°_{cell} = E°_{reduction} + E°_{oxidation}$ $E°_{cell} = (-0.14V) + (+0.76V) = 0.32$ volts

4. A galvanic cell was constructed between zinc and copper. The electrodes were massed before and after the reaction. Describe how the masses of each electrode will change as the cell operates. Be sure to specify which metal will be the anode and which metal will be the cathode. Give a reasonable explanation for your observations.

- In a cell between zinc and copper; zinc serves as the anode and copper serves as the cathode. Using the table of standard reduction potentials, zinc is more easily oxidized than copper.
- The zinc anode will lose some mass during the course of the experiment as zinc ions fall into solution — $Zn \rightarrow Zn^{2+} + 2e-$

- The copper cathode should actually weigh a bit more since copper ions from solution are reduced to copper metal on the surface of the electrode — $Cu^{2+} + 2e- \rightarrow Cu$

5. Rechargeable batteries are often referred to as NiCad batteries. A proposed reaction for the production of these batteries is: $Ni_{(s)} + Cd^{2+}_{(aq)} \rightarrow Ni^{2+}_{(aq)} + Cd_{(s)}$
   a. Will this reaction occur spontaneously as written? Explain.
      No, the reaction will not occur as written. In this reaction, cadmium is reduced since it has a more positive reduction potential, thus nickel is oxidized.

   b. Calculate the voltage produced by this cell.
      - $Ni_{(s)} \rightarrow Ni^{2+}_{(aq)} + 2e-$ (+ 0.25V)
      - $Cd^{2+}_{(aq)} + 2e- \rightarrow Cd_{(s)}$ (-0.40V)
      - $E°_{cell} = E°_{reduction} + E°_{oxidation}$ $E°_{cell} = (-0.40V) + (+0.25V) = -0.15$ volts (this would not be spontaneous—nothing would happen)

6. You visit a close relative during your spring break and are asked to help clean the silver that has tarnished to an almost black color due to the presence of Ag_2S. Your relative explains that the tarnish can be removed by boiling the silver in an aluminum pan with some baking soda. The relative states that they have no idea why this works. Being the expert chemistry student, explain the chemistry behind this process — in all the gory chemistry details including equations along with standard reduction potentials.
   - $Ag^+ + e- \rightarrow Ag_{(s)}$ (+ 0.80 volts)
   - $Al_{(s)} \rightarrow Al^{3+} + 3e-$ (+ 1.66 volts)
   - The reaction is definitely spontaneous with a voltage of 2.46. The baking soda provides an ionic solution where electrons are able to flow.

7. Given the following cell in standard cell notation,
   $Zn_{(s)} /Zn(NO_3)_{2(aq)}(1.0M) //Pb(NO_3)_{2(aq)}(1.0M) /Pb_{(s)}$

   a. Write the overall balanced chemical reaction.

      $Zn_{(s)} + Pb(NO_3)_{2(aq)} \rightarrow Zn(NO_3)_{2(aq)} + Pb_{(s)}$

   b. Assuming standard temperature, calculate the $E°$ for the cell.
      - $Zn \rightarrow Zn^{2+} + 2e-$ (+0.76V)
      - $Pb^{2+} + 2e- \rightarrow Pb_{(s)}$ (-0.13V)
      - $E°_{cell} = E°_{reduction} + E°_{oxidation}$ $E°_{cell} = (-0.13V) + (+0.76V) = + 0.63$ volts

8. Describe similarities and differences between galvanic cells and electrolytic cells.
   Similarities:
   - Anode = oxidation; cathode = reduction
   - Electrons flow from anode to cathode
   Differences:
   - Galvanic — uses chemical energy to produce electrical energy; electrolytic uses electrical energy to form pure elements
   - Galvanic occurs in two separate containers; electrolytic occurs in one cell
   - Galvanic uses different metals as electrodes; electrolytic usually uses inert electrodes
   - Galvanic is a battery; electrolytic needs a battery

9. An entrepreneurial student, after studying a bit of electrochemistry, decided to go into the business of electroplating. He decides to begin his business by electroplating semi-precious metals (cheap jewelry) with silver. Design an apparatus that would allow him to carry out this procedure. Be sure to label the following: anode, cathode, electron flow, the electrode's composition, the type and concentration of solution to be used, and the battery. Write the half-cell reaction that would occur at the cathode.

   - The cathode reaction is where the reduction takes place.
   - $Ag^+_{(aq)} + e- \rightarrow Ag_{(s)}$

10. A student performed an electrolysis experiment with an aqueous solution of potassium sulfate, K_2SO_4. She added a few drops of bromothymol blue indicator to her solution. Bromothymol blue indicator is an acid-base indicator that is green in a neutral solution; blue in a base; and yellow in an acid. Explain what she will observe as the electrolytic cell operates.

   a. Bubbles of hydrogen gas form at one electrode as the solution turns blue at this electrode.

   b. Bubbles of oxygen gas form at the other electrode and the color of the solution turns yellow.
      - The student is seeing the oxidation and reduction of water in this reaction. The reactions that she is observing are the following:
      - $2 H_2O_{(l)} + 2e- \rightarrow 2 OH^-_{(aq)} + H_{2(g)}$ (Cathode) The production of OH^- causes blue color.
      - $2 H_2O_{(l)} \rightarrow 4 H^+_{(aq)} + O_{2(g)} + 4e-$ (Anode)   The production of H^+ causes yellow color.
      - Water is more easily oxidized and reduced than either component of the salt in solution.

11. How many grams of chromium could be expected to plate out after 45 minutes in a cell that had 4.0 amps of current flowing through?

$$45 \, min \times \frac{60 \, s}{1 \, min} \times \frac{4.0 \, coul}{s} \times \frac{1 \, mole^-}{96,500 \, coul} \times \frac{1 \, mol \, Cr}{3 \, mole^-} \times \frac{52.00 \, g \, Cr}{1 \, mol \, Cr} = 1.9 \, g \, Cr$$

TEACHER PAGES

It's Electrifying!
Comparing Galvanic and Electrolytic Cells

There are many oxidation-reduction reactions that occur around us everyday. Currently, in this classroom, you are probably using at least one device that depends on the use of electrochemistry.

PURPOSE
In this lesson you will learn about the two types of electrochemical cells. You will be able to calculate reduction potentials for galvanic cells and use Faraday's law for calculations involving electrolytic cells.

MATERIALS
    student notes page
    calculator

CLASS NOTES: ELECTROCHEMISTRY
Electrochemistry involves two main types of cells: *Galvanic* and *electrolytic*. Galvanic cells are spontaneous chemical reactions for the production of electrical energy (batteries). Electrolytic cells are non-spontaneous chemical reactions that require electrical energy to produce pure forms of elements. Before exploring either of these cells, we must become familiar with a table of standard reduction potentials.

STANDARD REDUCTION POTENTIALS
Previously you learned how to balance redox reactions using the half-reaction method. Now we can use that knowledge of half-reactions to calculate overall cell potentials. A few important points follow:
- Each half-reaction has a cell potential
- Each potential was measured against a standard hydrogen electrode. The hydrogen electrode consists of a piece of inert platinum wire that is bathed by hydrogen gas at 1 atmosphere of pressure. Notice that the hydrogen electrode is assigned a value of ZERO volts.

Reading the reduction potential chart
- Elements that have the most positive reduction potentials are easily reduced — these substances would be good oxidizing agents. Generally speaking, non-metals are easily reduced.
- Elements that have the least positive reduction potentials are easily oxidized — these substances would be good reducing agents. Generally speaking, metals are easily oxidized.
- The reduction potential table can also be used as an activity series. Metals having less positive reduction potentials are more active and will replace metals with more positive potentials. Standard reduction potential tables may be written in different orders, be sure that you study the reduction potential values.

TABLE OF REDUCTION POTENTIALS

Half-reaction			$E°$ (V)
$F_2(g) + 2e^-$	$\rightarrow$	$2\,F^-$	2.87
$Co^{3+} + e^-$	$\rightarrow$	Co^{2+}	1.82
$Au^{3+} + 3e^-$	$\rightarrow$	$Au(s)$	1.50
$Cl_2(g) + 2e^-$	$\rightarrow$	$2\,Cl^-$	1.36
$O_2(g) + 4H^+ + 4e^-$	$\rightarrow$	$2\,H_2O(l)$	1.23
$Br_2(l) + 2e^-$	$\rightarrow$	$2\,Br^-$	1.07
$2\,Hg^{2+} + 2e^-$	$\rightarrow$	Hg_2^{2+}	0.92
$Hg^{2+}\ 2e^-$	$\rightarrow$	$Hg(l)$	0.85
$Ag^+ + e^-$	$\rightarrow$	$Ag(s)$	0.80
$Hg_2^{2+} + 2e^-$	$\rightarrow$	$2\,Hg(l)$	0.79
$Fe^{3+} + e^-$	$\rightarrow$	Fe^{2+}	0.77
$I_2(s) + 2e^-$	$\rightarrow$	$2\,I^-$	0.53
$Cu^+ + e^-$	$\rightarrow$	$Cu(s)$	0.52
$Cu^{2+} + 2e^-$	$\rightarrow$	$Cu(s)$	0.34
$Cu^{2+} + e^-$	$\rightarrow$	Cu^+	0.15
$Sn^{4+} + 2e^-$	$\rightarrow$	Sn^{2+}	0.15
$S(s) + 2H^+ + 2e^-$	$\rightarrow$	$H_2S(g)$	0.14
$2H^+ + 2e^-$	$\rightarrow$	$H_2(g)$	0.00
$Pb^{2+} + 2e^-$	$\rightarrow$	$Pb(s)$	-0.13
$Sn^{2+} + 2e^-$	$\rightarrow$	$Sn(s)$	-0.14
$Ni^{2+} + 2e^-$	$\rightarrow$	$Ni(s)$	-0.25
$Co^{2+} + 2e^-$	$\rightarrow$	$Co(s)$	-0.28
$Tl^+ + e^-$	$\rightarrow$	$Tl(s)$	-0.34
$Cd^{2+} + 2e^-$	$\rightarrow$	$Cd(s)$	-0.40
$Cr^{3+} + e^-$	$\rightarrow$	Cr^{2+}	-0.41
$Fe^{2+} + 2e^-$	$\rightarrow$	$Fe(s)$	-0.44
$Cr^{3+} + 3e^-$	$\rightarrow$	$Cr(s)$	-0.74
$Zn^{2+} + 2e^-$	$\rightarrow$	$Zn(s)$	-0.76
$Mn^{2+} + 2e^-$	$\rightarrow$	$Mn(s)$	-1.18
$Al^{3+} + 3e^-$	$\rightarrow$	$Al(s)$	-1.66
$Be^{2+} + 2e^-$	$\rightarrow$	$Be(s)$	-1.70
$Mg^{2+} + 2e^-$	$\rightarrow$	$Mg(s)$	-2.37
$Na^+ + e^-$	$\rightarrow$	$Na(s)$	-2.71
$Ca^{2+} + 2e^-$	$\rightarrow$	$Ca(s)$	-2.87
$Sr^{2+} + 2e^-$	$\rightarrow$	$Sr(s)$	-2.89
$Ba^{2+} + 2e^-$	$\rightarrow$	$Ba(s)$	-2.90
$Rb^+ + e^-$	$\rightarrow$	$Rb(s)$	-2.92
$K^+ + e^-$	$\rightarrow$	$K(s)$	-2.92
$Cs^+ + e^-$	$\rightarrow$	$Cs(s)$	-2.92
$Li^+ + e^-$	$\rightarrow$	$Li(s)$	-3.05

By using the reduction potential table, it is relatively simple to identify which substance will be oxidized and which substance will be reduced in a spontaneous chemical reaction. Using the reduction potential table given in these notes, the elements that are the most easily reduced have the most positive reduction potentials and those that are most easily oxidized have the smaller values. Let's try using the table to make some predictions:

1. Between copper and zinc, which would more likely be oxidized and which is more likely reduced?
   - Copper is +0.34 volts and zinc is -0.76. Copper is more easily reduced since it has the more positive value while zinc is more easily oxidized.

2. What would happen if your gold class ring accidentally dropped into a container of hydrochloric acid in the laboratory?
   - Nothing would happen to your gold ring. Why?
   - This question is a bit more difficult. First, determine which element is oxidized and which element is reduced using the table of reduction potentials.
   - Gold is +1.50 volts and hydrogen is 0.00 volts. Gold will be reduced (since it has a more positive reduction potential) while hydrogen will be oxidized.
   - If the gold ring were to react with HCl, the gold would lose electrons (act like a metal) to become a positive ion (cation) or in other words undergo oxidation. Hydrogen ion will gain an electron to become hydrogen gas or in other terms, undergo reduction. Since what we expect to see does not match the order listed on the table of reduction potentials the reaction will not take place.

Making predictions about reactions is just one benefit of using the reduction potential chart. Another use of the chart is to calculate the reduction potential for a galvanic cell. We will first explore the components of the galvanic cell and revisit this table for voltages a bit later.

GALVANIC CELLS

Galvanic or voltaic cells are spontaneous chemical reactions. Energy changes take place in all types of chemical reactions. The energy changes that take place in the redox reactions we are about to study often release energy in the form of heat. But if we separate the reactants, place them in solution and connect them via a wire, we get energy in the form of electricity. You know this type of electrochemical cell well — batteries have been, and continue to be, a major part of your life. Let's look at the basic design for a galvanic cell.

Parts of the voltaic or galvanic cell:
- Anode — the electrode where oxidation occurs. After a period of time, the anode may appear to become smaller as it reacts to form ions in solution. If the mass of the anode were measured before and after the reaction, the mass of the anode should decrease as metal atoms are converted into metal ions in solution.
- Cathode — the electrode where reduction occurs. After a period of time it may appear larger, due to ions from solution plating onto it. If the mass of the cathode were measured before and after the reaction, the mass of the cathode should increase as metal ions from solution are converted into metal atoms.
- Inert electrodes — used when a gas is involved or the conversion is ion to ion, such as Fe^{3+} being reduced to Fe^{2+}. These are usually composed of platinum or graphite.

- Salt bridge — a device used to maintain electrical neutrality in a galvanic cell. This may be a u-shaped tube filled with agar which contains a neutral salt or it may be replaced with a porous cup. Cations move from the salt bridge to the cathode to replace any positive ions that have been reduced and anions from the salt bridge move to the anode to balance the extra positive ions that are falling into solution. Without a salt bridge in place, ion charges would build up and the cell would quickly reach equilibrium and produce a voltage of zero. (What we call a dead battery!)
- Electron flow — electrons always move through the wire from anode to cathode.
- Standard cell notation (line notation) — anode/solution// cathode solution/ cathode
  Ex. Zn/Zn^{2+} (1.0 M) // Cu^{2+} (1.0M) / Cu   In this example zinc metal is the anode and is placed in a 1.0 M solution of zinc ions connected by a salt bridge to a piece of copper metal which serves as the cathode in a 1.0 M solution of copper (II) ions.
- Voltmeter — measures the reduction potential or the electromotive force. Usually this force is measured in volts. A voltmeter does not have to be a part of our cell — it is just for our convenience to measure the voltage.

A few mnemonics to help you remember some of this:

AN OX — oxidation occurs at the anode

RED CAT — reduction occurs at the cathode (the more positive reduction potential gets to be reduced)

FAT CAT — The electrons in a voltaic or galvanic cell ALWAYS flow
From the **A**node **T**o the **CAT**hode

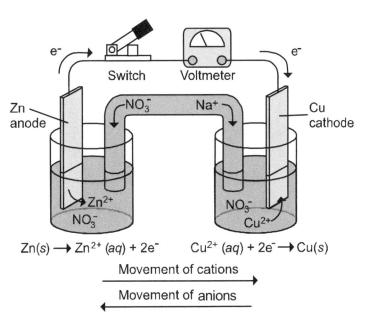

Fig. 1

Applications of Galvanic Cells:

Batteries: Batteries are cells that are connected in series where the potentials add together to give a total voltage. This is the most common application of galvanic cells.
Some common batteries include:
> Lead-storage batteries (car) — Pb anode, PbO_2 cathode, H_2SO_4 electrolyte
> Dry cell batteries
>> Acid versions: Zn anode, C cathode; MnO_2 and NH_4Cl paste
>> Alkaline versions: some type of basic paste, ex. KOH
>> Nickel-cadmium — anode and cathode can be recharged
> Fuel cells — Reactants continuously supplied (spacecraft –hydrogen and oxygen)

Corrosion prevention:
> Galvanizing of iron is one common method of corrosion prevention. A zinc coating is placed on the iron (galvanizing). Since zinc is a more active metal than iron, it will oxidize much more rapidly. Thus, zinc is a "sacrificial" coating on the steel and protects it for years. Other metals such as magnesium and titanium have also been used as protective coatings for hulls of ships and pipelines.

Calculating standard cell potential: $(E°_{cell})$ $E°_{cell} = E°_{reduction} + E°_{oxidation}$
Standard conditions for a galvanic cell consist of 1.0 M solutions, 25°C and 1.0 atm of pressure. The naught (°) on the E indicates these standard conditions.

To calculate the standard cell potential:
1. Write down the two half-reactions with voltages exactly as they are written in the table of standard reduction potentials.

2. The half-reaction with the more positive reduction potential will be reduced. Reverse the reaction with the smaller voltage and change the sign on the voltage. This should now be your oxidation half-cell.

3. If needed, multiply the terms of the half reactions by an integer to make the number of electrons lost equal the number of electrons gained. Do not multiply the voltages!

4. Add the reactions together and total the voltage.

Example: Calculate the voltage for a cell constructed using aluminum foil and iron nails in the appropriate solutions at standard conditions.

1. Copy reactions:
> $Al^{3+} + 3 e- \rightarrow Al$ -1.66 V
> $Fe^{2+} + 2 e- \rightarrow Fe$ -0.44 V (This number is more positive. That means Fe will be the cathode since a reduction takes place there — think RED CAT)

2. Reverse: $Al \rightarrow Al^{3+} + 3 e-$ +1.66 V (This will be the anode since it is an oxidation—think AN OX)

3. Multiply: the least common multiple between 2 and 3 is 6, therefore the reactions become:
$$3Fe^{2+} + 6e- \rightarrow 3Fe \qquad -0.44 \text{ V}$$
$$2Al \rightarrow 2Al^{3+} + 6 e- \qquad +1.66 \text{ V}$$

4. Find the sum the voltage:
$$3Fe^{2+} + 2Al \rightarrow 3Fe + 2Al^{3+} \quad E^\circ_{cell} = (-0.44) + (+1.66) = 1.22 \text{ volts}$$

ELECTROLYTIC CELLS

Electrolytic cells are non-spontaneous chemical reactions.

Electrolysis is a common term used to signify the use of electricity to bring about a chemical change. The literal translation means to "split with electricity". Water can be electrolyzed (split with electricity) into the elements hydrogen and oxygen. The components of an electrolytic cell appear basically the same as a voltaic cell but the purposes of these cells differ.

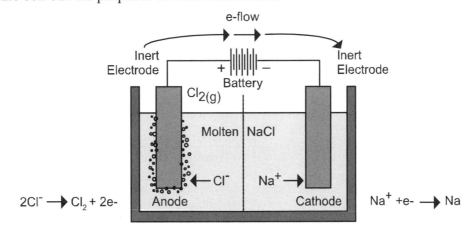

Fig. 2

Applications of electrolytic cells:

Production of pure forms of elements from mined ores:

Aluminum from Hall-Heroult process

Separation of sodium and chlorine (Down's cell)

Purify copper for wiring used in homes

Electroplating — applying a thin layer of an expensive metal to a less expensive one (The cathode is where reduction occurs, so the object to be electroplated must be attached to the cathode in order for this process to occur.)

Jewelry — 14 K gold plated

Bumpers on cars — Chromium plated

Charging a battery — For example, the alternator in your car recharges the battery as you drive.

Calculations Involving Electrolytic cells:

Faraday's Law states that the amount of a substance being oxidized or reduced at each electrode during electrolysis is directly proportional to the amount of electricity that passes through the cell. Some very practical problems can be solved using simple dimensional analysis and the following conversion factors.

For example, if you wanted to know how much metal could be plated out after a particular time period or how long it might take to plate a certain amount of metal you could solve both of these problems by combining some of the following conversion factors. You should memorize these.

1 Volt = 1Joule/Coulomb

1 Amp = 1Coulomb/second

Faraday's constant = 96,500 Coulomb/mole of e-

Balanced redox reactions give the number of moles of electrons transferred

The gram formula mass found on the periodic table is gives g/mol

Example: How many grams of gold could be expected to plate out after 60 minutes in a cell that had 2.0 amps of current flowing through?

$$60\,min \times \frac{60\,s}{1\,min} \times \frac{2.0\,coul}{s} \times \frac{1\,mol\,e^-}{96,500\,coul} \times \frac{1\,mol\,Au}{3\,mol\,e^-} \times \frac{196.97\,g\,Au}{1\,mol\,Au} = 2.5\,g\,Au$$

In Summary:

Galvanic Cells	Electrolytic Cells
Produce electrical energy from chemical energy	Use electrical energy to produce pure elements
Occur in two containers	Occur in one container
Spontaneous	Non-spontaneous
Usually electrodes are metals in appropriate metallic solutions	Usually electrodes are inert
Is a battery	Needs a battery
Electrons flow FAT CAT	Electrons flow FAT CAT
AN OX and RED CAT	AN OX and RED CAT

Complete the questions and calculations found on the student answer pages.

Name _____

Period _____

It's Electrifying!
Comparing Galvanic and Electrolytic Cells

ANALYSIS AND CONCLUSION QUESTIONS

1. List the two main types of electrochemical cells and briefly state the purpose of each.

2. Refer to the Table of Standard Reduction Potentials to answer the following questions.
   a. State the general location of metals and nonmetals on the table.

   b. Arrange the following metals in order of increasing activity. Explain your reasoning.

      Fe, Mg, Na, Au, Ag

   c. Write the reduction half-reaction for water with its voltage.

   d. Write the oxidation half-reaction for water with its voltage.

e. What would you expect to see if copper wire were placed into a solution of silver nitrate? Support your answer with appropriate equations.

f. Combining which of the following would yield the greatest cell potential? Support your answer.

Ni, Fe, Cu, Li, Ag, Al, Au

3. Sketch and label a diagram of a galvanic cell composed of tin and zinc. Both electrodes are in 1.0 M solutions of an appropriate ionic salt containing the metal ion and the experiment is carried out at 25°C. Be sure to include and label the following in your apparatus: anode, cathode, electron flow, salt bridge, ion flow, and voltmeter.

a. Write the half-cell reactions occurring at each electrode.

b. Write the balanced overall cell reaction.

c. Calculate the standard reduction potential for this reaction.

4. A galvanic cell was constructed between zinc and copper. The electrodes were massed before and after the reaction. Describe how the masses of each electrode will change at the cell operates. Be sure to specify which metal will be the anode and which metal will be the cathode. Give a reasonable explanation for your observations.

5. Rechargeable batteries are often referred to as NiCad batteries. A proposed reaction for the production of these batteries is: $Ni_{(s)} + Cd^{2+}_{(aq)} \rightarrow Ni^{2+}_{(aq)} + Cd_{(s)}$
   a. Will this reaction occur spontaneously as written? Explain.

   b. Calculate the voltage produced by this cell.

6. You visit a close relative during your spring break and are asked to help clean the silver that is tarnished to an almost black color, Ag_2S. Your relative explains that the tarnish can be removed by boiling the silver in an aluminum pan with some baking soda. The relative states that they have no idea why this works. Being the expert chemistry student, explain the chemistry behind this process — in all the gory chemistry details including equations along with standard reduction potentials.

7. Given the following cell in standard cell notation,
   $Zn_{(s)} /Zn(NO_3)_{2(aq)}(1.0M) //Pb(NO_3)_{2(aq)}(1.0M) /Pb_{(s)}$

   a. Write the overall balanced chemical reaction.

   b. Assuming standard temperature, calculate the E° for the cell.

8. Describe similarities and differences between galvanic cells and electrolytic cells.

9. An entrepreneurial student, after studying a bit of electrochemistry, decided to go into the business of electroplating. He decides to begin his business by electroplating semi-precious metals (cheap jewelry) with silver. Design an apparatus that would allow him to carry out this procedure. Be sure to label the following: anode, cathode, electron flow, the electrode's composition, the solution, and the battery. Write the half-cell reaction that would occur at the cathode.

10. A student performed an electrolysis experiment with an aqueous solution of potassium sulfate, K_2SO_4. She added a few drops of bromothymol blue indicator to her solution. Bromothymol blue indicator is an acid-base indicator that is green in a neutral solution; blue in a base; and yellow in an acid. Explain what she will observe as the electrolytic cell operates.

11. How many grams of chromium could be expected to plate out after 45 minutes in a cell that had 4.0 amps of current flowing through?

Deposition
Determining the Size of an Atom by Silvering Glass

OBJECTIVE

Students will identify purposes of deposition processes and describe the process involved in oxidation-reduction. The students will also learn how to deposit metallic silver on glass to make a mirror. Lastly the students will learn about Tollen's reagent and test.

LEVEL

Chemistry

NATIONAL STANDARDS

UCP.2, UCP.3, B.3, E.2, G.1, G.2, G.3

TEKS

1(A), 1(B), 2(B), 2(E), 3(C), 3(D), 4(A), 8(C), 10(A), 10(B)

CONNECTIONS TO AP

AP Chemistry:

    III. Reactions A. Reaction types 3. Oxidation-reduction reactions a. Oxidation number b. The role of the electron in oxidation-reduction

TIME FRAME

40-50 minutes

MATERIALS

(For a class of 28 working individually)

28 safety glasses
28 small clean test tubes with stoppers
1 gallon of distilled water
2 pipets or 2 syringes or 2 repipetors capable of delivering 0.5 mL
250 mL 5% silver nitrate solution (0.29M)
500 mL 0.1 M glucose solution
4-liter waste bottle

28 pairs of vinyl or latex gloves
pipet or syringe or repipetor capable of delivering 1 mL
pipet or syringe or repipetor capable of delivering 3 mL
250 mL 10% sodium hydroxide (2.5M)
250 mL 1:4 ammonium hydroxide solution (3.7M)

TEACHER NOTES

Caution: Chemicals may be hazardous if not handled properly.

If the repipetors are used, make sure the dispensing spouts do **NOT** point at the students or adults.

The waste solution should be diluted with concentrated nitric acid, neutralized with a base and then poured down the drain with copious amounts of water.

Suggested teacher script:

In many semiconductor fabrication processes, it is important to have a metal on the surface. Sometimes it is needed as a conductor of electrical current; sometimes it is needed to protect the surface from scratches; and sometimes it is placed on the surface just to make it shiny. One way in which metal is placed on surfaces is through deposition processes.

Deposition is a critical process in the fabrication of semiconductors. Fabricating a semiconductor product is similar in many ways to building a skyscraper. In a skyscraper, each level is built on top of the one below it. Semiconductor integrated circuits are fabricated in much the same way. The circuitry is built layer by layer. All the layers are built on a foundation, but the foundation is a silicon wafer instead of concrete. Instead of wood or steel construction, each layer is composed of materials such as silicon, silicon dioxide, copper, or silicon nitride. The layers are constructed in carefully designed patterns that will result in microscopic transistors, capacitors, diodes and resistors connected to each other by additional layers of metal.

Since integrated circuits are microscopic, they are not suitable for us to study. However, by studying the deposition process used to make mirrors, we can understand how the process can be employed in fabricating semiconductor products such as integrated circuits.

The production of a mirror requires a uniform deposit of a highly reflective material on a smooth surface. This exercise takes silver metal and *physisorbs* (physical adsorption) the silver onto a glass surface. *Physical adsorption* means that there is a mechanical bonding between the silver and the glass surface as opposed to a *chemical bond* (chemisorbed). The metal is *conformal*, that is, it takes the shape of the surface of the glass. This process used to be the way in which mirrors were commonly made.

To prepare the solutions:

250 mL 5% silver nitrate solution (0.29M):
> Measure 12.5 grams of silver nitrate, place into a 250. mL volumetric flask and fill to the line with distilled water.

500 mL 0.1 M glucose solution:
> Measure 9.00 grams of glucose, place into a 500. mL volumetric flask and fill to the line with distilled water.

250 mL 10% sodium hydroxide (2.5M):
Measure 25.00 grams of NaOH and place into a 250. mL volumetric flask and fill to the line with distilled water.

250 mL 1:4 ammonium hydroxide solution (3.7M):
> Measure 50.0 mL of concentrated (14.8 M) ammonia and place into a 250 mL volumetric flask, add distilled water to the line.

Background Information:

Silver (Ag^+) is in ionic form with a nitrate ion (NO_3^-).

In aqueous solution this salt is slightly acidic.

The sodium hydroxide (NaOH) is added to increase the pH (basic) and make the Ag^+ precipitate out as a solid mass of silver oxide.

$$2AgNO_{3(aq)} \; + \; 2NaOH_{(aq)} \; \rightarrow \; Ag_2O_{(s)} + 2\,NaNO_{3(aq)} \; + \; H_2O_{(l)}$$

Ammonium hydroxide is added to create complex Ag^+ ions in solid form that are placed in solution to produce Tollen's reagent. A complex ion is one that is composed of two or more ions surrounding and protecting the central metal ion. The silver ammonia complex allows the Ag^+ ion to collect free electrons from the reducing sugar, and reduce to silver metal. Remember a "*reduction*" is a gain of electrons and "*oxidation*" is a loss of electrons.

TOLLEN'S REAGENT

Tollen's Reagent = $Ag(NH_3)_2{}^+$

$$\begin{array}{c} O \\ \| \\ R\text{-}C\text{-}O^- \end{array} + Ag(NH_3)_2{}^+ + OH^- \longrightarrow \begin{array}{c} O \\ \| \\ R\text{-}C\text{-}O^- \end{array} + Ag_{(s)}$$

Reaction: The aldehyde is oxidized to a carboxylate anion and the Ag^+ is reduced to silver metal.

Reducing Sugars:

$$\begin{array}{c} CH_2OH \\ | \\ C{=}O \\ | \\ H\text{-}O\text{-}C\text{-}H \\ | \\ H\text{-}C\text{-}O\text{-}H \\ | \\ H\text{-}C\text{-}O\text{-}H \\ | \\ CH_2OH \end{array}$$

D-Fructose

$$\begin{array}{c} O \\ \| \\ C\text{-}H \\ | \\ H\text{-}C\text{-}O\text{-}H \\ | \\ H\text{-}O\text{-}C\text{-}H \\ | \\ H\text{-}C\text{-}O\text{-}H \\ | \\ H\text{-}C\text{-}O\text{-}H \\ | \\ CH_2OH \end{array}$$

D-Glucose

The D-fructose can be reduced even though it is not an aldehyde because the fructose is in equilibrium with two diastereomeric aldehydes through an enediol tautomeric intermediate.

$$CH_2OH(CHOH)_4\text{-}C\text{-}H + 2\ Ag(NH_3)_2{}^{1+} + 3\ OH^{1-} \longrightarrow 2\ Ag + CH_2\ OH(CHOH)_4\text{-}C\text{-}O + 4\ NH_3 + 2\ H_2O$$

dextrose
(an aldehydic sugar)

Alcohol   Intermediate   Aldehyde

Silvering Mirrors

A mirror may have a polished nickel or chrome coating on a steel or iron plate, a solution of tin in mercury (called an amalgam) pressed against a glass plate, or, most frequently, a deposit of finely divided silver metal. The most straightforward method of delivering this silver metal to a surface is to deposit it atom-by-atom, by means of electrolysis or by an oxidation-reduction reaction. In this exercise, the oxidation-reduction reaction will be used.

In an oxidation-reduction (also called a redox) reaction, one of the agents is *reduced* (a net gain of electrons) while the other is *oxidized* (a net loss of electrons). Silver is readily obtained as a *cation* (positive ion) in a salt such as silver nitrate. To change the positive ion into free or metallic silver, an electron supplier or reducing agent must be added. Certain organic compounds (aldehydes and some alcohols and ketones) can supply these electrons, especially when the silver is found in solution as its ammonium complex. The metallic silver is deposited as a thin film on the surface of the glass.

The recipe:

Organic chemists use this reaction to detect the presence of aldehydes. It is known as Tollen's test or the silver mirror test. The recipe is an old one for the silvering of mirrors known as Brashear's Process. The reducing agent is a common sugar, glucose.

Data Table:

Mass of the clean, dry test tube with stopper: <u>12.75 grams</u>

Mass of the silvered test tube with stopper: <u>12.95 grams</u>

Inside diameter of the container: <u>12 mm</u>

Length of tube: <u>100. mm</u>

Analysis: Show all work for full credit!
- Note: Rounding of answers did not occur until step 10 in calculation of the radius of the silver atom. Intermediate answers were left to four significant figures. You may wish to have students round at each step. However, answers may vary a bit more.

1. Calculate the mass of silver metal deposited inside the tube.
   - 12.95 g – 12.75g = 0.20 grams Ag

2. Convert the mass of silver metal to moles of silver atoms deposited.
   - $0.20 \text{ g Ag} \times \dfrac{1\,\text{mol Ag}}{107.87\,\text{g Ag}} = 0.001854\,\text{mol Ag}$

3. Calculate the number of atoms of silver deposited.
   - $0.001854\,\text{mol Ag} \times \dfrac{6.02\times10^{23}\,\text{atoms Ag}}{1\,\text{mol Ag}} = 1.116\times10^{21}\,\text{atoms Ag}$

4. Calculate the surface area of the cylindrical portion of the silver mirror.

   (Surface area of a cylinder = $2\pi r h$)
   - Remember that the radius is ½ of the diameter.
   - (2) (π) (6mm) (100 mm) = 3770. mm^2

5. Calculate the surface area of the half-sphere on the bottom of the test tube. (If your cylindrical container has a flat bottom and can stand alone on the lab counter, omit this step.)

   (Surface area of a sphere = $4\pi r^2$)
   - (4) (π) (6mm)^2 = 452.4 mm^2
   - ½ of the sphere = 226.2 mm^2

6. Calculate the total surface area of the silver deposited.
   - 3770. mm^2 + 226.2 mm^2 = 3996.2 mm^2

7. Calculate the volume in milliliters of the silver deposited. (The density of silver is 10.5 g/cm^3)
   - $D = \dfrac{\text{mass}}{\text{volume}}$

     $V = \dfrac{\text{mass}}{\text{density}}$
   - $V = \dfrac{0.20\,\text{g Ag}}{10.5\,\text{g/cm}^3} = 0.01905\,\text{cm}^3$

8. Calculate the thickness of the silver mirror. (Divide the volume by the total area.)

   - Thickness $= \dfrac{19.05 \, \text{mm}^3}{3996 \, \text{mm}^2} = 0.004767 \, \text{mm}$

   - Don't forget to convert mm and cm to obtain the same units!

9. Calculate the volume of a single silver atom using the volume of silver and the number of silver atoms.

   - $\dfrac{19.05 \, \text{mm}^3}{1.116 \times 10^{21}} = 1.707 \times 10^{-20} \, \text{mm}^3$

10. Calculate the radius of a single silver atom by using the volume of a silver atom (sphere). (Remember: $4/3 \, \pi \, r^3$)

    - $r^3 = \dfrac{V}{4/3 \, \pi} = \dfrac{1.707 \times 10^{-20} \, \text{mm}^3}{4/3 \, \pi} = 4.075 \times 10^{-21} \, \text{mm}^3$

      $r = 1.60 \times 10^{-7} \, \text{mm} = 1.60 \times 10^{-8} \, \text{cm}$

11. Determine the thickness of the silver layer in your mirror and report it as "number of atoms thick". Use the calculated diameter (radius x 2) of a silver atom.
    - $2(1.597 \times 10^{-7}) = 3.19 \times 10^{-7}$ atoms thick

12. The actual value for the radius of a silver atom is 1.75×10^{-8} cm. Determine your percent error.

    - $\% \, \text{error} = \dfrac{\left[1.60 \times 10^{-8} - 1.75 \times 10^{-8} \right]}{1.75 \times 10^{-8}} = 8.57\%$

POSSIBLE ANSWERS TO THE CONCLUSION QUESTIONS

1. What are the purposes of deposition processes?
   - To form conductive layers of metal on top of layers of other material
   - To form protective layers of materials on top of layers of other material
   - To form shiny layers of materials on top of layers of other material

2. Describe the process involved in an oxidation-reduction reaction.
   - One of the agents is reduced (gains electrons) while the other agent is oxidized (loses electrons).

3. Why does the surface that is going to receive the deposited metal need to be absolutely clean?
   - If there are damp areas or oily areas, the metal cannot adhere to the surface.
   - If there are particles on the surface, they might make holes in the deposited metal or come loose later and lift some of the metal off of the surface.

4. What is the usual purpose of Tollen's test?
   - Tollen's test is used by organic chemists to detect the presence of aldehydes.

REFERENCES:

Destination Digital Brochure, 2002, and http://www.destinationdigital.org
http://micron.com/K-12/lessonplans

P. Van Zant, *Microchip Fabrication, Fourth Edition*, McGraw Hill Publishing, 2000, glossary

Deposition
Determining the Size of an Atom by Silvering Glass

Semiconductor fabrication is one of the names for the mass production of integrated circuits. Fabrication includes all of the processes that can be done in wafer form. Many integrated circuits now have internal dimensions that are sub-micron in width. To put this in perspective, the human hair is about 100 microns in diameter. In order to produce integrated circuits with predictable yields, everything that comes near the integrated circuits prior to top-surface protective coating must essentially be free of particles and other contaminants. In addition, all of the layers formed during fabrication must be highly controlled in purity and must conform as closely as possible to the surface beneath.

In many semiconductor fabrication processes, it is important to have a metal on the surface. Sometimes it is needed as a conductor of electrical current; sometimes it is needed to protect the surface from scratches; and sometimes it is placed on the surface just to make it shiny. One way in which metal is placed on surfaces is through deposition processes.

Deposition is a critical process in the fabrication of semiconductors. Fabricating a semiconductor product is similar in many ways to building a skyscraper. In a skyscraper, each level is built on top of the one below it. Semiconductor integrated circuits are fabricated in much the same way. The circuitry is built layer by layer. All the layers are built on a foundation, but the foundation is a silicon wafer instead of concrete. Instead of wood or steel construction, each layer is composed of materials such as silicon, silicon dioxide, copper, or silicon nitride. The layers are constructed in carefully designed patterns that will result in microscopic transistors, capacitors, diodes and resistors connected to each other by additional layers of metal.

Since integrated circuits are microscopic, they are not suitable to study for this deposition exercise. However, by studying the deposition process used to make mirrors, it is possible to understand how the process can be employed in fabricating semiconductor products such as integrated circuits.

The production of a mirror requires a uniform deposit of a highly reflective material on a smooth surface. This exercise takes silver metal and *physisorbs* (physical adsorption) the silver onto a glass surface. *Physical adsorption* means that there is a mechanical bonding between the silver and the glass surface as opposed to a *chemical bond* (chemisorbed). The metal is *conformal*, that is, it takes the shape of the surface of the glass. This process used to be the way in which mirrors were commonly made.

In an oxidation-reduction (also called a redox) reaction, one of the agents is *reduced* (a net gain of electrons) while the other is *oxidized* (a net loss of electrons). Silver is readily obtained as a *cation* (positive ion) in a salt such as silver nitrate. To change the positive ion into free or metallic silver, an electron supplier or reducing agent must be added. Certain organic compounds (aldehydes and some alcohols and ketones) can supply these electrons, especially when the silver is found in solution as its ammonium complex. The metallic silver is deposited as a thin film on the surface of the glass.

39 *Deposition*

Organic chemists use the recipe used today to detect the presence of aldehydes. It is known as Tollen's test or the silver mirror test. The recipe is an old one used for the silvering of mirrors. The reducing agent is common sugar, glucose.

PURPOSE

In this activity you will be able to identify the purposes of deposition processes and to describe the process involved in oxidation-reduction. You will learn how to deposit metallic silver on glass to make a mirror. You will also learn about Tollen's test.

GLOSSARY

Aldehyde: An organic compound with a carbon bound to a $-(\overset{\overset{\textstyle H}{|}}{C}=O)$ group

Cation: A positively charged ion

Chemical Adsorption or Chemisorb: Formation of a chemical bond between two separate layers

Chip (or microchip): An individual integrated circuit built in a tiny, layered rectangle or square on a silicon wafer. There may be as many as hundreds of these chips on a single wafer.

Clean Room: A manufacturing facility where integrated circuits are fabricated. The air inside these rooms is cleaner than a typical surgical operating room.

Conformal: The deposited layer takes the shape of the layer on which it is deposited.

Contaminant: A general term used to describe unwanted material that adversely affects the physical or electrical characteristics of a semiconductor product

Deposition: A process in which a thin film of material is formed on top of the surface of other materials

Integrated Circuit: An electronic circuit containing as many as millions of microscopic transistors that work together to perform specific functions. All elements of the circuit are fabricated and interconnected in and on a single chip of semiconducting material. Also called chip or microchip or microcircuit.

Micron (or micrometer): One-millionth of a meter (10^{-6} meter); symbol is μ or μm.

Oxidation-Reduction Reaction: A process during which one of the agents is reduced (a net gain of electrons) while the other agent is oxidized (a net loss of electrons)

pH scale: A scale that indicates the strength or weakness of an acid (0 to 7) or base (7 to 14)

Physical Adsorption or Physisorb: Formation of a mechanical bonding between two layers

Semiconductor: A material that can be an electrical conductor or insulator. Silicon is the most common semiconductor used to manufacture integrated circuits.

Silicon: A basic element in the Periodic Chart. Sand is the primary source of silicon (Si).

Wafer: A thin slice of silicon, or other semiconductor material, in and on which multiple integrated circuits of the same design are fabricated

Laying the Foundation in Chemistry

MATERIALS

safety goggles
clean test tube with stopper
5% silver nitrate solution (0.29M)
0.1 M glucose solution

pair of vinyl or latex gloves
10% sodium hydroxide (2.5M)
1:4 ammonium hydroxide solution (3.7M)

Safety Alert

1. Chemicals may be hazardous if not handled properly.
2. Wear goggles and aprons at all times.
3. Silver nitrate will stain clothing and skin. Flush with water if it comes in contact with your skin.
4. The sodium hydroxide solution and the ammonia are fairly concentrated and should also be flushed with copious amounts of water if either comes in contact with skin.
5. Dispose of the waste as designated by your teacher.

PROCEDURE

1. Put on safety goggles and gloves.

2. Obtain a clean test tube with a stopper. Note that the inside glass surface must be absolutely clean. DO NOT PUT YOUR FINGERS OR ANYTHING ELSE INSIDE THE VIALS UNTIL INSTRUCTED TO DO SO.

3. Mass the test tube and stopper to the nearest 0.01 grams and record this in the data table on the student answer page.

4. Carefully add 1 ml of 5% aqueous silver nitrate solution to the container.

5. Carefully add 0.5 ml of a 10% sodium hydroxide solution. A gray precipitate (solid material) of silver oxide will form.

6. Carefully add 0.5 ml of the ammonium hydroxide solution and swirl the vial. Continue to add the ammonium hydroxide one drop at a time (3 drops should be adequate) until the precipitate ALMOST dissolves. YOU DO NOT WANT THE PRECIPITATE TO ENTIRELY DISSOLVE. The solution contains silver ions in an ammonium complex and is poised and ready to gain electrons from somewhere.

7. Add 3.0 ml of reducing solution, 0.1 M glucose, to the vial. ADD IT ALL AT ONCE TO THE SILVERING SOLUTION IN THE VIAL.

8. Stopper the tube and swirl, gently but continuously, and then roll the tube. A silver mirror deposit should cover the inside of the tube.

9. You should spend about 5 minutes of gentle mixing, and then pour the remaining solution and loose silver metal into the WASTE container as directed by your teacher.

10. Rinse the tube with tap water and then apply a quick acetone rinse and allow to dry. The mirror in the tube could last for many years.

11. Mass the tube with the silver deposit and stopper. Record this mass in the data table on the student answer page.

12. Measure the inside diameter of the tube to the nearest 0.1 cm and record.

13. Measure the length of the silver cylinder portion of the tube up to the point at which the end begins to curve and record.

14. Complete the ANALYSIS and CONCLUSION questions.

Name _____

Period _____

Deposition
Determining the Size of an Atom by Silvering Glass

DATA AND ANALYSIS

Data Table:

Mass of the clean, dry container with lid: _____

Mass of the silvered container with lid: _____

Inside diameter of the container: _____

Length of tube: _____

Analysis: Show all work for full credit!

1. Calculate the mass of silver metal deposited inside the tube.

2. Convert the mass of silver to moles of silver deposited.

3. Calculate the number of atoms of silver deposited.

4. Calculate the surface area of the cylindrical portion of the silver mirror. (Surface area of a cylinder $= 2\pi \, r \, h$)

5. Calculate the surface area of the half-sphere on the bottom of the tube. (Surface area of a sphere $= 4 \, \pi \, r^2$)

6. Calculate the total surface area of the silver deposited.

7. Calculate the volume in cm^3 of the silver deposited. (The density of silver is 10.5 g/cm^3)

8. Calculate the thickness of the silver mirror. (Divide the volume by the total area.)

9. Calculate the volume of a single silver atom using the volume of silver and the number of silver atoms.

10. Calculate the radius of a single silver atom by using the volume of a silver atom (sphere). (Remember: 4/3 πr^3)

11. Determine the thickness of the silver layer in your mirror and report it as "number of atoms thick". Use the calculated diameter (radius x 2) of a silver atom.

12. The actual value for the radius of a silver atom is 1.75 x 10^{-8} cm. Determine your percent error.

CONCLUSION QUESTIONS

1. What are the purposes of deposition processes?

2. Describe the process involved in an oxidation-reduction reaction.

3. Why does the surface that is going to receive the deposited metal need to be absolutely clean?

4. What is the usual purpose for Tollen's test?

Write Your Notes and Ideas Here!

Long, Long Chains
Identifying Polymers

TEACHER PAGES

OBJECTIVE
Students will learn the structures and uses of various recyclable polymers. They will use the physical and chemical characteristics of each to identify unknowns. Students will also become aware of how polymers can be recycled.

LEVEL
Chemistry

NATIONAL STANDARDS
UCP.1, UCP.5, A.1, B.2, E.2, E.4, F.3

TEKS
1(A), 1(B), 2(A), 2(B), 2(D), 3(B), 4(B)

CONNECTIONS TO AP
AP Chemistry:
   IV. Descriptive Chemistry 3. Introductions to organic chemistry: hydrocarbons and functional groups (structure, nomenclature, chemical properties).

TIME FRAME
45 minutes

MATERIALS
(For a class of 28 working in pairs)

samples of recyclable polymers 1-6
42 beakers (250-mL)
14 forceps
14 hot plates (student pairs may share if necessary)
700 mL acetone
14 glass stirring rods
14 watch glasses to cover the 250-mL beakers
14 permanent marking pens
6 containers to hold polymer samples and 6 containers to hold unknown samples.

14 Bunsen burners
14 beakers (100-mL)
14 crucible tongs
14 copper wires (10 cm long, 20 gauge)
1 quart isopropyl (rubbing) alcohol (70%)
14 watch glasses to cover 100-mL beakers
labels to label beakers
14 hole punches
14 permanent marking pens
paper towels

TEACHER NOTES

This activity is best done toward the end of a unit on organic chemistry. The week prior to the lab you should call students' attention to the recycle symbols on plastic containers and ask them to find at least one example of each of the six major recycled polymers. Have them bring the samples to class two days before you plan to do the lab. This will heighten student awareness of materials that can be recycled as well as provide an ample supply of material for the experiment.

Music is one way to improve retention of material and student interest. Michael Offutt has two tapes/CDs that are excellent interest builders. His song "Long, Long Chains" from Chemistry Songbag II is useful in teaching about polymers. This tape/CD may be ordered from Mike Offutt, 1020 Fox Trails Dr. S., Cary, IL 60013.

This lab will introduce the students to the use of a flow chart to identify an unknown. Ask students to develop the flow chart before the lab. You should check the flow charts or go over their plans before the lab is to be done to be sure that they understand the concept. The following flowchart is provided for you. After going over the flowchart with the students, you may want to put this chart on the board to avoid mistakes.

Flow Chart

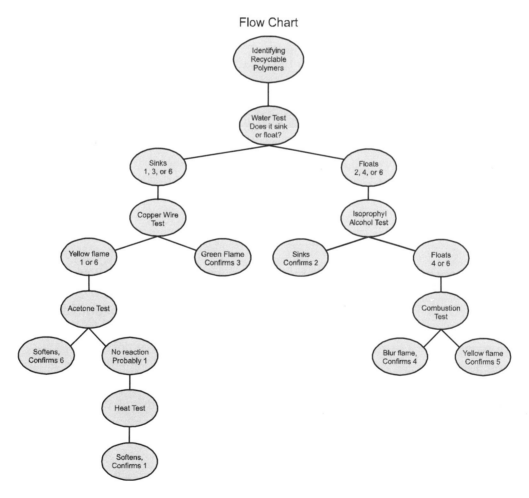

LAB SET-UP

Each group will need the following items:

400 mL beaker containing distilled water (Initial density test)

100 mL beaker containing 30 mL acetone— covered with a watch glass and set far away from open flames.

400 mL beaker containing mixture of isopropyl alcohol. See instructions below.

10 cm length of copper wire

250 mL beaker containing distilled water and placed on hot plate for heat test.

forceps or crucible tongs

Bunsen burner

Each student group will need a mixture of isopropyl alcohol and water. This is the trickiest set-up to do. Put about 500 mL distilled water in a large beaker. Float a sample of HDPE (2) and a sample of LDPE (4) in the water. Use a stirring rod to knock off any adhering bubbles. Gradually, with stirring, add isopropyl alcohol (2-propanol) to the water until the HDPE sinks and the LDPE floats. Distribute the solution to the student groups in 250-mL beakers covered with watch glasses.

Cut the polymer samples that the students have brought in into small pieces approximately 1 cm^2 in area. It is not important to worry too much about shape or size. Place the samples in labeled containers for the students to acquire before the lab.

Cut additional polymer samples and place them in containers labeled A-F. You may use the following key or make up one of your own.

Key	Polymer
A	4
B	3
C	5
D	6
E	1
F	2

Students can select three of these unknowns or you can assign three unknowns to analyze.

POSSIBLE ANSWERS TO THE CONCLUSION QUESTIONS AND SAMPLE DATA

1. In the introduction you are given three examples of addition polymers and three examples of condensation polymers. Give one additional example of each type of polymer.
   a. Addition polymer:
      - Student answers may vary. They may need to go to the Internet or do some additional research to answer this question. Addition polymers must have a double bond. The most likely answer will be polypropylene, as it is the one common recyclable polymer that is not listed in the introduction.
   b. Condensation polymer:
      - Student answers will vary, but common answers may include cellulose, proteins, or DNA.

2. You are in a boat which is sinking. You are not a good swimmer. In the boat are six large solid plastic blocks labeled 1, 2, 3, 4, 5, 6. You only have two arms. Which two blocks would you grab? Explain your choice.
   - You should choose low density polyethylene, #4,(density=0.92-0.94 g/mL) and polypropylene, #3, (density=0.92-0.94 g/mL). These polymers have the lowest density and would float and buoy you up better.

3. You are on a camping trip on an island on the gulf coast. You have run out of wood. Sadly, there is lots of plastic that has washed up on shore, having been discarded by boats. Which polymer would you use to cook your food? Why?
   - You should use Low Density Polyethylene (LDPE) because it burns with the hottest flame.

4. You wish to make a polymer handle for a cooking pan out of recycled plastic. Which polymer should you not use? Explain.
   - You should not use polyethylene terephthalate because it softens with heat.

5. You decide to redecorate your bathroom and transfer your fingernail polish remover into a new plastic container. The next day you find a messy blob instead of your stylish container. What type of polymer was probably used to make your container?
   - The fingernail polish remover most likely contained acetone and the container was made of polystyrene which softens in acetone.

6. When you are performing tests for density, why is it important to dislodge any bubbles from the samples?
   - The bubbles adhering to the surface of the polymer will increase the effective volume of the polymer and buoy it up so that its apparent density is less than its actual density.

7. From what you observed during the isopropyl alcohol test, estimate the approximate density of the alcohol solution.
   - The HDPE sinks (density=0.95-0.97g/mL) and the LDPE floats (density-0.92-0.94 g/mL). Therefore the density of the solution must be between 0.94 g/mL and 0.95 g/mL.

Ideas for this activity came from an article in CHEM 13 NEWS/January 1994. The article was written by high school student, Christopher S. Kollman from Kirkwood High School, Kirkwood, MO.

Long, Long Chains
Identifying Polymers

Generally when people think of polymers, they think of plastics. And, indeed, plastics are polymers. The term polymer comes from the Greek words *polus* and *mer* meaning "many parts". Polymers are giant molecules made up of repeating units called monomers. There may be hundreds or even thousands of these monomers in a molecule.

Many polymers are naturally occurring. For example, starch is a polymer made up of repeating units of glucose.

Other examples of natural polymers are cellulose, chitin, proteins, and the nucleic acids of RNA and DNA.

In 1868 the first modified natural polymer was developed. It was cellulose nitrate which was derived from wood. The new material could be molded into various shapes or made into a film called celluloid. Early movie film was made of celluloid. In 1884 a fiber which was marketed as "artificial silk" was developed. It was actually a form of rayon. In the 1920's and 1930's there was a rapid growth in the development of new polymers and the term "plastic" was coined. These decades saw the development of synthetic rubbers, polyvinyl chloride, and nylon. Today our world is filled with polymers that have been developed.

Polymers are formed from monomers in two ways. In the first method, the monomer has a double bond which allows monomers to add together when it is broken. These are called **addition polymers**. Examples of addition polymers include polyethylene, polyvinyl chloride, and polystyrene. In the other method, the repeating units come together by removing a molecule of water. These are known as **condensation polymers**. Examples include starch, polyethylene terephthalate, and nylon.

Two major issues in our use of polymers involve the use of non renewable resources to produce the materials and the amount of landfill that is being used as we discard these materials. We need to remember the 3 R's—Reduce, Reuse, and Recycle.

Many of our commonly used polymers are easily recyclable. In order to separate them, they are coded with a number on the plastic item. Start looking at the bottoms of containers that are made of polymers to find the codes. The week before you do this lab, you will have been asked to bring in samples of each of the six common polymers that have been used in packaging. Use the chart below to learn the names of each of the polymers.

Major Polymers and Their Uses

Code	Name	Monomer(s) Used to form polymer	Common Uses	Examples of recycled products
1	Polyethylene Terephthalate (PET)		Soft drink bottles, mouthwash bottles, peanut butter jars, salad dressing bottles	Liquid soap bottles, strapping, fiberfill for winter coats, surfboards, paint brushes, fuzz on tennis balls, soft drink bottles, skis, carpets, boats
2	High Density Polyethylene (HDPE)		Milk, water, and juice containers, grocery bags, toys, liquid detergent bottles	Soft drink bottles, flower pots, drain pipes, signs, stadium seats, trash cans, recycling bins, traffic barrier cones, golf bag liners, detergent bottles
3	Polyvinyl Chloride (PVC)		Clear food packaging, shampoo bottles	Floor mats, pipes, hoses, mud flaps
4	Low Density Polyethylene (LDPE)		Bread bags, frozen food bags, grocery bags	Garbage can liners, multi-purpose bags
5	Polypropylene (PP)		Ketchup bottles, yogurt containers, margarine tubs, medicine bottles	Manhole steps, paint buckets, video cassette storage cases, ice scrapers, fast food trays, lawn mower wheels, automobile battery parts

| 6 | Polystyrene (PS) | | Video cassette cases, compact disc jackets, coffee cups, cutlery, cafeteria trays, egg cartons, grocery store meat trays, fast food containers | License plate holders, septic tank drainage systems, desk top accessories, hanging files, food service trays, flower pots, trash cans |

Density Chart

Substance	Density (g/mL)
Water	1.00
1. PET	1.38-1.30
2. HDPE	0.95-0.97
3. PVC	1.16-1.35
4. LDPE	0.92-0.94
5. PP	0.90-0.91
6. PS	1.05-1.07

PURPOSE

Using samples of the six common recyclable polymers, you will develop a flow chart to identify the polymers. You will then use the flow chart to identify unknown samples of the polymers.

MATERIALS

samples of recyclable polymers 1-6
3 beakers (250-mL)
forceps or crucible tongs
hot plate
50 mL acetone in a small beaker covered with
   a watch glass
glass stirring rods
Sharpie pen
paper towels

Bunsen burner
1 beaker (100-mL)
hole punch
copper wires (10 cm long, 20 gauge)
mixture of isopropyl alcohol and water
   prepared by your teacher.
6 samples of known polymers
3 unknown samples of polymers

PROCEDURE

1. Before the polymer activity, search for and bring samples of the six common recyclable polymers to class.

2. Read the entire procedure before beginning.

3. The following statements will describe some the physical and chemical properties of the six polymers you will analyze. Read them carefully and then develop a flow chart of tests to identify each polymer.
   a. Some of the polymers will float in water, and some of them will sink. Divide the polymers into two groups.
   b. Within the group that floats, one of them will sink in the mixture of isopropyl alcohol and water. The other two will float. Which of the polymers will sink?
   c. Of the remaining two that float, one will burn with a blue flame and the other will burn with a yellow flame. Read the individual procedures for each test to determine which one burns with a blue flame.
   d. Within the group of polymers that sinks, one of them contains a halogen. The halogen will react with copper to give a green flame.
   e. Of the remaining two polymers, one will soften in acetone. The other will soften in hot water.

4. After you have developed your flow chart, obtain a sample of each polymer. Use a permanent marker to label each polymer with its recycle code number.

5. Use the following individual procedures to confirm your flow chart. Record your observations on the student answer page.
   a. **H₂O Test:** Place all of the samples in the beaker of distilled water. Use a stirring rod to knock off any excess bubbles. Note which samples sink and which samples float. Use the density chart to predict which samples are in each group. Confirm your predictions. Remove the samples from the water with forceps and dry with paper towels. Divide your samples into two groups based on the outcome of the H₂O test.
   b. **Copper Wire Flame Test:** You will want to perform tests b, c, and d to the sample(s) that sank in the water test. Using the forceps, hold the copper wire in the Bunsen burner flame until it is red hot. Remove from the flame and carefully push the hot wire through the sample. Place the wire back in the flame and observe the color of the flame that comes from the wire. The halogens will react with copper metal to create copper(II) ions which will produce a green flame.

c. **Acetone Test:** Place the samples in the beaker of acetone. *Remember to keep the acetone away from flames.* Allow the sample to sit in the acetone for 10 seconds. Remove the sample with forceps and press firmly between your fingers. The polymer chains made from styrene will "loosen up" in acetone. This is known as swelling. The surface will become soft and impressionable. Other polymers will not do this.

d. **Heat Test:** Using forceps, plunge the samples into boiling water and hold it there for 30 seconds. Polyethylene terephthalate has a relatively low softening point and should soften in the hot water. The other polymers, having a higher softening point will not be affected by the boiling water.

e. **Isopropyl Alcohol Test:** Place the three samples that floated in the initial water test in the solution of isopropyl alcohol that has been provided by your teacher. Use the stirring rod to knock off any bubbles. One of the samples should sink and the other two should float. Using the density chart, determine which sample sinks.

f. **Combustion Test:** To the remaining two samples, use a hole punch to punch a small circular sample of polymer. Using forceps hold the piece of polymer in the flame of the Bunsen burner until it begins to burn. Remove it from the flame, but hold it over the lab counter. Note the color of the flame as the sample continues to burn. When the reaction is complete, quench the sample in a beaker of water. Low Density Polyethylene burns slowly allowing oxygen to get to the sample. The flame will be a hot blue flame. The Polypropylene burns very quickly so not as much oxygen gets to the sample. Therefore the flame is cooler and is yellow.

6. Obtain three unknown samples from your teacher. The samples will be labeled with letters A-F. Record your unknown letters in the data table. Use the flow chart and the tests above to determine the identity of your unknowns. Complete the unknown data table.

Name _____

Period _____

Long, Long Chains
Identifying Polymers

FLOW CHART

Diagram the flow chart for this lab procedure:

OBSERVATIONS

1. Water test:

2. Copper flame test:

3. Acetone test:

4. Heat test:

5. Isopropyl alcohol test:

6. Combustion test:

ANALYSIS

Unknown Data Table:

Letter of Unknown	Confirming tests	Identity of Unknown

CONCLUSION QUESTIONS

1. In the introduction you are given three examples of addition polymers and three examples of condensation polymers. Give one additional example of each type of polymer.
   a. Addition polymer:

   b. Condensation polymer:

2. You are in a boat which is sinking. You are not a good swimmer. In the boat are six large solid plastic blocks labeled 1, 2, 3, 4, 5, 6. You only have two arms. Which two blocks would you grab? Explain your choice.

3. You are on a camping trip on an island on the gulf coast. You have run out of wood. Sadly, there is lots of plastic that has washed up on shore, having been discarded by boats. Which polymer would you use to cook your food? Why?

4. You wish to make a polymer handle for a cooking pan out of recycled plastic. Which polymer should you not use? Explain.

5. You decide to redecorate your bathroom and transfer your fingernail polish remover into a new plastic container. The next day you find a messy blob instead of your stylish container. What type of polymer was probably used to make your container?

6. When you are performing tests for density, why is it important to dislodge any bubbles from the samples?

7. From what you observed doing the isopropyl alcohol test, estimate the approximate density of the alcohol solution.

Biochemistry
Exploring Macromolecules

OBJECTIVE

Students will gain a better understanding of organic chemistry in a biological setting and focus on the properties and functions of the four main classes of macromolecules.

LEVEL

Chemistry

NATIONAL STANDARDS

UCP.1, UCP.5, B.2, C.2

TEKS

8(A), 8(B), 8(D), 11(A)

CONNECTIONS TO AP

AP Chemistry:

I. Structure of Matter B. Chemical bonding 1. Binding forces a. Types: ionic, covalent, metallic, hydrogen bonding, van derWaals(including London dispersion forces) b. Relationships to states, structure, and properties of matter 3. Geometry of molecules, ions, structural isomerism of simple organic molecules and coordination complexes; dipole moments of molecules; relation of properties to structure

IV. Descriptive Chemistry 3. Introduction to organic chemistry: hydrocarbons and functional groups (structure, nomenclature, chemical properties)

AP Biology:

I. Molecules and Cells A. Chemistry of Life 1. Water 2. Organic molecules in organisms
4. Enzymes
II. Heredity and Evolution B. Molecular Genetics 1. RNA and DNA structure and function

TIME FRAME

45 minutes

MATERIALS

models and/or examples of the 4 classes

TEACHER NOTES

Biochemistry is an important concept to teach the students before they enter the AP Biology course. This lesson connects bonding and molecular shapes in chemistry with structure and function in biology. The lesson begins with an introduction to organic chemistry, the chemistry of carbon, which is essential to the chemistry of life. This lesson should follow a bonding lesson and will be a nice introduction to Protein Properties. This lesson will prove challenging, but is a worthwhile endeavor.

POSSIBLE ANSWERS TO THE CONCLUSION QUESTIONS AND SAMPLE DATA

1. Organic chemistry is the study of carbon compounds. What unique bonding characteristics does the carbon atom have that allows for so many organic compounds to exist?
   - Carbon has four valence electrons so it has four available bonding sites.
   - Carbon can form single, double and triple covalent bonds with itself and other atoms.

2. List four elements that compose most of the molecules in living systems.
   - Carbon, hydrogen, oxygen and nitrogen

3. List five other elements that are essential to life on this planet. (Hint: use the periodic table given in the notes—there are many more than five!)
   - Student lists will vary—check the periodic table to make sure that they listed appropriate elements
   - One possible list might be: Na, K, S, P, Fe

4. Define functional group. Name and draw three functional groups that are common in biomolecules.
   - A functional group is a side chain or group of atoms that attach to a carbon backbone. A functional group both identifies the molecule and provides a site for chemical reactions.
   - Hydroxyl –OH
   - Carboxyl –COOH
   - Amine –NH_2

5. What is a polymer?
   - A polymer is series of linked repeating units known as monomers.

6. How are polymers formed? How are they broken apart?
   - Polymers are formed when two monomers join together by means of a covalent bond in a condensation or dehydration synthesis reaction. When the two monomers bond a water molecule is released as a product.
   - Adding a water molecule across the bond between two units in a polymer is a hydrolysis reaction. This reaction allows the polymer to be broken apart.

7. Name the monomer unit for each of the following biomolecules:
   - Proteins
   - Amino acids
   - Carbohydrates
   - Monosaccharides
   - Nucleic acids
   - Nucleotides

8. Describe how a peptide bond is formed. What is meant by the term dipeptide?
   - When two amino acids covalently bond to each other by means of a dehydration synthesis reaction, the bond formed is a peptide bond. It is formed between the amine group of one amino acid and the carboxylic acid group of the second amino acid.
 - A dipeptide consists of two amino acids joined by a peptide bond.

9. Discuss and describe the four levels of structural organization within a protein.
   • The primary structure of a protein is the sequence of amino acids that are joined together. The slightest change in this order may change the function of the protein.
   • Secondary structure may either be alpha helix or beta pleated sheet. The R groups that compose the protein will determine the type of folding and intermolecular forces that will occur within the protein. Emphasize that all of this folding and coiling is due to intermolecular forces whether attractive or repulsive.
   • Other complex structures may exist. They are not listed in this lesson but include tertiary structure, quaternary structure, and the globular structure [joining of 2 or more polypeptide chains] of proteins.

10. Explain the term disaccharide. Give one common example and tell how it is formed.
   • A disaccharide consists of two monosaccharides joined together. They join by means of a covalent bond known as a glycosidic bond which occurs due to a dehydration synthesis reaction.
   • One common example of a disaccharide is sucrose. It is formed when glucose and fructose join together by a dehydration synthesis reaction.

11. Glucose monomers are responsible for many carbohydrate polymers. List three and describe each of their functions.
   • Glycogen (the way animals store energy in the liver and muscles—branched chains)
   • Starch (used by plants to form glucose)
   • Cellulose (probably the most common organic compound on earth—straight chains linked together to form all types of fibers like cotton fibers)

12. Distinguish between the terms hydrophobic and hydrophilic? Structurally, how do molecules of these two groups differ?
   • Hydrophobic means water fearing. Hydrophilic means water loving.
   • Molecules that are hydrophobic tend to be largely nonpolar, unlike the water molecule.
   • Hydrophilic molecules are largely polar and are thus very attracted to the water molecule.

13. Describe what is meant when a lipid is described as saturated or unsaturated? Give one common example for each category.
   • Saturated compounds have all single bonds with the maximum number of hydrogens. One example is animal fat or lard.
   • Unsaturated compounds have some type of multiple bonding and less than the maximum number of hydrogens. One example is vegetable oil.

14. Explain why biological membranes are composed largely of lipids?
   • Biological membranes are composed largely of phospholipids—this gives them the unique characteristic of one polar end which will attract to water and one nonpolar end which will repel the water.

15. Describe three differences between DNA and RNA.
   - DNA is double stranded; RNA is single stranded
   - DNA has the sugar deoxyribose; RNA has the sugar ribose
   - DNA has four bases—ATGC; RNA has four bases—AUGC
   - Some students may cite the purpose—DNA carries the genetic code; RNA directs the building of proteins

16. Examine the DNA fingerprint analysis. Is Jeff Jones the father or not? Justify your answer.
   - M is the mother, Jennifer Jones.
   - C is the child, George Jones.
   - AF is the alleged father, Jeff Jones.
   - Remember, to conclude that the alleged father is truly the father, every band in the child's fingerprint that does not match in the mother's fingerprint must match in the father's.
   - Jeff Jones is the father. The child has matching second and third bands to the father while bands one and four match the mother.

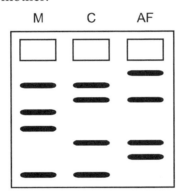

Biochemistry
Exploring Macromolecules

Biochemistry is the study of chemical reactions in living systems. The very name implies a connection between biology and chemistry. There are many types of macromolecules (very big molecules) that play a vital role in living systems.

PURPOSE
In this activity you will examine the fundamentals of organic chemistry and relate it to the four main classes of macromolecules.

CLASS NOTES
* While you are reading the notes pay close attention to properties, uses and examples of molecules in each class. You will be asked to answer questions related to reading.

Organic Chemistry:
Organic chemistry is the study of the chemistry of carbon. When studying bonding you may remember that carbon has four valence electrons available for bonding. Carbon is unique in that it may form single, double and triple covalent bonds with other carbon atoms. There are more organic compounds than inorganic (without carbon) compounds. Carbon chemistry forms the basis of the chemistry of living things. The molecules that make up living systems are predominantly composed of the elements carbon, oxygen, nitrogen and hydrogen. The periodic table shown in Figure 1 highlights the elements essential to life on this planet.

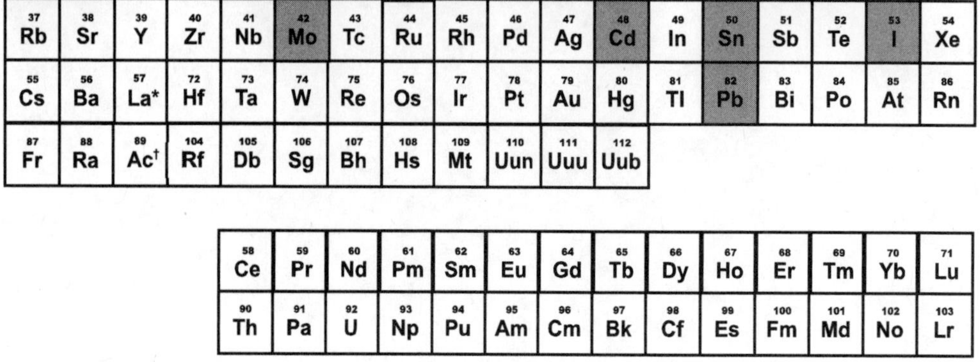

Fig. 1

As carbon atoms join together and form chains, there are often side chains that form along the carbon backbone. We refer to these, clusters of atoms as *functional groups*. The functional group is the actual building block of the compound which determines the characteristics of that compound. For example, one common functional group is the hydroxyl group, -OH. When this group is found on a carbon chain it makes the molecule become slightly polar like the water molecule. Long carbon chains without functional groups have little or no attractive forces and are therefore nonpolar molecules. For example, when crude oil spills into the ocean it does not mix with the water since the oil is nonpolar and the water is polar. An old saying to remember this is, "like dissolves like". The functional groups added to the carbon compounds in living systems are extremely vital since most living organisms are composed largely of water, a polar compound. Though there are many functional groups that exist, there are three that are essential to understanding the structure of the biomolecules that we will study in this lesson. They are:

- hydroxyl group: -OH. This group attaches by taking an H off of the carbon chain and replacing it with an –OH group. This creates something known as an alcohol. Figure 2 shows the structure for ethanol.

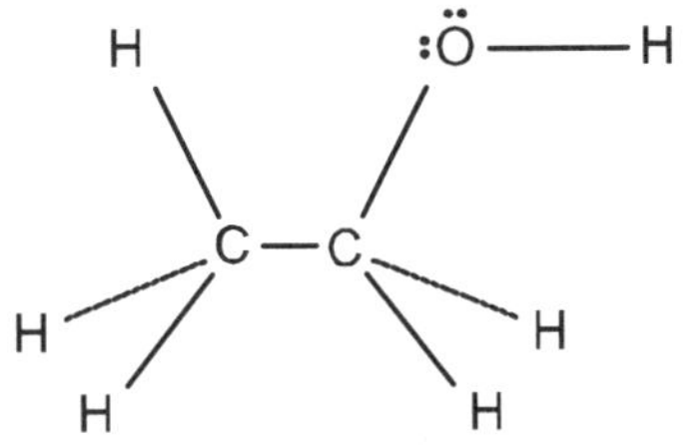

Fig. 2

- carboxylic acid group:-COOH. The C in this group is doubly bonded to one oxygen and singly bonded to the OH group leaving one bonding site open to attach to the main carbon chain. The compound shown in Figure 3 is ethanoic acid or more commonly referred to as acetic acid or vinegar.

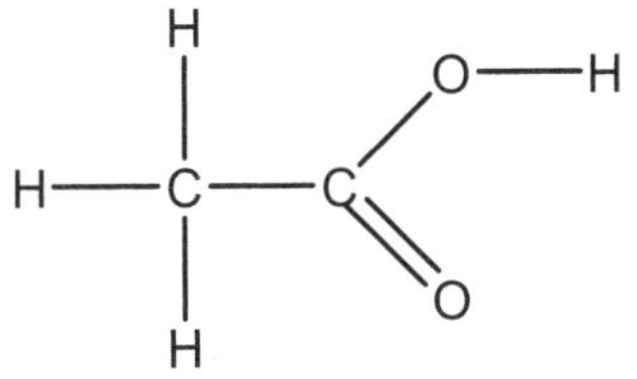

Fig. 3

- amine group: -NH$_2$. This group attaches to the main carbon chain in much the same way as the –OH group. A hydrogen is removed from the main carbon chain allowing the nitrogen to attach to the carbon chain. The R in Figure 4 represents "radical" which is the "rest" of the carbon chain.

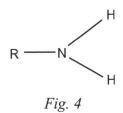

Fig. 4

- Adding any atom or group of atoms to a carbon chain, like the ones listed above, will change the molecules shape, which changes its function, physical properties and intermolecular attractive forces.

Large carbon molecules are known as macromolecules. They are actually large polymers. A polymer is composed of repeating units of smaller carbon units known as monomers. (Hence the prefixes, mono = one; di = two; tri = three; poly = many) The monomers are joined together in a chemical reaction known as a condensation reaction or a dehydration synthesis because water molecules drop out of the chain when the C—C bond forms linking monomers to form polymers. Breaking a polymer down into monomers occurs by an opposite reaction known as hydrolysis, which means "splitting with water" or adding water to break bonds. Remember that forming bonds releases energy, while breaking bonds requires energy.

There are four main classes of macromolecules or polymers that are essential for life:
- Proteins
- Carbohydrates
- Lipids
- Nucleic acids

Proteins:
Proteins are macromolecules that are composed of many amino acids and can provide energy to living systems. All amino acids have a carboxyl group, -COOH, and an amine group, -NH$_2$, attached to a central carbon atom. There are 23 amino acids, 20 of which are common to all living organisms. Look at the structures for the amino acids in Figure 5. Notice that the difference between them is the R group. The diagram shows how amino acid molecules are joined together. The amino acids are the monomers and when joined by a condensation reaction a water molecule drops out and a bond is formed between the amine group of one amino acid and the carboxyl group of the other amino acid. The bond linking the amino acid molecules together is known as a peptide bond. Proteins are nothing more than chains of amino acids. The 20 amino acids can be joined and rearranged in many ways, much like the 26 letters of the alphabet. Most proteins are series of hundreds of amino acids. The sequence of amino acids determines the protein's structure and function. Even the slightest change in the amino acid sequence will change the protein. For example, the difference between normal hemoglobin and hemoglobin in individuals with sickle-cell anemia is only two amino acids!

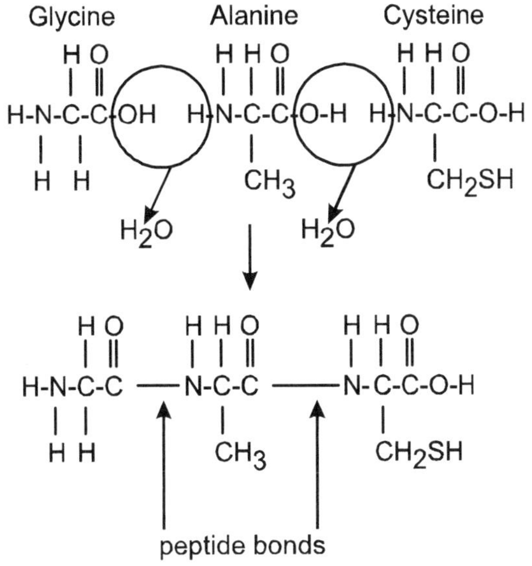

Fig. 5

Some examples of common materials composed of proteins include:

- Hair
- Tendons
- Ligaments
- Silk
- Hormones (transport substances throughout the body and fight infection—insulin is a hormone produced in the pancreas)
- Enzymes (control the body's chemical reactions—pepsin is a common digestive enzyme)

There are many levels of structure to most proteins. The primary structure is the basic amino acid sequence—the order in which they are joined. However, the R groups that are attached to the amino acid often form intermolecular attractive forces with other proteins or with themselves which leads to folding, bending or joining of many proteins together to give some very complex structures. The two most common types of intermolecular attractions that exist within and between proteins are disulfide bridges between two adjacent sulfur atoms and hydrogen bonding between the H on one molecule and the F, O, or N on an adjacent molecule. Remember even though they are called hydrogen bonds, they are merely an attraction, not a true bond! These intermolecular attractions often cause twisting and folding of the protein. Two common structures are shown in Figure 6 and Figure 7; the α-helix (like a spring) (Figure 6) and the β-pleated sheet (like a paper fan) (Figure 7). The shape of the protein is easily altered by change in temperature or pH. We call this alteration of shape, denaturing the protein. When an egg is boiled, the clear liquid of the egg turns white—cooking is a denaturing of the protein.

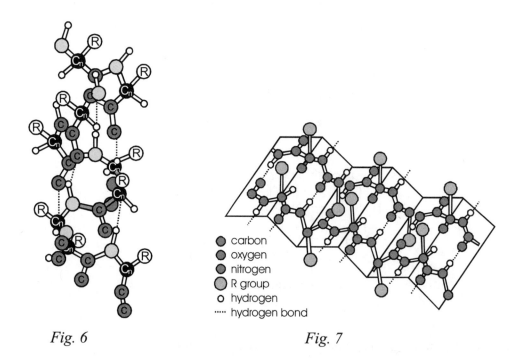

Fig. 6   *Fig. 7*

- ⬤ carbon
- ⬤ oxygen
- ⬤ nitrogen
- ⬤ R group
- ○ hydrogen
- ⋯⋯ hydrogen bond

Carbohydrates:

Carbohydrates have the general formula, $C_n(H_2O)_n$, thus the name is appropriate—hydrated (water) carbons. Carbohydrates like proteins also provide energy to living systems. Carbohydrates exist as monosaccharides, disaccharides, and polysaccharides. The monomer unit of a carbohydrate is a monosaccharide or simple sugar. Glucose and fructose are examples of monosaccharides that you may already know. Most naturally occurring carbohydrates contain more than one monosaccharide unit. Monosaccharides link together in much the same way that amino acid molecules link in the protein—by condensation or dehydration synthesis reactions. The bond formed between two monosaccharides is known as a glycosidic bond. Sucrose, common table sugar, is a disaccharide formed by linking the two monosaccharides glucose and fructose. See Figure 8.

Fig. 8

Three important polysaccharides that are made from glucose monomers are:

- Glycogen (the way animals store energy in the liver and muscles—branched chains)
- Starch (used by plants to form glucose)
- Cellulose (probably the most common organic compound on earth—straight chains linked together to form all types of fibers like cotton fibers)

Lipids:

Lipids are large organic molecules that are nonpolar, and thus not very soluble in water. Lipids, like proteins and carbohydrates serve as energy for living systems. Lipids are very efficient energy storage molecules, storing about two times the amount of energy as proteins and carbohydrates. The simplest lipids are known as fatty acids. A fatty acid is composed of a long, straight carbon chain with a –COOH, carboxyl group, at one end. The two ends of the fatty acid have very different properties! The long, straight carbon chain is nonpolar and will repel water, *hydrophobic* (water fearing), while the carboxyl group on the opposite end is polar and is very attracted to water, *hydrophilic* (water loving). Refer to Figure 9, the diagram of the fatty acid and the glycerol.

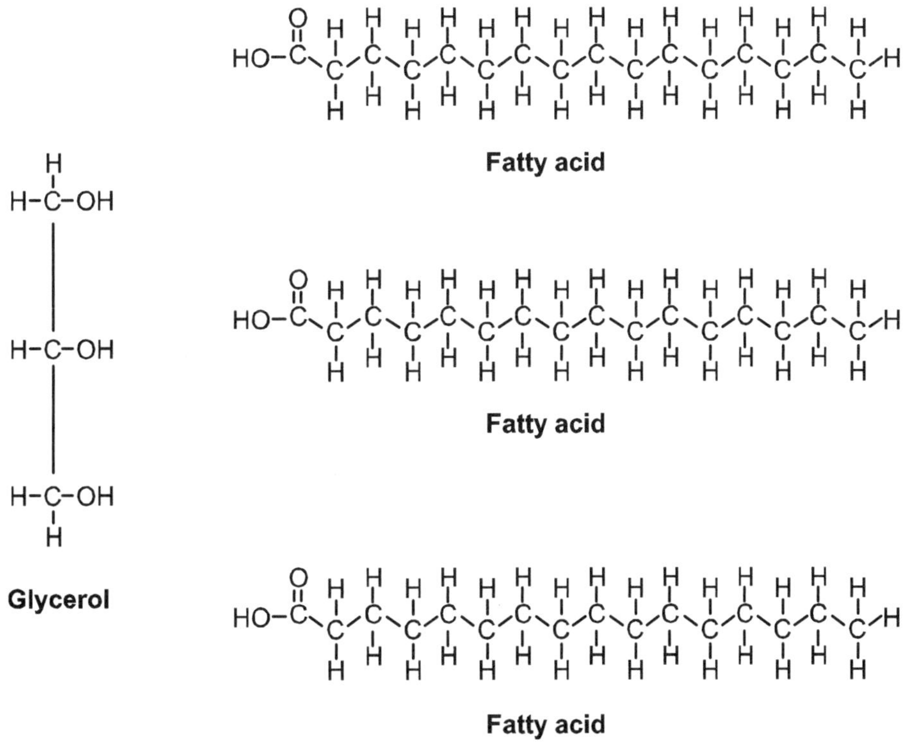

Fig. 9

This unique structure of one polar end and one nonpolar end allow the fatty acid to form membranes when dropped into water—one end attracts to the water, while the other end repels the water! This property is what gives soaps and detergents their cleaning power. (See Figure 10 and Figure 11)

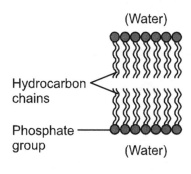

(Water)

Hydrocarbon chains

Phosphate group

(Water)

Fig. 10

Fig. 11

Fatty acids may be saturated or unsaturated fats. Saturated fats contain all single bonds between carbons—they are said to be "saturated" with hydrogen. Unsaturated fats contain some multiple bonding. The multiple bond replaces some of the hydrogen atoms.

Lipids may be divided into three basic categories according to their structure:

- Triglycerides (As the name implies—3 fatty acids joined to a glycerol. Saturated triglycerides are usually solids at room temperature while unsaturated triglycerides are usually liquids at room temperature. Shortening, animal fats and oils belong in this category.)
- Phospholipids (Two molecules of fatty acid joined to a glycerol and a phosphate, PO_4^{3-}, group. Refer to Figure 11. Cell membranes are composed of two layers of phospholipids, making a lipid bilayer.)

Polar head

Nonpolar tails

Fig. 11

- Waxes (A long fatty acid chain joined to a long alcohol chain. These compounds are highly waterproof and form protective coatings on plants and animals. For example, earwax helps prevent bacteria from entering into the middle ear.)

Steroids are a special class of lipids which are composed of four fused carbon rings with various functional groups attached to them. One of the most familiar steroids is cholesterol which is necessary for nerve cells and other body cells to function normally. The male hormone, testosterone, and the female hormone, estrogen, also belong to this category. Some typical steroids are shown in Figure 12.

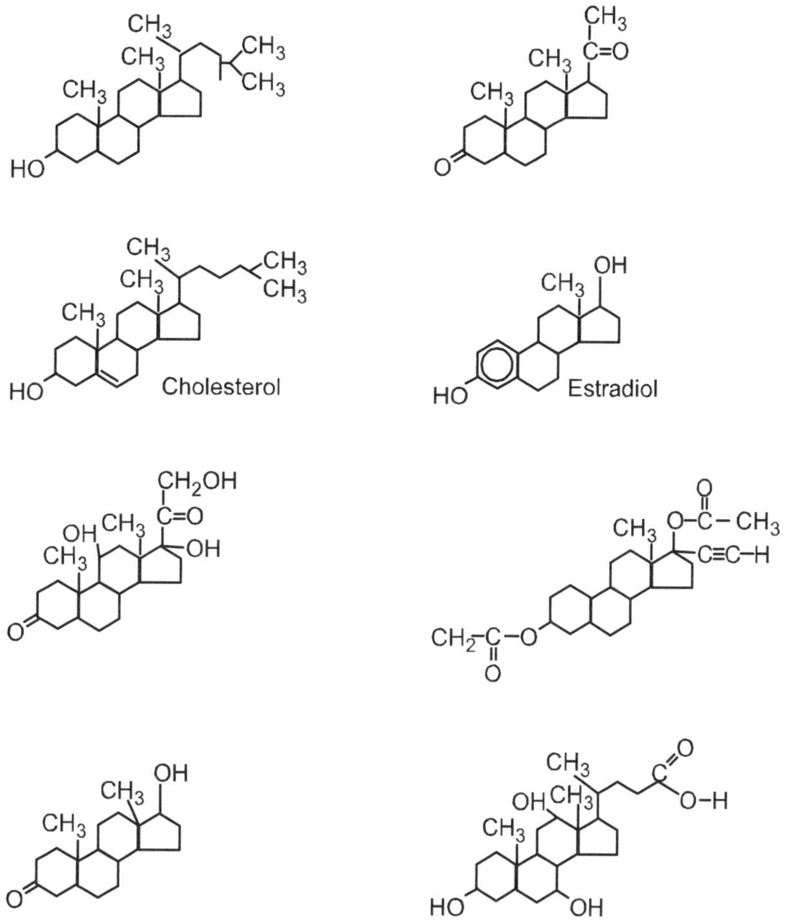

Fig. 12

Nucleic Acids:

Nucleic acids are the group of macromolecules that transmit genetic information. The two nucleic acids are DNA and RNA. DNA, deoxyribonucleic acid, contains the information necessary to pass on the genetic code from generation to generation. The monomer in DNA is a nucleotide. A nucleotide consists of a sugar (deoxyribose), a phosphate group, and a base. (Refer to Figure 13.)

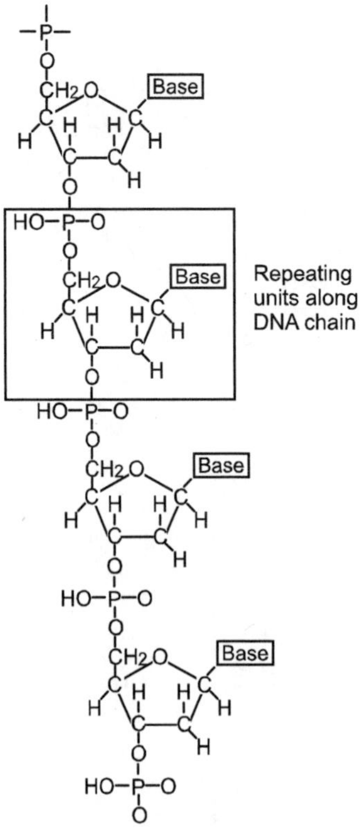

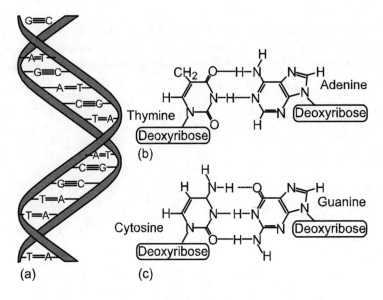

Fig. 13

There are only four bases in DNA that are sequenced in different orders to code for everything! Human DNA consists of over five billion base pairs. The four bases are adenine, thymine, guanine and cytosine. DNA is a double stranded helical-shaped molecule. The strands are held together by intermolecular attractions known as hydrogen bonds. DNA is found in the nucleus of every cell. (Refer to Figure 14)

Fig. 14

RNA, ribonucleic acid, is single stranded and is also composed of monomers which are known as nucleotides. DNA in the nucleus provides the code for creating RNA. The 2 main differences between the DNA and the RNA nucleotides are that RNA has a different sugar—DNA has deoxyribose while RNA has ribose and RNA's four bases are adenine, **uracil** (instead of thymine), guanine, and cytosine. RNA is responsible for carrying the genetic code from the nucleus to the cytoplasm of the cell where it codes for protein synthesis.

The bases in DNA and RNA are shown in Figure 15:

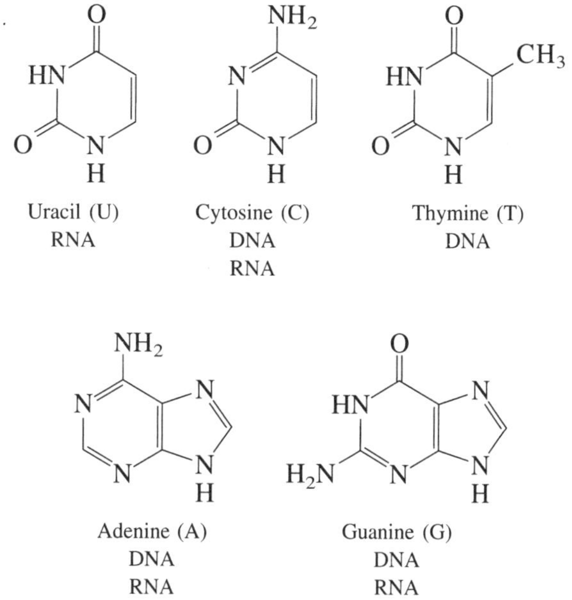

Fig. 15

The sugars in DNA and RNA are shown in Figure 16:

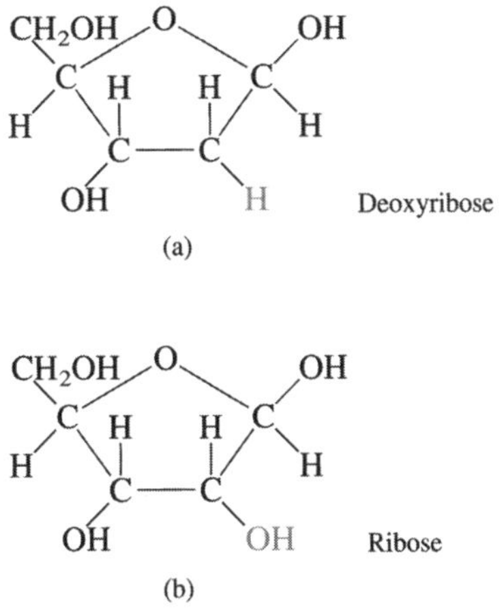

Fig. 16

Much has been learned about DNA in the last century. One really useful tool that is in widespread use is DNA fingerprinting. With only a small sample of blood, hair, skin tissue, etc. crimes have been solved and fathers have been identified. How does this work?

- Enzymes cut the DNA into fragments. Enzymes recognize specific base sequences so they know where to cut.
- The fragments will be of different lengths for different people.
- The fragments are loaded into a gel (yes, like Jello) and subjected to an electric field.
- DNA is negative so it is attracted to the positive terminal of the field.
- Heavy fragments move slowly through the matrix of the gel and do not travel as far as smaller fragments.
- If a person's fragments are loaded in one part of the gel and a sample from the crime scene or the potential father is loaded onto the SAME gel, comparisons can easily be made. (see Figure 17) If the fragments match the samples most likely came from the same person. There is only about a 1 in 5 billion chance that the DNA came from two different sources. Pretty impossible considering there are only about 6 billion people on the planet!

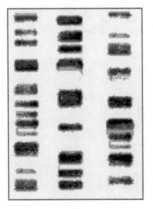

Fig. 17

Name _____

Period _____

Biochemistry
Exploring Macromolecules

CONCLUSION QUESTIONS: ANSWER COMPLETELY

1. Organic chemistry is the study of carbon compounds. What unique bonding characteristics does the carbon atom have that allows for so many organic compounds to exist?

2. List four elements that compose most of the molecules in living systems.

3. List five other elements that are essential to life on this planet. (Hint: use the periodic table given in the notes—there are many more than five!)

4. Define functional group. Name and draw three functional groups that are common in biomolecules.

5. What is a polymer?

6. How are polymers formed? How are they broken apart?

7. Name the monomer unit for each of the following biomolecules:
   a.   Proteins

   b.   Carbohydrates

   c.   Nucleic acids

8. Describe how a peptide bond is formed. What is meant by the term dipeptide?

9. Discuss and describe the four levels of structural organization within a protein.

10. Explain the term disaccharide. Give one common example and tell how it is formed.

11. Glucose monomers are responsible for many carbohydrate polymers. List three and describe each of their functions.

12. Distinguish between the terms hydrophobic and hydrophilic? Structurally, how do molecules of these two groups differ?

13. Describe what is meant when a lipid is described as saturated or unsaturated? Give one common example for each category.

14. Explain why biological membranes are composed largely of lipids?

15. Describe three differences between DNA and RNA.

16. Examine the DNA fingerprint analysis. Is Jeff Jones the father or not? Justify your answer.

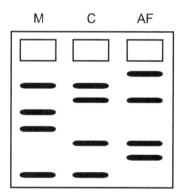

Protein Properties
Using Electrophoresis to Determine Net Charge

OBJECTIVE

Students will observe proteins migrating through an agarose gel and determine the net charge of the protein.

LEVEL

Chemistry

NATIONAL STANDARDS

UCP.2, UCP.3, UCP.5, B.2, C.2, C.5, G.1, G.2, G.3

TEKS

1(A), 2(A), 2(B), 2(D), 2(E), 3(C), 3(D), 8(B), 8(D)

CONNECTIONS TO AP

AP Biology:

    I. Molecules and cells A. Chemistry of Life, Organic molecules in organisms II. Heredity and Evolution B. Molecular Genetic, Nucleic acid technology and applications

AP Chemistry:

    I. Structure of Matter B. Chemical Bonding 1. Binding forces b. Relationships to states structure, and properties of matter c. Polarity of bonds 3. Geometry of molecules and ions, structural isomerism of simple organic molecules; relation of properties to structure

TIME FRAME

45 minutes

MATERIALS
(For a class of 28 working in pairs)

electrophoresis gel boxes (enough for 14 stations)

1.2 % melted agarose solution (enough for 14 casting trays)

several rolls of masking tape

sufficient buffer for electrophoresis boxes

14 microcentrifuge tubes with myoglobin

14 microcentrifuge tubes with serum albumin

DC power supply (enough for 14 stations)

14 micropipets

14 disposable beral pipets

14 microcentrifuge tubes with cytochrome c

14 microcentrifuge tubes with hemoglobin

14 microcentrifuge tubes with unknown protein

14 rulers

Optional
Coomassie blue stain

14 staining trays

destain

TEACHER NOTES
This laboratory activity is appropriate when studying protein structure, protein synthesis, or biotechnology. Students need to understand the underlying principles of electrophoresis and the effect of the net charge of proteins on their migration through an agarose gel.

If a paper chromatography activity has been done prior to this lab you can use that as an analogy for the mechanism of electrophoresis.

Preparations
Before beginning solution preparations, read the instruction manual that accompanies your electrophoresis equipment to determine the capacity of your chamber and gel plate. Use these values to figure out the total amount of buffer solution and agarose that will be needed for your class(es).

If money is not a consideration, there are molecular kits that can be purchased for a protein separation. The kit will include the proteins and buffer solution. They may or may not include the agarose. The following includes directions for preparing this lab without a kit. Once the lab is prepared, the protein solutions will last for some time.

Electrophoresis buffer should be made first since it is needed make the agarose gels. It will also be the buffer solution used to fill the electrophoresis chamber.

To prepare 1.0 L of buffer:
Mix 2.8 g of glycine and 0.4 g Tris base in 500 mL of distilled water. Bring the volume up to one liter of solution. Make enough for number of boxes that will be used during the lab. Electrophoresis buffer can be used for as many as three runs.

To prepare 1.2% agarose:
Use graduated cylinder and water to determine the volume of melted agarose needed per casting tray. Be sure to tape the ends securely and fill the tray until it reaches half-way up the comb. Multiply the volume required for one tray by the number of trays needed. Add 10% to this value to account for errors

and sealing the wells. This volume of buffer solution will be mixed to the powdered agarose to make the agarose solution. Pour the determined amount of buffer solution into a beaker capable of holding at least twice the required volume to avoid boiling over during heating. For example use a 500 mL beaker to hold 200 mL of buffer.

To determine the mass of powdered agarose required for the 1.2% solution, multiply the volume of electrophoresis buffer to be used by .012. Weigh out the appropriate number of grams of agarose and add it to the beaker containing the buffer. To melt the agarose put in a microwave and heat on high until all of the agarose has melted. This may take 2-4 minutes depending on the amount of agarose being melted and the power of the microwave. Check and stir every 30 seconds until the agarose has completely melted.

Sample Preparation Calcluation

- One tray holds 30 mL of agarose
- $\left(14 \text{ trays} \times \dfrac{30 \text{ mL}}{\text{tray}}\right) + 10\% = (420 \text{ mL}) + 42 \text{ mL} = 462 \text{ mL}$ *of melted agarose needed*
- *Round to 475 mL of melted agarose needed*
- $475 \text{ mL of buffer} \times \dfrac{0.012 \text{ g agarose}}{\text{mL of buffer}} = 5.7 \text{g of powderd agarose needed}$
- Add 5.7 g of powdered agarose to 475 mL of electrophoresis buffer.
- Put solution in a 1000 mL beaker to avoid boiling over in the microwave.
- Put in a microwave and heat on high until all the agarose has dissolved. Check and stir every *30* seconds until the agarose is completely dissolved.

Protein preparation

The proteins can be purchased from Sigma-Aldrich

>Sigma-Aldrich
>3050 Spruce Street
>St. Louis, Mo 63103
>(800)-325-5052
>www.sigma-aldrich.com

These proteins can be stored in the freezer for an indefinite time. The amount needed to make the protein solutions is minimal and the purchased protein can last for years.

The proteins used are

Protein	Color	Net charge in buffer 8.6
cytochrome c	orange	positive
myoglobin	rusty red	negative
hemoglobin	rusty red	negative
serum albumin	* blue	very negative
* Serum albumin is blue because bromophenol blue has been added to the sample and it binds to serum albumin.		

Protein solutions once made, can be stored in the freezer and are good for several years as long as they are kept frozen. It is recommended that you make 10 mL of each protein solution. This is enough to dispense 0.5 mL into 14 1 mL microcentrifuge tubes or 1.5 mL microcentrifuge tubes. This should supply 14 groups working in pairs. It will also allow you to make some unknown protein tubes.

To prepare the protein solutions:
1. Make 50 mL of a 50% glycerol solution by adding 25 mL of glycerol to 25 mL of water mix.

2. Protein solution #1: Add 10 mg of cytochrome c to 10 mL of 50% glycerol solution. Dispense 0.5 mL into each of 14 microcentrifuge tubes and label the tubes cytochrome c. Save the rest to make solutions of unknown proteins.

3. Protein solution #2: Add 10 mg of myoglobin to 10 mL of 50% glycerol solution. Dispense 0.5 mL into each of 14 microcentrifuge tubes and label the tubes myoglobin. Save the rest to make solutions of unknown proteins.

4. Protein solution #3: Add 10 mg of hemoglobin to 10 mL of 50% glycerol solution. Dispense 0.5 mL into each of 14 microcentrifuge tubes and label the tubes hemoglobin. Save the rest to make solutions of unknown proteins.

5. Before making this protein solution, take 10 mL of 50% glycerol and add .01 g of bromophenol blue. This will dye the serum albumin. Protein solution #4: Add 10 mg of serum albumin to 10 mL of 50% glycerol/bromophenol blue solution. Dispense 0.5 mL into each of 14 microcentrifuge tubes and label the tubes serum albumin. Save the rest to make solutions of unknown proteins.

6. To make unknown protein solutions, use the left over proteins. You may want to combine proteins so that there are two proteins that will separate out. Be sure you make a key so that you can check your student's answers.

The selected proteins are colored and can be observed without staining, however, you may want to stain them so that students can observe this process. Proteins are stained with Coomassie brilliant blue R stain. The staining and destaining process takes several hours so students will have to view the gels the following day. After the gels are stained, they are then destained to make the protein bands more visible.

To prepare the stain solution:

Combine the following
       440 mL methanol
       480 mL distilled water
       80 mL glacial acetic acid
       2.5 g Coomassie brilliant blue R

To prepare the destaining solution:

Combine the following:
       100 mL methanol
       100 mL glacial acetic acid
       800 mL distilled water

Let the gel stain for approximately 10 minutes, pour off stain and apply destaining solution. The destaining solution should completely cover the gels. You may have to destain several times and the destain may have to be changed after several hours. Destain until the proteins are visible. View the gels on a light box or overhead projector. Use plastic wrap to avoid staining the light source.

Be sure to instruct the students on the proper micropipeting technique. This protocol recommends filling the gels outside of the box and then sealing the wells with some melted agarose. This will save a a significant amount of time. If time is not a factor, you may want to demonstrate to the students how to load under buffer.

POSSIBLE ANSWERS TO THE CONCLUSION QUESTIONS AND SAMPLE DATA

Data Table			
Protein	Color of Protein	Distance Migrated Through Gel	Charge of the Protein
cytochrome c	orange	2.8	negative
myoglobin	rusty-orange	0.8	negative
hemoglobin	rusty-orange	0.6	negative
serum albumin	blue	-2.4	positive
unknown protein (s)			

7. What are the building blocks or monomers that make up the proteins? Explain how these molecules are structured and how they are different from one another.
   - Proteins are made from monomers called amino acids.
   - An amino acid has a central carbon and attached to the central carbon are: hydrogen, an amino group, a carboxyl group and a variable R group. There are twenty amino acids and each one has a different variable.

8. Explain what is meant by a protein's primary, secondary, tertiary and quaternary structure.
   - A protein's primary structure is the sequencing of its amino acids. Its secondary structure is hydrogen bonding between one amino acid's amino group and a neighbor amino acid's carboxyl group. This can take on several structures like the alpha helix and the beta pleated sheets. The tertiary structure is how the protein folds on itself to form three-dimensional structure. The quaternary structure involves a protein that has more than one polypeptide chain and how they fit with one another.

9. Explain why the four proteins used in this experiment moved at different rates through the gel?
   - The proteins moved at different rates because they have different charges. The net charge of the protein is due to the R groups. Different proteins have different number of amino acids with different R groups.

10. What can be determined by comparing the unknown proteins with the known proteins in the gel?
    - The charges of the unknown proteins can be determined and compared to the proteins with their known charges.

11. Explain why in most protein separations, the final step is the staining of the gels.
    - Most proteins do not have color and by staining the proteins they can be visualized.

REFERENCES:

Anderson, John, *A Laboratory Course in Molecular Biology, Overview of Proteins*, Instructors Manual. Dayton. Modern Biology. pg 1-5

Campbell, Neil, Lawrence Mitchell, and Jane Reece. *Biology*. Menlo Park. Benjamin/Cummings, 1999. pg 68-76.

Greenberg, Jon Revision Editor. *Biological Science, A Molecular Approach*. Chicago: Everyday Learning, 20001. pg 38-40

Gattorzzi, Linda, Earl Hagstrom, Marie Rediess, Mark Salminen, Dr. Clarence Suelter. "Constructing and Electrophoretic Gel Box for Running Miniature Agarose Submarine Gels", *Natural Science*, Fall 1988:13-14

TEACHER PAGES

Protein Properties
Using Electrophoresis to Determine Net Charge

Proteins are the molecules that carry out the "business of living". Humans can synthesize over 50,000 different proteins. Genes composed of DNA direct the synthesis of protein. The list below gives one an idea of the importance of proteins and how diverse their functions are.

Type of protein	Function / Examples
Structural Proteins	(hair, horns, nails, etc.)
Storage Proteins	(albumin, casein)
Transport Proteins	(permeases used for active transport)
Hormonal Proteins	(insulin used to lower sugar in blood)
Contractile Proteins	(found in muscles)
Antibodies	(used to fight off foreign invaders)
Enzymes	(digestive enzymes)

Proteins are relatively large polymers made from chemical building blocks or monomers known as amino acids. Twenty amino acids are used to build proteins. An amino acid is composed of a central carbon and bonded to this central carbon are four groups: hydrogen, an amine group, a carboxyl group, and a variable group (R group). Below are two amino acids, glycine and alanine. Notice that the difference between the two molecules lies in their R groups. There are twenty amino acids and each is different from one another in their R groups. It should be noted that some R groups are non-polar, some are polar and some are charged. The charged R groups are either an acid or a base.

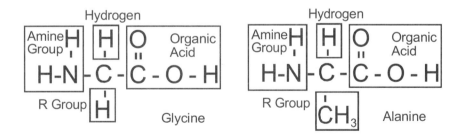

A polypeptide chain is a single chain of amino acids that are bonded together. A protein may contain one or more polypeptide chains. The formation of a polypeptide chain is called protein synthesis and occurs on the ribosome found in the cytoplasm. When two amino acids join together a molecule of water is removed in a dehydration reaction.

$$H_2O$$

$$\text{H-N-C-C-O-H} + \text{H-N-C-C-O-H} \longrightarrow \text{H-N-C-C} \text{—} \text{N-C-C-O-H}$$

The R groups of a given protein will influence the properties of that protein. Proteins in general usually range from 40-500 amino acids long. The particular sequencing of the amino acids in a polypeptide chain is called the primary structure of that protein. In addition to this sequencing, the primary structure may also include disulfide bonds. These are covalent bonds, which will cause the polypeptide to fold at a certain site and in a certain manner. Disulfide bonds are the result of a condensation reaction between two amino acids of cysteine. Cysteine has an S-H as part of its R Group. Two molecules of cysteine will bond forming an S-S bond and removing H_2. These disulfide bonds are very strong and will hold the shape of the polypeptide chain.

Proteins very seldom remain as linear molecules and generally fold on themselves to form three-dimensional shapes. Hydrogen bonding can occur between the hydrogen on the amine group and the oxygen on a neighboring amino acid's carboxyl group giving the protein secondary structure. This hydrogen bonding can produce a helical structure called an α-helix or a pleated sheet called a β-pleated sheet. These structures can then fold on themselves forming a three-dimensional structure termed the tertiary structure. If a protein has more than one polypeptide chain, then the protein can exhibit quaternary structure.

As a protein folds on itself to obtain its three-dimensional structure, the R groups of the amino acids play an important role. The polar R groups or that have a charge are usually found on the outside of the protein because these R groups are attracted to water. The R groups that are non-polar are buried inside the protein away from the water environment. A protein can be overall positive or negative, depending on whether the protein has more positive R groups or more negative R groups.

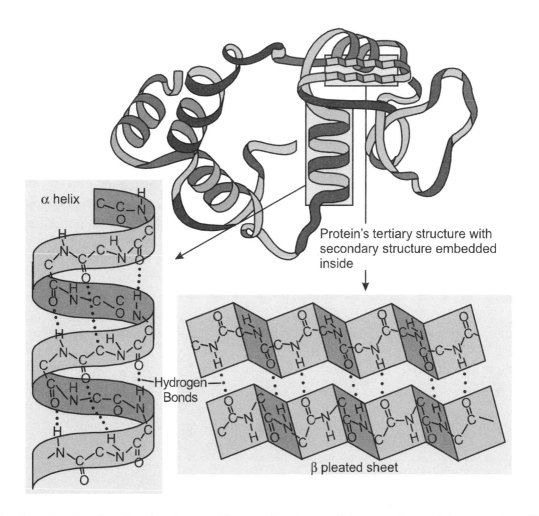

α helix

Hydrogen Bonds

β pleated sheet

Protein's tertiary structure with secondary structure embedded inside

A protein's function is related to its shape. Change the shape of the protein, and the protein will most often lose its function. There are certain physical factors that can also affect the shape of the protein. These include heat and pH. The factors can disrupt the H-bonds and thereby disrupt the protein's three-dimensional shape.

There are times when it is necessary or desirable to separate a mixture of proteins. One method for separating proteins involves electrophoresis. Electrophoresis is the separation of molecules through an agarose gel that is placed in an electric field. The agarose gel has preformed wells that the proteins are loaded into. The gel is porous and serves as a medium allowing the proteins to migrate. This migration is caused by the fact that proteins are positively or negatively charged. The rate that the proteins migrate depends on their relative charges and not their size as the pores are large compared to the proteins. Positive proteins will migrate to the negative pole and negative proteins will migrate to the positive pole. Proteins that have a greater charge will migrate faster than those proteins that have a lesser charge. Two proteins that have the same charge will migrate at the same rate.

Some proteins are colored and can be observed migrating through an agarose gel. Most proteins are colorless, however, and after the migration is complete the proteins must be dyed to determine their positions on the gel. A tracking dye is also used to make sure that the proteins do not migrate too far and run off the gel.

The proteins that will be used in this activity are equine cytochrome c, bovine myoglobin, bovine albumin, and rabbit hemoglobin.

Cytochrome c is protein found in the mitochondrion. It is a part of the electron transport chain during the production of ATP. This is a single polypeptide chain that has a heme group as a side group. The heme group contains iron, which gives this protein its characteristic orange color.

Myoglobin is a protein found in muscles and hemoglobin is the protein found in red blood cells. Myoglobin is closely related to hemoglobin. Both of these molecules can combine with oxygen and cells use these molecules for oxygen storage and transport. Both of these molecules contain an iron-based heme group. Because of the iron atoms, the proteins are rust colored. Myoglobin is found in muscle cells and stores oxygen. This oxygen is released to the muscle when the muscle demands an increase in oxygen. Hemoglobin is a much larger, complicated protein with four polypeptide chains and four heme groups. This protein is found in red blood cells and transports oxygen from the lungs to other cells in the body.

Serum albumin is the major protein found in blood plasma. It is used to transport a number of smaller molecules in the blood. Serum albumin is a colorless protein so it will need to be dyed with bromophenol blue in order to be visible on the gel.

PURPOSE
In this activity you will investigate the separation of proteins using electrophoresis. This method will determine whether a given protein has a net negative or positive charge.

MATERIALS
electrophoresis gel box	DC power supply
1.2 % melted agarose solution	micropipets
masking tape	disposable beral
buffer	cytochrome c
myoglobin	hemoglobin
serum albumin	unknown protein
ruler	250 mL beaker

Optional - If the gels are to be stained.

Coomasie blue stain	destain
staining tray	

Safety Alert
1. **Caution:** Coomassie blue stain may stain clothing and skin. Wear gloves when handling the gels and stain.
2. **Caution:** Follow the safety instructions on the proper use of eletrophoresis equipment to minimize electrical shock.

PROCEDURE

1. Make argarose gels by following the directions below:
    a. Obtain a casting tray, eletrophoresis box, and masking tape.
    b. Seal both ends of the gel casting tray with masking tape.
    c. Press the tape firm to get a tight seal.
       - *This is to insure that the melted agarose will not leak out of the tray.*
       - *Note some trays have their own dams, and do not need to be taped.*
       - *If this is the case, follow your teacher's instructions on the handling of the casting trays.*

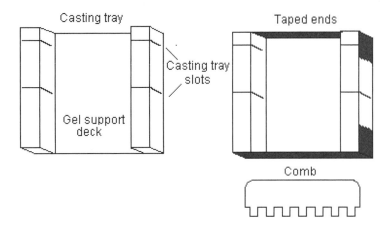

Fig. 1

2. Insert the comb at the midway point. Most casting trays are made so that the comb may be inserted either at the end or in the middle of the tray. If the tray does not allow you to insert the comb in the middle of the tray, you will have to watch and time your electrophoresis more carefully so that the positive proteins do not run off the short end of the tray.

3. Pour enough melted agarose in the casting tray so that the agarose reaches about half-way up the teeth of the comb.

4. Allow the agarose to solidify.
    - *The agarose changes appearance from clear to cloudy.*
    - *It should take about 10-15 minutes to solidify.*

5. Once the gel has solidified, carefully remove the comb by pulling straight up. Remove the tape from the ends of the casting tray.

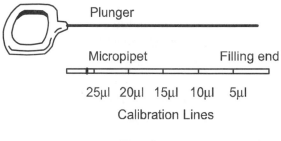

Fig. 2

6. Figure 2 shows an example of a micropipet. Above is an example of a micropipet. Insert the plunger into the glass micropipet at the end with the thick white line. There are other types of micropipets. If the pipet you are using is different from the one above. If this is the case, follow your teachers' instructions on the handling of the use of the micropipet.

7. Put the micropipet between the thumb and middle finger. The plunger can be pulled up with the index finger. Place the filling end of the micropipet into the solution to be measured. Measure out 20 mL of solution.

8. Load 10-20 mL of each of the four colored proteins and your unknown into the sample wells as indicated below.

Sample Well Number	Protein Sample
1	cytochrome c
2	myoglobin
3	hemoglobin
4	serum albumin
5	unknown protein

9. Rinse the pipet out between measurements by drawing up and expelling water three times by the pipet.

10. Seal your wells with some of the left over melted agarose using a disposable plastic pipet.

11. Transfer your gel to the electrophoretic box making sure you note the position of your gel in the electrophoresis chamber. Pour the Tris-buffer into the electrophoresis box until is just covers the gel.

12. Put of lid onto the electrophoresis box. Connect the box to the DC power supply. Turn the power to the voltage as directed by your teacher. Different boxes will require different voltage settings. Let protein move through the gel for 8 minutes. After 8 minutes, turn off and disconnect the power supply. Observe the color of the various proteins and record this information in the data table on the student answer page.

13. Remove the lid of the electrophoresis unit and record the relative position of the four proteins as compared to their starting point in the sample wells. Determine which proteins are positive and which are negative.

14. Resume electrophoresis until the bromophenol blue in the serum albumin sample has migrated to within 1 cm of the positive electrode end of the gel. Remove the gel from the casting tray unit and measure the distance of each protein (in cm) from the sample well.

Optional - If the gels are to be stained follow the directions below.

1. Carefully remove the casting tray from the electrophoresis box.

2. Slide the gel into a staining tray or weigh boat.

3. Add enough stain to cover the gel and allow it to sit undisturbed 1-24 hours.

4. When staining is complete, pour off the stain into the designated container.

5. Add about 100 ml of destaining solution. After several hours decant the destaining solution and replace with 100 ml of new destaining solution. Continue destaining until the proteins are visible when the gel is placed on a light source such as an overhead projector.

6. Measure the distance of each protein (in cm) from the sample well. Record this information on the student answer page.

Name _____

Period _____

Protein Properties
Using Electrophoresis to Determine Net Charge

HYPOTHESIS

DATA AND OBSERVATIONS

	Data Table		
Protein	Color of Protein	Distance Migrated Through Gel	Charge of the Protein
cytochrome c			
myoglobin			
hemoglobin			
serum albumin			
unknown protein (s)			

CONCLUSION QUESTIONS

1. What are the building blocks or monomers that make up the proteins? Explain how these molecules are structured and how they are different from one another.

2. Explain what is meant by a protein's primary, secondary, tertiary and quaternary structure.

3. Explain why the four proteins used in this experiment moved at different rates through the gel

4. What can be determined by comparing the unknown proteins with the known proteins in the gel?

5. Explain why in most protein separations, the final step is the staining of the gels.

Assessment

Assessment Foreword

Assessment in the foundational science courses is a critical component in the development of a successful AP program. Assessment can be used as a tool for measuring student understanding, measuring program success, and building students' test taking skills for use in a variety of settings. Well written, properly implemented assessment items play a valuable role in preparing students for success in subsequent AP courses and, ultimately, successful performance on AP exams.

Effective science programs expose students to a variety of assessment measures well in advance of the AP experience. For example, in foundational science courses, student performance can be measured through the use of laboratory practicals, laboratory reports, multiple choice questions, open-ended questions, free response items and other assessment tools. Formal and informal assessment should occur regularly and vary in complexity. Students should be exposed to questions of similar format and rigor as those that are used on the AP exams. Specific attention needs to be given to the development of test taking skills that will allow students to confidently approach a variety of multiple choice question styles and produce thorough and logical free responses.

Collaboration with members of a vertical team allows for a progression in the development and sophistication of test-taking skills. A cohesive science vertical team can implement strategies to ensure that students have sufficient exposure and opportunities to master these skills.

The assessment section of the *Laying the Foundation* series is designed to provide background information on the AP exams in science, offer samples of appropriate questions for use in the foundational courses, and suggest strategies that can be used to build student skills necessary for different question types. The types of multiple choice and free response items typically found on the exams are identified and described. Similar questions are presented at a level appropriate for use in foundational first year science courses. Additionally, this section contains a sample unit assessment and sets of multiple choice questions that can be used to assess student understanding of selected activities found within this book.

Hopefully, this section provides sufficient insight to allow you to confidently select or construct assessment items of the appropriate level of rigor and format to empower students.

AP TEST STRATEGIES AND INFORMATION

Course	Section	# of Questions	Relative Points	Time Allowed	Formula Sheet Provided?	Calculator Allowed
AP Biology	Multiple Choice	120 * will be 100 in 2004	60%	90 minutes	No	No
	Free Response (Problem/Essay)	4 (1 on molecules and cells, 1 on heredity and evolution, 2 on organisms and populations) At least one is lab-based.	40%	90 minutes	No	No
AP Chemistry	Multiple Choice	75	45%	90 minutes	No	No
	Free Response	2 mathematical problems (1 required; choose 1 of 2 others)	55%	40 minutes	Yes	Yes
		4 1 set of net ionic equations, 3 essay (2 mandatory, at least one of which is lab-based; choose 1 of 2 others)		50 minutes	Yes	No
AP Environmental Science	Multiple Choice	100	60%	90 minutes	No	No
	Free Response	4 (1 data, 1 document-based, 2 synthesis)	40%	90 minutes	No	No
AP Physics B	Multiple Choice	70	50%	90 minutes	Yes	Yes
	Free Response	7 (At least one lab-based)	50%	90 minutes	No	
AP Physics C (Electricity and Magnetism)	Multiple Choice	35	50%	45 minutes	No	No
	Free Response	3 (At least one lab-based)	50%	45 minutes	Yes	Yes
AP Physics C (Mechanics)	Multiple Choice	35	50%	45 minutes	No	No
	Free Response	3 (At least one lab-based)	50%	45 minutes	Yes	Yes

Overview of the AP Chemistry Exam

DESCRIPTION

*Multiple Choice Section

Number of questions – 75

Relative points – 45%

Time Allowed – 90 minutes

No formula sheet provided

No calculator allowed

Periodic Table and symbol definitions are included

*Free Response Section

Number of questions – Students answer 6

Breakdown – 2 mathematical calculations: 1 required equilibrium; 1 choice any topic; 1 set of net ionic equations; 3 "essay" or short answer questions: 2 required (at least one of these is lab oriented); 1 choice any topic

Relative points – 55%

The multiple choice portion of the exam is graded by machine while the free response portion is graded by a group of trained "readers" consisting of college professors and high school AP chemistry teachers. Each free-response question is scored based on a standard that is developed from input provided by not only the developers of the question but also those charged with the responsibility of grading the essays. For a complete description of the grading process go to AP Central at http://apcentral.collegeboard.com and follow the exam links to "exam scoring".

The scoring of the free response section is based on a positive point system in which the student starts out with zero points and collects up to 10 points (points vary with section of the exam) by writing correct mathematical relationships and/or ideas in response to the prompt. Points are not deducted for incorrect or inaccurate ideas but points may be deducted for incorrect significant figures. Students need acclimation to the positive grading process in which points are collected. This process has a liberating effect in comparison to the process of deducting points for mistakes. The positive point system encourages students to write more generously and share the details of their understanding of a concept.

Strategies for Developing Test Taking Skills

Success on the Advanced Placement exam in chemistry is dependent not only on the student's understanding of the content but also upon the student's test taking skills.

This segment specifically addresses developing multiple choice testing skills and describes the attributes of the free response segment of the test.

Students who know what types of questions to expect on the exam and how to approach these questions will perform more successfully. You can help your students develop these skills during the first year course by exposing them to a variety of types of multiple choice questions and periodically giving attention to test taking strategies. The skills and confidence your students develop under your guidance will serve them throughout their lives. The following strategies are offered to assist you in this skill development process.

STRATEGIES YOU CAN IMPLEMENT TO DEVELOP MULTIPLE CHOICE TEST TAKING SKILLS:

1. Teach students how to approach difficult multiple choice questions by encouraging them to:

    a. First read the entire question rather than simply scanning the words.

    b. Think about what the question is asking and then think of an answer before reading the choices. This will make it easier to recognize the correct answer when it is encountered.

    c. Round quantities and constants if it is a computational question for estimation of answers.

    d. Eliminate obvious wrong answers.

    e. Make an educated guess if one or two of the choices can be eliminated.

2. Provide frequent situations in which students are timed as they answer multiple choice questions. This is particularly important for students who have grown accustom to state level tests that have no time limitations.

3. Allow students the opportunity to bubble answers on an answer document during a timed test. Students should be taught to answer a question and immediately bubble the answer on the answer sheet rather than waiting to bubble at the end of the test.

4. Use multiple choice questions as warm-ups to provide skill development opportunities in situations outside the normal testing environment.

5. Allow the students opportunities to witness your approach to difficult multiple choice questions by modeling your thought processes aloud. This metacognition process allows students the opportunity to witness effective, logical, and mature thinking and is particularly important for computational questions. It also helps them to see that answers can be derived rather than memorized.

6. Include multiple choice items on major tests that are based on or specifically related to the lab activities performed in your course. This will encourage students to pay attention to both content and process during the lab experiences.

7. Inform students about the penalty for guessing that is often used on standardized tests, including the AP exams. Have students calculate what their test score would be on any given test if you were to deduct ¼ point penalty for each wrong answer.

8. Include material and multiple choice questions from previous "units" or topics on your tests. This helps motivate students to make an effort to retain the information rather than simply "for the test".

9. Give students feedback regarding their performance on multiple choice questions as quickly as possible. Prompt feedback allows student the opportunity to rethink the question and learn from their mistakes.

10. Include multiple choice questions that require a calculation but do not allow students to use their calculator. It is often difficult for students to learn this estimation process so give them plenty of practice.

11. Create several multiple choice test items for each exam that are similar to the AP exam questions. Use AP terminology as often as possible. For a list of common AP terminology go to: www.tealighthouse.org and choose Science, then scroll to the bottom of the page and choose Common Multiple choice terminology.

12. Expose your students to a variety of styles of multiple choice questions. Several types of multiple choice questions are described in the following section.

Examples of the Various Types of Multiple Choice Questions Found on the AP Chemistry Exam:

EXCEPT QUESTIONS

Multiple choice questions contain "EXCEPT" are sometimes difficult for students because they have to look for the answer that contains the **wrong** information. Since they have been trained to look for correct information in most questions, searching for the **wrong** information can be challenging. Help your students overcome this by including "EXCEPT" questions on your quizzes and tests.

Example from the 1994 AP Chemistry Exam:

All of the following statements concerning the characteristics of the halogens are true EXCEPT:

(A) The first ionization energies (potentials) decrease as the atomic numbers of the halogens increase.
(B) Fluorine is the best oxidizing agent.
(C) Fluorine atoms have the smallest radii.
(D) Iodine liberates free bromine from a solution of bromide ion.
(E) Fluorine is the most electronegative of the halogens.

Answer: D

Example suitable for use in a first year course:

All of the following statements concerning the characteristics of the halogens are true EXCEPT:

(A) The first ionization energies (potentials) decrease as the atomic numbers of the halogens increase.
(B) Fluorine is the most reactive member of the halogen family.
(C) Fluorine atoms have the smallest radii.
(D) Iodine reacts with sodium bromide to form sodium iodide and bromine liquid.
(E) Fluorine is the most electronegative of the halogens.

Answer: D

CLASSIFICATION SET QUESTIONS

In classification types of questions students are given a set of answers that are to be used to answer several questions. Students need practice with this type of question because frequently the answers can be used more than once or not at all. Students must be taught to **avoid** the tendency to think that if there are five questions following the answer choices that all five questions must be used. Frequently, on the AP Chemistry exam, one or more of the answer choices in an answer set will not be used at all.

Example from the 1994 AP Chemistry Exam:

Questions 8–10 refer to the following diatomic species.

(A) Li_2 (B) B_2 (C) N_2
(D) O_2 (E) F_2

  8. Has the largest bond—dissociation energy

  9. Has a bond order of 2

10. Contains 1 sigma (σ) and 2 pi (π) bonds

Answers: 8. c, 9. d, 10. c

Example suitable for use in a first year course:

Questions 1-3 refer to the following diatomic species.

(A) Li_2 (B) B_2 (C) N_2
(D) O_2 (E) F_2

  1. Requires the largest amount of energy to break the bond

  2. Contains a double bond

  3. The major component of our atmosphere

Answers: 1. c, 2. d, 3. c

TIERED-STEM QUESTIONS (MULTIPLE-MULTIPLE CHOICE)

Students may struggle with tiered-stem, multiple choice questions because they require more involved thought processes than do standard multiple choice questions. In these questions students are required to think about several items simultaneously. Exposure and practice with tiered-stem questions will help students develop confidence.

Example from the 1994 AP Chemistry Exam:

Correct procedures for a titration include which of the following?
  I.   Draining a pipet by touching the tip to the side of the container used for the titration.
  II. Rinsing the buret with distilled water just before filling it with the liquid to be titrated.
  III. Swirling the solution frequently during the titration.

(A) I only   (B) II only
(C) I and III only   (D) II and III only
(E) I, II, and III

Answer: C

Example suitable for use in a first year course:

Correct procedures for a titration include which of the following?
  I.   Draining a pipet by touching the tip to the side of the container used for the titration
  II. Rinsing the buret with the solution to be placed into it just before filling it with the solution to be titrated
  III. Swirling the solution frequently during the titration

(A) I only   (B) II only
(C) I and III only   (D) II and III only
(E) I, II, and III

Answer: E

LAB SET QUESTIONS

Students need extensive practice answering multiple choice questions that are based on experimental data or scenarios describing experimental procedures. This type of question can be used to assess science process skills as well as understanding of the scientific basis for collected data. Students will be expected to interpret diagrams, data tables and graphs. Additionally, they will be expected to apply their understanding of the concept to the data observed and recognize acceptable explanations of the trends observed.

Example from the 2002 AP Chemistry Exam:

36. A sample of a solution of an unknown acid was treated with dilute hydrochloric acid. The white precipitate formed was filtered and washed with hot water. A few drops of potassium iodide solution were added to the hot water filtrate and a bright yellow precipitate was produced. The white precipitate remaining on the filter paper was readily soluble in ammonia solution. What two ions could have been present in the unknown?

   (A) $Ag^+_{(aq)}$ and $Hg_2^{2+}_{(aq)}$

   (B) $Ag^+_{(aq)}$ and $Pb^{2+}_{(aq)}$

   (C) $Ba^{2+}_{(aq)}$ and $Ag^+_{(aq)}$

   (D) $Ba^{2+}_{(aq)}$ and $Hg_2^{2+}_{(aq)}$

   (E) $Ba^{2+}_{(aq)}$ and $Pb^{2+}_{(aq)}$

   Answer: B

Example suitable for use in a first year course:

A sample of an unknown clear solution was tested in the laboratory for the presence of several ions. Upon the addition of a dilute solution of hydrochloric acid, the solution turned white and cloudy. The white precipitate that formed was filtered and rinsed with distilled water. The sample was allowed to dry on the filter paper overnight in the classroom. The next class session students noticed that the sample had changed from a white solid to a grayish-purple solid. Which ion could have been present in the unknown?

   (A) $Ag^+_{(aq)}$

   (B) $Pb^{2+}_{(aq)}$

   (C) $Ba^{2+}_{(aq)}$

   (D) $Na^+_{(aq)}$

   (E) $Ca^{2+}_{(aq)}$

   Answer: A

FREE RESPONSE

In order to develop strong free response skills, students will need exposure to the types of questions found on the AP Chemistry exam in the first year chemistry course as well as the AP Chemistry course. The free response section is divided into several sections that require students to consider an expansive scope of the course content. The breadth of the questions make them challenging for the average student. In this section we will look at examples of each type of free response question encountered on the AP Chemistry Exam and how those questions may look in the first year course.

I. COMPUTATIONAL QUESTIONS

The first two questions encountered on the AP Chemistry exam require the use of calculations. The first of these questions is mandatory and has traditionally encompassed some type of equilibrium calculation. The second calculation allows the student a choice between two very different types of calculations and could address any topic or mixture of topics. For example the first part of a question may require knowledge of electrochemistry while the second part of the question may require knowledge of thermodynamics. Students are allowed use of a calculator on this portion of the exam only.

Equilibrium example from the 2002 AP Chemistry Exam:

---

1. $HC_3H_5O_{3(aq)} \rightleftharpoons H^+_{(aq)} + C_3H_5O_3^-_{(aq)}$

   Lactic acid, $HC_3H_5O_3$, is a monoprotic acid that dissociates in aqueous solution, as represented by the equation above. Lactic acid is 1.66 percent dissociated in 0.50M $HC_3H_5O_{3(aq)}$ at 298K. For parts (a) through (d) below, assume the temperature remains at 298K.

   (a) Write the expression for the acid-dissociation constant, K_a, for lactic acid and calculate its value.

   (b) Calculate the pH of 0.50 M $HC_3H_5O_3$.

   (c) Calculate the pH of a solution formed by dissolving 0.045 mole of solid sodium lactate, $NaC_3H_5O_3$, in 250. mL of 0.50 M $HC_3H_5O_3$. Assume that volume change is negligible.

   (d) A 100. mL sample of 0.10 M HCl is added to 100. mL of 0.50 M $HC_3H_5O_3$. Calculate the molar concentration of lactate ion, $C_3H_5O_3^-$, in the resulting solution.

Equilibrium example suitable for use in a first year course:

$$HC_3H_5O_{3(aq)} \rightleftarrows H^+_{(aq)} + C_3H_5O_3^-_{(aq)}$$

Lactic acid, $HC_3H_5O_3$, is a monoprotic acid that dissociates in aqueous solution, as represented by the equation above. Lactic acid is 1.66 percent dissociated in 0.50M $HC_3H_5O_{3(aq)}$ at 298K. For parts (a) through (d) below, assume the temperature remains at 298K.

(a) Write the expression for the acid-dissociation constant, K_a, for lactic acid and the expression for the % dissociation of lactic acid.

(b) Calculate the value of the hydrogen ion and the value lactate ion in solution.

(c) Calculate the Ka value for lactic acid.

(d) Calculate the pH of 0.50 M $HC_3H_5O_3$.

(e) Calculate the pH of a solution formed by dissolving 0.045 mol of solid sodium lactate, $NaC_3H_5O_3$, in 250. mL of 0.50 M $HC_3H_5O_3$. Assume that volume change is negligible. (Hint: common ion)

Choice computational example from the 2002 AP Chemistry Exam:

3. Consider the hydrocarbon pentane, C_5H_{12} (molar mass 72.15 g).

(a) Write the balanced equation for the combustion of pentane to yield carbon dioxide and water.

(b) What volume of dry carbon dioxide, measured at 25°C and 785 mmHg, will result from the complete combustion of 2.50 g of pentane?

(c) The complete combustion of 5.00 g of pentane releases 243 kJ of heat. On the basis of this information, calculate the ΔH for the complete combustion of one mole of pentane.

(d) Under identical conditions, a sample of an unknown gas effuses into a vacuum at twice the rate that a sample of pentane gas effuses. Calculate the molar mass of the unknown gas.

(e) The structural formula of one isomer of pentane is shown below. Draw the structural formulas for the other two isomers of pentane. Be sure to include all atoms of hydrogen and carbon in your structure.

$$H_3C \overset{\overset{H_2}{C}}{\diagup} \overset{C}{\underset{H_2}{}} \diagdown \overset{\overset{H_2}{C}}{} CH_3$$

n-pentane

Choice computational example suitable for use in a first year course:

Consider the hydrocarbon pentane, C_5H_{12} (molar mass 72.15 g).

(a) Write the balanced equation for the combustion of pentane to yield carbon dioxide and water.

(b) Calculate the number of moles of carbon dioxide that would result from the complete combustion of 2.50 g of pentane?

(c) What volume of dry carbon dioxide gas, measured at 25°C and 785 mmHg, will result from the complete combustion of pentane above?

(d) The complete combustion of 5.00 g of pentane releases 243 kJ of heat. On the basis of this information, calculate the ΔH for the complete combustion of one mole of pentane.

(e) Under identical conditions, a sample of an unknown gas effuses into a vacuum at twice the rate that a sample of pentane gas effuses. Calculate the molar mass of the unknown gas.

(f) The structural formula of one isomer of pentane is shown below. Draw the structural formulas for the other two isomers of pentane. Be sure to include all atoms of hydrogen and carbon in your structure.

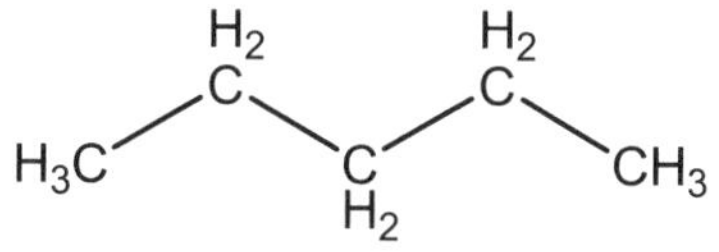

n-pentane

II. NET IONIC EQUATIONS

The equation portion of the AP Chemistry exam offers a chance for students to demonstrate basic descriptive chemistry awareness. The reactions given in this section are typical of reactions that students should have encountered in their high school laboratory experience. Students must have a working knowledge of solubility rules and formulas to be successful on this portion of the exam.

Example from the 2003 AP Chemistry Exam:

4. Write the formulas to show the reactants and the products for any FIVE of the laboratory situations described below. Answers to more than five choices will not be graded. In all cases, a reaction occurs. Assume that solutions are aqueous unless otherwise indicated. Represent substances in solution as ions if the substances are extensively ionized. Omit formulas for any ions or molecules that are unchanged by the reaction. You need not balance the equations.

Example: A strip of magnesium is added to a solution of silver nitrate.

$$\text{Ex.} \qquad Mg + Ag^+ \rightarrow Mg^{2+} + Ag$$

(a) A solution of potassium phosphate is mixed with a solution of calcium acetate.

(b) Solid zinc carbonate is added to 1.0 M sulfuric acid.

(c) A solution of hydrogen peroxide is exposed to strong sunlight.

(d) A 0.02 M hydrochloric acid solution is mixed with an equal volume of 0.01 M calcium hydroxide.

(e) Excess concentrated aqueous ammonia is added to solid silver chloride.

(f) Magnesium ribbon is burned in oxygen.

(g) A bar of strontium metal is immersed in a 1.0 M copper (II) nitrate solution.

(h) Solid dinitrogen pentoxide is added to water.

Example suitable for use in a first year course:

Write the formulas to show the reactants and the products for any FOUR of the laboratory situations described below. Answers to more than four choices will not be graded. In all cases, a reaction occurs. Assume that solutions are aqueous unless otherwise indicated. Show the total balanced formula equation, the total ionic equation and finally, the balanced net ionic equation. Represent substances in solution as ions if the substances are extensively ionized.

Example: A strip of magnesium is added to a solution of silver nitrate.

Ex. $Mg + 2\,AgNO_3 \rightarrow Mg(NO_3)_2 + 2Ag$

$Mg + 2\,Ag^+ + 2\,NO_3^- \rightarrow Mg^{2+} + NO_3^- + 2\,Ag$

$Mg + 2\,Ag^+ \rightarrow Mg^{2+} + 2\,Ag$

(a) A solution of potassium phosphate is mixed with a solution of calcium acetate.

(b) Solid zinc carbonate is added to 1.0 M sulfuric acid.

(c) A solution of hydrogen peroxide is exposed to strong sunlight.

(d) A 0.02 M hydrochloric acid solution is mixed with an equal volume of 0.01 M calcium hydroxide.

(e) Magnesium ribbon is burned in oxygen.

(f) A bar of strontium metal is immersed in a 1.0 M copper (II) nitrate solution.

(g) Solid dinitrogen pentoxide is added to water.

III. LABORATORY BASED QUESTIONS

Typically, at least one of the free response questions on the AP Chemistry exam is based on a laboratory activity. Often the lab based prompt has connections to one of the twenty-two recommended experiments listed in the Acorn booklet. In these questions the students are frequently asked to demonstrate science process skills such as graphing, making inferences, designing experiments, predicting experimental outcomes and critiquing experimental design.

Example from the 2003 AP Chemistry Exam:

$$H^+_{(aq)} + OH^-_{(aq)} \rightarrow H_2O_{(l)}$$

5. A student is asked to determine the molar enthalpy of neutralization, ΔH_{neut}, for the reaction represented above. The student combines equal volumes of 1.0 M HCl and 1.0 M NaOH in an open polystyrene cup calorimeter. The heat released by the reaction is determined by using the equation $q = mc\Delta T$.

Assume the following.

- Both solutions are at the same temperature before they are combined.

- The densities of all the solutions are the same as that of water.

- Any heat lost to the calorimeter or to the air is negligible.

- The specific heat capacity of the combined solutions is the same as that of water.

(a) Give appropriate units for each of the terms in the equation $q = mc\Delta T$.

(b) List the measurements that must be made in order to obtain the value of q.

(c) Explain how to calculate each of the following.

    i. The number of moles of water formed during the experiment

    ii. The value of the molar enthalpy of neutralization, ΔH_{neut}, for the reaction between $HCl_{(aq)}$ and $NaOH_{(aq)}$

(d) The student repeats the experiment with the same equal volumes as before, but this time uses 2.0 M HCl and 2.0 M NaOH.

    i. Indicate whether the value of q increases, decreases, or stays the same when compared to the first experiment. Justify your prediction.

    ii. Indicate whether the value of the molar enthalpy of neutralization, ΔH_{neut}, increases, decreases, or remains the same when compared to the first experiment. Justify your answer.

(e) Suppose that a significant amount of heat were lost to the air during the experiment. What effect would this have on the calculated value of the molar enthalpy of neutralization, ΔH_{neut}? Justify your answer.

Example suitable for use in a first year course:

This entire question is appropriate for a first year course.

$$H^+_{(aq)} + OH^-_{(aq)} \rightarrow H_2O_{(l)}$$

A student is asked to determine the molar enthalpy of neutralization, ΔH_{neut}, for the reaction represented above. The student combines equal volumes of 1.0 M HCl and 1.0 M NaOH in an open polystyrene cup calorimeter. The heat released by the reaction is determined by using the equation $q = mc\Delta T$.

Assume the following.

- Both solutions are at the same temperature before they are combined.

- The densities of all the solutions are the same as that of water.

- Any heat lost to the calorimeter or to the air is negligible.

- The specific heat capacity of the combined solutions is the same as that of water.

(a) Give appropriate units for each of the terms in the equation $q = mc\Delta T$.

(b) List the measurements that must be made in order to obtain the value of q.

(c) Explain how to calculate each of the following.

    i.   The number of moles of water formed during the experiment

    ii.   The value of the molar enthalpy of neutralization, ΔH_{neut}, for the reaction between $HCl_{(aq)}$ and $NaOH_{(aq)}$

(d) The student repeats the experiment with the same equal volumes as before, but this time uses 2.0 M HCl and 2.0 M NaOH.

    iii.   Indicate whether the value of q increases, decreases, or stays the same when compared to the first experiment. Justify your prediction.

    iv.   Indicate whether the value of the molar enthalpy of neutralization, ΔH_{neut}, increases, decreases, or remains the same when compared to the first experiment. Justify your answer.

(e) Suppose that a significant amount of heat were lost to the air during the experiment. What effect would this have on the calculated value of the molar enthalpy of neutralization, ΔH_{neut}? Justify your answer.

IV. CONTENT BASED QUESTIONS
The content based questions on the AP Chemistry exam come from an array of topics. Below is one example of a typical question that might be found in this section of the exam.

Example from the 2003 AP Chemistry Exam:

6. Use the principles of atomic structure and/or chemical bonding to explain each of the following. In each part, your answer must include references to <u>both</u> substances.

   (a)   The atomic radius of Li is larger than that of Be.

   (b)   The second ionization energy of K is greater than the second ionization energy of Ca.

   (c)   The carbon-to-carbon bond energy in C_2H_4 is greater than it is in C_2H_6.

   (d)   The boiling point of Cl_2 is lower than the boiling point of Br_2.

Example suitable for use in a first year course:

The entire example is suitable for a first year course.

Use the principles of atomic structure and/or chemical bonding to explain each of the following. In each part, your answer must include references to <u>both</u> substances.

   (a)   The atomic radius of Li is larger than that of Be.

   (b)   The second ionization energy of K is greater than the second ionization energy of Ca.

   (c)   The carbon-to-carbon bond energy in C_2H_4 is greater than it is in C_2H_6.

   (d)   The boiling point of Cl_2 is lower than the boiling point of Br_2.

The computational, the content questions and the laboratory based questions require the students to think across the "unit boundaries". The students are required to weave together content from all parts of the course curriculum to demonstrate a foundational understanding of major concepts. This level of thinking requires practice and incremental skill development. In addition to presenting the concepts of the course, you should provide multiple opportunities for students to develop their free response skills.

While some of the AP level questions in chemistry are appropriate for the first year course, many will incorporate concepts only appropriate for the second year course. The implication is that you will need to modify the AP level prompts or write your own prompts. In either case you should provide the students with questions or problems that require them to make choices, have multiple parts in the answer, incorporate multiple concepts and require students to make connections among a range of items.

You can have a strong, positive impact on your students test taking skills through the intentional incorporation of these assessment strategies. The next three segments of the assessment section provide you with an example of a unit test written at an appropriate level incorporating the various types of questions as well as 7 sets of multiple choice assessment questions along with a laboratory performance assessment to serve as resources as you apply these strategies.

Gas Law Unit Exam

Directions: Each set of lettered choices below refers to the numbered questions or statements immediately following it. Select the one lettered choice that best answers each question or best fits each statement and then fill in the corresponding oval on the answer sheet. A choice may be used once, more than once, or not at all in each set.

Use the following answers for questions 1-5:

 (A) Charles' Law

 (B) Boyle's Law

 (C) Gay-Lussac's Law

 (D) The Ideal Gas Law

 (E) Graham's Law

1. At constant temperature, the volume occupied by a definite mass of a gas is inversely proportional to the applied pressure.

2. At constant pressure, the volume occupied by a definite mass of a gas is directly proportional to the absolute temperature.

3. The relative rates of effusion of two gases at the same temperature and pressure are given by the inverse ratio of the square toots of the masses of the gas particles.

4. $PV = nRT$

5. Is commonly used in high school laboratories to experimentally determine the value of absolute zero.

   Directions: Each of the questions or incomplete statements below is followed by five suggested answers or completions. Select the one that is best in each case and then fill in the corresponding oval on the answer sheet.

6. Which of the following substances would have the greatest molar volume?

   (A) H_2O (s)

   (B) H_2O (ℓ)

(C) H_2O (g)

(D) CO_2 (s)

(E) CO_2 (ℓ)

7. Which of the following indicates the greatest pressure measurement?

(A) 95 kPa

(B) 375 torr

(C) 0.75 atm

(D) 425 mm Hg

(E) 785 Pa

8. Absolute temperature is the temperature at which

(A) a graph of V vs. 1/P intersects at the 1/P axis.

(B) gaseous hydrogen liquefies.

(C) the straight line graph of V vs. T intersects at the T-axis.

(D) a graph of P vs. 1/V intersects the 1/V-axis.

(E) a graph of T vs. 1/P intersects at the 1/T-axis.

9. Each of the following statements is a component of the kinetic molecular theory EXCEPT

(A) The volume occupied by gas particles is only significant at very low pressures.

(B) Gas molecules expand upon heating.

(C) The particles of a gas move in random straight line paths until a collision occurs.

(D) The collisions that occur between gas particles are considered to be "elastic" collisions.

(E) At a given temperature, all gas molecules within a sample possess the same average kinetic energy.

10. In a sample of an ideal gas, the average kinetic energy is directly proportional to the

(A) The pressure that the molecules exert on the container walls.

(B) Density of the gas sample.

(C) The absolute temperature (Kelvin) of the gas sample.

(D) The volume that the sample occupies.

(E) The molecular weight of the molecules in the sample.

11. A 1.0 L sample of gas at 25°C and 1.0 atm pressure is subjected to an increase in pressure and a decrease in temperature. The density of the gas

 (A) increases.

 (B) decreases.

 (C) remains the same since density is independent of temperature or pressure.

 (D) remains the same since the gas is behaving ideally.

 (E) becomes zero since the gas liquefies.

12. What is the molar mass of a gas if 0.214 g of the gas occupies 100. mL at STP?

 (A) $(0.214)(0.1)(0.0821)(273)$

 (B) $\dfrac{(0.214)(0.0821)(25)}{0.1}$

 (C) $\dfrac{(2.14)(0.0821)(273)}{760}$

 (D) $\dfrac{(0.214)(0.0821)(273)}{0.1}$

 (E) $\dfrac{(0.1)(0.0821)(273)}{0.214}$

13. Ten moles of a gas are contained in a ten liter container at 273K. Calculate the pressure of the gas.

 (A) 2.24 atm

 (B) 4.48 atm

 (C) 11.2 atm

 (D) 15.7 atm

 (E) 22.4 atm

14. What is the volume occupied by 34.0 g of gaseous ammonia at 227°C and 1520 torr?

 (A) 10.3 L

 (B) 18.6 L

 (C) 38.7 L

 (D) 41.1 L

 (E) 82.2 L

15. What is the average molar mass of air at 25°C and 1 atm pressure?

    (A) 28.0 g/mol

    (B) 28.8 g/mol

    (C) 32.0 g/mol

    (D) 32.4 g/mol

    (E) 60.0 g/mol

16. A gaseous compound contains 72% C, 12% H, and 16% O by mass. At STP 11.2 L of the gas weighs 100. g. What is the molecular formula of the gas?

    (A) $C_6H_{14}O$

    (B) $C_6H_{12}O$

    (C) $C_{12}H_{24}O_2$

    (D) $C_{12}H_{20}O$

    (E) $C_{14}H_{28}O_2$

17. A flask at 200°C has a total pressure of 2.5 atm and contains a mixture of carbon dioxide and carbon monoxide. If the partial pressure in the flask due to carbon monoxide is 1.3 atm, what is the partial pressure of carbon dioxide?

    (A) 0.50 atm

    (B) 0.60 atm

    (C) 1.2 atm

    (D) 2.6 atm

    (E) 3.8 atm

18. A mixture of 16.0 g of helium, 56.0 g of nitrogen and 64.0 g of oxygen at 25°C is in a 125 L container. What are the mole fractions of each of these three gases?

	X_{helium}	$X_{nitrogen}$	X_{oxygen}
(A)	0.11	0.41	0.48
(B)	0.33	0.33	0.33
(C)	0.50	0.25	0.25
(D)	0.57	0.14	0.29
(E)	0.25	0.50	0.25

19. A 50.0 mL sample of hydrogen was collected over water at 27°C on a day when the barometric pressure was 748 atm. What volume will the dry hydrogen occupy at STP? The vapor pressure of water is 27 torr at 27° C.

(A) $\dfrac{(748)(50.0)(273)}{(760)(300)}$

(B) $\dfrac{(721)(50.0)(27)}{(760)(300)}$

(C) $\dfrac{(721)(50.0)(300)}{(760)(273)}$

(D) $\dfrac{(721)(50.0)(273)}{(760)(300)}$

(E) $\dfrac{(760)(50.0)(300)}{(721)(273)}$

20. One half liter of element X (a diatomic gas) reacts with one liter of hydrogen gas to form one liter of a gaseous compound. All gas volumes are measured at the same conditions of temperature and pressure. What is the formula of the gaseous compound formed in this reaction?

(A) HX

(B) H_2X

(C) HX_2

(D) H_2X_3

(E) H_3X

21. What volume of hydrogen gas at STP would be required to react completely with 1 mole of Fe_3O_4 to produce solid iron metal and water vapor?

(A) 11.2 L

(B) 22.4 L

(C) 33.6 L

(D) 67.2 L

(E) 89.6 L

22. What is the order of increasing rate of effusion for the following gases?

$$Ar, CO_2, He, N_2$$

(A) $N_2 < Ar < CO_2 < He$

(B) $Ar < CO_2 < He < N_2$

(C) $Ar < He < CO_2 < N_2$

(D) $N_2 < CO_2 < Ar < He$

(E) $CO_2 < Ar < N_2 < He$

23. At a given temperature and pressure, the velocity of a helium molecule would be ___ times the velocity of a CH_4 molecule.

(A) one

(B) two

(C) three

(D) four

(E) five

Questions 24 & 25: A sample of propane undergoes complete combustion to produce carbon dioxide and water according to the *unbalanced* reaction below:

$$C_3H_{8\ (g)} + O_{2\ (g)} \rightarrow CO_{2\ (g)} + H_2O_{\ (g)}$$

24. If 6.0 moles of propane reacts with 15.0 moles of oxygen, how many moles of carbon dioxide are formed at STP?

(A) 3 moles

(B) 6 moles

(C) 9 moles

(D) 15 moles

(E) 45 moles

25. What volume of gas at STP is left unreacted?

(A) 11.2 L

(B) 22.4 L

(C) 44.8 L

(D) 67.2 L

(E) 78.4 L

Free Response: (You may get your graphing calculator once you have handed in the multiple choice portion of your exam.)

1. A student performed a laboratory investigation involving Charles' Law. A capillary tube that was closed at one end was fastened to a temperature probe and marked as shown in Figure 1. The following data was collected:

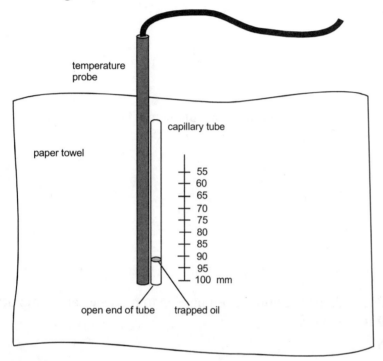

Data Table	
Length of trapped gas sample (mm)	Temperature (°C)
100	
90	121.23
85	108.79
80	93.861
75	62.142
70	41.617
67	28.246

(A) Perform a linear regression on the data paste the regression equation into ⟨Y=⟩ on your calculator.

    i.   What is the value of the correlation coefficient, "r" for this data? Express your answer to 4 decimal places.

    ii.   What is the value of the y-intercept for this data rounded to the nearest integer?

    iii.   What does this number represent?

    iv.   Why are we able to use length as a relative measure of volume in this experiment?

(B)   Sketch a graph of the data on the axes provided. Be sure to label the graph properly.

     i.   What would the temperature of the trapped air sample be when the oil drop is at the 55mm mark? Express your answer to the nearest tenth of a degree.

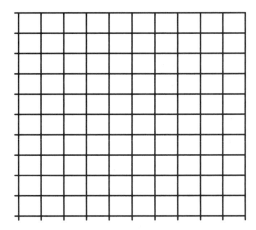

2. A rigid 2.00 L cylinder contains 14.0 g of nitrogen gas and 16.0 g of oxygen gas.

   (A)   Calculate the total pressure in atm, of the gas mixture in the cylinder at 298 K.

   (B)   The temperature of the gas mixture in the cylinder is decreased to 250 K. Calculate each of the following.

     i.   The mole fraction of oxygen gas in the cylinder.

     ii.   The partial pressure, in atm, of nitrogen gas in the cylinder.

   (C)   If the cylinder develops a pin-hole sized leak and some of the gaseous mixture escapes, would the ratio $\dfrac{\text{moles of } N_2(g)}{\text{moles of } O_2(g)}$ in the cylinder increase, decrease or remain the same? Justify your answer.

Multiple Choice Answers to Unit Test on Gas Laws					
1.	B	11.	A	21.	E
2.	C	12.	D	22.	E
3.	E	13.	E	23.	B
4.	D	14.	D	24.	C
5.	A	15.	B	25.	D
6.	C	16.	C		
7.	A	17.	C		
8.	C	18.	B		
9.	B	19.	D		
10.	C	20.	B		

Free Response Answers:

(A) Perform a linear regression on the data paste the regression equation into $\boxed{Y=}$ on your calculator.

    i.   0.9908

    ii.   -253°C

    iii. Absolute zero or the temperature when the volume of the trapped gas is theoretically zero.

    iv. The tube has a constant diameter, so that keeps the data relative to volume. The volume of the cylinder can be calculated as $V = \pi r^2 \ell$ where pi and r are constant for the tube so $V \propto \ell$.

(B) Sketch a graph of the data on the axes provided. Be sure to label the graph properly.

    i.   -20.4°C

    Answer graph:

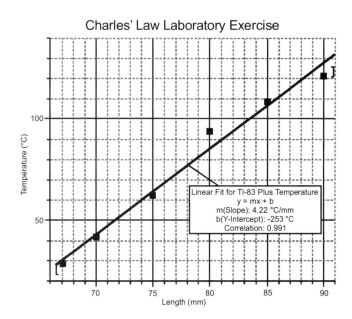

2. A rigid 2.00 L cylinder contains 14.0 g of nitrogen gas and 16.0 g of oxygen gas.

  (A) There are 0.500 moles of each gas for a total of one mole of gas. Using the ideal gas equation with one total mole of gas it is calculated that 12.2 atm is the total pressure.

  (B) i. $\dfrac{0.500 \text{ mole nitrogen gas}}{1.00 \text{ total moles}} = 0.500$

      ii. Using the ideal gas equation again, the new total pressure at the lower temperature is equal to 10.3 atm, and ½ of that is due to nitrogen, therefore the P_{nitrogen} = 5.13 atm.

  (C) Decrease. Nitrogen will effuse faster than oxygen since it has a smaller molar mass. Since nitrogen escapes at a faster rate than oxygen, more moles of oxygen will be present which will decrease the ratio since oxygen is in the denominator.

Why Do They Call It a Periodic Table?
Investigating and Graphing Periodic Trends

MULTIPLE CHOICE:
This type of multiple choice question is referred to as a "classification set question". It is a typical type of question that is used on the AP Chemistry Exam. It is important that students not view this as a matching question. Note that one answer may be used more than once, and some answer may not be not be used at all.

The following answers should be used to answer questions 1-5:

(a) Br

(b) K

(c) Ba

(d) Cs

(e) Ca

1. The element with the lowest first ionization energy.

2. The smallest atom

3. The atom with the lowest electron affinity.

4. The alkaline earth element with the smallest atomic radius.

5. The element that reacts most vigorously with water.

6. Which of the following has the highest electronegativity?
   a. Carbon
   b. Oxygen
   c. Nitrogen
   d. Fluorine
   e. Argon

7. Which of the following atoms would be most likely to have an oxidation state of 3+ when found in compounds?

    a. F

    b. P

    c. Al

    d. Zn

    e. Ag

Answer Key:

1. (D) Cs

   - The factors affecting ionization energy are the number of protons in the nucleus, the distance the valance electron is from the nucleus, and the screening from the inner electrons.

   - The combination of nuclear pull and screening from the core electrons is often referred to as effective nuclear charge and is symbolized by Z_{eff}.

   - Given the choices, the valence electrons in Cs are further away with more core screening and therefore would be easier to remove.

   - Cs and Ba are in the same period, however, the Z_{eff} is greater for Ba than for Cs.

2. (A) Br

   - Given the choices, K, Ca, and Br are all in the same period. However, the Z_{eff} is greater for Br due to a greater number of protons in the nucleus.

3. (D) Cs

   - The same reasons that would cause cesium to have a low first ionization potential would also cause it to have a low electron affinity. If it doesn't take much energy to remove an electron, it stands to reason that it would not better its energy situation to accept an extra electron.

4. (E) Ca

   - The student must know which atoms belong in the alkaline earth family, IIA.

   - There are only two choices, Ca and Ba.

   - Clearly, Ca has fewer principal energy levels than Ba.

5. (D) Cs

   - Of the choices, K, Ba, Cs, and Ca all react with water. However, the question asks which metal reacts most vigorously with water.

   - The alkali metals react with water more vigorously than alkaline earth metals. The reactivity increases as the electron becomes easier to remove.

   - Therefore, cesium has the most vigorous reaction with water since its valence electron is less tightly held.

6. (D) F

    - Due to its small size and high Z_{eff}, fluorine has the highest electronegativity of any atom.

7. (C) Al

    - Aluminum has three valence electrons which, when lost in forming an ionic bond, would give the atom an oxidation state of 3+.

Molecular Geometry
Investigating Molecular Shapes with VSEPR

MULTIPLE CHOICE:

Use the following answers for questions 1-5:

    (A)   linear

    (B)   trigonal planar

    (C) sp^3 hybridization

    (D) sp^3d^2 hybridization

    (E)   pi bonding

1. The nitrite ion

2. Selenium hexafluoride

3. Carbon tetrachloride molecule

4. The ammonia molecule

5. Triiodide ion

Answer Key:

1. E

2. D

3. C

4. C

5. A

Limiting Reactant
Exploring Molar Relationships

MULTIPLE CHOICE:

Questions 1-5:

A student performs the limiting reactant lab using 100 mL of copper (II) chloride solution of an unknown concentration and recorded the following data:

	Mass of Aluminum (g)
Initial	1.40 g
Final	0.50 g
Mass of aluminum consumed	

1. Which of the following represents a correctly balanced chemical equation for this reaction?

   (A) $3 \, Cu_2Cl_{(aq)} + Al_{(s)} \rightarrow AlCl_{3 \, (aq)} + 6 \, Cu_{(s)}$

   (B) $CuCl_{(aq)} + Al_{(s)} \rightarrow AlCl_{(aq)} + Cu_{(s)}$

   (C) $2 \, Cu_2Cl_{(aq)} + 2 \, Al_{(s)} \rightarrow 2 \, AlCl_{2 \, (aq)} + 4 \, Cu_{(s)}$

   (D) $3 \, CuCl_{2(aq)} + 2 \, Al_{(s)} \rightarrow 2 \, AlCl_{3 \, (aq)} + 3 \, Cu_{(s)}$

2. How many moles of aluminum reacted?

   (A) 0.014   (B) 0.019   (C) 0.033   (D) 0.050

3. How many moles of copper (II) chloride reacted?

   (A) 0.014   (B) 0.019   (C) 0.033   (D) 0.050

4. What is the molarity of the 100. mL sample of copper (II) chloride the student used?

   (A) 0.14 M   (B) 0.19 M   (C) 0.33 M   (D) 0.50 M

5. Which of the following are correct statements?

   I. If a student forgets to dry the aluminum strip before taking its final mass, the calculated molarity of the copper (II) chloride solution will be too high.

II. If a student fails to rinse all of the measured copper (II) chloride solution from the graduated cylinder into the reaction vessel, the calculated molarity of the copper (II) chloride solution will be too low.

III. If a student collects all of the copper formed in this reaction, dries and masses it and uses that mass to determine the initial molarity of the copper (II) chloride solution, the calculated molarity will be too high.

(A) I only

(B) I and II only

(C) II and III only

(D) I, II, and III

Answer Key:

1. D
2. C
3. D
4. D
5. B

The Iodine Clock Reaction
Investigating Chemical Kinetics: A Microscale Approach

MULTIPLE CHOICE:

Questions 1-3:

A student performs the iodine clock reaction laboratory and obtained the following data:

Data Table								
Well #	1	2	3	4	5	6	7	8
Drops A	4	4	3	3	2	2	1	1
Drops B	1	1	1	1	1	1	1	1
Drops C	4	4	4	4	4	4	4	4
Drops H_2O	0	0	1	1	2	2	3	3
Time required for blue color to appear:	14	16	21	22	32	31	58	58

Use this graph of the data to help you answer questions 1-3.

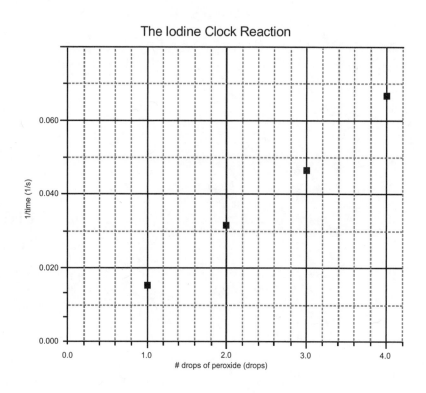

1. What is the order of the of $[H_2O_2]$ in this reaction?

   (A) zero   (B) 1st (C) 2nd (D) 3rd

2. Which of the following served as controls in this experiment?

   I. The number of drops of Solution A.

   II. The number of drops of solution B.

   III. The number of drops of solution C.

   (A) I only

   (B) II only

   (C) II and III only

   (D) I, II and III

3. Approximately how long would it take for 3.1 drops to react?

   (A)   0.05s   (B) 0.058s   (C) 18 s   (D) 20 s

   Use the following data collected from a different kinetics experiment to answer questions 4 and 5:

Experiment	Initial [A] (mol L^{-1})	Initial [B] (mol L^{-1})	Initial Rate of Formation of C (mol L^{-1} s^{-1})
1	0.05	0.10	0.058
2	0.10	0.10	0.230
3	0.10	0.05	0.115

4. The initial-rate data in the table above were obtained for the reaction represented below.

   $$A_{(g)} + 2\,B_{(g)} \rightarrow 2\,C_{(g)}$$

   What is the experimental rate law for the reaction?

   (A) rate $= k[A]\,[B]$
   (B) rate $= k[A]\,[B]^2$
   (C) rate $= k[A]^2\,[B]$
   (D) rate $= k[A]^2\,[B]^2$
   (E) rate $= k[A]\,/\,[B]$

5. The correct value for the rate constant, k, calculated for trial two is:

   (A) $2.3 \times 10^{-3}\,s^{-1}$ (B) $2.3 \times 10^{-4}\,M^{-2}\,s^{-1}$ (C) $23\,M^{-1}\,s^{-1}$ (D) $230\,M^{-2}\,s^{-1}$

Answer Key:

   1.   B
   2.   C
   3.   D
   4.   C
   5.   D

Disturbing Equilibrium
Observing Le Chatelier's Principle

MULTIPLE CHOICE:

1. In the following equilibrium system, when heat was added

    $$KNO_{3(s)} \rightleftharpoons KNO_{3(aq)}$$

    I.   The reaction shifted to the reactants

    II. KNO_3 precipitated out of solution

    III. KNO_3 dissolved into solution

    (A) I only

    (B) II only

    (C) III only

    (D) I and II only

    (E) I, II and III

2. The potassium nitrate system represented above is

    (A) endothermic.

    (B) exothermic.

    (C) constantly in a static equilibrium.

    (D) unaffected by a common ion.

    (D) extremely unstable.

A balanced equation for the equilibrium that existed after iron (III) and thiocyanate ions were combined in the beaker is:

$$Fe^{+3}_{(aq)} + SCN^{-}_{(aq)} \rightleftharpoons [FeSCN]^{2+}_{(aq)}$$

3. When more iron (III) nitrate was added to the system above,

    (A) the color of the solution became darker.

    (B) the color of the solution became lighter.

    (C) iron precipitates out of solution.

    (D) the system remains unchanged.

    (E) more SCN^{-} is formed.

Use the following system at equilibrium to answer questions 4 and 5:

$$N_{2(g)} + 3\,H_{2(g)} \rightleftharpoons 2\,NH_{3(g)} + heat$$

4. Which of the following would shift the equilibrium to produce more ammonia?

   I. Increase the pressure

   II. Decrease the temperature

   III. Decrease the amount of hydrogen

(A) I only

(B) II only

(C) I and II only

(D) III only

(E) I, II and III

5. All of the following are true for the reaction above EXCEPT

(A) Adding more nitrogen would shift the equilibrium to the products.

(B) The addition of a catalyst would shift the equilibrium to the reactants.

(C) Decreasing the pressure would cause the reaction to shift to the reactants.

(D) Adding argon gas would have no effect on the equilibrium.

(E) Removing ammonia as soon as it forms causes the reaction to continue in the forward direction.

Answer Key:

1. C

2. A

3. A

4. C

5. B

How Hot Is a Candle?
Determining the Heat of Combustion of Paraffin

MULTIPLE CHOICE:

1. During the day of the lab computing the heat of combustion of paraffin, the soda cans got wet. How would a wet can affect the value of the molar heat of combustion?

   a. The fact that the can was wet before the lab started will make the calculated heat of combustion lower than it should be because some of the heat will be absorbed by the extra water.

   b. The fact that the can was wet before the lab will have no effect on the calculated molar heat of combustion.

   c. The fact that the can was wet before the lab started will make the calculated heat of combustion higher than it should be because some of the heat will be absorbed by the extra water.

   d. The fact that the can was wet before the lab will cause the change in temperature to be greater than it should be.

2. Susie Student blew her candle out rather vigorously, splashing a small amount of paraffin on the lab counter. How will this affect the value of the molar heat of combustion?

   a. Since the CBL had stopped taking data, the loss of the paraffin will have no effect on the molar heat of combustion.

   b. The loss of the paraffin will cause the molar heat of combustion to be higher than it should be because there are more grams of paraffin involved.

   c. The loss of the paraffin will cause the molar heat of combustion to be lower than it should be because it appears that more paraffin has been burned that actually has been.

   d. The loss of the paraffin will change the specific heat of the paraffin and cause the molar heat of combustion to rise.

3. The principle of calorimetry is based upon

   a. Conservation of mass.

   b. Conservation of energy.

   c. Bond energies.

   d. Intermolecular forces.

4. The calculated value of the molar heat of combustion of paraffin is always lower than the actual value for the molar heat of combustion of paraffin. The reason for this is:

   a. Some of the heat from the candle is lost to the surrounding air.

   b. Some of the heat from the candle is absorbed by the aluminum can.

   c. Some of the heat from the candle is absorbed by the thermometer or probe.

d. All of the above.

5. The mass of the aluminum can was 18.0 g. The temperature of the system went from 5.0°C to 31°C. The specific heat of aluminum is 24.3 J/mol °C. How much heat was absorbed by the aluminum can?

a. 6.78×10^4 J

b. 421 J

c. 1.10×10^4 J

d. 43.3 J

Answer Key:

1. A

- Extra water in the can will absorb more heat and the change in temperature will be lower than it should be. That will make the heat of combustion lower than it should be.

2. C

- The lost paraffin will make it seem that more paraffin had been burned than actually had been burned and will make the calculated heat of combustion lower than it should be.

3. B

- Calorimetry is based upon the principle that energy is conserved. Then energy released by the candle is absorbed by the calorimeter.

4. D

- All of the above. Because the system is not isolated from the surroundings, heat is lost to many sources in the surroundings.

5. B

$$\frac{18.0 \text{ g}}{} \times \frac{\text{mol}}{27 \text{ g}} \times \frac{24.3 \text{J}}{\text{mol} °C} \times \frac{26°C}{} = 421 \text{J}$$

This problem can be solved by estimation without a calculator as follows:

Reduce: $\frac{18}{27} = \frac{2}{3}$; Round: 24.3 to 25 and 26 to 25

This simplifies: $2 \times \frac{25 \times 25}{3} = 416$ (only one answer is in this range)

Hess's Law
Determining the Heat of Formation of Magnesium Oxide

MULTIPLE CHOICE: (A CALCULATOR WILL BE NEEDED)
Questions 1-3:

A student performs a lab to determine the heat of formation of magnesium oxide and obtained the following data:

	Mg + HCl	**MgO + HCl**
Volume of 1.0 M HCl (mL)	60.0 mL	60.0 mL
Mass of solid (grams)	0.65 g	1.05 g
Initial temperature of HCl (°C)	22.5 °C	22.5 °C
Highest temperature of mixture (°C)	54.5 °C	27.0 °C

1. Which of the following represents a correctly balanced equation for the reaction that occurs in the first experiment?

    (A) $Mg_{(s)} + HCl_{(aq)} \rightarrow MgCl_{(aq)} + H_{(g)}$

    (B) $Mg_{(s)} + HCl_{(aq)} \rightarrow MgCl_{2(aq)} + H_{(g)}$

    (C) $Mg_{(s)} + 2HCl_{(aq)} \rightarrow MgCl_{2(aq)} + H_{(g)}$

    (D) $2Mg_{(s)} + 2HCl_{(aq)} \rightarrow 2MgCl_{(aq)} + H_{2(g)}$

    (E) $Mg_{(s)} + 2HCl_{(aq)} \rightarrow MgCl_{2(aq)} + H_{2(g)}$

2. How many moles of magnesium were consumed in the first experiment between magnesium and hydrochloric acid?

    (A) 0.010 mol Mg

    (B) 0.027 mol Mg

    (C) 0.040 mol Mg

    (D) 0.65 mol Mg

    (E) 2.53 mol Mg

3. Calculate the amount of heat given off by the reaction between magnesium and hydrochloric acid. The density of the solution is 1.04 g/mL and the specific heat of water is 4.184 J/g °C.

    (A) 32.0 J

    (B) 87.0 J

    (C) 1940 J

(D) 6500 J

(E) 8442 J

4. All of the following are true in this experiment EXCEPT

(A) The reaction between magnesium oxide and hydrochloric acid was endothermic.

(B) Hess' Law was used to determine the heat of formation for magnesium oxide using the following three reactions: Mg + HCl; MgO + HCl and the formation of water from its elements.

(C) Hydrochloric acid was the excess reagent in each experiment.

(D) The total mass of the solution is found by multiplying the volume of the solution by the density of the solution then adding the mass of the solid sample.

(E) The proper units for the heat of formation of MgO are kJ/mol.

5. The assumption that heat was not lost to the surroundings during this experiment was not truly accurate. Which of the following would be true if heat was lost to the surroundings?

   I. The experimental value for ΔH_f for MgO will be lower than expected.

  II. The % error will be higher than expected

 III. The experimental value for ΔH_{rxn} for Mg +HCl will be higher than expected.

(A) I only

(B) II only

(C) I and II only

(D) II and III only

(E) I, II, and III

Answer Key:

1. E

2. B

3. E

4. A

5. C

It's Not Easy Being Green
Making Solutions

OBJECTIVE
This activity assesses student's ability to make solutions of specified concentration.

LEVEL
Chemistry

NATIONAL STANDARDS
UCP.3, A.1, A.2, B.6, E.1, E.2.

TEKS
2(A), 2(B), 2(C) 2,(D), 13(B)

CONNECTIONS TO AP
AP Chemistry:
   II. States of Matter, C. Solutions, 2. Methods of expressing concentration.
   AP Chemistry Labs: 17. Colorimetric or spectrophotometric analysis.

TIME FRAME
45 minutes for students
Teacher prep time for making solutions and standard curve: 30 minutes.

MATERIALS
(For a class of 28 working in pairs)

14 graphing calculators
14 volumetric flasks with stoppers, caps, or parafilm
   (assorted volumes from 10-mL to 100-mL)
14 scoops
500 g $NiCl_2·6H_2O$
1 syringe, 20 mL capacity
14 wash bottles filled with distilled water
1 spectrometer (Spectronic-20, CBL colorimeter,
   or other similar device)

student answer pages and pencils
balances
14 condiment cups with lids
50 mL of nickel(II) stock solution
5 25-mL volumetric flasks
assorted beakers
24 cuvettes
assorted graduated cylinders

TEACHER NOTES
This lesson may be used as a classroom laboratory activity. However, a better use it to allow this activity to serve as an assessment for a unit on solutions. Before performing this activity, be very certain that students have ample experience in calculating solution concentrations, particularly molarity. Also, students should have been exposed to the proper use of volumetric flasks, and the proper method of massing and transferring solids.

Proper method of massing solids and using volumetric flasks: Mass a waxed weighing paper. If an electronic balance is available, the tare feature may be used. Using a scope, transfer the desired amount of the nickel salt on the weighing paper. Record the mass. Fold the weighing paper slightly to assist in transferring the compound to the volumetric flask. Using a wash bottle, wash the weighing paper into the volumetric flask. Add distilled water to the volumetric flask until the bottom of the meniscus is at the etched line on the flask. Mix thoroughly.

Students will be asked to prepare a specific volume of nickel(II) chloride hexahydrate of a specified molarity. The volumes will be dictated by the assortment of volumetric glassware that is available in your lab. If you are short volumetric glassware, you should check with your local hospitals or pathology labs. They often discard volumetric glassware because of minor flaws like a small scratch. This "flawed" equipment works perfectly well in a high school laboratory. The concentrations should be between [0.05] and [0.10]. Use the following table to record the student assignments and the absorbances that they produce.

Student pair ID #	Volume (mL)	Concentration (M)	% Transmittance
1			
2			
3			
4			
5			
6			
7			
8			
9			
10			
11			
12			
13			
14			

To prepare standard solutions:

Prior to the lab, make a stock solution of $NiCl_2$. Using an analytic balance, mass between 2.2 and 2.4 grams $NiCl_2\cdot6H_2O$ on a weighing paper. Record the exact mass. Transfer the compound to a 100-mL volumetric flask. Use the wash bottle to wash any residue on the paper into the flask. Allow the solid to dissolve. Dilute to the mark on the volumetric flask with distilled water and mix thoroughly. Calculate the molarity of your stock solution. Use the 25.0 mL volumetric flasks to make dilutions. You may use any serial dilution scheme that you wish. Use the syringe to measure the stock solution and dilute each solution with distilled water. Cover and mix each solution thoroughly. This is sample data for 2.2962 g nickel(II) chloride hexahydrate in 100 mL solution.

Sample #	Concentration (mol/L)	%Transmittance	Absorbance
1	0.106 Stock solution	27.8	-0.556
2	0.0847 20.0 mL stock solution diluted to 25.0 mL	36.0	-0.444
3	0.0636 15.0 mL stock solution diluted to 25.0 mL	46.0	-0.337
4	0.0424 10.0 mL stock solution diluted to 25.0 mL	58.1	-0.236
5	0.0212 5.00 mL stock solution diluted to 25.0 mL	77.8	-0.109

SAMPLE CALCULATION:

$M_1V_1=M_2V_2$

Example #2: $(0.0966\,M)(20.0\,mL)=(x)(25.0\,mL)$

Sample standard curve for data.

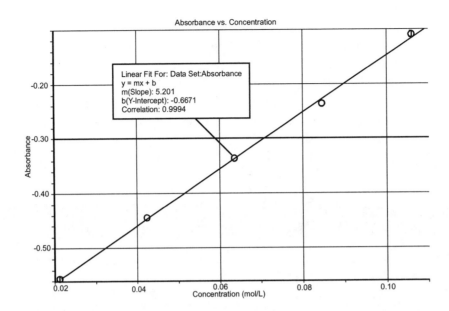

Use this table to record your own data.

Sample #	Concentration (M)	% Transmittance	Absorbance Absorbance = –log(%Transmittance/100)
1			
2			
3			
4			
5			

Post your data on the board so that students may enter it in their calculators.

Spectroscopic analysis: Turn on the spectrometer and allow it to warm up for 30 minutes. Set the wavelength at 395 nm. Zero the spectrometer with no sample in the sample holder. Fill one cuvette with distilled water and the other 5 cuvettes with solutions 1-5. Each of the cuvettes should be rinsed with the solution that it will contain. Insert the cuvette containing the distilled water into the sample holder and adjust the instrument to 100% transmittance. Remove the distilled water and put in the cuvette filled with sample #1. Read and record the %transmittance. Do the same with each of the other four solutions.

Making the standard curve: Enter the concentration in L_1 and the %Transmittance in L_2. Arrow up so the cursor is on L_2 and type [(-)] [LOG] [(] [2nd] [2] [÷] [1][0][0][)] [ENTER]. This will convert %Transmittance to Absorbance. Beer's Law says that absorbance is directly proportional to concentration. (A=abc where a=the molar absorptivity and b=the length of the light path and c=concentration). Once you have the standard curve, run a linear regression. Go to [STAT] Edit [ENTER] [4]:LinReg [VARS][▶]Y-VARS [ENTER]

ENTER ENTER. Go to ZOOM 9 to look at the graph. Your calculator is now set up to easily check the concentrations of each of the student's solutions. Go to Y= Arrow down to Y_2. Enter (-) LOG (2nd 2 ÷ 1 0 0). Go back to ZOOM 9. Press 2nd TRACE 5 ENTER ENTER ENTER and you can read the concentration at the bottom of the screen.

Depending upon the students' expertise with calculators, you may need to go over this concept with students before the lab. You should check each student's calculations with your own standard curve.

Sample scoring guidelines

1. (5 points) Students are wearing proper eye protection.

2. (10 points) Calculations on lab reporting sheet are correct.

3. (3 points) Students use a weighing paper or weighing cup to mass solid $NiCl_2 \cdot 6H_2O$.

4. (2 points) Students wash the weighing paper or cup into volumetric flask.

5. (10 points) Students mix solution in a volumetric flask—not a beaker or graduated cylinder.

6. (10 points) Present error is less than 5%. (2 points penalty for each 5% over initial 5%)

POSSIBLE ANSWERS TO DATA AND OBSERVATIONS

1. Write the formula for nickel(II) chloride hexahydrate.

   - $NiCl_2 \cdot 6H_2O$

2. Calculate the grams of nickel(II) chloride hexahydrate needed to make your solution. Show your work.

   - Student answers will vary. Calculations shown for 50.0 mL of 0.025 M solution.
   - $$\frac{237.7\,g\,NiCl_2 \cdot 6H_2O}{mol} \times \frac{0.025\,mol}{L} \times \frac{0.0500\,L}{} = 0.30\,g$$

POSSIBLE ANSWERS TO CONCLUSION QUESTIONS

1. Explain any error in this lab.

   - Student answers will vary, but most common errors involve not thoroughly mixing the solution and errors in massing the compound. Also, students may over-shoot the mark when diluting the solution.

2. Explain how the following errors would affect the molarity of your solution.

   a. You forget to add the water of hydration in calculating the molar mass.

      - Your calculated molar mass would be low and therefore you would not use enough compound to make your solution. Your molarity would be low.

   b. You forget to thoroughly mix the solution before giving it to your teacher for measurement.

      - The majority of the nickel(II) chloride is at the bottom of the flask. When the solution is poured into the cuvette, it is mostly water and the molarity will be very low.

It's Not Easy Being Green
Making Solutions

In this exercise, you will be asked to make up a solution with a specified volume and molarity. Your teacher will then use an instrument called a spectrophotometer to measure the concentration of the solution that you made. Your grade will depend upon using safe and correct laboratory techniques as well as accuracy in producing a solution with the correct molarity.

PURPOSE
You will produce a solution of nickel(II) chloride with a specific molarity.

MATERIALS

graphing calculator

assorted glassware as provided by your teacher

scoop

$NiCl_2 \cdot 6H_2O$ in small container provided by
   your teacher

student answer pages and pencils

balance

weighing paper or weighing boat

wash bottle containing distilled water

Safety Alert
1. Wear eye protection.
2. Return solution to your teacher for disposal.
3. Wash your hands before leaving the lab.

PROCEDURE
1. Your task in this exercise is to make a certain volume of nickel(II) chloride solution. Your teacher will assign you a specified volume and concentration to prepare.

2. Calculate the correct amount of nickel(II) chloride hexahydrate needed to prepare your solution. Show your calculations clearly on the student answer page.

3. Prepare the solution using the materials available in your lab. If you feel that you need something that is not readily available, you may ask your teacher.

4. When your solution is complete, take it to your teacher who will use a spectrophotometer to measure the concentration. A portion of your grade will depend upon the accuracy of your concentration.

5. Your teacher will provide you with data for the standard curve. Enter concentration in L_1 and absorbance in L_2. You will use this curve to determine the actual concentration of your solution.

Name _____

Period _____

It's Not Easy Being Green
Making Solutions

DATA AND OBSERVATIONS

You are to make _____ mL of _____ M nickel(II) chloride.

1. Write the formula for nickel(II) chloride hexahydrate.

2. Calculate the grams of nickel(II) chloride hexahydrate needed to make your solution. Show your work.

3. Describe your plan for making your solution.

4. Prepare your solution.

5. Take your solution to your teacher.

6. Percent transmittance of solution _____.

7. Absorbance of solution. Absorbance= -log(%Transmittance/100) _____

ANALYSIS

1. Use the standard curve provided by your teacher to calculate the actual molarity of your solution.
   The actual concentration of my solution is _____.

2. Calculate the % error of your solution.

   $$\% \text{ Error} = \frac{|\text{actual concentration-assigned concentration}|}{\text{assigned concentration}} \times 100\%$$

CONCLUSION QUESTIONS

1. Explain any error in this lab.

2. Explain how the following errors would affect the molarity of your solution.

   a. You forget to add the water of hydration in calculating the molar mass.

   b. You forget to thoroughly mix the solution before giving it to your teacher for measurement.

Appendixes

A Brief History of Science

LEVEL
All levels of science

NATIONAL STANDARDS
Science in Personal and Social Perspectives; History and Nature of Science

OBJECTIVE AND INTRODUCTION
Science is a human endeavor, and the discoveries in science, whether accidental or deliberately pursued, always occur within the contexts of culture and social environment, curiosity and necessity, and the availability of technology. The following chart which chronicles some of the major developments in science and technology over thousands of years is not meant to be complete, but is intended to convey how some of the scientific concepts were discovered and how they developed from one generation to the next. The discoveries which are highlighted were chosen because of their direct relevance to the content typically covered in middle school and high school science courses.

It should also be noted that even though modern science is often partitioned into narrow, specific areas of concentration, as science continues to progress we continue to see how one science is dependent on another. For example, the discovery of the structure of the DNA molecule is certainly one of the most important discoveries of the modern era, and could not have been possible without the cooperation and contributions of physicists, chemists, and biologists. If one of the sciences doesn't have any events highlighted during a particular time period in the chart, it should not be assumed that there were no significant developments during that time period. Often progress in one area of science ultimately reveals itself in a major development in another area of science.

It is important to help the students realize that the discoveries listed on this chart did not happen overnight. Scientists struggled for years, decades, centuries, and sometimes millennia to grasp an understanding of the difficult riddles the natural world presents us. Science teachers should continually remind students that because these concepts took considerable time and effort to discover and understand, it will likely take time and effort for the students to understand them.

Perhaps this chart can help teachers put the science they are exploring with students into perspective as it relates to time, culture, and available technology. If you like to have the students write reports on scientists and their discoveries, you might choose a particular time period or topic and have the students do further research and present their findings in a creative way, such as posters, timelines, oral and written reports, or a Power Point presentation.

The primary sources for this chart can be found at http://www.sciencetimeline.net.

	10000 BC - 800 BC	800 BC - 400 BC	400 BC - 200 BC
Biology	10,000 BC - 6500 BC Animals were probably first domesticated. 2000 BC Egyptians considere the souring of wine comparable to the souring of milk. 1600 BC Egyptian papyrus list many diagnoses of head and neck injuries and their treatment. 1000 BC Horse breeders experiment with cross-breeding of horses and donkeys.	580 BC Thales of Miletus suggests that water is the fundamental component to all life. 510 BC Almaeon of Crotona locates the seat of perception in the brain by dissection. 500 BC Xenophanes examines fossils and speculates on the evolution of the earth. 400 BC Hippocrates of Cos, maintains that diseases have natural causes.	250 BC Erasistratus of Alexandria dissects the brain and distinguishes between the cerebrum and the cerebellum.
Physics	4800 BC Astronomical calendar stones used in Egypt. 3300 BC Numerals first used in Sumerian and Egyptian hieroglyphics. 3200 BC First evidence of wheeled vehicles in Uruk. 1600 BC Documents maintain the Earth was a globe and the Earth circled the Sun. 1500 BC Babylonians understand right-triangle relationships.	From 747 BC, a continuous record of solar and lunar eclipses was kept in Mesopotamia. 585 BC Thales predicts and eclipse 530 BC Pythagoras discovers musical intervals in strings depends on length and tension. 425 BC Herodotus writes the first scientific history. 400 BC Arrow-shooting catapult developed at Syracuse.	400 BC Babylonian astronomers can predict the occurence of lunar eclipses. 370 BC Eudoxus of Cnidus invents a model of concentric spheres by which he was able to predict the motions of the moon, sin, and planets. 335 BC Aristotle writes *Physics*, which becomes the standard for science for 2000 years. 300 BC Euclid writes *Elements*, providing the basis for geometry. 260 BC Aristarchus of Samos suggests the sun-centered model of the universe, and calculates the Earth-Sun distance to the Earth-moon distance; Archimedes formulates buoyancy principles.
Chemistry	4000 BC Copper smelting is introduced in Mesopotamia. 2500 BC Smelting of bronze in Sumeria. 1200 BC Smelting of iron in Armenia.	450 BC Empedocles of Agrigento divides matter into the four elements: earth, water, air, fire. 440 BC Leucippus of Miletus suggests the existence of atoms.	250 BC 'Zero' appears in the Babylonian place-value system.

	200 BC - 200 AD	200 - 1000	1000 - 1300
Biology	1st Century AD Pedanius Dioscorides publishes recommendations as to the medicinal use of specific plant extracts. 170 Claudius Galen uses pulse-taking as a diagnostic, performs numerous animal dissections, and writes treatises on anatomy.	900 Abu Bakr al-Razi, distinguishes smallpox from measles in the course of writing several medical books in Arabic.	1200 Medical doctors, especially in Italy, begin writing case-histories, describing the symptoms and courses of numerous diseases. 1266 Hugh and Theodoric Borgogoni advocate putting surgical subjects to sleep with narcotic-soaked sponges.
Physics	134 BC Hipparchus of Rhodes measures the year with great accuracy and builds the first comprehensive star chart with 850 stars and a luminosity scale. 45 BC Sosigenes designs the calendar adopted by Julius Caesar. 100 AD Hero of Alexandria explains that the four elements consist of atoms. 141 AD Claudius Ptolemy publishes *Almagest*, the standard book for astronomy for 1500 years.	517 John Philoponus determines that falling objects fall with the same acceleration. 530 Simplicius of Cilicia writes a commentary in Greek on Aristotle's writings on 'gravity'. 1000 Ibn al-Haitam, or al-Hazen, in *Opticae Thesaurus*, introduces the idea that light rays emanate in straight lines in all directions from every point on a luminous surface.	1054 Chinese astronomers at the Sung national observatory at K'ai-feng observe the explosion of a supernova in the Crab Nebulae, visible in daylight for twenty-three days. 1268 Roger Bacon publishes proposals for educational reform, arguing for the study of nature, using observation and exact measurement, and asserting that the only basis for certainty is experience, or verification.
Chemistry	1st century AD Titus Lucretius Carus maintains that the universe came into being through the working of natural laws in the combining of atoms.	800 Jabir ibn Hayyan, later known as Geber, bases his chemical system on sulfur and mercury. 850 Moors in Spain prepare pure copper by reacting its salts with iron, a forerunner of electroplating.	1100 Alchemists develop the art of distillation to the stage at which distillates could be captured by cooling in a flask. 1260 Albertus writes a book in which he geology into a coherent theory. He was the first to produce arsenic in a free form.

	1300 - 1400	1400 - 1500	1500 - 1550
Biology	1316 Mondino of Luzzi publishes *Anatomia*, introducing the practice of public dissections for teaching. 1360, Guy de Chauliac, recommends extending fractured limbs with pulleys and weights, and replacing lost teeth with bone fastened to the sound teeth with gold wire.	1410 Benedetto Rinio publishes an herbal which contains 450 paintings of plants, botanical notes, citations of authorities used, and the names of the plants in various languages. 1482 Leonardo da Vinci begins his notebooks on dissections of the human body, the impossibility of perpetual motion, dynamics, statics, and numerous machines.	1541 Giambattista Canano publishes illustrations of each muscle and its relation with the bones. 1546 Fracastoro publishes the idea that diseases were caused by disease-specific seeds which could be contagious.
Physics	1304 Theodoric of Freiberg shows that rainbows could be explained through experiments with hexagonal crystals and spherical crystal balls. 1323 William Ockham introduces the distinction between 'being in motion' and 'being moved,' that is, as it is now called, between dynamic motion and kinematic motion. 1364 Giovanni di Dondi builds a complex clock which kept track of calendar cycles and computed the date of Easter by using various lengths of chain.	1420 Felipe Brunelleschi draws panels in scientifically-accurate perspective. 1437 Johann Gutenberg becomes the first in Europe to print with movable type cast in molds.	1543 Copernicus publishes *De revolutionibus orbium coelestium*, detailing the sun-centered model of the solar system. 1572 Tycho Brahe observes a supernova in the constellation Cassiopeia, now known as Tycho's star. 1583 Galileo Galilei discovers by experiment that the oscillations of a swinging pendulum take the same amount of time regardless of their amplitude. In 1590 Zacharias and Hans Janssen combines double convex lenses in a tube, producing the first telescope.
Chemistry	1300 Giles of Rome puts forward an atomic theory based on Avicebron's theory of matter.		Theophrastus Bombastus von Hohenheim (Paracelsus) suggests the chemical properties are combustibility, fluidity, and changeability, solidity, and permanence.

	1600 - 1625	1625 - 1650	1650 - 1700
Biology	1627 William Harvey confirms his observation that the blood circulates throughout the body.	1645 Marc Aurelio Severino discovers the heart of the higher crustacea, recognizes the respiratory function of fish gills, and recognizes the unity of vertebrates. 1650 Francis Glisson publishes an account of infantile ricketts.	1651 Harvey publishes the concept that all living things originate from eggs. 1652 Thomas Bartholin discovers the lymphatic system and determines its relation to the circulatory system. 1655 Thomas Sydenham promotes the idea that diseases are organisms inside a host. 1665 Robert Hooke names and gives the first description of cells. 1674 Anton van Leeuwenhoek reports his discovery of protozoa.
Physics	1600 William Gilbert, in *De Magnete*, holds that the earth behaves like a giant magnet with its poles near the geographic poles. 1604 Johannes Kepler and many other astronomers witness the outburst of a supernova in the constellation Serpens. 1605 Francis Bacon, with the *Advancement of Learning*, begins the publication of his philosophical works, in which he urges collaboration between the inductive and experimental methods of proof. 1609 Kepler publishes his 1st and 2nd laws of planetary motion. 1610 Galileo observes the moons of Jupiter, phases of Venus, craters on the moon. 1621 Willibrord Snell discovers the law of refraction.	1633 Galileo is placed under house arrest for his heliocentric views published in his book *Dialogue on the Two Chief World Systems*. 1638 Galileo publishes *Discourses on Two New Sciences*, outlining his theory of motion. 1644 Blaise Pascal builds a five digit adding machine. 1644 Evangelista Torricelli devises the mercury barometer and creates an artificial vacuum. 1648 Pascal shows that barometric pressure results from atmospheric pressure and that pressure applied to a confined fluid is transmitted equally to all areas and at right angles to the surface of the container.	1666 Isaac Newton discovers the essentials of calculus, the law of universal gravitation, and that white light is composed of all the colors of the spectrum. 1669 Newton circulates a manuscript containing the first notice of his calculus. 1676 Ole Roemer proves that light travels at a finite speed by repeated observations of eclipses of Jupiter's moon, Io. 1684 Gottfried Wilhelm von Leibniz publishes his system of calculus, developed independently of Newton. 1687 Newton publishes the *Principia*, a summary of his discoveries in motion, gravitation, and calculus. 1693 Edmund Halley discovers the formula for the focus of a lens. 1694 Rudolph Jakob Camerarious reports the existence of sex in flowering plants.

Chemistry	1600 - 1625	1625 - 1650	1650 - 1700
		1630 Jean Rey states that the slight increase in weight of lead and tin during their calcination could only have come from the air. 1644 Evangelista Torricelli devises the mercury barometer and creates an artificial vacuum. 1648 Jean Baptiste van Helmont concludes that plants derive their sustenance from water, demonstrates that physiological changes have chemical causes, coined the name 'gas' from the Greek *chaos*, distinguishes gases as a class with liquids and solids.	1661 Robert Boyle gives the first precise definitions of a chemical element, a chemical reaction, chemical analysis, made studies of acids and bases, and shows that pressure and volume of a gas are inversely proportional. 1670 Boyle produces hydrogen by reacting metals with acid. 1679 Denis Papin demonstrates the influence of atmospheric pressure on boiling points.

Laying the Foundation in Chemistry

	1700 - 1750	1750 - 1800	1800 - 1850
Biology	1715 Thomas Fairchild produces the first artificial hybrid plant. 1745 Maupertius proposes the notion of descent from a common ancestor. 1749 Buffon begins the publication of the 44 volumes of *Histoire Naturelle*, in which he draws attention to vestigial organs and asserted that species are mutable. 1788 Jean Senebier demonstrates that it is light, not heat, from the sun that is effective in photosynthesis. 1791 Luigi Galvani shows that it is possible to control the motor nerves of frogs using electrical currents, i.e., that the nerves transmitted electricity.	1752 James Lind calls attention to the value of fresh fruit in the prevention of scurvy. 1753, Carl Linné publishes *Species plantarum*, in which he distinguished plants in terms of genera and species, and later applying the system to animals. 1762 Marcus Antonius Plenciz says that living agents are the cause of infectious diseases.	1800 Karl Friedrich Burdach introduces the term 'biology,' which replaces 'natural history,' which traditionally had three components, zoology, botany, and mineralogy. 1809 Jean-Baptiste Monet de Lamarck states that heritable changes in 'habits,' or behavior, could be brought about by the environment. 1820 Lamarck describes the origin of living things as a process of gradual development from matter. 1827 Robert Brown notices random movement of microscopic particles contained in the pollen from plants when suspended in fluid (Brownian movement). 1831 Brown discovers the cell nucleus in the course of a microscopic examination of orchids. 1833 Marshall Hall describes the mechanism by which a stimulus can produce a response independent of both sensation and volition, and coins the term 'reflex.' 1837 Heinrich Gustav Magnus determines that carbon dioxide released in the lungs had been carried there by blood and that more oxygen and less carbon dioxide was contained in arterial than in venous blood; René Dutrochet observes that chlorophyll is necessary for photosynthesis; Hugo von Mohl describes 'chloroplasts' as discrete bodies within the cells of green plants. 1838 Mattias Jakob Schleiden puts forward the theory that plant tissues are composed of cells, and recognizes the significance of the nucleus. 1839 Mohl describes the appearance of the cell plate between the daughter cells during cell division, or 'mitosis.' 1846 William Morton demonstrates the effective use of ether as an anesthesia. 1848 Louis Pasteur discovers molecular dissymmetry, or chirality, and coins the distinction between users and non-users of oxygen, 'aerobic' and 'anaerobic.' 1801 Thomas Young observes that light passing through a double-slit recombines to create light and dark areas, and measures the wavelength of light using this pattern.

Physics	1700 - 1750	1750 - 1800	1800 - 1850
	1704 Newton, in *Opticks*, presents his discoveries using light and elaborates his theory that it is composed of particles.	1751 Benjamin Franklin publishes *Experiments and Observations on Electricity* after several years of experiments.	1807 Young coins the word 'energy' for the fundamental quantity created by the heat which moved particles in Bernoulli's kinetic theory.
	1705 Halley recognizes the orbit of the comet that bears his name and predicts its reappearance in 1758.	1752 Thomas Melvill notices that the spectra of flames into which metals or salts have been introduced show bright lines characteristic of the metal or salt.	1814 Joseph von Fraunhofer devises a primitive spectroscope by allowing light to pass through a narrow slit and then a prism.
	1718 Halley states that stars move, since they had changed position since Ptolemy's *Almagest*.	1756 Franz Ulrich Theodosius Aepinus, realizes that the causes of magnetic and electrical phenomena were extremely similar.	1816 Augustin Jean Fresnel shows that diffraction, interference, and polarization can be explained in terms of the transverse wave theory of light.
	1738 Daniel Bernoulli asserts the principle that as the speed of a moving fluid increases, the pressure within the fluid decreases, inventing the kinetic theory of gases.	1759 The return of Halley's comet confirms Newton''s mechanics.	1820 Hans Christian Øersted initiates the study of electromagnetism by placing a needle parallel to a wire conducting electric current and discovering that this produces a magnetic field that curls around the wire.
	1746 Andreas Cunaeus invents the 'Leyden jar,' a form of capacitor.	1759 Aepinus fathers the action-at-a-distance/localization of charge theory of electricity and magnetism.	1824 Sadi Carnot shows that even under ideal conditions a steam engine cannot convert into mechanical energy all the heat energy supplied to it.
	1798 Cavendish constructs a torsion balance by which he measured the mean density of the Earth.	1768 Euler proposes that the wavelength of light determines its color.	1827 Georg Simon Ohm discovers that the ratio of the potential difference between the ends of a conductor and the current flowing through it is constant, and is the resistance of the conductor.
		1783 Carnot specifies the optimal and abstract conditions for the operation for all sorts of actual machines.	1831 Faraday discovers the means of producing electricity from magnetism, i.e., electromagnetic induction.
		1785 Charles Augustin de Coulomb formulates the inverse square law for the force between electric charges.	1841 Julius Robert Mayer, working with established experimental results, derives the general relationship between heat and work, which is the first law of thermodynamics, a form of the law of conservation of energy.
			1842 Christian Doppler develops the theory that the frequency of energy in the form of the form of waves changes depending on the motion of either the sender or the receiver.
			1843 James Prescott Joule demonstrates experimentally the equivalence of the heat produced and the mechanical work spent in the operation.
			1846 Johann Gottfried Galle discovers the planet Neptune where Urbain Jean Joseph Le Verrier and, independently, John Couch Adams had predicted that a planet would be found.

	1700 – 1750	1750 – 1800	1800 – 1850
Physics (Continued)		1792 Volta discovers he could arrange metals in a series in such a way that chemical energy is converted into electrical energy.	1847 Hermann von Helmholtz formulates the law of the conservation of energy in an equation which expresses the most general form of the principle. 1850 Jean Foucault, using a rotating mirror, determines the speed of light in the air as 298,000 km/s.
Chemistry	1709 Gabriel Daniel Fahrenheit constructs an alcohol thermometer and, five years later, a mercury thermometer. 1742 Celsius develops the centigrade temperature scale which carries his name.	1754 Joseph Black heats calcium carbonate which separates into calcium oxide and carbon dioxide and then recombines back into calcium carbonate. 1757 Black discovers latent heat, distinguishing between heat and temperature. 1774 Priestly discovers sulphur dioxide, ammonia, and 'dephlogisticated air,' later named oxygen by Lavoisier. 1780 Lavoisier and Laplace develop a theory of chemical and thermal phenomena based on the assumption that heat is a substance, which they called 'caloric' and deduced the notion of 'specific heat.' 1787 Charles determines that the volume of a fixed mass of gas at constant pressure is proportional to its temperature. This was published by Joseph Louis Gay-Lussac in 1802. 1789 Lavoisier proves that mass is conserved in chemical reactions and created the first list of chemical elements.	1803 Dalton applies atomic theory to a table of atomic weights. 1808 Dalton publishes *A New System of Chemical Philosophy*, launching chemical atomic theory; Gay-Lussac ennunciates the 'Law of combining volumes,' which states that when gases combine they do so in small whole number ratios. 1811 Berzelius simplifies chemistry through his suggestion that they be represented by the first letter of each element's Latin name, with the addition of the second letter when necessary. Proportions in a compound were indicated with appropriate number as subscript. 1811 Amedeo Avogadro proposes that equal volumes of gases at the same temperature and pressure contain the same number of molecules. 1815 William Prout proposes that the atomic weights of elements are multiples of that for hydrogen. 1825 Faraday discovers benzene. 1834 Faraday states that the amount of chemical change produced is proportional to the quantity of electricity passed and the amount of chemical change produced in different substances by a fixed quantity of electricity is proportional to the electrochemical equivalent of the substance. 1835 Berzelius suggests the name 'catalysis' for reactions which occurred only in the presence of some third substance. 1839 Christian Swann discovers the existence of ozone. 1848, W. Thomson proposes what became known as the 'Kelvin scale,' after the title bestowed on him by the British government. 1850 Runge demonstrates the separation of inorganic chemicals by their differential adsorption to paper. This is forerunner of chromatographic separations.

Biology	1850 - 1875	1875 - 1900	1900 - 1925
	1852 Georges Newport observes the penetration of the vitelline membrane of a frog egg by sperm.	1876 Robert Koch devises the method of employing aniline dyes to stain microorganisms, isolating pure cultures of bacteria and showing the bacterial origin of many infectious diseases.	1900 Mikhail Tsvet discerns three green pigments, chlorophyll a, b, and c, differing in color, fluorescence, and spectral absorption.
	1857 Pasteur demonstrates that lactic acid fermentation is carried out by living bacteria; Albert von Kolliker describes what were later named 'mitochondria' in the nucleus of muscle cells.	1879 Walther Flemming names 'chromatin' and 'mitosis,' made the first accurate counts of chromosome numbers, and discerned the longitudinal splitting of chromosomes.	1902 Karl Landsteiner found that human blood was one of four types, A, B, A-B, and O, thus making transfusions safe; Fischer proposes that proteins consist of chains of amino acids; Ivan Pavlov combines associative learning with reflex acts, postulating the existence of associated stimuli, or 'conditioned responses.'
	1858 Darwin's friends, arrange for the simultaneous announcement of Wallace's and Darwin's idea of natural selection.	1883 Wilhelm Roux suggests that the filaments within the cell's nucleus carry the hereditary factors.	1903 Tsvet develops methods in chromatography.
	1859, Darwin, in *Origin of Species* asserts all life had a common ancestor.	1885 Hertwig and Strasburger develop the conception that the nucleus is the basis of heredity.	1905 Edmund Beecher Wilson discovers that the X chromosome is linked to the sex of the bearer.
	1862 Pasteur publishes the 'germ theory': Infection is caused by self-replicating microorganisms, and that attenuated viral cultures granted immunity. These beneficent antigens he named 'vaccines' in honor of Jenner and his vaccinia virus.	1890 Hans Driesch separates two cells of a fertilized sea urchin egg by shaking with very different results than Roux: From a single cell arose an entire sea urchin; Richard Altmann reports the presence within cells of organisms which live as intracellular symbionts, later named mitochondria.	1908 Godfrey Harold Hardy works out the equilibrium formula for a population heterogenous for a single pair of alleles.
	1865 O.F.C. Deiters proposes the image of the nerve cell which is accepted today: cell body with its nucleus, multiple, branching dendrites, and a single axon; Lister, using carbolic acid as antiseptic and sterilizing his instrument, proved the efficacy of antiseptic surgery.	1894 H.J.H. Fenton discovers a reaction now considered to be one of the most important mechanisms of oxidative damage in living cell.	1910 Konstantin S. Mereschovsky publishes an essentially modern view of the bacterial origin of what later came to be called eukaryotic cells.
	1866 Gregor Mendel interprets heredity in terms of a pairing of dominant and/or recessive unit characters.		1911 Alfred Henry Sturtevant, an undergraduate student of Morgan's, constructs the first rudimentary map of the fruit fly chromosome, establishing that genes are real.
	1871 Darwin, in *The Descent of Man*, suggests that there is no sharp discontinuity between the evolution of humans and animals.		1913 Lawrence Joseph Henderson proposes that the concept of fitness be extended to the environment. This has ramifications for the origin of life.

	1850 - 1875	*1875 - 1900*	*1900 - 1925*
Biology (Continued)	1873 Anton Schneider describes chromosomes during the process of mitosis during cell division.		1921 Victor Jollos hypothesizes that the disappearance of environmentally-induced acquired traits, even after hundreds of generations, indicates that their acquisition should be assigned to the cytoplasm rather than the nucleus; Muller raises the question of the relationship of genes to viruses, or 'naked genes'. 1922 Walter Garstang shows that phylogeny is not the cause but the product of different ontogenies. 1923 Robert Feulgen discovers a selective staining technique for DNA localization, which is still in use; Jean Piaget maintains that child development proceeds in the same sequence of genetically determined stages.
Physics	1851 Foucault demonstrates that a pendulum's swing, seen relative to the Earth, would gradually precess, evidence of the Earth's rotation. 1859 Kirchhoff proves a theorem about blackbody radiation, namely, the energy emitted E depends only on the temperature and the frequency of the emitted energy. 1861, Maxwell announces his discovery that some of the properties of the vibrations in the magnetic medium are identical with those of light, and predicts the speed of light theoretically; Anders Jonas Ångström, using a spectroscope, confirms the presence of hydrogen in the Sun.	1876 Alexander Graham Bell invents the telephone. 1879 Crookes attempts to determine the paths of the 'lines of molecular pressure,' or cathode rays, in an evacuated glass tube through which two electrodes are passed. 1879 Albert Michelson determines the speed of light to be 186,350 miles per second. 1881 Venn represents logical propositions diagrammatically.	1900 Planck introduces 'quantum theory' to explain a formula, $E=hf$, where E is energy, f is frequency, and h is a new constant; Rutherford identifies a third type of radiation, which he calls 'gamma radiation.' 1903 Orville and Wilbur Wright achieve flight in a manned, gasoline power-driven, heavier-than-air flying machine. 1904 Lorentz formulates the so-called 'Lorentz transformation,' which describes the increase in mass, the shortening of length, and the time dilation of a body moving at speeds close to that of light. 1904 Hantaro Nagaoka proposes a 'Saturn model' of the atom with a nucleus and many electrons in a ring around it.

	1850 - 1875	1875 - 1900	1900 - 1925
Physics (Continued)	1865 Maxwell publishes his four equations of electromagnetism based on the work of Coulomb, Gauss, Ampere, and Faraday. 1871 Crookes creates a vacuum of about one millionth of an atmosphere which made possible the discovery of X-rays and the electron.	1887 Michelson and Edward W. Morley, using an interferometer to investigate whether the speed of light depends on the direction the light beam moves, fail to detect the motion of the Earth with respect to the aether, thereby refuting the hypothesis that the aether exists; Heinrich Hertz produces electromagnetic radio waves. 1888 Nicola Tesla patents his invention of alternating electric current. 1892 Lorentz proposes a theory in which a body carries a charge if it has an excess of positive or negative particles, and an electric current in a conductor is a flow of particulate particles. 1895 Wilhelm Conrad Röntgen, using a Crookes' tube, observes a new form of penetrating radiation, which he named X-rays. 1897 Joseph John Thomson, using a Crookes' tube, demonstrates that cathode rays consisted of units of electrical current made up of negatively charged particles of subatomic size (electrons). 1899 Ernest Rutherford characterizes 'alpha rays' and 'beta rays'; Becquerel shows that radioactivity in uranium consists of charged particles that are deflected by a magnetic field.	1905 Albert Einstein publishes three papers describing his explanation of the photoelectric effect, Brownian movement, and his theory of special relativity. 1907 Einstein deduces the expression for the equivalence of mass and energy, $E=mc^2$. 1908 Robert Andrews Millikan determines the probable minimum unit of an electrical charge, that is, of an electron. 1909 Hans Geiger and E. Marsden, under Rutherford's direction, scatter alpha particles with thin films of heavy metals, providing evidence that atoms possess a discrete nucleus. 1911 Heike Kamerlingh Onmes discovers 'superconductivity,' the ability of certain materials at low temperatures to carry electric current without resistance; Einstein postulates that light is bent by gravity. 1911-13 Hertzsprung and Russell publish graphs plotting color or spectral class against the absolute magnitude of stars. These are now called HR diagrams and are the basis of the theory of stellar evolution. 1913 Niels Bohr, applying the Planck quantum hypothesis to Rutherford's atomic model, places electrons in discrete energy levels, and postulating the quantum model of the atom; Einstein and Marcel Grossman investigate curved space and time as it relates to a theory of gravity. Einstein contributed the physics and Grossman the mathematics.

	1850 - 1875	1875 - 1900	1900 - 1925
Physics (Continued)			1919 E. Rutherford discovers the proton, which contains the positive charge within the nucleus of an atom, and publishes the first evidence of artificially-produced splitting of atomic nuclei; Eddington and Frank W. Dyson measure the bending of starlight by the gravitational pull of the sun, thus confirming Einstein's general theory of relativity. 1920 E. Rutherford postulates the existence of the neutron, required in order to keep the positively-charged protons in the nucleus from repelling each other. 1922 Arthur Compton demonstrates an increase in the wavelengths of X-rays and gamma rays when they collide with loosely bound electrons, verifying the quantum theory since the effect requires the rays be treated as particles, not waves. 1923 Louis de Broglie hypothesizes that a moving electron particle has wave-like properties.
Chemistry	1855 David Alter described the spectra of hydrogen and other gases. 1858 Friedrich August Kekulé von Stradonitz suggests that carbon atoms are formed in chains. 1859 Robert Wilhelm Bunsen discovers that each element produces its own characteristic set of lines in the spectrum. 1866 Alfred Nobel patents dynamite in Sweden.	1879 Stefan conjectures that that the radiant energy emitted by an enclosure equivalent to a black body is proportional to the fourth power of the body's temperature. 1884 Jacobus van't Hoff explains the principle of equilibrium in chemical dynamics and osmotic electrical conductivity. 1894 Strutt and William Ramsay discover and isolate argon in the process of explaining the discrepancy between the weight of nitrogen obtained from the air and from ammonia.	1905 Arrhenius expresses concern about global warming as a result of burning fossil fuels. 1913 Frederick Soddy discovers that different forms of the same element were, in fact, groups of elements with the same chemical character, but varying in their masses (isotopes), and that radioactive decay is accompanied by the transmutation of one element to another. 1916 Gilbert Newton Lewis states that the chemical bond consists of two electrons held jointly by two atoms.

	1850 - 1875	1875 - 1900	1900 - 1925
Chemistry (Continued)	1869 Dmitri Mendeléev and, independently, Julius Lother Meyer formulate the 'Periodic law.' Mendeléev placed the chemical elements in seven rows in an order where those elements having similar chemical properties were aligned vertically. 1869 John Hyatt produces 'celluloid,' the first synthetic plastic to be put into wide use.	1896 Eduard Buchner discovers a chemical in yeast, which he called zymase. He noted that the crushed yeast, that is, cell-free yeast, fermented sugar. This observation opened the era of modern biochemistry. 1896 Antoine Henri Becquerel discovers radioactivity in uranium. 1897 Felix Hoffman synthesizes a form of acetysalicylic acid that enabled the mass production of aspirin two years later. 1898 Marie Sklodowska Curie and P. Curie discover and isolate radium and polonium, and clarify that radiation is an atomic property. M. Curie coins the term 'radioactive.' 1898 J. Thomson shows that neon gas consists of two types of charged electrons, or ions, each with a different charge, or mass, or both. This raised the possibility that varieties of a single element might exist with the same atomic number but differ in mass; Wien identifies a positive particle equal in mass to the hydrogen atom, which later was named the 'proton'; Ramsey and Morris Travers discover neon, krypton, and xenon; James Dewar liquefies hydrogen.	1925 Wolfgang Pauli puts forth the principle that no two electrons in the atom can be in the same quantum state.

	1925 - 1950	1950 - 1975	1975 - 2000
Biology	1928 Alexander Fleming discovers penicillin, a relatively innocuous antibiotic. 1929 It was found that deoxyribonucleic acid (DNA) is located exclusively in the chromosomes, whereas ribonucleic acid (RNA) is located mainly outside the nucleus; Fisher provides a mathematical analysis of how the distribution of genes in a population will change as a result of natural selection, and maintained that once a species' fitness is at a maximum, any mutation will lower it. 1930 Phoebus Aaron Levene elucidate the structure of mononucleotides and showed them to be the building blocks of nucleic acids. 1931 Harriet B. Creighton and Barbara McClintock, working with maize, and Curt Stern, working with Drosophila, provide the first visual confirmation of genetic 'crossing-over.' 1935, William Cumming Rose recognizes the essential amino acid 'threonine.' 1937, Krebs discovers the citrus acid cycle, also known as the tricarboxylic acid cycle and the Krebs cycle. 1938 Hans Spemann proposes the concept of cloning and insists that cell differentiation is the outcome of an orderly sequence of specific stimuli, namely, chemical inductive agents, which were predominantly cyto-plasmic in operation; Warren Weaver coins the term 'molecular biology.' 1940 Ernst Boris Chain and Howard Walter Florey extract and purify penicillin and demonstrate its therapeutic utility.	1953 James Watson and Francis Crick build a model of DNA showing that the structure is two paired, complementary strands, helical and anti-parallel, associated by secondary, noncovalent bonds. Maurice H. F. Wilkens' and Rosalind Franklin's X-ray crystallographs of DNA supported the discovery of the structure. 1954 Salk develops an injectable killed-virus vaccine against poliomyelytis, the incidence of which began to decline after mass immunization began the following year. 1968 Norman Geschwind and Walter Levitsky show that in male and female humans there are characteristic anatomical differences, e.g., the size of the planum temporale in the hemispheres of the brain. 1973 Mertz, Davis, Lobban, Berg, Boyer, Cohen, and Morrow, animal genes were spliced into the small rings of DNA, thus beginning recombinant cloning and launching of the biotechnology industry. 1975 E. M. Southern devises an extension of gel electrophoresis, known as 'Southern blotting,' which greatly aids cloning by enabling the identification and sizing of DNA fragments.	1977 Jack Corliss, in a diving bell 2600 meters below the surface of the Pacific Ocean, observes boiling, lightless deep-sea thermal vents with hundreds of species, including a nine-foot tube worm, most of them new to science. 1978 Mary Leaky announces the discovery of fossilized human footprints from about 3.5 million years ago. 1984 Richard Leaky and Alan Walker excavate a Homo erectus skeleton, dated 1.6 million years ago; Alec John Jeffreys discovers 'genetic fingerprinting,' the pattern of nonfunctional repetitions unigue to each individual's DNA. 1990 Teams led by Robin Lovell-Badge and Robin Goodfellow isolate the testis-determining factor gene SRY, the master switch for mammalian sex determination.

	1925 - 1950	*1950 - 1975*	*1975 - 2000*
Biology (continued)	1941 Astbury establishes that DNA has a crystalline structure. 1943 Thomas Francis and Jonas Edward Salk develop a formalin-killed-virus vaccine against type A and B influenzas. 1944 Oswald T. Avery, Colin MacLeod, and Maclyn McCarty establish that the material of heredity is deoxyribonucleic acid; Archer John Porter Martin and Richard Synge devise 'paper partition chromatography.' 1948 William Howard Stein and Stanford Moore isolate amino acids by passing a solution through a chromatographic column filled with potato starch. 1949 Sven Furberg draws a model of DNA, setting sugar at right angles to base, with the correct three-dimensional configuration of the individual nucleotide; Frederick Sanger claims that proteins are uniquely specified, the implication being that, as there is no general law for their assembly, a code was necessary. 1950 Ernst L. Wynder and Evarts A. Graham publish a survey indicating a strong correlation between contracting lung cancer and smoking tobacco.		

	1925 - 1950	*1950 - 1975*	*1975 - 2000*
Physics	1926 Erwin Schrödinger initiates the development of the final quantum theory by describing wave mechanics, which predicted the positions of the electrons, vibrating as Bohr's standing waves. 1927 Heisenberg states that electrons do not possess both a well-defined position and a well-defined momentum simultaneously. 1927 George P. Thomson diffracts electrons by passing them in a vacuum through a thin foil, thus verifying de Broglie's wave hypothesis; Davisson and Germer measure the length of a de Broglie wave by observing the diffraction of electrons by single crystals of nickel. 1929 Robert van de Graaf develops an electro-static particle accelerator; Hubble observes that all galaxies are moving away from each other. 1930 Ernest O. Lawrence publishes the principle of the cyclotron which uses a magnetic field to curl the particle trajectory of a linear accelerator into into a spiral. 1931 Pauli, in order to solve the question of where the energy went in beta decay, predicts the existence of a 'little neutral thing,' the 'neutrino.' 1932 Irène Curie and Frédéric Joliot bombard nonradioactive beryllium with alpha particles, transmuting it briefly into a radioactive element; James Chadwick isolates the neutron, the first particle discovered with zero electrical charge. 1934 I. Curie and Joliot announce the discovery of "artificial radiation obtained by bombarding certain nuclei with alpha particles."	1950 Hoyle claims to have coined 'big-bang' for the primal fireball, disparaging the notion that such ever occurred. 1956 Leon Cooper shows that in superconductivity the current is carried in bound pairs of electrons, or 'Cooper pairs.' This led to the BCS theory of superconductivity the following year. 1957 John Backus leads the team which creates 'Fortran,' the Formula Translation language for the IBM 704 computer. 1957 The United States government forms the Advanced Research Agency, or ARPA, in response to the Soviet Union's Sputnik, the first artificial satellite. 1958, Jack Kilby builds the first integrated circuit. 1959, James A. Van Allen, Carl E. McIlwain, and George H. Ludwig establish the existence of geometrically trapped electrons and protons in two belts above the Earth, later called the Van Allen Belts. 1960 Theodore H. Maiman describes the first laser, which used a synthetic ruby rod as the lasing medium.	1978 Elementary particle physicists begin speaking of the 'Standard Model' as the basic theory of matter. 1979 The spacecraft Voyager 1 photographs Jupiter's rings, and subsequently visits Saturn, Uranus, and Neptune. 1983 Carlo Rubbia and Simon van der Meer, using the CERN particle accelerator, confirmed the existence of the Z and Ws particles. 1986 Johannes Georg Bednorz and Karl Alexander Müller find a new class of layered materials which superconduct at much higher temperatures than any which had been found previously. 1987 A supernova, SN 1987A, explodes in the Large Magellanic Cloud, and was the nearest supernova to have been observed since the invention of the astronomical telescope. 1990 NASA and the European Space Agency (ESA) launch the Hubble Space Telescope, or HST. Servicing missions were carried out in 1993, 1997, and 2002; Tim Berners-Lee and CERN, The European Organization for Nuclear Research, implemented a hypertext system for information access for physicists.

	1925 - 1950	1950 - 1975	1975 - 2000
Physics (Continued)	1935 IBM introduced a punch card machine with an arithmetic unit based on relays which could do multiplication. 1938, Otto Hahn and Lise Meitner, with their colleague Fritz Strassman, bombard uranium nuclei with slow neutrons. Meitner, interprets the results to be 'nuclear fission,' the term fission being borrowed from biology. 1942 Fermi creates the first controlled, self-sustaining nuclear chain reaction. 1945 The first atomic bombs are exploded over Hiroshima, Japan, then, three days later, over Nagasaki. 1946 John Mauchly and John Presper Eckert demonstrate ENIAC, or Electronic Numerical Integrator and Computer. Its components were entirely electronic.	1963 Murray Gell-Mann and, independently, George Zweig, invent the notion of a more fundamental particle than neutrons and protons which Gell-Mann named the 'quark.' 1965 Arno Allan Penzias and Robert Woodrow Wilson discover cosmic background radiation. The implication is that intergalactic space is above absolute zero, or about 3 degrees K, leading to a drastic shift of the consensus to favor acceptance of the big-bang cosmology. 1967 Steven Weinberg, and Abdus Salam complete the observation of Glashow that the weak and electromagnetic forces result from the same fundamental force. 1968 ARPA , under Lawrence G. Roberts, contracts with BBN to build ARPANET, the prototype of the computer internet. 1970 Stephen Hawking and Penrose prove that the universe must have had a beginning in time, on the basis of Einstein's theory of General Relativity, i.e., mathematically, the big-bang must have arisen from a singularity. 1972 Ray Tomlinson creates the first electronic mail program.	1992 the United States' COBE, or 'Cosmic Background Explorer,' astronomical satellite detects very small variations, or ripples or lumps, in the background cosmic radiation which are thought to be imprints of quantum fluctuations from the early universe, or, in other words, the seeds of later giant structures; CERN releases to the public their hypertext for physicists, naming it the World Wide Web. 1995 Michel Mayor and Didier Queloz detect the first extra-solar planet using the 'wobble technique.' 1997 Ian Wilmut and Keith Campbell clone a sheep, 'Dolly,' from adult cells.

Chemistry	*1925 - 1950*	*1950 - 1975*	*1975 - 2000*
	1927 Walter Heitler and Fritz London show that chemical bonding, the force which holds atoms together, is electrical, and a consequence of quantum mechanics. 1931 Pauling details the rules of covalent bonding. 1944 Seaborg proposed a second 'lanthanide group' as an addition to the periodic table of the elements, as well as existence of a similar series, 90 through 103, or 'actinide group.' 1950 Leo Rainwater combines the liquid drop and shell models of the atomic nucleus.	1950 Leo Rainwater combines the liquid drop and shell models of the atomic nucleus. 1963 Stephanie Louise Kwolek synthesizes polybenzamide, or PBA, a liquid crystalline polymer, used in lightweight body armor. 1969 Calvin publishes *Chemical Evolution* in which he gives several autocatalytic scenarios for the origin of life. 1970 Woodward and Hoffman, in *The Conservation of Orbital Symmetry*, design a set of rules for postulating the areas around atoms where it is most probable that electrons will be found.	1977 Mandelbrot publishes *The Fractal Geometry of Nature* in which complex curves are reduced to straight lines, or fractals, and undergo invariant scaling.

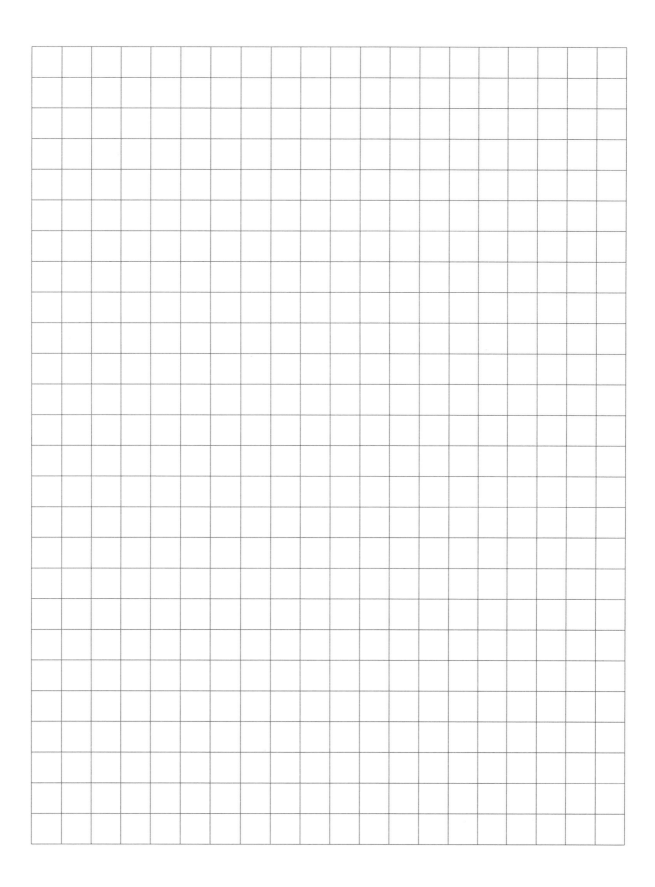

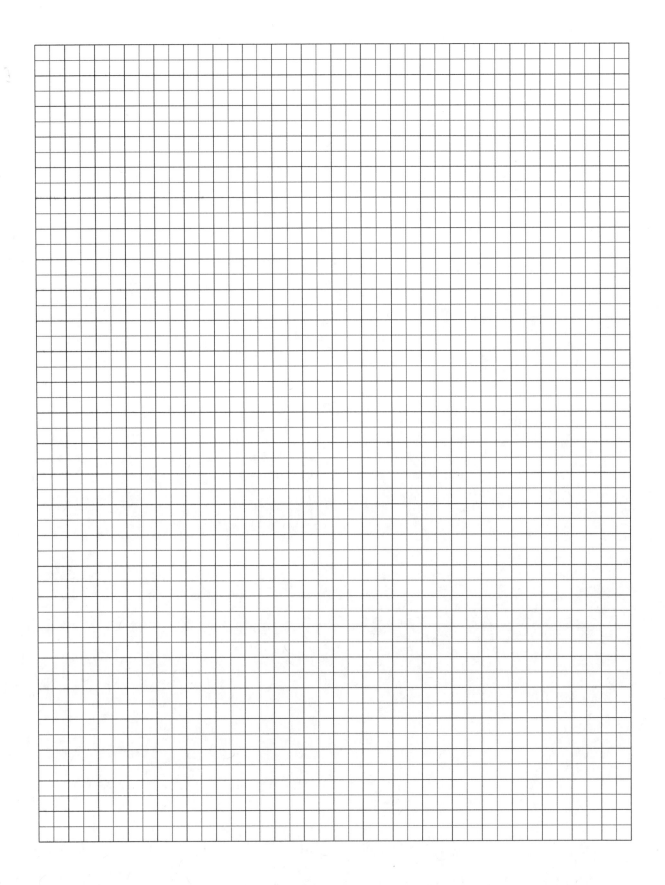

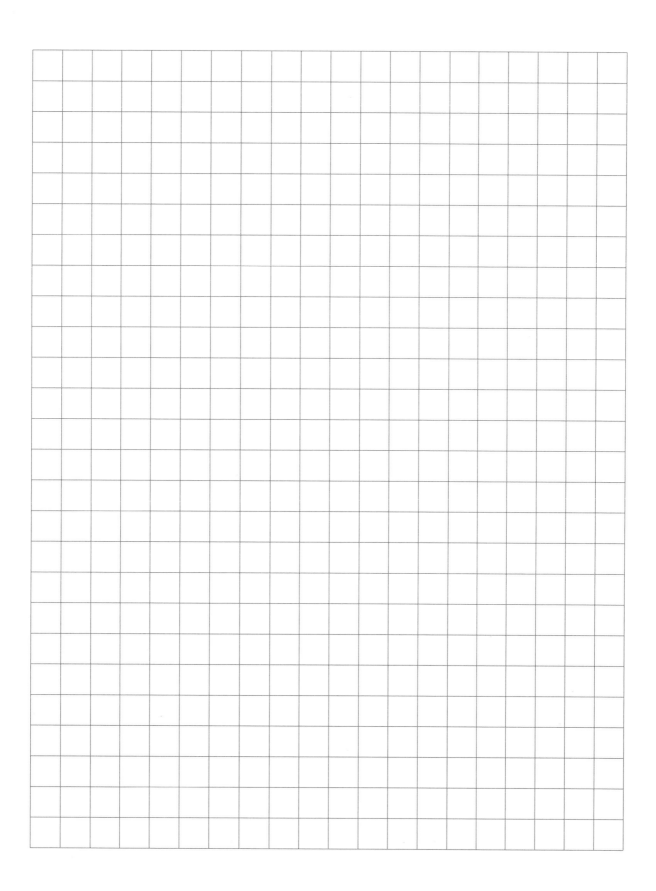

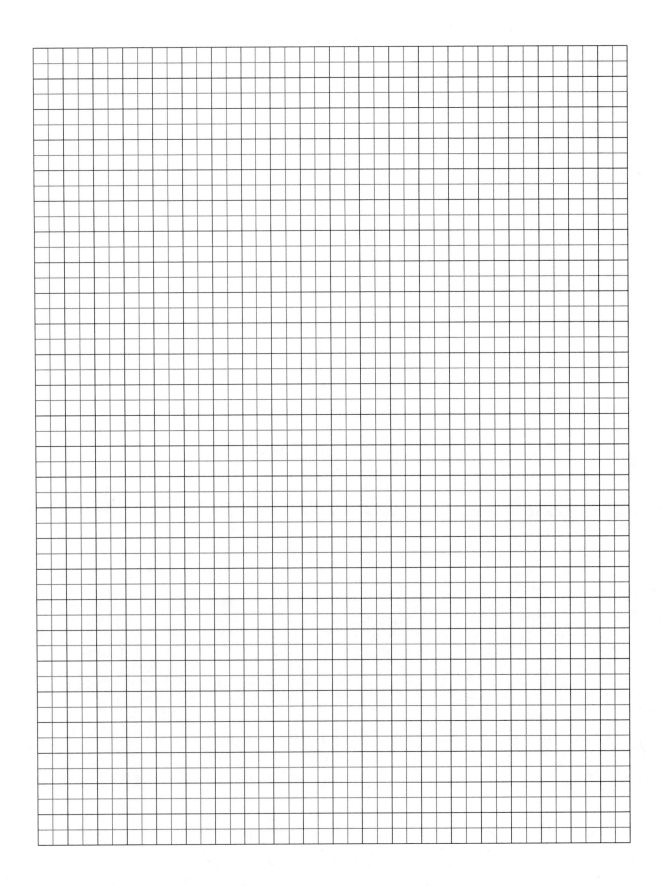

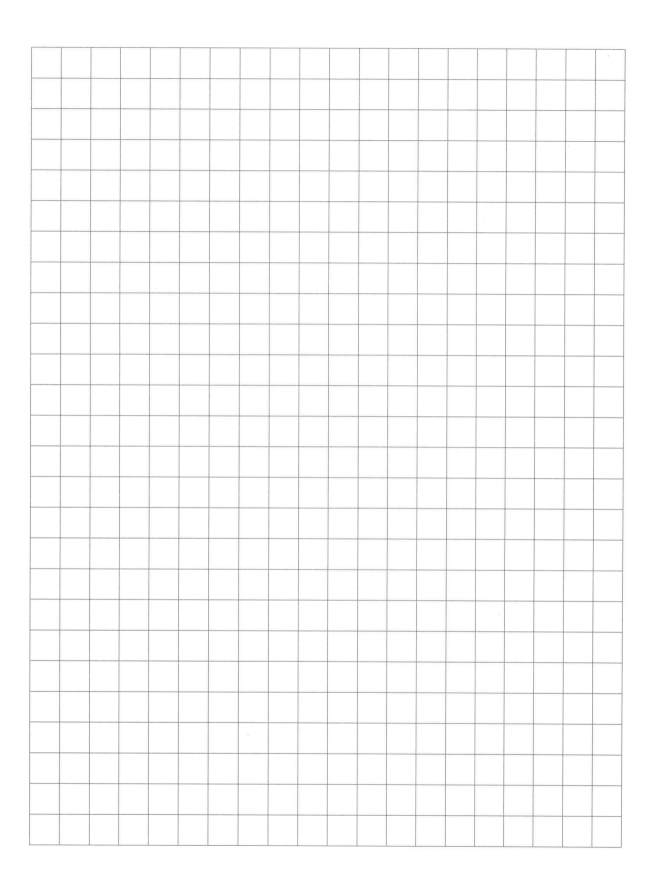

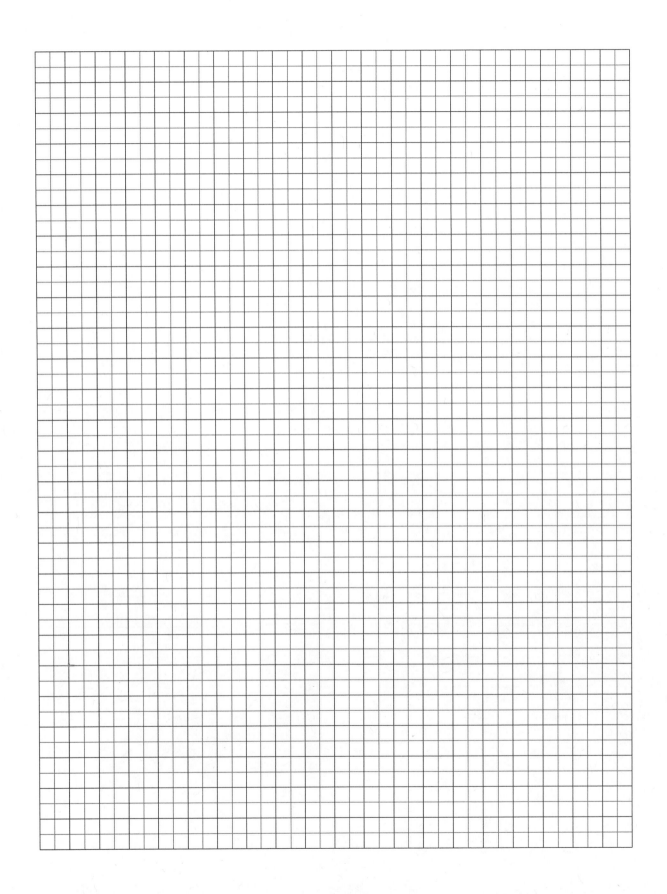

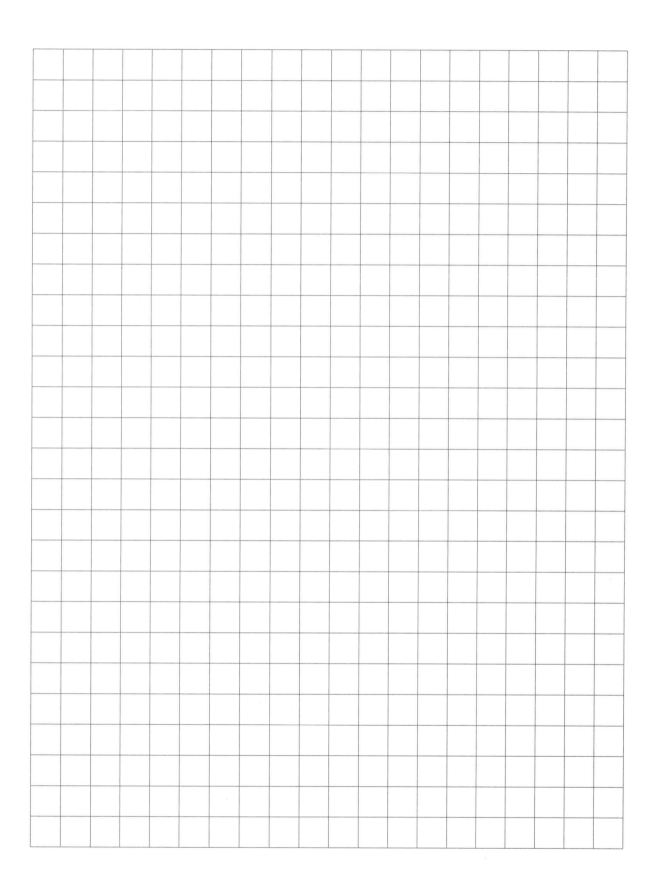

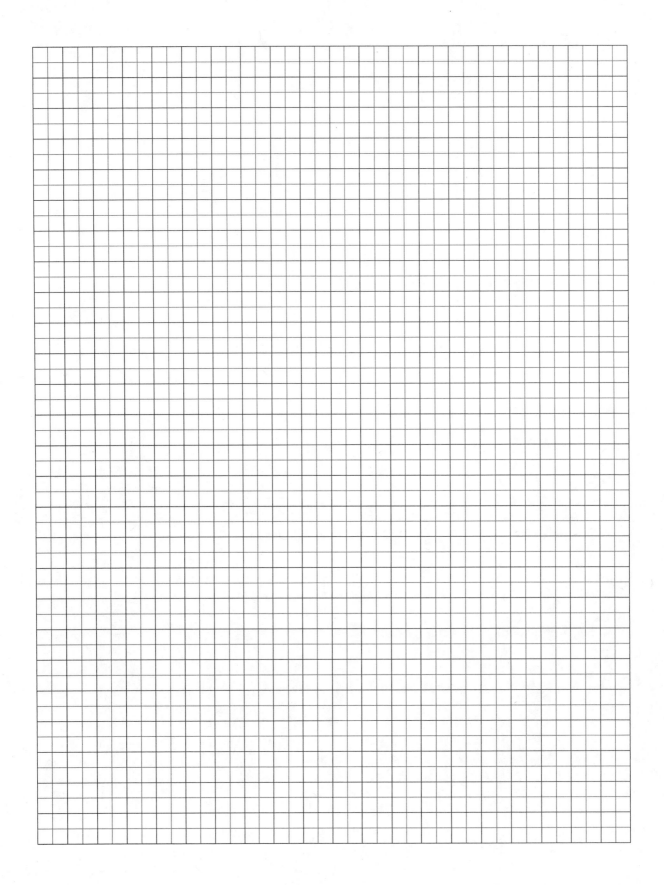